PACIFIC SYMPOSIUM ON BIOCOMPUTING 2006

PACIFIC SYMPOSIUM ON BIOCOMPUTING 2006

Maui, Hawaii
3–7 January 2006

Edited by

Russ B. Altman
Stanford University, USA

A. Keith Dunker
Indiana University, USA

Lawrence Hunter
University of Colorado Health Sciences Center, USA

Tiffany Murray
Stanford University, USA

Teri E. Klein
Stanford University, USA

NEW JERSEY • LONDON • SINGAPORE • BEIJING • SHANGHAI • HONG KONG • TAIPEI • CHENNAI

Published by

World Scientific Publishing Co. Pte. Ltd.

5 Toh Tuck Link, Singapore 596224

USA office: 27 Warren Street, Suite 401-402, Hackensack, NJ 07601

UK office: 57 Shelton Street, Covent Garden, London WC2H 9HE

British Library Cataloguing-in-Publication Data
A catalogue record for this book is available from the British Library.

PACIFIC SYMPOSIUM ON BIOCOMPUTING 2006

ISBN 981-256-463-2

Printed in Singapore by Mainland Press

PACIFIC SYMPOSIUM ON BIOCOMPUTING 2006

The stunning advances in stem cell research by Drs. Woo Suk Hwang and Shin Yong Moon of the Seoul National University have captured the world's attention. Meanwhile, stem cell research in the U.S. is being impeded for a complex of reasons. Bioethicist Eric Meslin, Director of the Indiana University Center for Bioethics and the Assistant Dean for Bioethics of the Indiana University School of Medicine, will address this timely issue in his keynote lecture. We are looking forward to this extremely important presentation and the discussion to follow, not only for the insight it will provide on this particular issue, but also for illuminating trends in the development of U.S. science policy over the last several years.

Michael Ashburner, Professor of Biology at the University of Cambridge and former Joint-Head of the European Bioinformatics Institute (EBI), will deliver the second keynote lecture. His presentation will describe the use of ontologies in biological research. The importance of ontologies for all of our research has been reflected in past sessions on this topic at PSB.

The 13th International Congress on Genes, Gene Families, and Isozymes was held in Shanghai, China, from September 17-21, 2005 (http://www.cafs.ac.cn/page/cafs/guanggao/jiyin/show.asp). This meeting is but one of many international meetings that are being held with increasing frequency in Asia. The significant increases in investment in research as evidenced by the stem cell advances, the improved overall scientific infrastructure as evidenced by the international meetings, and the general increases in the quality of life are encouraging significant numbers of Western-trained Asian scientists, including many PSB participants, to consider accepting research positions in the Far East. We anticipate that these trends will lead to increasing numbers of PSB participants from Pacific Rim countries in the coming years. Such a result would be entirely consistent with the long range goals of the PSB organizers.

Thanks to the effort of Jie Sun and Tiffany Murray of Indiana and Stanford Universities, respectively, PSB articles are now linked with PubMed. Next time your PubMed query leads to a PSB article, our logo (PSB Full Text) will appear above the article's abstract, which of course provides a direct link to the article itself. If you haven't discovered this already, we invite you to try it out. So far only PDF versions of the articles are available. This effort was no small task. The various electronic forms of many different PSB articles had to be converted one-by-one into PDF format and then checked for fidelity. In the early years, PSB articles were submitted on paper, and in many instances the authors never followed up with electronic versions of their published articles. In still other cases, the electronically formatted articles exhibited differences from the actual articles in the PSB proceedings. In all such cases, the paper versions were scanned electronically and PDF-formatted manuscripts were created from the scans. Thus, except for the session introductions and the prefaces to each year's book, the articles in PSB can now be reached automatically from PubMed. We believe that this will be a significant gain for all of you who have published with us or who use articles from PSB in your research. Cudos to Jie Sun and Tiffany Murray for carrying out this project. The PSB online proceedings web site has also introduced a Google-style search facility which makes it much easier to find articles based on simple word searches.

At each PSB meeting, the seeds for the following PSB are planted as participants begin to think about possible sessions for the next meeting. Over the years by discussions with the various participants, each of the organizers has helped to nurture ideas into session proposals. We would like to call to your attention that the process of organizing a session is a great way for a young faculty member to gain visibility for her- or himself, and just as important, this is a great way to stimulate interest in and focus on newly emerging areas of study. We are proud that many areas in biocomputing received their first significant focused attention at PSB. It is no coincidence that many of these sessions were conceived and executed by some of our younger participants. If you have an idea for a new session, we the organizers are available to talk with you, either at the meeting or later by e-mail.

Again, the diligence and efforts of a dedicated group of researchers has led to an outstanding set of sessions. This year, in addition to the sessions, there will also be survey tutorials. These tutorials of one hour for each session are intended to provide key background information for the upcoming presentations. These organizers and their sessions are as follows:

Bobbie-Jo Webb-Robertson, Bill Cannon, Joshua Adkins, and Deborah Gracio,
Computational Proteomics

Andrew G. Clark, Andrew Collins, Francisco M. De La Vega, Kenneth K. Kidd,
Design and Analysis of Genetic Studies After the HapMap Project

Kevin Bretonnel Cohen, Olivier Bodenreider, and Lynette Hirschman,
Linking Biomedical Information Through Text Mining

Maricel Kann, Yanay Ofran, Marco Punta, and Predrag Radivojac,
Protein Interactions in Disease

Robert Stevens, Olivier Bodenreider, and Yves A. Lussier,
Sematic Webs for Life Sciences

In addition to the sessions and survey tutorials, this year's program will include two in depth tutorials of three hours each. The presenters and titles of these tutorials are as follows:

Giselle M. Knudsen, Reza A. Ghiladi, and D. Rey Banatao,
Integration Between Experimental and Computational Biology for Studying Protein Function

Jotun Hein, Mikkel Schierup, and Thomas Mailund,
Association Mapping: Fundamental Principles and Applications

The Department of Energy and the National Institutes of Health are thanked again for their continuing support of this meeting. Their support both enables the infrastructure of the meeting and also provides travel grants to many of the participants. Applied Biosystems and the International Society for Computational Biology continue to sponsor PSB, and as a result, we are able to provide travel grants to many meeting participants.

PSB is now receiving far more submitted manuscripts than can be accommodated in our program. As a result, many excellent papers cannot be included, and as a further result, the task of reviewing the papers has become increasingly difficult. Thus, we would like to especially acknowledge the many busy researchers who found the time to review the submitted manuscripts on a very tight schedule. The partial list following this preface does not include many who wished to remain anonymous, and of course we apologize to any who may have been left out by mistake.

Aloha!

Pacific Symposium on Biocomputing Co-Chairs September 27, 2005

Russ B. Altman
Department of Genetics, Stanford University

A. Keith Dunker
Department of Biochemistry and Molecular Biology, Indiana University School of Medicine

Lawrence Hunter
Department of Pharmacology, University of Colorado Health Sciences Center

Teri E. Klein
Department of Genetics, Stanford University

Thanks to the reviewers...

Finally, we wish to thank the scores of reviewers. PSB requires that every paper in this volume be reviewed by at least three independent referees. Since there is a large volume of submitted papers, paper reviews require a great deal of work from many people. We are grateful to all of you listed below and to anyone whose name we may have accidentally omitted or who wished to remain anonymous.

Patrik Aloy
Gitte Andersen
Aida Andres
Dominik Aronsky
Lan Aronson
John Belmont
Jadwiga Bienkowska
Christian Blaschke
Olivier Bodenreider
Stefan Böhringer
Phil Bradley
Carol Bult
Martha L. Bulyk
Evelyn Camon
Michael Cantor
Stephen Chanock
Nitesh Chawla
Ming Chen
Hao Chen
Cheng Cheng
Aaron Cohen
Nigel Collier
Marc Colosimo
Rob Culverhouse
Mehmet Dalkilic
Mariza de Andrade
Seth Dobrin
Mary Dolan
Andreas Doms
Tarazona, Eduardo
Sarah Ennis
Piero Fariselli
Jan Feng
Jose' Maria
Fernandez
Steven Finch
Alessandro Flammini
Sarel Fleishman
Federico Fogolari
Kristofer Franzen
Tim Frayling
Iddo Friedberg
Katheleen Gardiner
John Gennari
Carole Goble
Sheng Gu
Dan Gusfield
Bjarni Halldorsson
Inbal Halperin
Midori Harris
Frank Hartel
Mark Hoffman
Fiona Hyland
Lilia Iakoucheva
Cliff Joslyn
Raja Jothi
Rachel Karchin
Vipul Kashyap
Toni Kazic
Sun Kim
Gad Kimmel
Mickey Kosloff
Patrick Lambrix
Ross Lazarus
Suzi Lewis
Lang Li
Jian Li
Hongfang Liu
Jane Lomax
Philip Lord
Esti Yeger Lotem
Joanne Luciano
Yves Lussier
Bob MacCallum
Tom Madej
Yael Mandel-
Gutfreund
Nikolas Maniatis
Elisa Margotti
Leonardo Marino-
Ramirez
Andrew McBride
Jacob McCauley
Alexa McCray
Robin McEntire
Brett McKinney
Cristian Micheletti
Joyce Mitchell
Madan Mohan
Sean Mooney
Jason Moore
Alex Morgan
Andrew Morris
Bickol Mukesh
Mark Musen
Murad Nayal

Michael Ng
David Page
Grier Page
Andrew Pakstis
Anna Panchenko
Helen Parkinson
Paul Pavlidis
Tzu Phang
Ron Y. Pinter
Bin Qian
Arun Ramani
Tom Rindflesch
Marylyn Ritchie
Andrea Rossi
Loic Royer
Indra Neil Sarkar
Mikkel Heide Schierup
Avner Schlessinger
Santiago Schnell
Michael Schroeder
Stefan Schulz
Steffen Schulze-Kremer
Laura J. Scott
Robert D. Sedgewick
Ben Shoemaker
Trevor Siggers
Mona Singh
Barry Smith
Einat Sprinzak
Robert Stevens
Daniel Stram
He Tan
Lorraine Tanabe
Haixu Tang
Nelson Tang
Will Tapper
Michael Thompson
Caroline Thorn
Tricia Thornton
Kevin Thornton
Silvio Tosatto
Anna Tramontano
Jun'ichi Tsujii
Daniele Turi
Vladimir Uversky
Vladimir Vacic
Alfonso Valencia
Karin Verspoor
Slobodan Vucetic
Thomas Waechter
Bonnie Webber
Patricia Whetzel
W. John Wilbur
David Wild
Jennifer Williams
Ian Wilson
Limsoon Wong
Chris Wroe
Cathy Wu
Rongling Wu
Momiao Xiong
Golan Yona
Yi-Kuo Yu
Chengfeng Zhao
Hongyu Zhao

CONTENTS

SEMANTIC WEBS FOR LIFE SCIENCES

DESIGN AND ANALYSIS OF GENETIC STUDIES AFTER THE HAPMAP PROJECT

COMPUTATIONAL APPROACHES FOR PHARMACOGENOMICS

LINKING BIOMEDICAL INFORMATION THROUGH TEXT MINING: SESSION INTRODUCTION

K. BRETONNEL COHEN
Center for Computational Pharmacology
University of Colorado
Denver, CO 80045, USA

OLIVIER BODENREIDER
National Library of Medicine
Bethesda, MD 20894, USA

LYNETTE HIRSCHMAN
The MITRE Corporation
Bedford, MA 01730, USA

This session is focused on text mining applications that link information from the biomedical literature to the growing array of structured resources available to researchers, such as protein databases (e.g., UniProt, PDB, PIR), model organism databases (e.g., FlyBase, MGI, SGD), ontologies (the Gene Ontology, as well as the growing number of ontologies in OBO – Open Biological Ontologies), and nomenclatures (HUGO, HUPO). To achieve this focus, there was an explicit requirement that submissions include both a text mining component and a mapping between at least two publicly available data sources. There were twenty papers submitted to this session, with nine papers accepted (7 for oral presentation).

This session builds on two threads of work that have been well represented at past PSB meetings, namely text mining and ontologies. There have been PSB sessions on text mining in 2000, 2001, 2002, and 2003. Many of the systems discussed in these earlier sessions focused on recognition of biomedical entities and relations in order to provide effective indexing into the literature. Other papers focused on topic-based document clustering, to provide tools to manage the vast biomedical literature at the document level. However, these systems were limited in that they did not link to resources outside the text collections (generally PubMed). The entity recognition systems identified entities or relations by simply pointing to substrings in the input text. Such outputs are of intrinsically limited value. For example, a system that produces a table of protein-protein interactions is potentially highly valuable if it refers to specific entities in PDB, but of much more limited utility if it outputs only a list of potentially ambiguous symbols and names.

The second relevant thread at PSB investigated the linguistic and semantic characteristics of a variety of publicly available biomedical data sources, including gene names and Gene Ontology terms. Much of this work was presented at PSB sessions on ontologies in 2003, 2004, and 2005 or the PSB sessions on biomedical language processing listed above. Of particular interest is the identification of various kinds of relations among biomedical entities, which can enrich existing ontologies and subsequently benefit text mining.

This 2006 session on linking biomedical information represents the logical next step. Our goal has been to solicit papers that follow through on the insights gained into the structure of available data sources and advances in text mining, to create language processing systems that not only locate information in texts, but also map it to these explicit knowledge models. Two recent competitive evaluation tasks from BioCreAtIvE (Critical Assessment of Information Extraction in Biology) showed that it is possible to create systems that produce grounded outputs and to perform principled evaluations of them. BioCreAtIvE Task 1b [1] involved mapping references to genes in abstracts to specific gene identifiers from the appropriate model organism database. BioCreAtIvE Task 2 [2] involved assigning Gene Ontology terms to proteins mentioned in journal articles. Taken together, these two tasks demonstrate that it is possible to link the literature to specific entities and to specific concepts. At the same time, they make it clear that there is considerable room for improvement in performance of these tasks.

The papers for this session demonstrate the progress that has been made in using text mining to link across resources and to anchor mentions of biological entities to accepted biological nomenclatures and ontologies. These papers tackle a number of biological problems using a variety of technologies:

- Four papers emphasize the linkage to ontologies. One paper (Johnson et al.) discusses lexically-based techniques for ontology alignment between GO and several other ontologies. The other three papers focus on annotation into an ontology: Höglund et al. produce improved results for subcellular localization by combining both sequence data and text mining; Lussier et al. describes PhenoGO, a system that maps from text into one of several anatomical ontologies; and Stoica and Hearst describe improved results for BioCreAtIvE

Task 2 (functional annotation of papers on human proteins) by using orthologous genes in Mouse.

- Two papers describe summarization applications: Lu et al. focus on generation of GeneRIFs based on overlap of GO annotations with PubMed abstracts; Ling et al. describe an algorithm for generation of summaries by identifying documents about a particular gene and then extracting the most relevant sentence(s) for six aspects of gene function to create the summary.
- The paper by Vlachos et al. uses relations from the Sequence Ontology to improve on named entity results for FlyBase genes and to support an ontology-based coreference resolution strategy for these genes and gene products.

Overall, the papers in this session reflect the growing maturity of text mining as a bioinformatics tool that can be used, often in conjuction with other bioinformatics tools, to extract knowledge from the biomedical literature and to integrate it effectively with other knowledge sources.

References

1. Hirschman L, Colosimo M, Morgan A, Yeh A. Overview of BioCreAtIvE task 1B: normalized gene lists. BMC Bioinformatics. 2005;6 Suppl 1:S11.
2. Blaschke C, Leon EA, Krallinger M, Valencia A. Evaluation of BioCreAtIvE assessment of task 2. BMC Bioinformatics. 2005;6 Suppl 1:S16.

EXTRACTION OF GENE-DISEASE RELATIONS FROM MEDLINE USING DOMAIN DICTIONARIES AND MACHINE LEARNING

HONG-WOO CHUN[1], YOSHIMASA TSURUOKA[1,2], JIN-DONG KIM[1,2], RIE SHIBA[3,4], NAOKI NAGATA[3], TERUYOSHI HISHIKI[3], AND JUN'ICHI TSUJII[1,2,5]

1. Tsujii Laboratory, Room 615, 7th Building of Science, University of Tokyo, Hongo 7-3-1, Bunkyo-ku, Tokyo, 113-0033, Japan
2. CREST, Japan Science and Technology agency, Hongo, Bunkyo-ku, Tokyo, 113-0033, Japan
3. Biological Information Research Center, National Institute of Advanced Industrial Science and Technology, AIST Waterfront Bio-IT Research Building, Aomi 2-42, Koto-ku, Tokyo, 135-0064, Japan
4. Integrated Database Team, Japan Biological Information Research Center, Japan Biological Informatics Consortium, AIST Waterfront Bio-IT Research Building, Aomi 2-42, Koto-ku, Tokyo, 135-0064, Japan
5. School of Informatics, University of Manchester POBox 88, Sackville St, MANCHESTER M60 1QD, UK
E-mail: {chun,tsuruoka,jdkim,tsujii}@is.s.u-tokyo.ac.jp, {rshiba,nnagata,t-hishiki}@jbirc.aist.go.jp

We describe a system that extracts disease-gene relations from *MedLine*. We constructed a dictionary for disease and gene names from six public databases and extracted relation candidates by dictionary matching. Since dictionary matching produces a large number of false positives, we developed a method of machine learning-based named entity recognition (NER) to filter out false recognitions of disease/gene names. We found that the performance of relation extraction is heavily dependent upon the performance of NER filtering and that the filtering improves the precision of relation extraction by 26.7% at the cost of a small reduction in recall.

1. Introduction

The continuing rapid development of the internet makes it very easy to quickly access large amounts of data online. However, it is impossible for a single human to read and comprehend a significant fraction of the available information, and there is a real need for the application of natural language processing techniques in many domains that would facilitate quick and easy retrieval of useful information. Genomics is not an exception. Databases such as *MedLine* have a vast amount of knowledge.

Our aim in this paper is to extract diseases and their relevant genes from *MedLine* abstracts, which we term *relation extraction*. There are some existing systems for relation extraction from biomedical literature. Arrow-Smith (Swanson 1986) [1] and BITOLA (Hristovski 2003) [2] extract relations between diseases and genes using background knowledge about the chromosomal location of the starting disease as well as the chromosomal location of the candidate genes from resources such as LocusLink, HUGO and OMIM. These systems are designed to discover new, potentially meaningful relations between diseases and genes which do not occur together in the same published article. If concept X and concept Y are related to each other, the systems assume that concepts Z and X have some relationship if Z is relevant to Y. Finally, the systems check whether X and Z appear together in the medical literature. If they do not appear together, this pair (X and Z) is considered as a potentially new relation. G2D (Perez-Iratxeta 2002) [3] also extracts relations by *Relative score*, which is calculated by co-occurrence information. G2D assumes that relevant terms occur together in many abstracts. An appealing feature of these three systems is that all outputs of these systems are terms used in publicly available biomedical data sources, which means these outputs are linked to such databases and can be used by other researchers. However, these approaches have some problems: Their results could conceivably contain a lot of false positives because they yield too many relations that are dependent only on the co-occurrence information; so many of their results may be unreliable. They have done only a preliminary analysis on the precision of the outputs.

There are some studies that employ various NLP techniques in order to obtain high-precision knowledge from biomedical literature. Proux (2000) [4] extracted gene-gene interactions by manually constructed predicate patterns, which they call scenarios. For example, '[gene product] *acts* as a [modifier] of [gene]' is a scenario of the predicate 'act', which can cover a sentence like: "Egl protein *acts* as a repressor of BicD". In this approach, they employed several techniques for linguistic analysis. Concerning the named entity recognition, they used a part-of-speech (POS) tagger that is based on finite state transducers (FST). This POS tagger contained tokenization and morphological analysis to provide possible POS tags. They used a Hidden Markov Model (HMM) for disambiguation and domain-specific corpora for correcting errors. They then attempted to identify entity names. After that, they did shallow parsing of local structures around verbs to analyze their subjects and objects and made a conceptual graph using a domain-specific ontology. Experimental results show 81% precision and 44% recall. Pustejovsky (2002) [5] also used predicate patterns. They did not build these patterns manually, but extracted patterns from a manually-constructed training corpus. Then they analyzed the subject and the object relation for a main verb to extract them as the arguments for a relation. In this approach, they attempted to recognize entity names by shallow parsing and identify semantic type using a domain ontology, and they dealt with acronym problems and anaphora resolution. Experimental results show 90% precision and 59% recall. The advantages of these approaches are that they considered various contextual features using NLP techniques. However, these approaches have a problem in terms of extracting practical and reusable biological knowledge. The outputs only provide information about relations among the "terms" appearing in text. In other words, the entities in the outputs are not explicitly linked to entities in biological databases. If the outputs provide links to explicit knowledge models, then the utility of these outputs will be increased for other researchers.

In this paper, we extract relations by named entity recognition that consists of two steps. The first step uses a dictionary-based longest matching technique. We create dictionaries constructed from public biomedical databases, which enables us to explicitly link extracted relations with the entries in such databases. Since dictionary-based matching produces many false positives, we filter them out by machine learning in the second step.

2. Relation Extraction using Dictionaries and Machine Learning

Figure 1 shows the architecture of our system. Our system first collects sentences that contain at least one pair of disease and gene names, using the dictionary-based longest matching technique. The system then attempts to extract a binary relation between the disease and gene names in each sentence [a].

In this work, we use machine learning to filter out false positives from the dictionary-based longest matching results.

[a] When a sentence contains more than one disease or one gene, the system makes copies of the sentence according to the number of disease-gene pairs. We call each of these copies $co-occurrence$, and regard these items as the input unit of our system. For example, if there are two gene names and one disease name in a sentence, then our system makes two co-occurrences for this sentence.

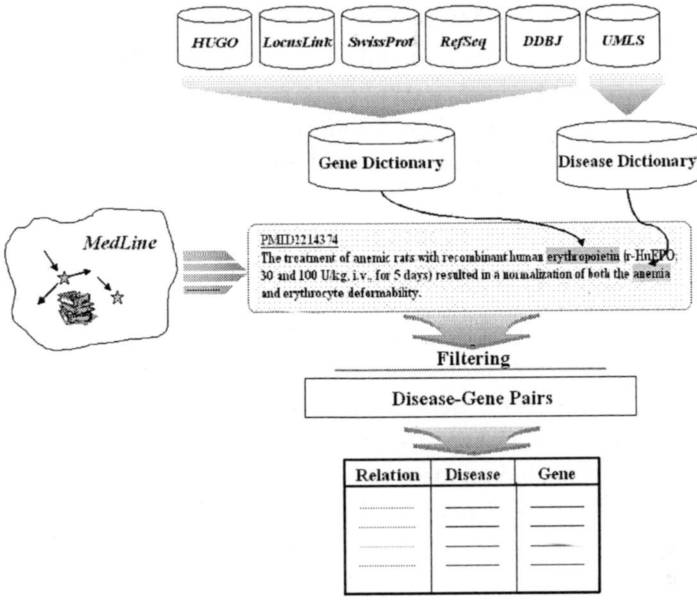

Figure 1. The system architecture

We have three types of false positives in the dictionary-based results:

- False gene names
- False disease names
- False relations

There are some existing studies in natural language processing aimed at filtering out the first two types of false positives. Tsuruoka and Tsujii [6] proposed a dictionary-based longest matching approach for protein name recognition where they employed a Naive-Bayes classifier to filter out false positives. However, since their dictionary was constructed from the training corpus, their experimental setting is different from the real situation where we have a dictionary constructed from biomedical databases. Furthermore, they used only local context as the features for filtering.

In the following sections, we explain our techniques including dictionaries, a corpus, and the NER filter in detail.

2.1. *Construction of the Gene and Disease Dictionaries*

In order for each output entry to be linked to publicly available biomedical data sources, we created a human gene dictionary and a disease dictionary by merging the entries of multiple public biomedical databases. These two dictionaries provide gene and disease-related terms and cross-references between the original databases.

2.1.1. *The gene dictionary*

A unique *LocusLink* identifier for genetic loci is assigned to each entry in the gene dictionary, which enables us to consistently merge gene information dispersed in different databases. Each entry in the merged gene dictionary holds all relevant literature information associated with a given gene. We used five public databases to build the gene dictionary: *HUGO*, *LocusLink*, *SwissProt*, *RefSeq*, and *DDBJ*(July 2004). Each entry consisted of five items: gene name, gene symbol, gene product, chromosomal band, and PubMed ID tags. Based on these principles, we created a database-merging system to automatically collect relevant gene information from biomedical data resources. The current version of the gene dictionary contains a total of 34,959 entries with 19,815 HUGO-approved gene symbols, 19,788 HUGO-approved gene names, and 29,470 gene products. It should be noted that there are numerous alias gene symbols and alias gene names in these entries. We found at least 202 approved gene symbols and 253 approved gene names that are used as aliases, in different entries, or entries without a LocusLink identifier. This tedious merging of data is a result of inconsistencies between databases that cannot be simply solved by combining data into one database. In addition, some words belong to multiple categories and cannot be easily classified into one category. We plan to address these problems in the near future by improving our algorithms. We also hope to improve the merging system to create other types of dictionaries that will allow comparative genome research.

2.1.2. *The disease dictionary*

We used the Unified Medical Language System (UMLS) to collect disease-related vocabulary. From the 2003AC edition of the UMLS Metathesaurus, we selected 12 TUIs (unique identifiers of semantic types) that correspond to diseases names, types of abnormal phenomena, or their symptoms (Table 1). From these TUIs, 431,429 SUIs (unique identifiers for strings) for

Table 1. Selected TUIs (Unique identifiers of semantic type)

T019	Congenital Abnormality
T020	Acquired Abnormality
T033	Finding
T037	Injury or Poisoning
T046	Pathologic Function
T047	Disease or Syndrome
T048	Mental or Behavioral Dysfunction
T049	Cell or Molecular Dysfunction
T050	Experimental Model of Disease
T184	Sign or Symptom
T190	Anatomical Abnormality
T191	Neoplastic Process

159,448 CUIs (unique identifiers for concepts) were extracted and stored as a disease-related lexicon.

2.2. *Annotation of Corpus*

The purpose of building an annotated corpus is to construct the training data for machine learning that will filter out false positives from the dictionary-based results.

To build training and testing sets, 1,362,285 abstracts were collected through a Medline search, using Medical Subject Headings (MeSH) terms. In this work, we used *"Diseases Category"[MeSH] AND ("Amino Acids, Peptides, and Proteins"[MeSH] OR "Genetic Structures"[MeSH])* as the keywords. From the resulting abstracts, we generated 2,503,037 co-occurrences using the dictionary-based longest matching technique. Each co-occurrence is a candidate of a relation between one disease and one gene. We chose 1,000 co-occurrences randomly[b], and they were annotated by one biologist.

Figure 2 shows an example of an annotation. Disease and gene candidates are highlighted: there are four candidates in two co-occurrences. *PRCC* and *PSA* are candidate genes and *renal cell carcinoma* and *BPH* are candidate diseases. These items were recognized by the dictionary-based longest matching technique. The check boxes labeled *correct gene* and *correct disease* are marked by a biologist if he considers the candidates to be correct gene (or disease) names[c].

As for the annotation on disease-gene relations, we considered the fol-

[b]We checked all the 1,000 co-occurrences and found that they were all different sentences

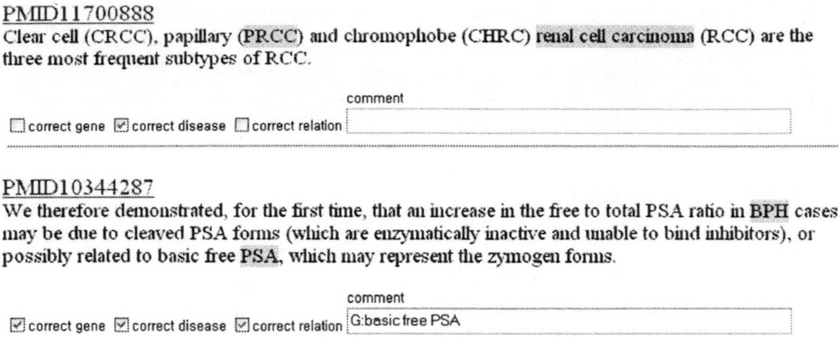

PMID11700888
Clear cell (CRCC), papillary (PRCC) and chromophobe (CHRC) renal cell carcinoma (RCC) are the three most frequent subtypes of RCC.

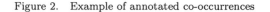

comment

☐ correct gene ☑ correct disease ☐ correct relation

PMID10344287
We therefore demonstrated, for the first time, that an increase in the free to total PSA ratio in BPH cases may be due to cleaved PSA forms (which are enzymatically inactive and unable to bind inhibitors), or possibly related to basic free PSA, which may represent the zymogen forms.

comment

☑ correct gene ☑ correct disease ☑ correct relation G:basic free PSA

Figure 2. Example of annotated co-occurrences

lowing three aspects. In other words, the annotator judged a co-occurrence as "correct" if any of the following three types of relations between the gene and disease was described in the sentence.

- Pathophysiology, or the mechanisms of diseases, containing etiology, or the causes of diseases.
- Therapeutic significance of the genes or the gene products, more specifically classified to their therapeutic use and their potential as therapeutic targets.
- The use of the genes and the gene products as markers for the disease risk, diagnosis, and prognosis.

Among 1,000 co-occurrences, 572 co-occurrences contained correctly identified diseases and genes by a biologist. The important observation was that 94% of the 572 co-occurrences were annotated as correct relations, which means that there are few false positives for relations if the disease and gene names are correct. Therefore, we did not perform filtering for relations in this work. Figure 3 shows an example of the remaining 6% of the 572 co-occurrences whose gene and disease were identified as correct but whose relation was incorrect.

---

and they all came from different abstracts.

[c]A name can be embedded in a different name. For example, the dictionary matching may find the disease name *APC* in the term *APC gene*, in which *APC* would be annotated as "incorrect". Embedded names are a major source of false recognitions of gene/disease names.

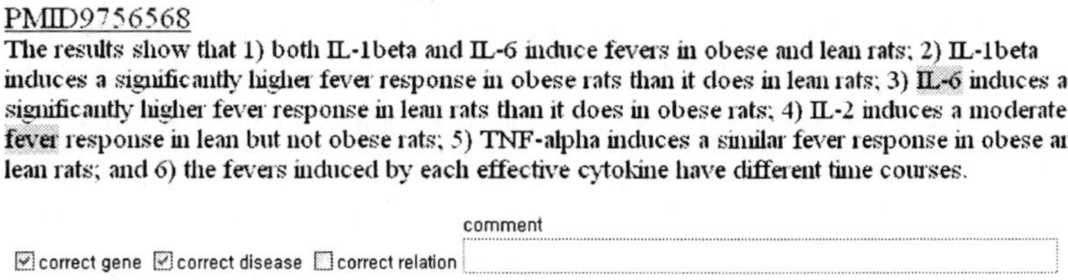

PMID9756568
The results show that 1) both IL-1beta and IL-6 induce fevers in obese and lean rats; 2) IL-1beta induces a significantly higher fever response in obese rats than it does in lean rats; 3) IL-6 induces a significantly higher fever response in lean rats than it does in obese rats; 4) IL-2 induces a moderate fever response in lean but not obese rats; 5) TNF-alpha induces a similar fever response in obese and lean rats; and 6) the fevers induced by each effective cytokine have different time courses.

comment

☑ correct gene ☑ correct disease ☐ correct relation

Figure 3. An example of an annotated co-occurrence whose gene and disease are identified as correct but relation as incorrect

2.3. *Filtering with a Maximum Entropy-based NER Classifier*

To improve the precision of recognizing gene and disease names, we propose the use of a maximum entropy model to filter out false positives. Maximum entropy models exhibited the best performance in the CoNLL-2003 Shared Task of NER, and are widely used in classification problems in natural language processing. For smoothing, we used Gaussian prior modeling and tuned this parameter with empirical experiments and set it to 300 for genes and 400 for diseases.

2.3.1. *Features for NER*

The feature sets used in our experiments are as follows:

- Candidate names and contextual terms:
The features we considered were the candidate name itself as well as unigrams and bigrams. A unigram refers to the word either before or after the candidate name; a bigram refers to the two adjacent words either before or after the candidate name.
- Head word information and the predicate:
We used the head word information (the word itself and its part-of-speech) of the maximal projection of the disease/gene name as a feature. This analysis is given by the deep-syntactic parser ENJU [7][d].
In addition, we expect that an important clue for NER is whether or not the candidate is used as an argument of a verb. This is because certain verbs in biomedical literature occur fre-

[d]ENJU achieved 87.85% precision and 86.85% recall on the Penn Treebank and the average parsing time was 360 ms [8].

quently and have a relationship with a disease/gene name; for example, *induce*, *activate*, *contain*, and *phosphorylate*. We named this kind of verb the predicate and considered it as a feature.

- The expanded form of an acronym:
One of the difficulties in term recognition from biomedical literature is the problem of *ambiguous acronyms*. One acronym can be used with different meanings. We can solve this problem if we have access to its full form. Thus, we tried to map the acronym of a candidate name to its full form by scanning the entire abstract. When coming across an acronym, the system searches for the full form of the acronym and uses the last word of the full form as a feature. In practice, an acronym and its full form usually occur simultaneously as *full form* (*acronym*) when they first appear in a document.
- Part-of-speech (POS) tags:
We considered the POSs of the candidate name and its surrounding words. To tag the words with POS labels, we used the *Genia Part-of-Speech Tagger* [9] which is trained on a combined set of the newswire corpus (Penn Treebank) and biological corpus (GENIA corpus [10]).
- Use of capitals and digits in the candidate term:
Capital characters and numbers frequently appear in biomedical terms. We considered whether candidate names contain capital characters and digits or not.
- Greek letters in the candidate term:
Greek letters (e.g. *alpha*, *beta*, *gamma*, etc.) are strong indicators of biomedical terms. These Greek letters appear in their original forms such as α, β, $\Gamma(\gamma)$.
- Affixes of the candidate term:
Prefixes and suffixes can be very important cues for terminology identification. We considered the 11 suffixes given in Table 2. These affixes are commonly used in biomedical terms.

3. Experimental Results

We conducted two sets of experiments for disease-gene relation extraction. One is an experiment without NER filtering and the other is an experiment with NER filtering.

Table 2. Affix feature

Prefix/Suffix	Examples
~cin	actinomycin
~mide	Cycloheximide
~zole	Sulphamethoxazole
~lipid	Phospholipids
~rogen	Estrogen
~vitamin	dihydroxyvitamin
~blast	erythroblast
~cyte	thymocyte
~peptide	neuropeptide
~ma	hybridoma
~virus	cytomegalovirus

3.1. *Experiments without Filtering (Baseline)*

Our baseline experiment is very simple: we assume that all disease-gene pairs recognized by dictionary matching indicate relations. The performance of this baseline experiment is shown in the first row of Table 3.

It should be noted that our dictionaries do not cover all disease/gene names, and thus we cannot calculate the *absolute* recall in this experiment. Instead, we use *relative recall* as a performance measure, and the relative recall given by the baseline method is 100% by definition. In this approach, our interest is in how precise our system is at correctly identifying the relations, rather than how often it misses other meaningful relations.

3.2. *Experiments with Filtering*

The second set of experiments made use of the maximum entropy-based NER filter. Table 3 lists the performance percentages of relation extraction. We found that NER filtering improves the precision of relation extraction by 26.7% at the cost of a small reduction in recall. This suggests that the performance of relation extraction is very much dependent upon the performance of NER. In this experiment, we used the best combination of features for NER (see Table 4):

- Recognition of Gene names:
  Contextual terms, capitalization, Greek letters, POS of disease/gene names and its head, words of predicate and head and full forms if candidate names are acronyms.
- Recognition of Disease names:
  Contextual terms, capitalization, POS of disease/gene names and unigram words and words of head.

Table 3. Relation extraction performance

	Precision(%)	Relative recall(%)
without filtering	51.8	100.0
with filtering	78.5	87.1

Table 4. NER performance

	Features											Precision	Relative recall
	1	2	3	4	5	6	7	8	9	10	11	(%)	(%)
	√	√										86.4	90.2
	√	√	√									85.9	90.2
G	√	√		√								86.2	90.6
E	√	√			√							86.0	90.2
N	√	√				√						86.3	89.4
E	√	√					√					85.9	90.2
	√	√		√		√		√				86.2	90.9
	√	√		√		√				√		86.5	90.5
	√	√		√		√		√	√	√	√	**89.0**	90.9
	√	√										88.5	97.8
D	√		√									88.5	97.9
I	√			√								88.6	98.1
S	√				√							88.6	98.1
E	√					√						88.5	96.0
A	√						√					89.8	95.5
S	√	√				√	√	√				**90.0**	96.6
E	√	√				√	√	√	√	√		89.6	96.6
	√	√				√	√	√			√	89.6	96.0

Note: 1: Candidate disease/gene names and Contextual terms; 2: Use of capitals in the candidate term; 3: Use of digits in the candidate term; 4: Greek letters in the candidate term; 5: Affixes of the candidate term; 6: POS of disease/gene names; 7: POS of disease/gene names and unigram; 8: Head word; 9: POS of head word; 10: Predicates of a candidate disease/gene name; 11: Expanded forms if candidate disease/gene names are acronyms.

All the experimental results for NER considered *contextual terms*. This is because this feature is the most powerful in recognizing candidate names. It leads to improved NER performance of 6.6% for genes and 2.1% for diseases.

4. Conclusion and Future work

The aim of this research was to build a system to automatically extract useful information from publicly available biomedical data sources. In particular, our focus was on relation extraction between diseases and genes. We found that named-entity recognition (NER) using ME-based filtering significantly improves the precision of relation extraction at the cost of a small reduction in recall.

We conducted experiments to show the performance of our relation extraction system and how it depends on the performance of the NER scheme. We could safely regard co-occurrences as containing correct relations if candidate disease and gene names were considered to be correct.

In this work, we did not address the problem of polysemous terms, which would cause difficulty in linking such terms with database entries. One solution would be to incorporate techniques for ambiguity resolution into our system. For example, S. Gaudan et al. proposed the use of SVMs for abbreviation resolution and achieved 98.9% precision and 98.2% recall.

The number of co-occurrences in the training and testing sets was rather small for the purpose of evaluating our system. Future work should encompass increasing the size of the annotated corpus and enriching annotation.

References

1. D.R. Swanson, Fish oil, Raynaud's syndrome, and undiscovered public knowledge, *Perspect Biol Med*, 30(1), pp.7–18 (1986).
2. D. Hristovski, B. Peterlin, J.A. Mitchell, and S.M. Humphrey, Improving literature based discovery support by genetic knowledge integration, *Stud. Health Technol. Inform.*, 95, pp.68–73 (2003).
3. C. Perez-Iratxeta, P. Bork, M.A. Andrade, Association of genes to genetically inherited diseases using data mining, *Nat Genet*, 31(3), pp.316–319 (2002).
4. D. Proux et al., A pragmatic information extraction strategy for gathering data on genetic interactions, *ISMB*, 8, pp.279–285 (2000).
5. J. Pustejovsky et al., Medstract : Creating Large-scale Information Servers for biomedical libraries, *Proceedings of the Workshop on Natural Language Processing in the Biomedical Domain*, pp.85–92 (2002).
6. Y. Tsuruoka and J. Tsujii, Boosting Precision and Recall of Dictionary-Based Protein Name Recognition, *Proc. of the ACL-03 Workshop on Natural Language Processing in Biomedicine*, pp.41–48 (2003).
7. Enju v1.0: http://www-tsujii.is.s.u-tokyo.ac.jp/enju/index.html (2004).
8. T. Ninomiya, Y. Tsuruoka, Y. Miyao, and J. Tsujii, Efficacy of Beam Thresholding, Unification Filtering and Hybrid Parsing in Probabilistic HPSG Parsing, *Proceedings of the 9th International Workshop on Parsing Technologies* (2005).
9. GENIA Part-of-Speech Tagger v0.3: http://www-tsujii.is.s.u-tokyo.ac.jp/GENIA/postagger/ (2004).
10. GENIA Corpus 3.0p: http://www-tsujii.is.s.u-tokyo.ac.jp/genia/topics/Corpus/3.0/GENIA3.0p.intro.html (2003).
11. S. Gaudan et al., Resolving abbreviations to their senses in Medline, *Bioinformatics*, 21(18), pp.3658–3664 (2005).

SIGNIFICANTLY IMPROVED PREDICTION OF SUBCELLULAR LOCALIZATION BY INTEGRATING TEXT AND PROTEIN SEQUENCE DATA

ANNETTE HÖGLUND[†], TORSTEN BLUM[†], SCOTT BRADY[‡], PIERRE DÖNNES[†], JOHN SAN MIGUEL[‡], MATTHEW ROCHEFORD[‡], OLIVER KOHLBACHER[†], HAGIT SHATKAY[‡*]

† Div. for Simulation of Biological Systems, ZBIT/WSI, University of Tübingen, Sand 14, D-72076 Tübingen, Germany

‡ School of Computing, Queen's University, Kingston, Ontario, Canada K7L 3N6

Computational prediction of protein subcellular localization is a challenging problem. Several approaches have been presented during the past few years; some attempt to cover a wide variety of localizations, while others focus on a small number of localizations and on specific organisms. We present a comprehensive system, integrating protein sequence-derived data and text-based information. It is tested on three large data sets, previously used by leading prediction methods. The results demonstrate that our system performs significantly better than previously reported results, for a wide range of eukaryotic subcellular localizations.

1. Introduction

In this paper we introduce a new system for computationally assigning proteins to their subcellular localization. By integrating several types of sequence-derived features and text-based information, the achieved performance is the best reported so far, in terms of sensitivity, specificity, and overall accuracy. Unlike several recent systems which focus on a few subcellular localizations or on a specific organism[1,2,3,4], our system is applicable to – and retains its good performance across – a wide variety of organisms and subcellular localizations. Moreover, we show that the integrated system, which combines sequence and text, performs significantly better than its individual components, based on each data source alone.

The task of protein subcellular localization prediction is important and well-studied[5,6]. Knowing a protein's localization helps elucidate its function, its role in both healthy processes and in the onset of disease, and its potential use as a drug target. Experimental methods for protein localization range from immunolocalization[7] to tagging of proteins using green fluorescent protein (GFP)[8]

*To whom correspondence should be addressed: shatkay@cs.queensu.ca. HS is supported by NSERC Discovery grant 298292-04.

and isotopes[9]. Such methods are accurate but, even at their best, are slow and labor-intensive compared with large-scale computational methods. Computational tools for predicting localization are useful for a large-scale initial "triage", especially for proteins whose amino acid sequence may be determined from the genomic sequence, but are hard to produce, isolate, or locate experimentally.

The past decade, and most notably the last five years, has seen much progress in computational prediction of protein localization from sequence data. Nakai and Kanehisa[10,11] introduced PSort, a rule-based expert system, which was later improved upon by a probabilistic[12] and by a K-nearest neighbor[13] classifier. Another pair of prominent systems, TargetP[1] and ChloroP[14], based on artificial neural networks, demonstrated a significantly higher accuracy when applied to a limited set of subcellular localizations in plant and animal cells. Other recent systems use a variety of machine learning techniques. Most of them focus on a few subcellular localizations and improve upon – or just meet – the state of the art on those[15,3,16].

Several recent publications have examined the possibility of using text to support subcellular localization. Specifically, Stapley *et al.*[17] represented yeast proteins as vectors of weighted terms from all the PubMed articles mentioning their respective genes. They then trained a support vector machine (SVM) on protein-text-vectors, to distinguish among subcellular localizations. The performance was favorable when compared to a classifier trained on amino acid composition alone, but it was not compared against any state-of-the-art localization system, and the reported results do not suggest an improvement over earlier systems. Moreover, while their text-based classifier performed better than an amino acid composition classifier, combining the two forms of data did not significantly improve performance with respect to the text-based classifier alone.

Nair and Rost[2] used the text taken from Swiss-Prot annotations of proteins to represent these proteins, and trained a subcellular classifier using this representation. They concentrate on a few subcellular localizations, and report results that are compatible – but do not improve upon – the state of the art at that time. Their work was elaborated upon by Eskin and Agichtein[18], who added subsequences from the protein's amino acid sequence as part of the terms considered in the text representation. The system was not tested against existing systems or data sets, and the reported results do not indicate improvement over previous systems.

The best performing comprehensive systems reported so far, which were tested on a large set of proteins, are PLOC[19] and, more recently, MultiLoc[20]. While they report the best accuracy until now, on a broad range of organisms and localizations, there is still room for improvement.

The work reported here, similarly to that reported by Nair and Rost[2], uses Swiss-Prot as a text source. Unlike them though, we use the PubMed abstracts referenced by Swiss-Prot, rather than the annotation text placed by Swiss-Prot curators. Furthermore, unlike Stapley *et al.* who use all abstracts that contain the gene name for the protein, we use only abstracts that are referenced by Swiss-Prot, and moreover, rather than use all the terms in them with a standard (TF*IDF[a]) weighting, as done by Stapley *et al.*, we select terms based on a *distinguishing* criterion described in Section 2, and apply a probability-based weighting scheme. We train an SVM as a text-based classifier, and combine it with a sequence-based classifier, to produce a comprehensive subcellular categorizer. Our integrated system is tested on a number of publicly available, extensive, homology-reduced, data sets which were used for evaluating earlier systems (TargetP, PLOC, and MultiLoc). For each system, we first conduct a comparison using the same data and the same subcellular localizations as reported in the paper published about that system. We then conduct a test using all the proteins in Swiss-Prot for which a subcellular annotation is assigned, among the 11 localizations: chloroplast, cytoplasm, endoplasmic reticulum, extracellular space, Golgi apparatus, lysosome, mitochondria, nucleus, peroxisome, plasma membrane, and vacuole. On each of the data sets our system performs better than the state-of-the-art systems in terms of overall prediction accuracy, and other standard measures.

The next section outlines the methods used, while in Section 3 we demonstrate the performance of our system. Section 4 concludes and outlines future work.

2. Methods

Our system combines five separate classifiers, four sequence-based and one text-based. Their output is integrated through a sixth classifier to produce an improved prediction of protein subcellular localization. The sequence-based classifiers have been successfully used before by the MultiLoc system[20] and are briefly described below. Section 2.2 then presents the novel text-based method, while Section 2.3 explains how all these classifiers are combined to form an integrated prediction system. Four of the five classifiers are based on support vector machines (SVMs), using the LIBSVM implementation[21]. The latter supports soft, probabilistic categorization for n-class tasks[22], assigning to each classified item an n-dimensional vector denoting its probability to belong to each of the n classes. Radial Basis Function kernels were used throughout this study. Further details are given below.

2.1. *Sequence-based methods*

Each of the sequence-based classifiers utilizes a different approach to derive biologically informative features that can be used to predict localization, and classifies the input protein sequence to its respective localization using these features.

[a] An acronym for Term Frequency, Times Inverse Document Frequency.

Three of these classifiers are SVM-based. The fourth scans the protein sequences for short sequence motifs indicative of structure and function. The four classifiers are briefly described below (see the MultiLoc paper[20] for further details).

SVMTarget – This classifier uses the N-terminal targeting peptide (TP) to predict a few subcellular categories. It distinguishes among four plant (chloroplast (*ch*), mitochondria (*mi*), secretory pathway (*SP*), and other (*OT*)) and three non-plant (*mi, SP, OT*) localizations. The targeting peptides are represented by their partial amino acid composition, motivated by the observation that TPs for specific localizations have a similar amino acid composition while their actual sequence may differ. Given an input protein, the classifier outputs a three-dimensional vector (four-dimensional for plant) of class probabilities. SVMTarget alone demonstrated a slightly better performance than TargetP[1] in a comparative study[20].

SVMSA – Some proteins of the secretory pathway carry a signal anchor (SA) that, unlike the targeting peptide, is usually located further away from the N-terminus and contains a longer hydrophobic component. SVMSA can predict secretory pathway (*SP*) proteins that are hard to detect using SVMTarget. It is a binary classifier, trained to distinguish proteins carrying SA from those that do not. It outputs, given an input sequence, its probability to contain a signal anchor.

SVMaac – This method uses the whole protein amino acid composition (*aac*), and categorizes proteins into any of the possible localizations. It combines a collection of binary classifiers, each trained to distinguish one class from all others, although one classifier in the collection was especially trained to distinguish cytosolic (*cy*) from nuclear (*nu*) proteins, as these are hard to separate using the one-against-all approach. Given an input protein, p, with n possible localizations, the classifier outputs an n-dimensional probability vector containing p's probability to belong to each localization.

MotifSearch – Proteins from several subcellular localizations can be characterized by a few types of short sequence motifs, such as Nuclear Localization Signal and DNA-binding domains. The motifs were obtained from the PROSITE[23] and from the NLSdb[24,25] databases. This classifier outputs a discrete, binary vector, representing the presence (1) or the absence (0) of each type of motif in the query protein sequence.

2.2. *Text-based method*

The idea underlying the text-based classifier is the representation of each protein as a vector of weighted text features. While text-based localization has been presented before[2,17], the key differences between the current work and previous ones is in the text source used, the feature selection, and the term weighting scheme.

First, for each protein the text comes from the abstracts curated for the protein in its Swiss-Prot entry. We used a script that scanned each protein in Swiss-Prot for all the PubMed identifiers occurring in its Swiss-Prot entry, and obtained the respective title and abstract[b] from PubMed. Each protein is thus assigned a set of PubMed abstracts, based on Swiss-Prot. This choice of abstracts is different from that of Stapley *et al.*[17] who used all the PubMed abstracts mentioning the gene's name, and from that of Nair and Rost[2] – who use Swiss-Prot annotation text rather than PubMed abstracts. The assigned abstracts are then tokenized into a set of terms, consisting of singleton and pairs of consecutive words, with a list of standard stop words excluded from consideration. The results reported here also include the application of Porter stemming[26] to all the words in the terms.

Second, from all the extracted terms, we select a subset of *distinguishing terms*. This is done by scoring each term with respect to each subcellular localization, where the score reflects the probability of the term to occur in abstracts that are associated with proteins of this certain localization. Intuitively, a term is *distinguishing* for a localization L, if it is much more likely to occur in abstracts associated with localization L than with abstracts associated with all other localizations. We formalize this idea in the following paragraphs.

Let t be a term, L a localization, and p a protein. If protein p is known to be localized in L, we denote this $p \in L$. We also define the following sets:

- The set of all PubMed abstracts associated with protein p according to Swiss-Prot, denoted D_p ;
- The set of all proteins known to be localized at L, denoted P_L ;
- The set of abstracts that are associated with a localization L, denoted D_L, is defined as: $D_L = \bigcup_{p \in P_L} \{d \mid d \in D_P\}$. It is the set of all the abstracts associated with the proteins that are in localization L. The number of documents in this set is denoted $|D_L|$.

The probability of a term t to be associated with a localization L, denoted Pr_L^t, is defined as the conditional probability of the term to appear in a document, given that the document is associated with the localization: $Pr_L^t = Pr(t \in d \mid d \in D_L)$. A maximum likelihood estimate for this probability is simply the proportion of documents containing t among all those associated with the localization: $Pr_L^t \approx (\#\text{ of documents } d \in D_L \ s.t.\ t \in d)/|D_L|$. For each term t and each localization L, the estimate for the probability Pr_L^t is calculated.

Based on this probability, a term t is called *distinguishing* for localization L, if and only if its probability to occur in localization L, Pr_L^t, is significantly different from its probability to occur in any other localization L', $Pr_{L'}^t$. The statistical test applied, uses the Z-score[27], which evaluates the difference between two binomial

[b]Without using any of the MeSH terms.

Table 1. Examples of distinguishing stemmed terms for several localizations

Localization	**Example Terms**
Nucleus	*bind, control, dna, histon, nuclear, promot, transcript*
Mitochondria	*coa (CoA), complex, cytochrom, dehydrogenas, mitochondri, oxidas, respiratori*
Golgi Apparatus	*acceptor, catalyt domain, fucosyltransferas, galactos, glycosyltransferas, golgi, transferas*
Endoplasmic Reticulum	*calcium, chaperon, disulfid isomeras, endoplasm, lumen, microsom, transmembran*

probabilities, Pr^t_L, and $Pr^t_{L'}$, as follows:

$$Z^t_{L,L'} \doteq \frac{Pr^t_L - Pr^t_{L'}}{\sqrt{\overline{P}\cdot(1-\overline{P})\cdot\left(\frac{1}{|D_L|}+\frac{1}{|D_{L'}|}\right)}}\ , \text{ where } \overline{P} = \frac{|D_L|\cdot Pr^t_L + |D_{L'}|\cdot Pr^t_{L'}}{|D_L|+|D_{L'}|}\ .$$

When $|Z^t_{L,L'}| \geq 1.96$, the hypothesis that the two probabilities Pr^t_L, $Pr^t_{L'}$ are different is accepted with a confidence level greater than 95%. Therefore, if the term t has a localization L such that for any other localization L' $|Z^t_{L,L'}| \geq 1.96$, t is considered *distinguishing for localization* L, and is included in the set of distinguishing terms. In our representation of proteins as term vectors, we use only *distinguishing terms*. In the experiments described in Section 3, using several different proteins sets, the average number of PubMed abstracts is on the order of $10,000$, while that of distinguishing terms is about 800. Some examples of distinguishing terms for several localizations are shown in Table 1.

Finally, once the collection of N distinguishing terms, denoted as T_N, was established, each protein p is represented as an N-dimensional vector, where the weight $W^p_{t_i}$ at position i, (where $1 \leq i \leq N$), is the conditional probability of the term t_i to appear in the abstracts associated with the protein p, given all the PubMed abstracts related to the protein, (the set D_p) This probability is estimated as the ratio between the total number of times the term t_i occurs in the abstracts associated with the protein p and the total number of all the occurrences of distinguishing terms in these same abstracts. Formally it is calculated as:

$$W^p_{t_i} = \frac{\sum\limits_{d\in D_p,\ s.t.\ t_i\in d} (\#\text{ of times } t_i \text{ occurs in } d)}{\sum\limits_{d\in D_p}\sum\limits_{t_j\in T_N} (\#\text{ of times } t_j \text{ occurs in } d)}\ ,$$

where the sums are taken over all the abstracts d in the set of abstracts associated with the protein p, D_p.

The representation of proteins as weighted term vectors, is then partitioned into training and test sets for each subcellular localization, and as before, an SVM is trained to classify these protein vectors into their respective localization. This classifier, like SVMacc described above, produces an n-dimensional probability vector denoting the probability of the protein to be in each of the n localizations.

2.3. *Integrated method*

The output from the five classifiers above, is a set of four probability vectors and one binary-valued vector (resulting from MotifSearch). These are all concatenated to form one integrated feature vector for each protein. Again, an SVM classifier is trained on these feature vectors to produce a prediction. This classifier consists of a set of one-against-one classifiers (each of which distinguishes between a pair of localizations) and its output, yet again, is a probabilistic vector, holding for each localization the probability of the protein to belong to it. Based on this final classification step, a protein is assigned to the localization with the highest probability value in the last output vector. The training and evaluation procedure uses strict five-fold cross-validation, where no test protein was used to train any of the classifiers comprising the system.

3. Experiments and Results

To train and to evaluate our integrated system, we used three different data sets, namely those used for training and testing TargetP, MutliLoc, and PLOC. These sets provide the basis for an extensive and sound comparison. The data sets, the evaluation procedure, and the results are described throughout this section.

3.1. *Experimental setting*

The data sets used in our experiments are the following:

TargetP – This data set[1] contains a total of 3,415 distinct proteins representing four plant (*ch*, *mi*, *SP*, and *OT*) and three non-plant (*mi*, *SP*, and *OT*) localizations. Homologs were removed from it by the TargetP authors. The *SP* category includes proteins from several localizations in the secretory pathway: endoplasmic reticulum (*er*), extracellular space (*ex*), Golgi apparatus (*go*), lysosome (*ly*), plasma membrane (*pm*), and vacuole (*va*). The *OT* category includes *cy* and *nu* proteins.

MultiLoc – The MultiLoc data set[20] contains a total of 5,959 protein sequences, which were extracted from the Swiss-Prot database release 42.0[28]. Animal, fungal, and plant proteins with an annotated subcellular localization[c] were grouped into eleven eukaryotic localizations: *cy*, *ch*, *er*, *ex*, *go*, *ly*, *mi*, *nu*, peroxisome (*pe*), *pm*, *va*. In the experiments reported here homologous proteins with identity higher than 80%, (the same threshold used by PLOC[19]), were excluded from the set, to avoid the occurrence of highly similar sequences in both the training and the test sets[d]. Further details about the data set extraction and the implications of homology reduction are available in the MultiLoc publication[20].

[c] Excluding proteins whose annotation was commented *by similarity* or *potential*.

[d] We also conducted experiments with a more lenient and more stringent homology constraints, of 90% and 40% identity, respectively (data not shown).

PLOC – The PLOC data set was used by Park and Kanehisa[19] and consists of proteins extracted from Swiss-Prot release 39.0, covering 12 localizations. In contrast to MultiLoc, (aside for the older Swiss-Prot version), this data set introduces an additional category within the *cy* proteins, namely, the cytoskeleton (*cs*). There are 41 *cs* proteins, compared to 1,245 *cy* proteins. The total number of sequences is 7,579 (max. sequence identity 80%). This set is larger than the MultiLoc data set due to a less restrictive data extraction, assigning proteins to localization even when the localization annotation includes the words "potential" or "by similarity".

Using these three data sets, the performance of our integrated system is compared to that of TargetP, PLOC, and MultiLoc[e]. In addition, we also compare the performance of the integrated system to that of an SVM classifier applied to the text data alone. Following previous evaluations[1,19], we consistently employ five-fold cross-validation. For comparison against the PLOC data set we use the same split as the one used by Park and Kanehisa[19]. For the TargetP data, as the split used by Emanuelsson *et al.*[1] was not provided, we ran the five-fold cross-validation procedure five times, each using a different randomized five-way split, to ensure robustness. The reported results are averaged over all the 5 folds, and over the 5 randomized splits when those are used.

Since the performance of previous systems[1,19] was evaluated using several different metrics, for a fair comparison we calculated these same performance measures. Thus, for each system and data set the performance is measured, for each localization, in terms of the sensitivity (*Sens*), specificity (*Spec*), and Matthews correlation coefficient (*MCC*)[29]. These are defined as:

$$Sens = \frac{TP}{TP+FN}, \quad Spec = \frac{TP}{TP+FP}, \quad \text{and}$$

$$MCC = \frac{TP \cdot TN - FP \cdot FN}{\sqrt{(TP+FN)\cdot(TP+FP)\cdot(TN+FN)\cdot(TN+FP)}},$$

where TP, TN, FP, FN denote the number of true positives, true negatives, false positives, and false negatives, respectively, with respect to a given localization. Like Park and Kanehisa[19] we also measure the *overall accuracy*, namely, $Acc = C/N$, where C is the number of correctly classified proteins over all the localizations, and N is the total number of classified proteins. They also measured the *average sensitivity*, over all the localizations, a metric they call *local accuracy*, which we calculate as well. This last measure, which we denote as *Avg*, gives an equal weight to the categorization performance on each localization, regardless of the number of proteins known to be associated with it.

[e]Comparison to PSort[11] is not included here, since MultiLoc has already demonstrated a higher prediction accuracy compared to this method[20].

3.2. *Results*

We present the results of running the sequence-based system, MultiLoc, the text-based classifier alone (denoted *Text*), and the integrated system (denoted *MultiLocText*), on all the three data sets. For completeness, we also present the results reported by the authors of PLOC[19] and of TargetP[1] on the respective data sets. These numbers were directly taken from the respective publications.

Table 2 summarizes the results, showing the overall accuracy (*Acc*) and the average local accuracy (*Avg*) for both the TargetP and the PLOC data sets. For TargetP the results are shown for plant and non-plant proteins, while for PLOC results are shown for plant, animal, and fungal proteins. Table 3 compares the performance of TargetP and PLOC with our integrated system, with respect to the individual subcellular localizations.

Table 2. An overview of the prediction results using the TargetP and PLOC data sets. Both the total (*Acc*) and the average (*Avg*) prediction accuracies are shown for all the methods. The highest values appear in bold. Standard deviations, (denoted ±) are provided where available.

Data set	Method	***Acc* [%] (± Standard Deviation) / *Avg* [%] (± Standard Deviation)**		
TargetP		**Plant**	**Non-Plant**	
	TargetP	85.3 (±3.5) / 85.6 (n/a)	90.0 (±0.7) / 90.7 (n/a)	
	MultiLoc	89.7 (±1.6) / 90.2 (±2.0)	92.5 (±1.2)/ 92.8 (±1.1)	
	Text	81.2 (±2.6) / 78.1 (±3.2)	88.7 (±1.1)/ 89.8 (±1.6)	
	MultiLocText	**94.7** (±1.5) / **94.4** (±1.6)	**96.2** (±0.8) / **96.7** (±0.9)	
PLOC		**Plant**	**Animal**	**Fungal**
	PLOC	78.2 (±0.9)/ 57.9 (±2.1)	79.6 (±0.9)/ 59.9 (±3.3)	79.5 (±0.9)/ 56.8 (±1.9)
	MultiLoc	73.6 (±0.7) / 71.3 (±2.8)	76.0 (±0.7) / 73.6 (±3.9)	75.8 (±0.8) / 72.5 (±2.5)
	Text	68.7 (±0.7) / 73.5 (±1.8)	70.2 (±0.7) / 75.5 (±2.7)	67.8 (±0.5) / 72.4 (±2.6)
	MultiLocText	**85.3** (±1.2) / **84.2** (±2.4)	**86.4** (±0.8) / **84.5** (±3.6)	**85.4** (±0.8) / **83.8** (±2.8)

Table 3. Localization specific results using the TargetP (left), and the PLOC (right) data sets. For both sets, the results reported in the respective papers are compared to results of our integrated system (MultiLocText). As PLOC localization-specific results are averaged over all three organisms, we show such averaged results for our system as well. Specificity and MCC values were not available for PLOC, hence only its *Sensitivity* is listed and compared with our sensitivity values. The highest compared values for each data set are shown in bold.

TargetP Data Set		
Loc	**TargetP**	**MultiLocText**
	Plant (*Sens Spec MCC*)	
ch	0.85 0.69 0.72	**0.93 0.89 0.89**
mi	0.82 0.90 0.77	**0.95 0.99 0.95**
OT	0.85 0.78 0.77	**0.95 0.87 0.89**
SP	0.91 0.95 0.90	**0.95 0.98 0.95**
	Non-Plant (*Sens Spec MCC*)	
mi	0.89 0.67 0.73	**0.97 0.88 0.91**
OT	0.88 0.97 0.82	**0.95 0.99 0.93**
SP	0.96 0.92 0.92	**0.98 0.96 0.96**

PLOC Data Set		
Loc	**PLOC**	**MultiLocText**
	Avg. *Sens*	**Avg. (*Sens Spec MCC*)**
ch	0.72	**0.84** 0.83 0.82
mi	0.57	**0.85** 0.85 0.83
cs	0.59	**0.83** 0.26 0.46
cy	0.72	**0.79** 0.78 0.74
er	0.47	**0.86** 0.71 0.78
ex	0.78	**0.88** 0.91 0.88
go	0.15	**0.82** 0.30 0.49
nu	**0.90**	0.88 0.94 0.88
pe	0.25	**0.81** 0.63 0.71
pm	**0.92**	0.89 0.98 0.91
va	0.25	**0.83** 0.28 0.48
ly	0.62	**0.81** 0.52 0.64

A comparison of the performance of our three systems (MultiLoc alone, Text alone, and the integrated MultiLocText) using five-fold cross-validation over the

5,959 proteins of the MultiLoc data set, is presented in Table 4. The sensitivity (*Sens*), specificity (*Spec*), and Matthews *MCC* values for the plant and animal versions are listed. (Similar results were obtained for the fungal version, and are not shown here due to space limitation).

The results in Tables 2, 3, and 4 clearly show that the combined classifier, which integrates text and sequence data, outperforms earlier prediction methods. It also outperforms its own text-based (Text) and sequence-based (MultiLoc) components, if taken separately. A significance test was performed to evaluate the differences between the values obtained from MultiLocText and those obtained from each of MultiLoc and Text alone, (Table 4). The improved performance values of MultiLocText are highly statistically significant ($p \ll 0.05$), for almost all the subcellular localizations. The only exceptions are the Golgi (*go*, animal and plant), where there is no significant difference in sensitivity with respect to text-alone, as well as the peroxisome predictions (*pe*, animal and plant), where MultiLocText does not outperform the text-alone system.

4. Discussion and Conclusion

The methods, experiments, and results presented here clearly demonstrate a significant improvement in the prediction of protein subcellular localization through the integration of sequence- and text-based methods. Table 4 shows that the two

Table 4. Prediction performance of MultiLoc, Text, and MultiLocText on the MultiLoc data set. Both localization-specific values (*sens, spec*, MCC) and overall results (*Acc* and *Avg*) are shown. Highest values appear in bold.

Loc	MultiLoc	Text	MultiLocText
		Plant (*Sens Spec MCC*)	
ch	0.88 0.85 0.85	0.89 0.70 0.78	**0.94 0.91 0.92**
cy	0.68 0.85 0.70	0.53 0.75 0.54	**0.81 0.91 0.82**
er	0.72 0.54 0.61	0.73 0.55 0.62	**0.82 0.63 0.71**
ex	0.68 0.81 0.70	0.74 0.80 0.73	**0.84 0.90 0.84**
go	0.75 0.41 0.54	0.82 0.42 0.57	**0.84 0.61 0.70**
mi	0.85 0.81 0.80	0.80 0.80 0.78	**0.90 0.88 0.88**
nu	0.82 0.75 0.75	0.80 0.72 0.72	**0.89 0.85 0.85**
pe	0.71 0.34 0.47	**0.88 0.71 0.79**	0.85 0.59 0.70
pm	0.74 0.89 0.77	0.80 0.91 0.82	**0.84 0.96 0.87**
va	0.70 0.20 0.36	0.59 0.15 0.29	**0.83 0.29 0.48**
***Acc* [%]**	74.6	73.1	**85.1**
***Avg* [%]**	75.2	76.0	**85.5**
		Animal (*Sens Spec MCC*)	
cy	0.67 0.85 0.68	0.51 0.77 0.53	**0.83 0.91 0.82**
er	0.68 0.56 0.60	0.74 0.48 0.58	**0.82 0.67 0.73**
ex	0.79 0.83 0.77	0.76 0.78 0.72	**0.86 0.90 0.86**
go	0.71 0.43 0.53	0.86 0.40 0.57	**0.87 0.65 0.74**
ly	0.69 0.36 0.48	0.75 0.32 0.47	**0.86 0.55 0.68**
mi	0.88 0.82 0.83	0.80 0.79 0.77	**0.93 0.91 0.91**
nu	0.82 0.73 0.73	0.84 0.71 0.73	**0.89 0.83 0.84**
pe	0.71 0.31 0.44	**0.93** 0.60 0.74	0.89 **0.68 0.77**
pm	0.73 0.90 0.76	0.80 0.91 0.81	**0.85 0.95 0.87**
***Acc* [%]**	74.6	72.5	**86.2**
***Avg* [%]**	74.1	77.5	**86.8**

types of methods distinctly complement each other. MultiLoc, which is based on sequence data, typically performs well predicting protein localizations that are directed by N-terminal signals such as the mitochondria and the chloroplast. The use of text information complements and significantly boosts its performance for localizations whose sequence-based signal is not as overt, including the peroxisome and localizations related to the secretory pathway such as the Golgi apparatus and the endoplasmic reticulum.

In this work we have demonstrated, using five-fold cross-validation, that our system can reproduce, with unprecedented sensitivity and specificity, localizations of proteins which were already annotated in Swiss-Prot. A natural next step is to apply the method to yet un-localized proteins. We are developing the means to predict subcellular localization of proteins for which PubMed reference exist in Swiss-Prot but no localization assigned, as well as for those with no curated PubMed reference. Our current use of "raw text" from PubMed abstracts (in contrast, for instance, to the use of Swiss-Prot annotation text as was done before[2]), is expected to make our approach amenable to such extensions. We are also investigating methods for the localization of proteins with no PubMed references, through the use of alternative data sources.

References

1. Emanuelsson, O., Nielsen, H., Brunak, S., von Heijne, G.: Predicting subcellular localization of proteins based on their N-terminal amino acid sequence. J Mol Biol. **300** (2000) 1005–1016
2. Nair, R., Rost, B.: Inferring sub-cellular localization through automated lexical analysis. Bioinformatics **18** (2002) S78–S86
3. Gardy, J.L., Spencer, C., Wang, K. *el al.*: PSORT-B: Improving protein subcellular localization prediction for gram-negative bacteria. Nucleic Acids Research **31** (2003) 137–140
4. Cai, Y.D., Chou, K.C.: Predicting 22 protein localizations in budding yeast. Biochem Biophys Res Commun. **323** (2004) 425–428
5. Schneider, G., Fechner, U.: Advances in the prediction of protein targeting signals. Proteomics **4** (2004) 1571–1580
6. Dönnes, P., Höglund, A.: Predicting Protein Subcellular Localization: Past, Present, and Future. Genomics, Proteomics, and Bioinformatics **2** (2004)
7. Burns, N., Grimwade, B., Ross-Macdonald, P., Choi, E., Finberg, K., GS, R., M, S.: Large-scale analysis of gene expression, protein localization and gene disruption in Saccharomyces cerevisiae. Genes and Development **8** (1994) 1087–1105
8. Hanson, M.R., Köhler, R.H.: GFP imaging: Methodology and application to investigate cellular compartmentation in plants. Journal of Experimental Botany **52** (2001)
9. Dunkley, T., Watson, R., Griffin, J., Dupree, P., Lilley, K.: Localization of organelle proteins by isotope tagging (LOPIT). Molecular and Cellular Proteomics **3** (2004)
10. Nakai, K., Kanehisa, M.: Expert system for predicting protein localization sites in gram-negative bacteria. Proteins: Structure, Function and Genetics **11** (1991) 95–110

11. Nakai, K., Kanehisa, M.: A knowledge base for predicting protein localization sites in eukaryotic cells. Genomics. **14** (1992) 897–911
12. Horton, P., Nakai, K.: A probabilistic classification system for predicting the cellular localization of proteins. In: Proc. of the Int. Conf. on Intelligent Systems for Molecular Biology (ISMB). (1996)
13. Horton, P., Nakai, K.: Better prediction of protein cellular localization sites with the k nearest neighbors classifier. In: Proc. of the Int. Conf. on Intelligent Systems for Molecular Biology (ISMB). (1997)
14. Emanuelsson, O., Nielsen, H., von Heijne, G.: Chlorop, a neural network-based method for predicting chloroplast transit peptides and their cleavage sites. Protein Science **8** (1999) 978–984
15. Bannai, H., Tamada, Y., Maruyama, O., Nakai, K., Miyano, S.: Extensive feature detection of N-terminal protein sorting signals. Bioinformatics. **18** (2002) 298–305
16. Nair, R., Rost, B.: Mimicking cellular sorting improves prediction of subcellular localization. J Mol Biol. **348** (2005) 85–100
17. Stapley, B.J., Kelley, L.A., Sternberg, M.J.E.: Predicting the subcellular location of proteins from text using support vector machines. In: Proc. of the Pacific Symposium on Biocomputing (PSB). (2002) 374–385
18. Eskin, E., Agichtein, E.: Combining text mining and sequence analysis to discover protein functional regions. In: Proc. of the 9th Pacific Symposium on Biocomputing (PSB). (2004) 288–299
19. Park, K.J., Kanehisa, M.: Prediction of protein subcellular location by support vector machines using compositions of amino acids and amino acid pairs. Bioinformatics. **19** (2003) 1656–1663
20. Höglund, A., Dönnes, P., Blum, T., Adolph, H., Kohlbacher, O.: Using N-terminal targeting sequences, amino acid composition, and sequence motifs for predicting protein subcellular localization. German Conference on Bioinformatics (GCB) 2005.
21. Chang, C.C., Lin, C.J.: LIBSVM: A library for support vector machines (2003) *http://www.csie.ntu.edu.tw/~clin/libsvm/*.
22. Wu, T.F., Linand, C.J., Weng, R.C.: Probability Estimates for Multi-class Classification by Pairwise Coupling. Journal of Machine Learning Research **5** (2004) 975–1005
23. Bairoch, A., Bucher, P.: PROSITE: recent developments. Nucleic Acids Res. **22** (1994) 3583–3589
24. Cokol, M., Nair, R., Rost, B.: Finding nuclear localization signals. EMBO Rep. **1** (2000) 411–415
25. Nair, R., Carter, P., Rost, B.: NLSdb: database of nuclear localization signals. Nucleic Acids Res. **31** (2003) 397–399
26. Porter, M.F.: An Algorithm for Suffix Stripping (Reprint). In: Readings in Information Retrieval. Morgan Kaufmann (1997) *http://www.tartarus.org/~martin/PorterStemmer/*.
27. Walpole, R.E., Myers, R.H., Myers, S.L. In: One- and Two-Sample Tests of Hypotheses. (1998) 235–335
28. Bairoch, A., Apweiler, R.: The SWISS-PROT protein sequence database and its supplement in TrEMBL in 2000. Nucleic Acids Res. **28** (2000) 45–48
29. Matthews, B.W.: Comparison of predicted and observed secondary structure of T4 phage lysozyme. Biochim Biophys Acta. **405** (1975) 442–451

EVALUATION OF LEXICAL METHODS FOR DETECTING RELATIONSHIPS BETWEEN CONCEPTS FROM MULTIPLE ONTOLOGIES*

HELEN L. JOHNSON,[†] K. BRETONNEL COHEN, WILLIAM A. BAUMGARTNER JR., ZHIYONG LU, MICHAEL BADA, TODD KESTER, HYUNMIN KIM, AND LAWRENCE HUNTER

Center for Computational Pharmacology
University of Colorado School of Medicine

We used exact term matching, stemming, and inclusion of synonyms, implemented via the Lucene information retrieval library, to discover relationships between the Gene Ontology and three other OBO ontologies: ChEBI, Cell Type, and BRENDA Tissue. Proposed relationships were evaluated by domain experts. We discovered 91,385 relationships between the ontologies. Various methods had a wide range of correctness. Based on these results, we recommend careful evaluation of all matching strategies before use, including exact string matching. The full set of relationships is available at compbio.uchsc.edu/dependencies.

1. Introduction

Lexical analysis of an ontology is a powerful tool for suggesting relationships between concepts within the ontology [25, 28, 29, 30, 32] or among multiple ontologies [7, 6]. However, there are many possible text types to match between (e.g. term names, synonyms, and definitions) and variations on matching techniques (e.g. stemming, case normalization, etc.), and there is no reason to expect similar, and equally valid, results for all of them. Most importantly, the mere existence of a match does not prove a valid relationship between concepts. In this paper, we systematically evaluate three text matching techniques in two text types, and use domain experts to evaluate the correctness of the resulting matches.

Recently, [7] demonstrated the utility of an external, publicly-available resource for finding within-ontology relationships. They hypothesized that

*This work is supported by NLM grant R01-LM00811 to Lawrence Hunter.
†Authors Johnson and Cohen contributed equally to the work reported here.

two GO terms that share a relationship to a single ChEBI term are related to each other. They detected 771,302 within-GO relationships by finding sets of GO terms and synonyms that lexically matched a ChEBI term or its synonyms. They noted that 55% of all GO terms contained 26% of all ChEBI terms, totalling 20,497 GO-ChEBI relationships. Implicit in their work is the assumption that relationships found between GO and ChEBI terms are valid and meaningful. More recently [6], they have proposed extending this technique to detect relationships between all OBO ontologies.

Various ontology working groups have become interested in integrating external ontologies into their own, and have pointed out some of the obstacles to doing this [15]. In this paper, we show that a variety of publicly available resources can be exploited for large-scale, automated suggestion of between-ontology relationships. We define a *relationship* as any direct or indirect association between two ontological concepts.

In this paper, we evaluate the following hypotheses:

(1) Valid relationships exist between concepts from GO and from other OBO ontologies.
(2) Gene Ontology definitions are a fruitful resource for discovering relationships between concepts in ontologies.
(3) Language processing techniques for discovering relationships have quantifiable and variable rates of correctness.

A novel aspect of this paper as compared to [6] and [7] is that we make use of the text of GO definitions. There is some history for using definitions in language processing applications, particularly in the word sense disambiguation task [21]. More recently, [23] shows the value of Gene Ontology definitions for predicting GeneRIFs. We also evaluate simple linguistic processing techniques for term detection and normalization. The findings may be useful for semi-automatically linking ontologies, whether to support reasoning tasks or annotation, and also for detecting terms from ontologies in natural language texts.

1.1. *Context and motivation*

This research falls into the general category of semantic integration (SI). Semantic integration is a currently active topic of research in the general computer science, artificial intelligence, Internet, and data mining communities [26]. It has crucial roles to play in areas as diverse as interoperability in Semantic Web Services [8], coreference resolution in free text [22], schema and data matching in databases [12], and communication between intelligent agents and resources [13]. There is much related work in the ontology community, e.g. [27] and [24] among many others. Within the biomedical ontology literature, closely related work includes the description-logic-based GONG project [33], in which GO metabolism terms were linked to biological-substance terms from MeSH using lexical tools and term synonyms of UMLS. [2], [19], [5] used various non-lexical techniques to find relations within GO.

Mapping vs. alignment of ontologies: Integration of multiple, independently produced ontologies is an important task in molecular biology. One well-studied aspect of this task is *mapping,* the identification of equivalent concepts in multiple ontologies [15, 20, 30]. This work has shown some of the difficulties of textual analysis of biological ontologies. For example, [30] points out that biological terminologies pose difficulties for standard normalization procedures, since they often contain alphanumeric modifiers. Other problems include synonymy and morphological variation [20].

Ontology alignment is the task of making overlapping concepts among multiple ontologies compatible. Although mapping may be a part of alignment, the alignment task requires finding meaningful relationships between non-identical concepts. The identification of such relationships may also be valuable within an ontology, e.g. in order to improve compositionality [25, 32, 28, 29] or in defining and populating novel relationships. The work reported here is relevant both to the mapping and to the alignment task.

Natural language processing: The relevance of locating concepts from an ontology in free text is clear from the inclusion of this task in recent "bake-off" competitions in the NLP community. The overall low performance on these tasks [4, 16, 9] demonstrates their difficulty. The work in this paper can be thought of as a step towards recognizing OBO concepts in free text: GO terms and definitions are themselves a type of semi-structured natural language, fitting the sublanguage model but having enough complexity to be a challenge, while not being as unstructured as the language of scientific abstracts.

2. Methods

Materials: We retrieved the current versions of the GO [1, 15], ChEBI [11], Cell Type [3], and BRENDA Tissue [31] ontologies from SourceForge. (In the remainder of this paper, when we say "(other) OBO ontologies," we mean the ontologies other than GO.) We chose these three other ontologies because we expected high degrees of subject-matter overlap between them and GO, and because they are in relatively advanced stages of development.

Table 1. Materials: ontologies, data files, and revisions.

Ontology	terms	synonyms	avg. syn./term	data file	revision_date
Gene Ontology	19,508	8,202	.42	gene_ontology.obo	09:06:2005 17:10
ChEBI	11,549	19,295	1.67	chebi.obo	25:05:2005 10:54
Cell Type	748	215	.29	cell.obo	24:05:2005 17:10
BRENDA	2,222	1,208	.54	BrendaTissue.txt	10:5:2005 13:49:02

Finding relationships: We used Lucene [14] to search for the OBO concepts in GO. Lucene is a Java information retrieval library[a]. We modeled the GO concepts as documents to be retrieved, and the other OBO concepts as search engine queries. We indexed the GO concepts, placing the terms and definitions in distinct fields, which allowed us to search them separately. We constructed Lucene *phrasal queries* from the other OBO concepts. This meant that for searches on multi-word OBO concepts, word order could not vary and no words could intrude. Synonym queries were done by constructing phrasal queries for each synonym, and then grouping the phrasal queries with Boolean *OR*. Both indexing and searching require a Lucene class called an *analyzer.* We used the WhiteSpace and Porter-Stemmer analyzers. Lucene gave us an efficient and robust framework for carrying out searches and for manipulating their results.

Evaluation: We drew a random sample from the relationships proposed by each technique for each ontology, for a total of 2,389 relationships. The sample is unevenly distributed across various categories of ontologies, linguistic manipulations, and GO terms vs. GO definitions, but covers all combinations of those categories. These 2,389 relationships were manually examined by domain experts. One domain expert (DE1) has considerable experience in ontologies, biology, and structural chemistry. The other domain expert (DE2) is a bioinformatics doctoral candidate with experience with GO and with protein function and subcellular localization. The experts were presented with (1) the ID and name of a concept from an OBO ontology, and (2) the ID, name, and definition of some concept from the GO. In addition, the experts had access to the definitions of the OBO concept, as well as any other helpful information found in the ontologies themselves.

[a]Previous applications of Lucene to text processing in the biomedical domain are reported in [10] and [18].

They were instructed to evaluate the output with the following question in mind: *Is this OBO term the concept that is being referred to in this GO term/definition?* They were permitted to classify all relationships as either true positive or false positive. We calculated *correctness* as the number of true positive relationships divided by the number of proposed relationships (similar to precision or specificity). All relationships are available for public inspection at compbio.uchsc.edu/dependencies.

Inter-annotator agreement (IAA): DE1 evaluated the majority of the output. A sample of 400 proposed relationships was also evaluated by DE2. Initial IAA between the two was 93.5% (374/400). After dispute resolution, the consensus IAA was 98.2% (393/400). For the remaining seven cases, DE1 had the deciding vote.

Linguistic manipulations: We queried by exact match to the OBO concept name. We also queried using synonyms of OBO concepts. Since all work in this area has observed moderate differences in concept name realization, such as pluralization, we also implemented the standard linguistic manipulations of stemming and stop word removal [17]. We evaluated the correctness of the resulting searches individually.

What we counted: For each ontology, we give data on the following:

- Relationships found by matches between the OBO ontology and GO terms (T)
- Relationships found by matches between the OBO ontology and GO definitions (D)
- The union of T and D ($T \cup D$)[b]
- The intersection of T and D ($T \cap D$)[c]
- The relative complement of T and D (T-D)[d]
- The relative complement of D and T (D-T)[e]

Gain, the magnitude of the increase in the number of relationships detected by examining definitions, rather than just terms, is the relative complement of D and T divided by the union of T and D ((D-T)/($T \cup D$)).

In addition, for each ontology, we calculated the analogous set relations

[b] $T \cup D$ gives the number of relationships that are found in terms or definitions. Some of its relationships are revealed by both. It equals ($T \cap D$) + D-T + T-D.
[c] $T \cap D$ is the number of relationships that are found in both terms and definitions.
[d] T-D is the number of relationships that can be found in terms, but cannot be found by examining definitions. It equals T - ($T \cap D$).
[e] D-T is the number of relationships that can be found in definitions, but cannot be found by examining terms. It equals D - ($T \cap D$).

for the various language processing techniques. This allows us to quantify the yield and the correctness of the various techniques with respect to the three ontologies.

For the Cell Type ontology, we filtered out all matches to the terms *cell* (CL:0000000, 3215 matches), *cell by organism* (CL:0000004, 96 matches), and *cell by function* (CL:0000144, 10 matches), since we realized early on that they were either content-free or incorrect.

3. Results

Finding relationships between ontologies: Our initial hypothesis was that there are relationships between GO and the various OBO ontologies. Table 2 summarizes the number of matches between GO and the three other ontologies and the average correctness calculated by manually examining a subset of the matches. Searching GO terms and definitions for terms from the other ontologies resulted in a total of 91,385 proposed relationships. The majority of these links (73,002) are between GO and ChEBI. The average correctness across the three ontologies is 80.62%. This is generally consistent with the precision reported for the mapping task by [30] (range from .36 for BLAST to .94 for exact match) and [20] (range from .25 for Chimaera to 1.0 for PROMPT). These data are consistent with the initial hypothesis, validating the goal expressed in [6], and gives an idea of the size of the set of potential relationships.

Table 2. Counts and correctness of proposed relationships between ontologies. Numbers in parentheses are the correct and total manually evaluated pairs.

Ontology	Relationships to GO	Avg. Correctness
ChEBI	73002	84.2% (977/1161)
Cell Type	1961	92.99% (584/628)
BRENDA	16469	60.83% (365/600)
TOTAL	91385	80.62% (1926/2389)

Correctness and Error Analysis: To assess the correctness of the matches, a random set of 2,389 was manually examined by domain experts. All results are given in Table 3. Note that although correctness is generally high, *some combinations of ontology and linguistic technique had quite low correctness.* This has important consequences for more ambitious efforts to detect relationships across all OBO ontologies, such as proposed by [6]: we cannot use any technique, including exact matching, without assessing its correctness for a particular pair of data sources.

Exact matching was the most accurate type of search, ranging from 76% to 100% correct. This is consistent with results reported for the mapping task. One source of false positives for exact matching was polysemy, or words with multiple meanings. For example, the word *group* (CHEBI:24433) also has a General English meaning, and often appeared with that sense in GO definitions. Similarly, the BRENDA term *joint* (BTO:0001686), which refers to an anatomical joint, appears as an adjective meaning *combined* in GO concepts. We found examples of false positives related to non-General-English, domain-specific terms as well, e.g. *reticulum* (BTO:0000347) incorrectly matching the definition of GO:0006614. Incorporating OBO term synonymy resulted in slightly lower correctness, ranging from 42% to 94%, with an average of 67.4% (397/589). Finally, the stemming/stop-word-removal searches show the lowest correctness, ranging from 7% to 92%.

Table 3. Correctness Rates (correct/evaluated)

Ontology	Exact	Synonyms	Stemming
ChEBI			
GO Term	99.5% (199/200)	42.0% (42/100)	73.0% (73/100)
GO Def	97.8% (451/461)	69.0% (138/200)	74.0% (74/100)
Cell Type			
GO Term	100% (200/200)	94% (44/47)	76% (41/54)
GO Def	98.7% (231/234)	50% (21/42)	92% (47/51)
BRENDA			
GO Term	76.0% (76/100)	83.0% (83/100)	7.0% (7/100)
GO Def	93.0% (93/100)	69.0% (69/100)	15.0% (15/100)

GO terms versus GO definitions: A novel hypothesis of this paper is that GO definitions are a fruitful resource for discovering relationships between GO and other ontologies. Table 4 addresses this hypothesis for BRENDA, and the corresponding data for the other ontologies are given on the website (compbio.uchsc.edu/dependencies). Searching for relationships in the GO definitions in addition to the GO terms had a large impact on the quantity of relationships found between ontologies. The table presents the number of links found in GO terms and in GO definitions, as well as the union, intersection and relative complements of these sets. The number of links found only in GO terms is given by the relative complement of terms and definitions (T-D), listed in the fifth column of the table. The number of links found only in GO definitions is given by the relative complement of definitions and terms (D-T), the sixth column. The final column in Table 4 describes the gain from searching in GO definitions for relationships. It is calculated by dividing the D-T by the union of D and T. For instance, in the first row of Table 4, which displays the number of links found in the GO using an exact BRENDA term search, a gain of 49.59% means that just under half of these between-ontology links could be found only by searching the GO definition. Note that for all ontologies and for all search strategies, the number of relationships is higher when definitions are considered. The gain is never lower than 24.3% (270 additional matches for exact matching of Cell Type concepts), and it is generally higher than 50% (43,146 additional matches just for the case of allowing stemming matches for ChEBI concepts). The correctness (see Table 3) of relationships detected by matches to definitions is comparable to the correctness of relationships detected by matches to terms.

Table 4. Relationships in GO terms vs. GO defs for BRENDA

Ontology	T	D	T∪D	T∩D	T-D	D-T	Gain
Exact	1465	2447	2906	1006	459	1441	49.59%
Exact+synonyms	1875	3093	3686	1282	593	1811	49.13%
Stemmed	3892	15409	15722	3579	313	11830	75.24%

3.1. *Linguistic techniques in relationship searches*

Using synonyms: Results for including the synonyms associated with BRENDA terms in the search string are given in Table 5; corresponding data for the other ontologies is on the website. *E* is *exact match, Syn* adds synonyms for the OBO concept, and the other columns are the union, intersection, and relative complements. Adding synonyms increased the yield of relationships by an average of 36% (23,300/64,987) over using only the exact OBO term query. The set E should be a proper subset of Syn, and the relative complement of E and Syn is the empty set. The yield of using OBO synonyms ranged from 9% (85/925) to 40.69% (8669/21300), and was generally quite similar for GO terms and for GO definitions. Synonyms allowed us to detect some relationships that could not have been found by any other technique, e.g. relating *adipose* (BTO:0000441) to *larval fat body development*(GO:0007504). Correctness for synonymy-based matches was sometimes low, ranging from 42 to 94% (see Table 3)—not surprising in the face of the history of query expansion attempts in information retrieval.

Table 5. Using BRENDA synonyms

GO	E	Syn	E∪Syn	E∩Syn	E-Syn	Syn-E	Gain
T	1465	1875	1875	1465	0	410	21.9%
D	2447	3093	3093	2447	0	646	20.9%
T∪D	2906	3686	3686	2906	0	780	21.2%

Stemming GO concepts and OBO terms: Results for stemming and stop word removal (labelled *Stem*) for BRENDA are given in Table 6; data for the other ontologies is on the website. Stemming garnered the greatest increase of proposed relationships, with an average gain of 78% (146,777/188,464). However, this increase comes at a price, with a lower average correctness rate of 51% (257/505). Note that the correctness of matching BRENDA to GO terms or definitions by stemming is extremely low (7-15%). Again, searching for stemmed OBO terms also returned the subset of relationships that the exact term searches returned, and E-Stem is the empty set.

Stemming allowed pluralized forms of the same term to be matched. It also picked up other morphological variation in terms, e.g. matching *neuron* (CL:0000540) to *neuronal* in the definition of *syntrophin* (GO:0016013). In a random sample of 97 relationships matched by stemming across the three ontologies, 57% was due to pluralization, 11% to adjectival derivation, and 32% to other morphological variation.

Table 6. Stemming and stop word removal: BRENDA

GO	E	Stem	E∪Stem	E∩Stem	E-Stem	Stem-E	Gain
T	1465	3892	3892	1465	0	2427	62.36%
D	2447	15409	15409	2447	0	12962	84.12%
T∪D	2906	15722	15722	2906	0	12816	81.52%

4. Discussion and conclusions

Our results are consistent with the hypotheses that there are many valid relationships between GO and other OBO ontologies, and that in addition to GO terms, GO definitions are an important source for detecting them.

Implications for ontology mapping: One implication of this study comes from the observation that correctness is almost never 100%: even exact string matches do not guarantee a valid match. Ontologists attempting to carry out the goal stated in [6] should not ignore these findings. The results on BRENDA are especially cautionary.

In contrast to work on the mapping task done by the ontology community, the evaluation of work such as ours and Burgun and Bodenreider's has been hampered by the lack of a curated gold standard. One important product of the work reported in this paper is a data set of 2,389 GO/other-OBO concept pairs that has been examined by at least one domain expert. This data set includes 1,926 true positive relationships and 463 known unrelated (i.e., the false positive) pairs. It is publicly available at compbio.uchsc.edu/dependencies, and will allow future researchers in this area to do principled automatic evaluations.

Furthermore, the set of known unrelated pairs can be used in future efforts to filter out terms that are known to produce high numbers of irrelevant, incorrect, or simply unrevealing matches. We suspect that a relatively small set of OBO terms contributed many of the errors, and that correctness can be improved by filtering them. This analysis continues.

Implications for ontology enrichment: One limitation for the application of these relationships to ontology enrichment (the addition of relationships among existing terms) is the fact that most of the relationships that we detect are indirect. For example, our techniques relate *T cell* (CL:0000084) to both *T cell proliferation* (GO:0042098) and *regulation of T cell proliferation* (GO:0042129), but an ontologist would likely prefer to find only the direct relationship from *T cell* to *T cell proliferation.* Another limitation for ontology enrichment is that our methods do not automatically differentiate between relation types (see e.g. [28]). Future work should attempt to differentiate between direct and indirect relationships, and to characterize the nature of the relations between concepts.

Implications for language processing: This study provides cautionary data on the limits of various techniques, even exact string matches. The data provides a list of terms that are likely to produce false-positive matches under conditions of exact match and specific linguistic manipulations; these lists can be used to filter results from any language processing system that seeks to recognize concepts from the ChEBI, Cell Type, and BRENDA ontologies. It also points us towards techniques that might allow us to predict which terms are likely to produce high rates of false positive matches, such as ones at high positions in an ontology (e.g. *cell* (CL:0000000)) and ones that are isomorphic with General English words (e.g. *groups* (CHEBI:24433)). Additionally, it highlights the importance of building biomedical-domain-specific preprocessing tools, such as stemmers.

References

1. Ashburner, M.; et al. (2000) Gene Ontology: tool for the unification of biology. *Nature Genetics* 25:25-29.
2. Bada, M.; D. Turi; R. McEntire; and R. Stevens. (2004) Using reasoning to guide annotation with Gene Ontology terms in GOAT. *SIGMOD Record* 33(2):27-32.
3. Bard, J.; S.Y. Rhee; and M. Ashburner. (2005) An ontology for cell types. *Genome Biology* 6:R21.
4. Blaschke, C.; E.A. Leon; M. Krallinger; and A. Valencia. (2005) Evaluation of BioCreative assessment of task 2. *BMC Bioinformatics* 6:(Suppl. 1):S16.
5. Bodenreider, O.; M. Aubry; and A. Burgun (2005) Non-lexical approaches to identifying associative relations in the Gene Ontology. *PBS 2005* pp. 104-115.
6. Bodenreider, O.; and A. Burgun. (2005) Linking the Gene Ontology to other biological ontologies. *ISMB Bio-ontologies SIG meeting.*
7. Burgun, A.; and O. Bodenreider. (2005) An ontology of chemical entities helps identify dependence relations among Gene Ontology terms. *Semantic mining in biomedicine.*
8. Burstein, M.H.; and D.V. McDermott. (2005) Ontology translation for interoperability among Semantic Web Services. *AI Mag.* 26(1):71-82.
9. Camon, E.B.; D.G. Barrell; E.C. Dimmer; V. Lee; M. Magrane; J. Maslen; D. Binns; and R. Apweiler. (2005) An evaluation of GO annotation retrieval for BioCreative and GOA. *BMC Bioinformatics* 6(Suppl. 1):S17.
10. Carpenter, B. (2004) Phrasal queries with LingPipe and Lucene: ad hoc genomics text retrieval. *NIST Special Publication: SP 500-261 The Thirteenth Text Retrieval Conference (TREC 2004).*
11. Degtyarenko, K. (2003) Chemical vocabularies and ontologies for bioinformatics. *Proc. 2003 Int. Chemical Info. Conf.*, Nimes, France.
12. Doan, A.; and A.Y. Halevy. (2005) Semantic-integration research in the database community: a brief survey. *AI Mag.* 26(1):83-94.
13. Gruninger, M.; and J.B. Kopena. (2005) Semantic integration through invariants. *AI Mag.* 26(1):11-20.
14. Gospodnetić, O.; and E. Hatcher. (2005) *Lucene in action.* Manning.
15. Hill, D.P.; J.A. Blake; J.E. Richardson; and M. Ringwald. (2002) Extension and integration of the Gene Ontology (GO): combining GO vocabularies with external vocabularies. *Genome Research* 12(12):1982-1991.

16. Hirschman, L.; A. Yeh; C. Blaschke; and A. Valencia. (2005) Overview of BioCreative: critical assessment of information extraction for biology. *BMC Bioinformatics* 6(Suppl. 1):S1.
17. Jackson, P.; and I. Moulinier (2002) *Natural language processing for online applications: text retrieval, extraction, and categorization.* John Benjamins.
18. Konrad, K.; R. Steinbach; and H. Stenzhorn (2005) Competitive intelligence with Lucene in XtraMind's XM-InformationMinder. In Gospodnetić and Hatcher (2005), pp. 344-350.
19. Kumar, A.; B. Smith; and C. Borgelt. (2004) Dependence relationships between Gene Ontology terms based on TIGR gene product annotations. *Proc. CompuTerm* pp. 31-38.
20. Lambrix, P.; and A. Edberg. (2003) Evaluation of ontology merging tools in bioinformatics. *PSB 2003* pp. 589-600.
21. Lesk, M. (1986) Automatic sense disambiguation using machine readable dictionaries: How to tell a pine cone from an ice cream cone. SIGDOC pp. 24-26.
22. Li, X.; P. Morie; and D. Roth. (2005) Semantic integration in text: from ambiguous names to identifiable entities. *AI Mag.* 26(1):45-58.
23. Lu, Z.; K.B. Cohen; and L. Hunter. (2006) Finding GeneRIFs via Gene Ontology annotations. *PSB,* this volume.
24. McGuinness, D.L.; R. Fikes; J. Rice; and S. Wilder. (2000) The Chimaera ontology environment. *AAAI 2000* pp. 1123-1124.
25. Mungall, C.J. (2004) Obol: integrating language and meaning in bio-ontologies. *Comparative and Functional Genomics* 5:509-520.
26. Noy, N.F.; A. Doan; and A.Y. Halevy. (2005) Semantic integration. *AI Mag.* 26(1):7-9.
27. Noy, N.F.; and M.A. Musen. (2000) PROMPT: Algorithm and tool for automated ontology merging and alignment. *AAAI 2000* pp. 450-455.
28. Ogren, P.V.; K.B. Cohen; G.K. Acquaah-Mensah; J. Eberlein; and L. Hunter. (2004) The compositional structure of Gene Ontology terms. *PSB 2004* pp. 214-225.
29. Ogren, P.V.; K.B. Cohen; and L. Hunter. (2005) Implications of compositionality in the Gene Ontology for its curation and usage. *PSB 2005* pp. 174-185.
30. Sarkar, I.N.; M.N. Cantor; R. Gelman; F. Hartel; and Y.A. Lussier. (2003) Linking biomedical language information and knowledge resources: GO and UMLS. *PSB 2003* pp. 427-450.
31. Schomburg, I.; et al. (2004) BRENDA, the enzyme database: updates and major new developments. NRA Vol. 32, D431-D433.
32. Verspoor, C.M.; C. Joslyn; and G.J. Papcun. (2003) The Gene Ontology as a source of lexical semantic knowledge for a biological natural language processing application. *Participant notebook of the ACM SIGIR '03 workshop on text analysis and search for bioinformatics* pp. 51-56.
33. Wroe, C.J.; R. Stevens; C. A. Goble; and M. Ashburner. (2003) A methodology to migrate the Gene Ontology to a description logic environment using DAML+OIL. *PSB 2003* pp. 624-635.

AUTOMATICALLY GENERATING GENE SUMMARIES FROM BIOMEDICAL LITERATURE*

XU LING, JING JIANG, XIN HE, QIAOZHU MEI
CHENGXIANG ZHAI, BRUCE SCHATZ

Department of Computer Science and Institute for Genomic Biology
University of Illinois at Urbana-Champaign Urbana, IL 61801
E-mail: {xuling,jiang4,xinhe2,qmei2,czhai,schatz}@uiuc.edu

Biologists often need to find information about genes whose function is not described in the genome databases. Currently they must try to search disparate biomedical literature to locate relevant articles, and spend considerable efforts reading the retrieved articles in order to locate the most relevant knowledge about the gene. We describe our software, the first that automatically generates gene summaries from biomedical literature. We present a two-stage summarization method, which involves first retrieving relevant articles and then extracting the most informative sentences from the retrieved articles to generate a structured gene summary. The generated summary explicitly covers multiple aspects of a gene, such as the sequence information, mutant phenotypes, and molecular interaction with other genes. We propose several heuristic approaches to improve the accuracy in both stages. The proposed methods are evaluated using 10 randomly chosen genes from FlyBase and a subset of Medline abstracts about Drosophila. The results show that the precision of the top selected sentences in the 6 aspects is typically about 50-70%, and the generated summaries are quite informative, indicating that our approaches are effective in automatically summarizing literature information about genes. The generated summaries not only are directly useful to biologists but also serve as useful entry points to enable them to quickly digest the retrieved literature articles.

1. Introduction

The rise of modern genomics in the 21st century is catalyzing the necessity for gene annotation of new organisms, which are not model genetic organisms and whose gene functions are largely unknown. There are already an order of magnitude more organisms whose sequences are known

*This work is in part supported by the National Science Foundation under award numbers 0425852 and 0428472.

than those whose genetics is known, and the number of such new organisms is growing rapidly. As part of the BeeSpace project at the University of Illinois (www.beespace.uiuc.edu), we are developing fully automatic annotation methods for model organisms beyond the genetic models, using computational methods. In particular, we are annotating genome data about the honey bee Apis mellifera using new text processing technologies on biomedical literature combined with existing model genetic databases, especially about the fruit fly Drosophila melanogaster. This paper describes a component software that supports automatic summarization of gene descriptions from biomedical literature.

The generated summary covers six aspects of a gene: (1) Gene products; (2) Expression location; (3) Sequence information; (4) Wild-type function and phenotypic information; (5) Mutant phenotype; and (6) Genetical interaction. Such a summary not only is itself very useful, but also can serve as useful entry points to the literature through linking each aspect to the supporting evidence in the literature, allowing biologists to more easily keep track of new discoveries occurring in the literature. If gene summaries can be automatically generated with decent accuracy, we would be able to curate the databases for other model organisms equivalently well as FlyBase[7] did, but much more efficiently.

To the best of our knowledge, this is the first attempt to automatically generate such a structured summary of a gene from biomedical literature. We present a two-step method, retrieving relevant articles then extracting informative sentences from these articles for each aspect. In the retrieval step, we propose several heuristics to address gene name variations to improve the retrieval accuracy. In the extraction step, we exploit training sentences in existing curated databases and score a sentence for each aspect based on its content, location, and the document containing the sentence.

We evaluate the proposed method using 10 randomly chosen genes from FlyBase and a subset of Medline abstracts about Drosophila. The precision of the top selected sentences in the 6 aspects is about $50 - 70\%$ and the generated summaries are quite informative, indicating that our approaches are effective in automatically summarizing literature about genes. Since our method is quite general, it is likely to work on other organisms as well.

2. Related Work

Most existing studies of biomedical literature mining focus on automated information extraction, using natural language processing techniques to identify relevant phrases and relations in text, such as protein-protein interactions[1] (see [2,3] for reviews of these works). The information we extract is at the sentence level, which allows us to cover many different aspects of a gene and extract information in a more robust manner.

A problem closely related to ours was addressed in the Genomics Track in the Text REtrieval Conference (TREC) 2003, where the task was to generate descriptions about genes from Medline records. The major differences between this task and ours are: (1) The generated descriptions do not organize the information into clearly defined aspects. In contrast, we define six reasonable aspects of genes and propose new methods for selecting sentences for specific aspects. (2) In genomics track, the existing GeneRIF in LocusLink (http://www.ncbi.nih.gov/entrez/query.fcgi?db=gene) can be used as training data, which makes the problem easier, while we are dealing with situations where no such resource is available.

Automatic text summarization, notably news summarization has also been extensively studied. According to the scheme given in a detailed review[4], our gene summarization task is a type of informative, query-oriented, multi-document extraction. Again, a distinctive feature of our work is that the generated summary has explicitly defined semantic aspects, whereas most news summaries are simply a list of extracted sentences. Despite this difference, our two-step process of generating a summary and some of our heuristics used in sentence selection are similar to what has been used for news summarization[5].

3. Automatic Gene Summarization

3.1. *Overview*

Our automatic gene summarization system mainly consists of two components: a Keyword Retrieval module that retrieves documents about a target gene, and an Information Extraction module that extracts sentences from the retrieved documents to summarize the target gene. The Information Extraction module itself consists of two components, one for training data generation, and the other for sentence extraction. The whole system is illustrated in Figure 1.

3.2. *Keyword Retrieval Module*

First, to identify documents that may contain useful information for the target gene, we use a dictionary-based keyword retrieval approach to retrieve all documents containing any synonyms of the target gene.

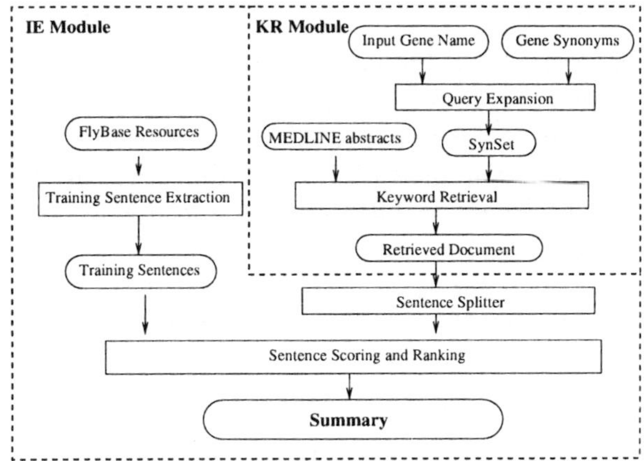

Figure 1. System Overview.

3.2.1. *Gene* SynSet *Construction*

Gene synonyms are very common in biomedical literature. It is important to consider all the synonyms of a target gene when searching for relevant documents about the gene. We used the synonym list for fly genes provided by BioCreAtIvE Task 1B[6] and extended it by adding names or functional information of proteins encoded by each gene from FlyBase's annotation. In the end, we constructed a set of synonyms and protein names (called *SynSet* here) for each known Drosophila gene.

To further improve the recall of retrieval, we investigated variations in gene name spelling. The following variations are identified and addressed in our system: (1) There are various ways to separate name constituents: they can be contiguous or separated by various separators such as white spaces, hyphens, slashes and brackets. (2) Gene names can be spelled in upper or lower case. To deal with these variations, our system uses a special tokenizer for both Medline abstracts and *SynSet* entries. The tokenizer converts the input text into a sequence of tokens, where each token is either a sequence of lowercase letters or a sequence of numbers. White spaces and all other symbols are treated as token delimiters. For instance, the different synonyms for gene *cAMP dependent protein kinase 2*, "PKA C2", "Pka C2", and "Pka-C2", are all normalized to the same token sequence "pka c 2" to allow them to match each other. A Medline abstract is considered as being relevant only if it matches the token sequence of a synonym *exactly*.

3.2.2. *Synonym Filtering*

Some gene synonyms are ambiguous, for example, the gene name "PKA" is also a chemical term with a different meaning. In these situations, a document containing the synonym with an alternative meaning would be retrieved. Our strategy of alleviating this problem is based on the observations that (1) the longer or full name of a gene is often unambiguous; (2) when a gene's short abbreviation is mentioned in a document, its full or longer name is often present as well. Therefore, we force all retrieved documents to contain at least one synonym of the target gene that is at least 5-character long.

3.3. *Information Extraction Module*

The information extraction module extracts sentences containing useful factual information about the target gene from the documents returned by the keyword retrieval module. To ensure the precision of extraction, we only consider sentences containing the target gene, which are further organized into the six general categories listed in Table 1, which we believe are important for gene summaries.

Table 1. Categories for Gene Summary

GP	Gene Product, describing the product (protein, rRNA, *etc.*) of the target gene.
EL	Expression Location, describing where the target gene is mainly expressed.
SI	Sequence Information, describing the sequence information of the target gene and its product.
WFPI	Wild-type Function & Phenotypic Information, describing the wild-type functions and the phenotypic information about the target gene and its product.
MP	Mutant Phenotype, describing the information about the mutant phenotypes of the the target gene.
GI	Genetical Interaction, describing the genetical interactions of the target gene with other molecules.

3.3.1. *Training Data Generation*

To help identify informative sentences related to each category, we construct a training data set consisting of "typical" sentences for describing each of the six categories using three resources: the *Summary* pages, the *Attributed data* pages, and the *references* of each gene in FlyBase.

The "Summary" Paragraph: FlyBase curators have compressed all the relevant information about a gene into a short paragraph, the text *Summary* in the FlyBase report. This paragraph contains good example sentences for each aspect of a gene. A typical paragraph contains information related to gene product, sequence information, genetical interaction, *etc.* More importantly, verbs such as "encode", "sequence" and "interact" in the text are very indicative of which category the sentence is related to. Based on the regular structure of these text summaries, we decompose each paragraph into our six categories with non-relevant sentences discarded.

However, since these sentences are generated from a common template by a curator, they are not good examples of typical sentences that appear in real literature. For instance, genetical interaction can be described in many different ways using verbs such as "regulate", "inhibit", "promote" and "enhance". In the "summary" paragraph, it is always described using the template "It interacts genetically with ...". Thus we also want to obtain good examples of original sentences from the literature.

The "Attributed Data" Report: One resource of original sentences is the "attributed data" report for each Drosophila gene provided by FlyBase. For some attributes such as "molecular data", "phenotypic info." and"wild-type function", the original sentences from literature are listed. These sentences seem to be good complements of the training data from the "summary" paragraph. In our system, we collect the sentences from "phenotypic info." and "wild-type function" as training sentences for the category *WFPI*.

The References: For categories such as "gene product" and "interacts genetically with", the "attributed data" reports only list the noun phrases related to the target gene, but do not show any complete sentences. In order to find the patterns of sentences containing such information, we exploit the links to the corresponding references given in the "attributed data" reports to find the PubMed ID of the reference. We then look for occurrences of the item, *i.e.,* a protein name in "gene product" or another gene name "interacts genetically with", in the abstract of the reference. We add the sentence containing both the item and the target gene to our training data. Inclusion of these sentences is useful because verbs such as "enhance" and "suppress" now appear in the training data.

3.3.2. *Sentence Extraction*

To extract sentences related to each category for a target gene, we first preprocess sentences by removing the stop words and stemming with a Porter stemmer. We then score each sentence as follows.

Category Relevance Score (S_c): We use the vector space model and cosine similarity function from information retrieval to assign a relevance score to each sentence *w.r.t.* each category. Specifically, For each category, we construct a corresponding term vector V_c using the training sentences for the category. Following a commonly used information retrieval heuristic, we define the weight of a term t_i in the category term vector for category j as $w_{i,j} = \text{TF}_{i,j} * \text{IDF}_i$, where $\text{TF}_{i,j}$ is the term frequency, i.e., the number of times term t_i occurs in all the training sentences of category j, and IDF_i is the inverse document frequency. IDF_i is computed as $\text{IDF}_i = 1 + \log \frac{N}{n_i}$, where N is the total number of documents in our document collection, and n_i is the number of documents containing term t_i. Intuitively, V_c reflects the usage of different words in sentences describing a category.

Similarly, for each sentence we can construct a sentence term vector V_s, with the same IDF and the TF being the number of times a term occurs in the sentence. The category relevance score is then the cosine of the angle between the category term vector and the sentence term vector: $S_c = \cos(V_c, V_s)$.

Document Relevance Score (S_d): A good sentence to be included in our summary should be both relevant to a category and informative. To measure the informativeness of a sentence, we compute a document relevance score for each sentence, which is the cosine similarity between the sentence vector V_s and the document vector V_d, which is computed similarly to the other vectors described above.

Location Score (S_l): A useful heuristic for news article summarization is to favor sentences at the beginning of a document. For scientific literature, however, the last sentence of an abstract is usually a summary of the experimental results or the discovery. Therefore, we also assign each sentence a location score, which is 1 for the last sentence of an abstract, and 0 otherwise.

Sentence Ranking and Summary Generation: The final score of a sentence S is a weighted sum of the three scores mentioned above with the weights set empirically: $S = 0.5S_c + 0.3S_d + 0.2S_l$. To ensure reliable association between sentences and categories, for each sentence, we rank all the categories based on S and keep only the top two categories. To generate a structured, category-based summary, for each category, we rank all the kept sentences according to S and pick the top-k sentences. Such a category-based summary is similar to the "attributed data" report in FlyBase. We also generate a paragraph-long summary by combining the top sentences of all the categories in the following way: We "grow" our paragraph summary by taking a top-ranked sentence from each category that is different from all the already included sentences in the paragraph summary. We impose an order on the categories so that the most specialized category would have a chance to contribute a sentence first.

4. Experiments and Evaluation

4.1. *Experiment Setup*

We retrieved 22092 Medline abstracts as our document collection using the keyword "Drosophila". We used the Lemur Toolkit[a] to implement the keyword retrieval module. Lemur is a C++ toolkit supporting a variety of information retrieval functions. We mainly exploited its indexing capability to quickly retrieve documents containing a given keyword. Our gene summarization algorithm runs very fast, taking only seconds to generate a summary on a Dell PowerEdge 2650 (3.06GHz CPU, 4GB Memory).

We used about 1/5 of the training data in FlyBase for training and randomly selected 10 genes from FlyBase for evaluation. For each gene, we ran three experiments. The first is a baseline run (*BL*), in which we randomly select k sentences. In the second run (*CatRel*), we use Category Relevance Score S_c to rank sentences. In the third run (*Comb*), we use combined score S to rank sentences.

4.2. *Evaluation and Discussion*

For each category of each gene, we generated top-k sentences from each run, and then asked two annotators with domain knowledge to judge the relevance. A sentence is considered to be relevant to a category if and only if it contains information on this aspect, regardless whether it contains any extra information. The evaluation metric is the precision of the top-k sentences for each category. The results are shown in Table 2. The average precisions of top-10 sentences for most categories by the two ranking methods are about $50 - 70\%$, while the average precision by random selection is typically about 20%. In most cases, combining all three scores performs only slightly better than using the Category Relevant Score alone. This could either be due to the fact that we use a simple function to combine the three scores and the parameters are not fully optimized, or suggest that those general text summarization heuristics may not be applicable to our problem.

[a] http://www.lemurproject.org/

We notice that the improvements over the baseline are most pronounced for categories *EL*, *SI*, *MP* and *GI*. This may be because these four categories are more specific and thus harder to detect by random selection.

Table 2. Precision of the top-k extracted sentences

		Avg. Precision					Avg. Precision		
cat.	top-k	*BL*	*CatRel*	*Comb*	cat.	top-k	*BL*	*CatRel*	*Comb*
	1	0.1	**0.9**	0.85		1	0.1	**0.6**	0.55
EL	2	0.1	**0.8**	0.73	MP	2	0.13	0.53	**0.55**
	5	0.14	**0.58**	0.58		5	0.13	0.36	**0.43**
	10	0.18	0.48	**0.51**		10	0.17	0.33	**0.45**
	1	0.45	**0.8**	0.75		1	0.1	**0.7**	0.7
GP	2	0.43	0.78	**0.8**	GI	2	0.13	**0.68**	0.65
	5	0.42	0.73	**0.75**		5	0.21	0.62	**0.67**
	10	0.4	0.57	**0.67**		10	0.23	0.56	**0.58**
	1	0.1	**0.85**	0.85		1	0.45	**0.6**	0.55
SI	2	0.05	0.78	**0.8**	WFPI	2	0.58	**0.78**	0.73
	5	0.12	0.63	**0.66**		5	0.6	**0.78**	0.77
	10	0.15	0.49	**0.54**		10	0.6	0.73	**0.75**

In Table 3, we show a sample structured summary generated for the well-studied gene *Abl*, in which all the extracted sentences are quite informative as judged by biologists. For comparison, we show the human-generated FlyBase summary of the same gene in Table 4.

To see how well our system performs on a less-studied gene, we show a sample structured summary generated for the less-studied gene *Camo\Sod* in Table 5. In this case, some sentences are not very relevant. However, by reading this summary, a biologist could still get some basic idea of the gene *Camo\Sod*. We cite one possible reconstruction of information based solely on our results in Table 5:

> *Camo\Sod* encodes the protein, CuZn superoxide dismutase, involved in superoxide production. In Drosophila, it is suggested that this gene is expressed in central nervous system. All the protein's important amino acids are conserved in related organisms. The mutation of this gene is known to be lethal.

The FlyBase summary for this gene is shown in Table 6, which is seen to be very short and barely informative. Considering that we have used no external information, the rich information content of our results is a strong indication of the usefulness of our system.

One problem of predefined categories is that not all genes fit into this framework. For instance, the gene *Amy-d* is an enzyme involved in carbohydrate metabolism and not typically studied by genetic means. As a

Table 3. Text summary of gene *Abl* by our system

GP	The Drosophila melanogaster abl and the murine v-abl genes encode tyrosine protein kinases (TPKs) whose amino acid sequences are highly conserved.
EL	In later larval and pupal stages, abl protein levels are also highest in differentiating muscle and neural tissue including the photoreceptor cells of the eye. abl protein is localized subcellularly to the axons of the central nervous system, the embryonic somatic muscle attachment sites and the apical cell junctions of the imaginal disk epithelium.
SI	The DNA sequence encodes a protein of 1520 amino acids with sequence homology to the human c-abl proto-oncogene product, beginning at the amino terminus and extending 656 amino acids through the region essential for tyrosine kinase activity.
MP	The mutations are recessive embryonic lethal mutations but act as dominant mutations to compensate for the neural defects of abl mutants.
GI	Mutations in the Abelson tyrosine kinase gene show dominant interactions with fasII mutations, suggesting that Abl and Fas II function in a signaling pathway that controls proneural gene expression.
WFPI	We have examined the expression of the abl protein throughout embryonic and pupal development and analyzed mutant phenotypes in some of the tissues expressing abl. abl protein, present in all cells of the early embryo as the product of maternally contributed mRNA, transiently localizes to the region below the plasma membrane cleavage furrows as cellularization initiates.

Table 4. Text summary of gene *Abl* from FlyBase

D. melanogaster gene ***Abl tyrosine kinase***, abbreviated as ***Abl***, is reported here. It has also been known in FlyBase as CG4032 and l(3)04674. It encodes a product with protein-tyrosine kinase activity (EC:2.7.1.112) involved in axon guidance which is localized to the axon; it is expressed in the embryo (embryonic central nervous system) and ovary (oocyte and ovary). It has been sequenced and its amino acid sequence contains a protein kinase, a SH2 motif, a tyrosine protein kinase, a SH3, a tyrosine protein kinase, active site and a protein kinase-like. It has been mapped cytologically to 73B1–4. It interacts genetically with Nrt, ena, fax, Lar, robo and 17 other listed genes. There are 28 recorded alleles: 15 in vitro constructs (none available from the public stock centers), 12 classical mutants (3 available from the public stock centers) and 1 wild-type. Amorphic mutations have been isolated which affect the central nervous system, the longitudinal connective, the commissure and 5 other listed tissues and are pupal recessive lethal, reduced (with Df(3L)st-j7) viable and neuroanatomy defective. *Abl* is discussed in 206 references (excluding sequence accessions), dated between 1981 and 2005. These include at least 30 studies of mutant phenotypes , 8 studies of wild-type function and 10 molecular studies . Among findings on *Abl* mutants, *Abl* mutants show phenotypes in somatic muscles and eye imaginal disks. Among findings on *Abl* function, *Abl* gene product may play a role in establishing and maintaining cell-cell interactions.

result, most sentences in *MP* and *GI* categories will be judged as irrelevant. Thus, the low precision in some occasions may simply be because there is little research on this topic. In general, the lack of information on some

Table 5. Text summary of gene *Camo\Sod* by our system

GP	Superoxide production by Drosophila mitochondria was measured fluorometrically as hydrogen peroxide, using its dependence on substrates, inhibitors, and added superoxide dismutase to determine sites of production and their topology.
EL	The aim of this study was to ascertain the status of CuZn superoxide dismutase (CuZn-SOD) expression in the central nervous system of Drosophila melanogaster.
SI	Comparison of the Drosophila Cu,Zn SOD amino acid sequences with the Cu,Zn SOD of Bos taurus and Xenopus laevis (whose three-dimensional structure has been elucidated) reveals conservation of all the protein's functionally important amino acids and no substitutions that dramatically change the charge or the polarity of the amino acids.
MP	The gene for cytoplasmic superoxide dismutase (cSOD) maps within this interval, as does low xanthine dehydrogenase (lxd).–Recessive lethal mutations were generated within the region by ethyl methanesulfonate mutagenesis and by hybrid dysgenesis.
GI	Drosophila orthologues of the mammalian Cu chaperones, ATOX1 (a human orthologue of yeast ATX1), CCS (copper chaperone for superoxide dismutase), COX17 (a human orthologue of yeast COX17), and SCO1 and SCO2, did not significantly respond transcriptionally to increased Cu levels, whereas MtnA, MtnB and MtnD (Drosophila orthologues of human metallothioneins) were up-regulated by Cu in a time- and dose-dependent manner.
WFPI	The 2.5 kb clone consists of a wild-type 1.84 kb EcoRI fragment containing the Cu,Zn SOD gene previously isolated in our laboratory, with an insertion of 0.68 kb derived (by an internal deletion) from an autonomous, 2.9 kb P element.

Table 6. Text summary of gene *Camo\Sod* from FlyBase

Superoxide dismutase, abbreviated as ***Camo\Sod***, is reported here. It has been sequenced . There is one recorded allele, which is wild-type. *Camo\Sod* is discussed in 4 references (excluding sequence accessions), dated between 1992 and 2001.

aspects of a query gene is not a major problem for our system in the sense that, if information about one aspect is missing, a biologist could infer that this aspect may have not been well studied or is not biologically interesting.

5. Conclusion and Future Work

In this paper, we proposed a novel problem in biomedical text mining: automatic generation of structured gene summaries. We developed a system which employed information retrieval and information extraction techniques to automatically summarize information about genes from PubMed abstracts. The system was tested on 10 randomly selected genes, and evaluated by domain experts. The promising results with an average precision above 50% indicate that the system is very effective in summarizing biomedical literature.

We realized that one obvious limitation of our approach was its dependence on the high-quality data in FlyBase. To address this issue, we will in the future incorporate more training data from databases of other model organisms and resources such as GeneRIF in Entrez Gene. We believe the mixture of data from different resources will reduce the domain bias and help build a general tool for gene summarization.

We employed many heuristic methods in our system, primarily because it is unclear at the beginning which computational strategy would be most suitable for our problem. A major future work is to explore more generic methods including probabilistic models for sentence selection. Our long-term goal is to extend our system so that it can be used by all biomedical researchers. Even though we used some fly-specific resources and tested mainly on fly genes, the general framework we proposed is independent of the actual biological domains. We will next be testing the methods on bee genes using the same training set on fly genes but extracting sentences from bee literature, to test applicability across insects. Eventually, we hope to produce automatic summarization of all genes in all organisms, using the entire biomedical literature for extraction and the entire set of model genetic databases for training.

References

1. I. Iliopoulos, A. Enright, C. Ouzounis, (2001) Textquest: document clustering of medline abstracts for concept discovery in molecular biology. *PSB*, 384-395.
2. L. Hirschman, J. C. Park, J. Tsujii, L. Wong, C. H. Wu, (2002) Accomplishments and challenges in literature data mining for biology. Bioinformatics **18(12)**:1553-1561.
3. H. Shatkay, R. Feldman, (2003) Mining the Biomedical Literature in the Genomic Era: An Overview. *JCB.*, **10(6)**:821-856.
4. D. Marcu, (2003) Automatic Abstracting. *Encyclopedia of Library and Information Science*, 245-256.
5. D. R. Radev, H. Jing, M. Sty, D. Tam, (2004) Centroid-based summarization of multiple documents. Inf. Process. Manage. 40(6):919-938.
6. L. Hirschman, M. Colosimo, A. Morgan, A. Yeh, (2005) Overview of BioCreAtIvE Task 1B: Normailized Gene Lists. *BMC Bioinformatics* 2005, 6(Suppl):S11.
7. R. A. Drysdale, M. A. Crosby and The FlyBase Consortium, (2005) FlyBase: genes and gene models. *Nucleic Acids Res.* **33**: 390-395.

FINDING GENERIFS VIA GENE ONTOLOGY ANNOTATIONS

ZHIYONG LU, K. BRETONNEL COHEN, AND LAWRENCE HUNTER

Center for Computational Pharmacology
School of Medicine, University of Colorado
Aurora, CO, 80045 USA
E-mail: {Zhiyong.Lu, Kevin.Cohen, Larry.Hunter}@uchsc.edu

A Gene Reference Into Function (GeneRIF) is a concise phrase describing a function of a gene in the Entrez Gene database. Applying techniques from the area of natural language processing known as automatic summarization, it is possible to link the Entrez Gene database, the Gene Ontology, and the biomedical literature. A system was implemented that automatically suggests a sentence from a PubMed/MEDLINE abstract as a candidate GeneRIF by exploiting a gene's GO annotations along with location features and cue words. Results suggest that the method can significantly increase the number of GeneRIF annotations in Entrez Gene, and that it produces qualitatively more useful GeneRIFs than other methods.

1. Introduction

The National Library of Medicine (NLM) started a Gene Indexing initiative on April 1, 2002, the goal of which is to link any article about the basic biology of a gene or protein to the corresponding Entrez Gene entry [1]. The result is an entry called a Gene Reference Into Function (GeneRIF) within the Entrez Gene[a] (previously LocusLink) database. Each GeneRIF is a concise phrase (limited to 255 characters in length) describing a function related to a specific gene, supported by at least one PubMed ID. For example, the GeneRIF *LATS1 is a novel cytoskeleton regulator that affects cytokinesis by regulating actin polymerization through negative modulation of LIMK1* is assigned to the human gene LATS1 (GeneID: 9113) and is associated with a citation titled *LATS1 tumor suppressor affects cytokinesis by inhibiting LIMK1* (PMID: 15220930) in PubMed/MEDLINE (see Table 2). In principle, GeneRIFs provide an up-to-date summary of facts

[a] `http://www.ncbi.nlm.nih.gov/entrez/query.fcgi?db=gene`

relevant to each gene, justified by specific literature citations. However, despite growing at a rate of about 35,000 per year, the *GeneRIF coverage*, i.e. the percentage of genes associated with at least one GeneRIF, remains quite modest — 1.3M Entrez genes have no GeneRIFs. Even in humans, the organism with the best GeneRIF coverage, only 26.8% of all genes are associated with at least one GeneRIF. Thus the main objective of this work is to increase the currently low GeneRIF coverage, which might be due to the time- and labor-intensive fully manual indexing process. Table 1 shows the current GeneRIF coverage for the four organisms with the largest number of GeneRIFs. Column 4 shows the number of genes with no GeneRIFs for which our method could potentially generate at least one GeneRIF and Column 5 shows the number of genes for which it could increase the number of GeneRIFs already present. The largest potential coverage increase is for mouse genes. In the current database, 12.6% of mouse genes (6,081/48,447) have already been associated with at least one GeneRIF. Meanwhile, 6,050 mouse genes (12.5%) do not have any GeneRIF, but they are associated with at least one Gene Ontology (GO) [2] annotation, and 4,919 (10.2%) more with one or more GeneRIFs could gain additional GeneRIFs by our method.

Table 1. The first four rows are organism-specific. The last row is for all Entrez genes regardless of species. Columns are: **Species** (the name of the organism); **Entrez Genes** (the number of genes currently in the Entrez database); **W/ GeneRIFs** (the number of genes having at least one GeneRIF); **GO Only** (the number of genes having at least one GO annotation (and its corresponding PubMed article) and no GeneRIFs); **GO And GeneRIFs** (the number of genes having both GeneRIFs and GO annotations supported by different PubMed articles). Col. 3 is a proper subset of Col. 5.

Species	Entrez Genes	W/ GeneRIFs	GO Only	GO And GeneRIFs
Homo sapiens	32,791	8,790 (26.8%)	2,225 (6.79%)	5,789 (17.7%)
Mus musculus	48,447	6,081 (12.6%)	6,050 (12.5%)	4,919 (10.2%)
Rattus norvegicus	28,665	3,143 (11.0%)	1,359 (4.74%)	1,604 (5.60%)
Drosophila melanogaster	20,763	1,274 (6.14%)	218 (1.05%)	10 (0.00%)
All Species	1.3M	22,352 (1.69%)	15,282 (1.15%)	12,267(0.92%)

We hypothesize that it is possible to use automatic summarization techniques [3] to automatically predict GeneRIFs by exploiting GO annotations associated with Entrez gene entries, in combination with automatic summarization techniques, to find sentences that would be good GeneRIF annotations. This approach links the Entrez Gene database, the Gene Ontology, and the Medline database by automatic summarization techniques. The method is based on two observations: the fact that GeneRIFs are in many ways similar to single-document summaries, and the fact that the subject matter of GeneRIFs often has considerable overlap with the semantic content of Gene Ontology terms. The method consists of calculating a score for the title and every sentence in an abstract, and then selecting the highest-scoring candidate as a GeneRIF. The score is calculated based on features known to be useful in selecting sentences for automatically-generated summaries, and crucially, based on similarity between the candidate and the Gene Ontology terms with which the gene is annotated.

Table 2. The first row (LATS1, Entrez Gene ID 9113) is an example of the 413 GeneRIFs that we used as the gold standard. The PMID 15220930 is the reference for the GO term *regulation of actin filament polymerization* and for the GeneRIF. The second row is an example of the 6,050 target genes. The PMID is the reference for both GO terms, but it is not the reference for any GeneRIF.

Gene	GO Term	PMIDs	GeneRIFs
LATS1	regulation of actin filament polymerization (IDA)	15220930	LATS1 is a novel cytoskeleton regulator that affects cytokinesis by regulating actin polymerization ...
	G2/M transition of mitotic cell cycle (IDA) sister chromatid segregation (IDA)	15122335	WARTS plays a critical role in maintenance of ploidy through its actions in both mitotic progression and the G(1) tetraploidy checkpoint
BST2	signal transducer activity (IMP) positive regulation of I-kappaB/NF-kappaB cascade (IMP)	12761501	N/A

In order to evaluate our system, we assembled a gold standard data set consisting of 413 GeneRIFs found in the current Entrez database. The data set consisted of all human genes (e.g. gene LATS1 in Table 2) that have both (1) a GeneRIF and its corresponding PubMed article, and (2) GO term(s) that are supported by the same PubMed article. The 413 GeneRIFs are associated with an average of three GO terms each. We only evaluate our system for human genes (the organism with the most GeneRIFs) in this paper, but the method is organism-independent and we have applied it to all four organisms in Table 1.

2. Related Work

GeneRIFs were first characterized and analyzed by Mitchell et al. 2003. Their prediction was the subject of the TREC 2003 competition. The secondary task of the TREC 2003 Genomics Track [4] was to reproduce GeneRIFs from MEDLINE records. Each contestant team was given 139 GeneRIFs. The results were later described in [4]:

> Most participants found that the GeneRIF text most often came from sentences in the title or abstract of the MEDLINE record, with the title being used most commonly ... The best approaches ([5] and [6]) used classifiers to rank sentences likely to contain the GeneRIF text. No groups [achieved] much improvement beyond using titles alone.

As shown below, our results are significantly better than this baseline.

3. System and Method

3.1. *Data*

We downloaded both GeneRIFs and Entrez Gene flat files on June 16, 2005 from NCBI's ftp[b] site.

3.2. *The relationship between GeneRIFs and their sources*

To understand why our method works, it is helpful to be familiar with the relationship between GeneRIFs and their source documents. Every GeneRIF annotation includes a PMID (PubMed identifier) that identifies a specific document that provides the literary evidence for the GeneRIF. GeneRIFs typically have an *extractive* relationship to their document, meaning that the GeneRIF is, to a large extent, "cut-and-pasted" from its source. Furthermore, GeneRIFs typically come from particular *locations* in the document, definable either by sentence position (e.g. first, second, penultimate, last) in the abstract or by being the document title. We investigated the extent to which these patterns hold by examining the 413 GeneRIFs that constituted our gold standard. Specifically, for each GeneRIF we first computed the "classic" Dice coefficient (a measure of overlaps in two strings, [4]) between the GeneRIF text and each of the abstract sentences and title of an article. Next, we selected a sentence or

---

[b] `ftp://ftp.ncbi.nlm.nih.gov/gene`

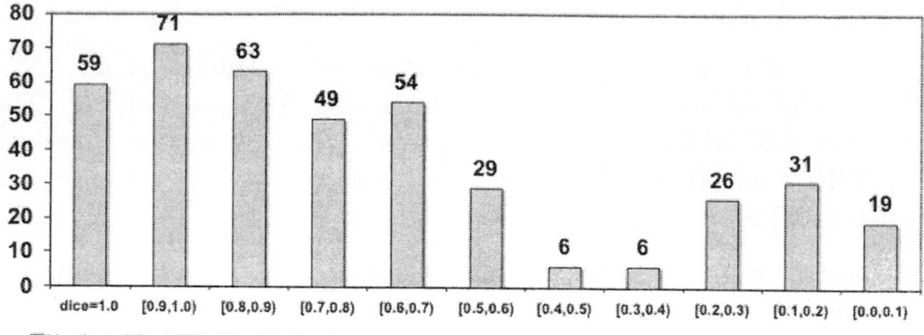

Figure 1. Distribution of 413 GeneRIFs according to their maximum Dice coefficients.

the title of an abstract that is most similar to the GeneRIF (i.e. the Dice coefficient between the selected one and the GeneRIF is the largest). Figure 1 shows the distribution of 413 GeneRIFs according to their maximal Dice coefficient. As can be seen, 59 GeneRIFs have a Dice coefficient of 1.0. That is, these 59 GeneRIFs are exact matches to either the title or a sentence of an abstract. Finally, we analyzed which sentence of an abstract is most similar to the GeneRIF. Data are shown in Table 3 with different Dice coefficient thresholds. We found that the ones most similar to the

Table 3. Best mappings of the 413 GeneRIF texts against their corresponding abstract titles and sentences under different Dice coefficient thresholds T. T $<$ 0.5 is not considered as an acceptable match.

matching	T = 0.5	T = 0.6	T = 0.7	T = 0.8	T = 0.9	T = 1.0
the title	25.4%	23.2%	19.9%	16.9%	12.3%	9.69%
the last sentence	26.1%	24.5%	20.6%	16.2%	9.93%	2.42%
the penultimate sentence	8.96%	7.51%	5.08%	4.36%	3.63%	0.97%
other sentences	17.7%	16.5%	13.0%	9.24%	5.65%	1.22%
total matching	78.7%	71.7%	58.6%	46.7%	31.5%	14.3%
no matching	21.3%	28.3%	41.4%	53.3%	68.5%	85.7%

GeneRIF are always the title, the last sentence or the penultimate sentence of an abstract. In addition, an acceptable match was found much more often in the title and the last sentence than in the penultimate sentence. For example, when the threshold was set to 0.5, 25.4% (105) GeneRIFs matched best to the title, 26.1% (108) to the last sentence, and 8.96% (37) to the penultimate sentence. As the Dice coefficient threshold increases (i.e. the matching criterion becomes stricter), there are fewer matches for those

GeneRIFs. There were only 22.3% (92) GeneRIFs not matching to the title or any sentence when the threshold was set to 0.5. But 85.7% (354) did not have a match when exact matches were required (i.e. T = 1.0). Table 3 gives us a baseline approach: picking the title or the last sentence of an abstract, depending on the Dice coefficient threshold. Since the numbers are approximately identical for both when the threshold is less than 0.8 and the numbers favor the title when the threshold equals 0.9 or 1.0, we used "picking the title" as the baseline.

3.3. *System*

The algorithm works by assigning each candidate a score based on the presence of GO terms, the candidate's position, and the presence of cue words. The highest-scoring candidate is suggested as a GeneRIF. The system architecture is shown in Figure 2. For any gene with GO annotations, we retrieve

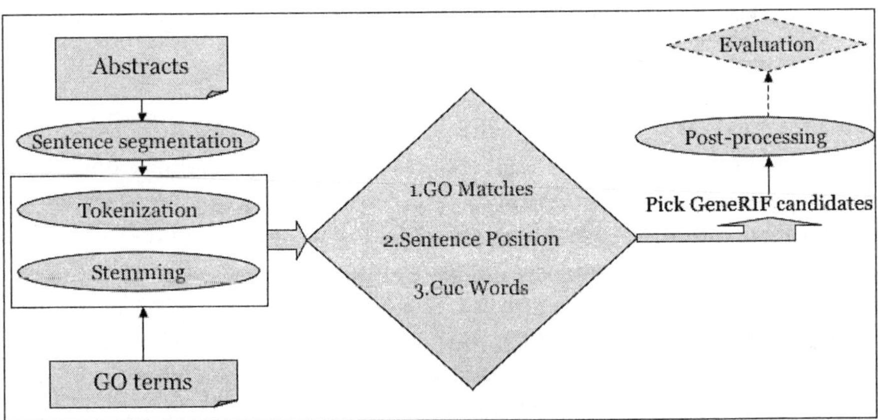

Figure 2. Architecture of the prediction method.

the abstracts associated with the Gene Ontology annotations. Input abstracts are segmented into individual sentences. Each sentence is tokenized and stemmed into a bag of stemmed tokens. Similarly, the set of GO terms associated with that gene is preprocessed via tokenization and stemming. Tokens from a stop list[c] are removed, and the set of unique tokens from the

[c] ftp://ftp.cs.cornel.edu/pub/smart/english.stop

Gene Ontology terms is assembled. Then all of these tokens are processed by the algorithm described below:

```
1: for every sentence S in an abstract A do
2:    for every unique sentence token ST in S do
3:       for every unique GO token GT in all GO terms do
4:          if ST equals GT then
5:             assign one point to S
6:          end if
7:       end for
8:    end for
9:    if S is the title or penultimate or last sentence of A then
10:      assign one point to S
11:   end if
12:   if S has a cue word match then
13:      assign one point to S
14:   end if
15:   if S is assigned more points than other sentences then
16:      generif_candidate ← S
17:   end if
18: end for
```

The pseudo-code above describes the three scoring procedures illustrated in the center diamond in Figure 2:

GO Matches — Pseudo-code lincs 2 to 8. We look for GO-term presence in the title or sentences in an abstract. Our search is based on string matching of stemmed tokens. For example, *GO:0030833 regulation of actin filament polymerization* was preprocessed into four stemmed tokens: "regul", "actin", "filament" and "polymer". The word "of" was dropped via the stop-word list during the process. Similarly, the title and the sentences in the abstract were tokenized and stemmed. After preprocessing, the last sentence of the abstract, *Our findings indicate that LATS1 is a novel cytoskeleton regulator that affects cytokinesis by regulating actin polymerization through negative modulation of LIMK1.* contained 14 unique stemmed tokens, three of which were identical to the ones in the GO term (i.e. "regul", "actin" and "polymer"). Thus, GO matching gives this sentence a score of three.

Sentence Position — Pseudo-code lines 9 to 11. Titles, penultimate sentences, and final sentences are each given one point. The example sentence is the last sentence, so one point is added to the score, for a total of four.

Cue Words — Pseudo-code lines 12 to 14. We found many words to be very indicative of GeneRIFs, such as "findings", "novel", "role", et al. In the automatic summarization literature, these are known as *cue words*—words or phrases that indicate that a sentence is likely to be a component of a good summary. A complete list of these cue words can be found at the paper supplementary website. Note that some of these keywords are often not seen directly in GeneRIFs because they are removed when a sentence is selected as a GeneRIF[d]. For example, the phrase "Our findings indicate that" was cut from the last sentence while the remainder was used as the GeneRIF. We assembled this keyword list mainly by human examination. In particular, we manually inspected those 59 GeneRIFs in Table 1 that are very similar to but not exactly the same as a title or sentence (i.e. 0.9 <= Dice coefficient < 1.0). We then verified our keywords with a list of words that have the highest mutual information [8] produced by a Naïve Bayes classifier. If a title or an abstract sentence contains any of these cue terms, it will be given a single point. The example sentence contains a cue word ("novel"), so one point is added to the score, for a total of five.

For each abstract, the sentence with the largest number of points is selected as the GeneRIF candidate. (Tie-breaking procedures are described on the supplementary website.) For the LATS1 example, the last sentence was given a total of five points. This is the highest score among the title and all abstract sentences, so it is the GeneRIF candidate for LATS1. In a post-processing step, we removed polarity-indicating words/phrases, since they are often omitted in GeneRIFs. For example, the phrase *Our findings indicate that* was removed from that last sentence in the LATS1 example. The complete set of predictions is posted at the paper supplementary website[e].

4. Results

We evaluated our system on the gold standard data set under different Dice coefficient thresholds. Figure 3 shows that the prediction result of our system is better than that of the baseline (i.e. picking the title or the last sentence) approach at all thresholds except 1.0. For example, our method has made a 21.3% (131 vs. 108) increase in producing correct GeneRIF candidates when the threshold was set to 0.5. Since there is no explicit definition for GeneRIF selection, in principle any sentence could

[d]These are often polarity-indicating phrases [7]. They are typically omitted in GeneRIFs.
[e]`http://compbio.uchsc.edu/Hunter_lab/Zhiyong/psb2006`.

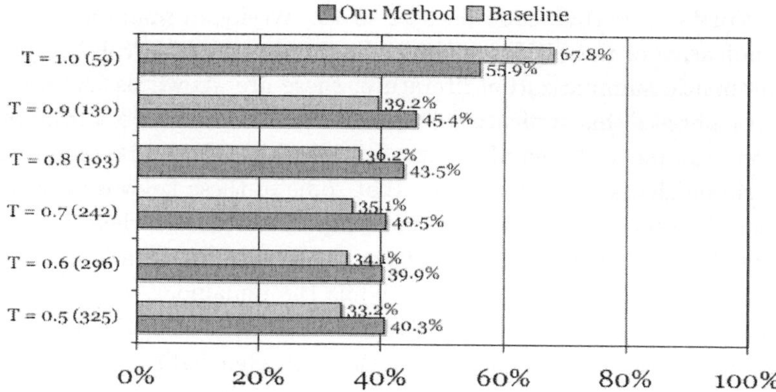

Figure 3. System performance of our prediction approach compares to the baseline method. **T** is the Dice coefficient. **(number)** is the number of abstracts (out of 413 totally) that has a Dice coefficient score equal or greater than the threshold **T** between one of its sentences and the corresponding GeneRIF. **Our Method** is the number of correct predictions made by our system. **Baseline** is the number of correct predictions made by the baseline method.

be a GeneRIF as long as it describes a gene function and is less than 255 characters long. Those GeneRIF candidates selected by our system that do not exact match GeneRIFs are not necessarily false positives. Further analysis shows that (1) many of our outputs are as meaningful and informative as the corresponding GeneRIFs. For example, the GeneRIF *ANGPTL3 stimulates endothelial cell adhesion and migration via integrin alpha vbeta 3 and induces blood vessel formation in vivo* (PMID: 11877390) for the human gene ANGPTL3 (GeneID: 27329) was not chosen by our method. Rather, a candidate *ANGPTL3 is the first member of the angiopoietin-like family of secreted factors binding to integrin alpha(v)beta(3) and suggest a possible role in the regulation of angiogenesis* based on the last sentence of the abstract was predicted. Not only does this sentence have more matches to GO terms, but it also summarizes three previous sentences (including the GeneRIF) in that abstract. Therefore, we argue that in this case, our method produced a better candidate than the current GeneRIF from this abstract. (2) Some candidates reflect information complementary to the current GeneRIFs. Since our outputs are based mainly on GO matches, our GeneRIF candidates mostly express gene functions in GO terms. For instance, the human gene BSCL2 (GeneID: 26580) has only one GO term *GO:0030176 integral to endoplasmic reticulum membrane* associated with a

PubMed article (PMID: 14981520). Our system suggests *seipin is an integral membrane protein of the endoplasmic reticulum (ER)* as the GeneRIF rather than the current one *Heterozygous missense mutations in BSCL2 are associated with distal hereditary motor neuropathy and Silver syndrome*, the actual GeneRIF (and the title of the paper). In this example, although the actual GeneRIF and the candidate one are not similar, they each describe an important functional aspect of this gene. Thus, we believe both should be included.

We also applied our method to all genes in column 4 of Table 1. For each gene, we produced one or more GeneRIFs (depending on the number of PubMed articles), each of which is associated with one or more GO terms.

5. Discussion

5.1. *Comparison with other features and methods*

As mentioned above, both teams in TREC 2003 used classification methods attempting to reproduce GeneRIFs. They both experimented with a number of different features and reported several useful ones including MeSH terms and Target_Gene (i.e. is the target gene mentioned?). We therefore experimented with these two, and their combinations, in our system. Additionally, we also extended GO terms to GO definitions. We stemmed and tested the MeSH terms and GO definitions in the same way as GO terms. Each match adds one point to the score. MeSH terms for each PubMed article were retrieved from Medline. GO definitions are parsed from the publicly available GO.defs file. For Target_Gene, both the gene official name and aliases are used. When the target gene is found in a sentence, one additional point is assigned. The combination of GO and MeSH matches are the intersection of each individual match, thus the same token in a sentence will not be credited twice.

Table 4. Performance comparison with other features and their combinations. ∪ represents the union of two features. TG stands for Target_Gene.

Various Matchings	T = 0.5	T = 0.6	T = 0.7	T = 0.8	T = 0.9	T = 1.0
GO only	**131**	118	98	84	59	33
GO Defs only	129	**125**	**105**	**89**	**61**	32
MeSH only	123	112	98	83	57	**37**
GO ∪ MeSH	127	114	95	80	60	35
GO ∪ TG	127	114	95	81	58	33
GO ∪ MeSH ∪ TG	124	112	94	79	**61**	35

Table 4 shows that by just using GO terms or definitions our system can achieve better performance in most cases. Other features (i.e. MeSH, Target_Gene and GO definitions) and their combinations with GO did not significantly enhance the system performance. This is possibly because (1) GO terms are as informative as MeSH terms and GO definitions, if not more so; and (2) although GeneRIFs are linked to particular genes, many GeneRIFs do not contain explicit gene mentions in their text. In addition, many abstract sentences include target gene names, which also makes the feature Target_Gene not very discriminative.

Inspired by the previous studies, we also experimented with machine learning (ML) algorithms implemented in WEKA [9]. In our experiments, we used three features: (1) sentence position (the title vs. the last sentence vs. the penultimate sentence vs. all others); (2) the number of GO matches; and 3) a binary feature indicating whether there is a cue word match. No better results were achieved by ML methods compared to our weighted voting system.

5.2. *GeneRIF prediction as automatic summarization*

GeneRIFs can be thought of as single-document summaries. As summaries, they are somewhat unusual, since the fact that they are often derived from abstracts (as that term is used by PubMed/MEDLINE) makes them in some sense summaries of summaries. However, they are clearly characterizable as summaries—a fact which we were able to exploit in predicting them. Specifically, GeneRIFs can be thought of as low-compression, single-document, extractive, informative, topic-focussed summaries of the abstracts from which they are derived, and these facts make them attractive targets for a summarization-based approach.

The fact that GeneRIFs can be thought of as summaries has practical implications for the design of systems that seek to predict GeneRIFs—specifically, we can use location features and cue words, standard parts of the summarization toolkit [3], to find them. These are the elements of the second and third part of our algorithm. Adding the knowledge of Gene Ontology annotations to our summarization system led to high performance and biologically relevant output.

6. Conclusion

NLM's Gene Indexing initiative results in a GeneRIF linking an Entrez Gene record to a specific PubMed article. However, the percentage of genes covered by at least one GeneRIF remains quite **low for all** species after three years of curation. We implemented a system that can automatically produce GeneRIF candidates from the title or abstract. These predicted GeneRIFs reflect the gene attributes represented in their corresponding GO terms. Our results show that we can (1) significantly improve the current GeneRIF coverage by adding more than 10,000 high-quality GeneRIFs to the current database; (2) produce qualitatively more useful GeneRIFs than previous approaches, and (3) continuously generate GeneRIFs when future GO annotations are included for genes which currently do not have any GO annotations. For example, it is estimated that approximately 1,500 such genes are processed each year at MGI[f].

7. Acknowledgments

This work was supported by NIH grant R01-LM008111 (LH).

References

1. J. A. Mitchell *et al.* Gene Indexing: Characterization and Analysis of NLM's GeneRIFs. In *Proceedings of AMIA 2003 Symposium*, (2003)
2. M. Ashburner *et al.* Gene Ontology: tool for the unification of biology. *Nature Genet* **25**, (2000)
3. I. Mani. Automatic Summarization. *John Benjamins Publishing Company*, (2001)
4. W. Hersh *et al.* TREC Genomics Track Overview. In *Proceedings of The Twelfth Text REtrieval Conference (TREC 2003), National Institute of Standards and Technology (NIST)*, (2003)
5. G. Bhalotia *et al.* BioText Report for the TREC 2003 Genomics Track. In *Proceedings of The Twelfth Text REtrieval Conference (TREC 2003), National Institute of Standards and Technology (NIST)*, (2003)
6. B. Jelier *et al.* Searching for GeneRIFs: concept-based query expansion and Bayes classification. In *Proceedings of The Twelfth Text REtrieval Conference (TREC 2003), National Institute of Standards and Technology (NIST)*, (2003)
7. H. Shatkay *et al.* Searching for High-Utility Text in the Biomedical Literature. *BioLINK SIG: Linking Literature, Information and Knowledge for Biology, Detroit, Michigan*, (2005)
8. T. Mitchell. Machine Learning. *McGraw-Hill*, (1997)
9. I. H. Witten *et al.* Data Mining: Practical machine learning tools and techniques. *Morgan Kaufmann*, (2005)

[f] `http://www.informatics.jax.org`

PHENOGO: ASSIGNING PHENOTYPIC CONTEXT TO GENE ONTOLOGY ANNOTATIONS WITH NATURAL LANGUAGE PROCESSING

YVES LUSSIER*[,1,2], TARA BORLAWSKY[§,1], DANIEL RAPPAPORT[§,1], YANG LIU[1], CAROL FRIEDMAN*[,1]

1- Department of Biomedical Informatics, Columbia Center for Systems Biology, 2- Department of Medicine ; Columbia University, New York, NY 10032

Natural language processing (NLP) is a high throughput technology because it can process vast quantities of text within a reasonable time period. It has the potential to substantially facilitate biomedical research by extracting, linking, and organizing massive amounts of information that occur in biomedical journal articles as well as in textual fields of biological databases. Until recently, much of the work in biological NLP and text mining has revolved around recognizing the occurrence of biomolecular entities in articles, and in extracting particular relationships among the entities. Now, researchers have recognized a need to link the extracted information to ontologies or knowledge bases, which is a more difficult task. One such knowledge base is Gene Ontology annotations (GOA), which significantly increases semantic computations over the function, cellular components and processes of genes. For multicellular organisms, these annotations can be refined with phenotypic context, such as the cell type, tissue, and organ because establishing phenotypic contexts in which a gene is expressed is a crucial step for understanding the development and the molecular underpinning of the pathophysiology of diseases. In this paper, we propose a system, PhenoGO, which automatically augments annotations in GOA with additional context. PhenoGO utilizes an existing NLP system, called BioMedLEE, an existing knowledge-based phenotype organizer system (PhenOS) in conjunction with MeSH indexing and established biomedical ontologies. More specifically, PhenoGO adds phenotypic contextual information to existing associations between gene products and GO terms as specified in GOA. The system also maps the context to identifiers that are associated with different biomedical ontologies, including the UMLS, Cell Ontology, Mouse Anatomy, NCBI taxonomy, GO, and Mammalian Phenotype Ontology. In addition, PhenoGO was evaluated for coding of anatomical and cellular information and assigning the coded phenotypes to the correct GOA; results obtained show that PhenoGO has a precision of 91% and recall of 92%, demonstrating that the PhenoGO NLP system can accurately encode a large number of anatomical and cellular ontologies to GO annotations. The PhenoGO Database may be accessed at the following URL: http://www.phenoGO.org

1 Introduction, Related Work and Background

In recent years, there has been a growing interest in automatic methods that annotate biomedical journals. Several methods that use Medline Abstracts in order to annotate genes to *Gene Ontology* (**GO**) terms[1] have been proposed and have yielded up to 10% to 20% recall and 61-99% precision[2,3]. However, to our knowledge, no method is available to automatically process text in order to map contextual pheno-

* Corresponding authors that have contributed equally to the work
§ These authors have contributed equally to the work

types to Gene Ontology Annotations. Establishing ***phenotypic contexts*** in which a gene is expressed is a crucial step for understanding the molecular underpinning of the pathophysiology of diseases. Since complete genomes of multicellular organisms are increasingly annotated in GO, phenotypic context annotations of these genes could serve as a basis for large scale comparative analyses of gene phenotype interactions (phenomics). For example, the specific cell type(s) in which a gene is expressed are very useful to establish the functional molecular networks of differentiated cells (e.g. "CD4+ T Lymphocytes", but not "CD8+", are responsible for murine "interleukin-2-deficient" colitis resembling ulcerative colitis in humans[4] [MGI: 96548 Il2 interleukin 2, GO:0005134 interleukin-2 receptor binding]). More particularly, bioinformatics methods in systems biology are based on the analysis of datasets relating multiple scales of biology together. In this paper we describe an automated system, PhenoGO, which combines NLP and knowledge-based methods to infer the anatomical and cellular context of existing associations between gene products and GO terms as specified in *GOA*[5]. GOA databases comprise gene-GO associations according to a reference (usually a *Pubmed identification*: **PMID**) and the curation process in GOA has a precision of 91%-100% and a recall of 72%[2]. In addition, we performed an evaluation of PhenoGO over the *Mouse Genome Database* (**MGI**) annotations of GO, and report on the results, which show high precision and recall.

1.1 Related Work

A key step for understanding the phenomes is to provide phenotype and genotype information. While GO provides some phenotypic information, other "orthogonal" ontologies have been developed such as *Cell Ontology* (**CO**)[6], the Adult Mouse Anatomy (**MA**)[7] and the *Mammalian Phenotype Ontology* (**MP**)[8]. In addition, more traditional ontologies and terminologies can be filtered to yield other complementary phenotypes, such as the *Unified Medical Language System* (**UMLS**)[9] and the *NCBI Taxonomy*[10].

Natural Language Processing (NLP). Since 1998 there has been an increasing amount of language processing research that involves extraction and mining of biomolecular information in journal articles. Some systems recognize and/or identify biomolecular entities, some detect relations among biomolecular entities, and some discover new knowledge by piecing together information from heterogeneous resources[11]. Krauthammer and Nenadic provide a review of entity recognition systems[12], Cohen and Hersh, and Hirschman and colleagues each provide an overview of relation extraction and text mining systems[13,14]. Until recently biological language processing systems generally extracted terms and relations, but did not map them to concepts in an established ontology. In the biological domain, it has recently been recognized that to achieve interoperability and improved comprehension, it is critical for text processing systems to map extracted information to ontological concepts. A number of researchers have developed systems mapping genes to GO codes[2,3,14,15,16]. Work in the medical domain involving the mapping of text to UMLS concepts has also been explored. For example, Aronson developed MetaMap, which consists of a mixture of statistics and linguistic methods[17], Nadkarni and colleagues[18] use a string matching approach, and Friedman and colleagues[19]use an *NLP system* called MedLEE. MedLEE differs from the other NLP coding systems in that the codes are shown with modifier relations so that concepts may be associated with temporal, negation, uncertainty, degree, and descriptive information, which affect the underlying meaning and are critical for accurate retrieval of subsequent medical applications. The PhenoGO system discussed in this paper utilizes an adaptation of the MedLEE engine, as discussed in the methods. While bioinformatics techniques have been developed to infer phenotypes from biological databases[20] (e.g. microarray experiments), to our knowledge, none have used NLP techniques over the literature to extract relations that associate anatomical or cellular phenotypes to genes and GO terms together.

Knowledge Management, MeSH Indexing and NLP. The *Medical Subject Headings* terminology (**MeSH**) is the National Library of Medicine controlled vocabulary thesaurus[21] and covers all biomedical concepts classes, including phenotypes. To index the main concepts in the Medline paper, MeSH headings are created manually by experts who read the entire Medline paper and then assign MeSH headers to relevant PMIDs. Of relevance to the proposed methods, MeSH terms have been mapped to other terminologies and organized in a semantic network in the UMLS. Recently, phenomic systems have been developed by Bodenreider and Lussier to relate phenotypes and genes relying exclusively on computational terminology methods[22]. The PhenoGO system reuses the knowledge of MeSH indexes and GOA to infer the phenotypic context of genes mapped to GO terms.

1.2 Background

***BioMedLEE NLP System*.** The NLP component of PhenoGO utilizes an existing NLP system, called BioMedLEE, which is under development by the Friedman language processing group. BioMedLEE extracts and encodes genotype-phenotype relations from information in text. An early version is described in Chen and colleagues[23], but differs substantially from the current one in that it extracted phenotypic information only, and did not map textual terms to codes. The BioMedLEE system is based on an adaptation of the MedLEE system[19], which

```
<genefunc v = "regulation" code = "GO:0050789^regulation of biological process">
<process v = "proliferation"><arg v = "target"></arg>
<cell v = "progenitor cell" code = "UMLS:C0038250^stem cell"></cell></process>
<gene_gproduct v = "MGI:98958^Wnt5a"><arg v = "agent"></arg></gene_gproduct> </genefunc>
```

Figure 1. XML output of BioMedLEE for "*Wnt5A regulates proliferation of progenitor cells.*"

extracts and encodes clinical information in patient reports. An important feature of MedLEE/BioMedLEE for PhenoGO is the flexible infrastructure for **mapping textual terms to codes** and is described in more detail by Friedman elsewhere[19].

A detailed description of the BioMedLEE system is being submitted as a separate publication. In this paper, we summarize the components critical to PhenoGO: 1) the first one, prepares the articles for processing by extracting relevant textual sections, and by handling parenthetical expressions, 2) **the entity tagging component** performs semantic tagging of certain entities, such as biomolecular entities 3) the next component identifies section and sentence boundaries, and performs lexical lookup using a lexicon that specifies the semantic and syntactic categories of terms that were not previously tagged. Many of the lexical entries were automatically generated using GO, MA, MP, cell ontology, and the UMLS, 4) the parser structures the sentence according to grammar rules which specify the relations among the concepts. A parse of the complete sentence is attempted first, and then, if unsuccessful, large segments are attempted in an effort to maximize the capture of relations, 5) the last stage performs encoding using a coding table and the structured output from the previous parsing stage to find the most specific codes, and to generate XML output. Codes in the table are represented as triples consisting of a prefix that identifies the specific ontology, an identifier within the ontology, and the name of the concept (e.g. GO:00507899^regulation of biological process). ***Figure 1*** illustrates a simplified form of the output that was generated by processing *Wnt5A regulates proliferation of progenitor cells.* The molecular function **genefunc** with a value attribute **regulation** is the main output structure; it has a code **GO:0050789** followed by the name of the concept. The function is associated with two arguments where one has a cellular modifier. One argument is a **process,** which has the value **proliferation**, but does not have a code. **Process** is the target of **regulation,** and thus, it has an **arg** tag with a value **target**. The **cell** tag with the value **progenitor cell** and **code UMLS:C0038250** modifies **process**.The other argument, the agent of **regulation**, is a gene whose code is **MGI:98958.** What is significant about the above output is that the MGI-GO-cell triplet is contained within one structure **genefunc**, signifying that BioMedLEE found a relation between the 3. In other relevant sentences, BioMedLEE may find a relation between the phenotype and only one of the MGI-GO pair, or may just find the context without any relation to MGI or GO.

Phenotype Organizer System (***PhenOS***) *Knowledge Management and Computational Terminology System.* As the focus is on NLP, and the knowledge management, and computational terminology methods have been published earlier in related papers, we provided a summary of PhenOS. *PhenOS* is a system under development by the Lussier research group with purpose of bridging the gap between heterogeneous biomedical terminologies. The system produces a directed acyclic graph from the UMLS, provides lexico-semantic and model theoretic methods that automatically map an ontology to another one independently of the UMLS and organizes and structures phenotypes across heterogeneous datasets [22,24,25]. Specific methods of PhenOS used in the current study were the integration of phenomic knowledge structures via structured terminology.[25,24]

2 Methods

The method for assigning phenotypic context to GOA was implemented as a system called PhenoGO. An overview of the overall components is illustrated in ***Figure 2.*** First, Medline abstracts and their gene-GO annotations are identified from GOA, and then obtained. Two distinct and independent processes extract phenotypic context from the abstracts. The design is such that PhenoGO can function with either or both of the processes. ***Component 1a*** consists of the **BioMedLEE NLP** system, which processes the title and abstract and generates structured output, which in this study identifies genes, GO codes, coded phenotypes (UMLS, CO, MA, MP), along with gene-GO-context relations. ***Process 1b*** simply obtains relevant phenotypic MeSH headings from the Medline abstracts. ***Component 2***, the PhenOS knowledge management system completes the contextual assignment of contextual phenotypes to GOA gene-GO terms pairs related to the same PMIDs.

2.1 Phenotypic Context Determination and Encoding Components

Process 1a- Determining Context using NLP (BioMedLEE). Abstracts are parsed by the NLP system BioMedLEE (which was not adapted for PhenoGO) according to 2 different abstract sections: (i) titles and (ii) body. BioMedLEE can extract about 50 distinct semantic types from biological text as well as generate codes from multiple ontologies, but this study focuses by design on the following 6 coding systems, 4 types of entities, and their associations: 1) MGI:genes, 2) GO: terms, 3) Cells coded in CO or UMLS and 4) anatomies above the cellular level coded in MA, MP, and UMLS. Two training sets of 50 PMIDs were selected, parsed by BioMedLEE and analyzed thoroughly. Consequently, ten UMLS encodings observed ambiguous in

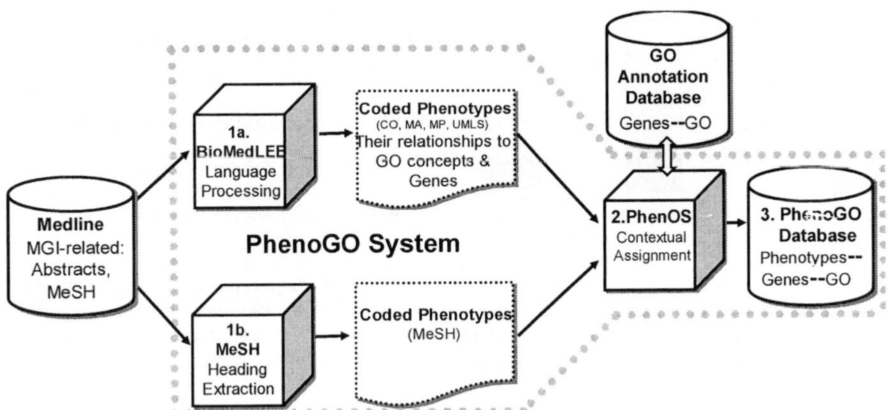

Figure 2: ***Diagram of the PhenoGO System showing Software and Database Components Involved in the Assignment of Phenotypic Context to Gene Ontology Annotations.***

the training set or meaningless for our aims are filtered out (e.g. *back, helix, tissue*).

Process 1b- Determining Context using Curated Knowledge (MeSH and PhenOS). MeSH terms considered useless in PhenoGO were filtered out during training (e.g. "cell", etc.), while the MeSH headings subsumed by the following *concepts of the UMLS semantic networks* (**TUI**) were selected: Anatomical Structure, Embryonic Structures, Body Part, Organ, Tissue, Body Substance, and Systems.

2.2 Phenotype Organizer System (PhenOS) Contextual Assignment

*The PhenOS system (**Component 2, Figure 2**)* performs the contextual assignment, and uses a different process depending on whether the coded phenotype came from processes ***1a** or **1b*** (***Figure 2***). First, component 2 obtains the gene-GO pair that was associated with the specific article in the GOA database. If the coded phenotype originated as a MeSH header that was selected by PhenOS, it is assigned as a contextual phenotype augmenting the gene-GO pair. *Our hypothesis is that when the MeSH header lists a contextual phenotype, it is relevant not only to the article but also to the gene-GO pair associated with the Medline article in GOA.*

If the coded phenotype was generated by the NLP system, a more complex procedure is followed. The assumption in this case is that context may be mentioned incidentally and picked up by the NLP system, but may not be related to the gene-GO pair. In order to achieve high precision, *we hypothesize that if there is a relation between the codes in a gene-GO-phenotype triple and the gene and GO codes match the corresponding ones in GOA, then the phenotype is highly likely to be related to the matched pair, and thus is likely to be the correct context.* The other extreme is that if the context does not match any gene or GO code in the pair asso-

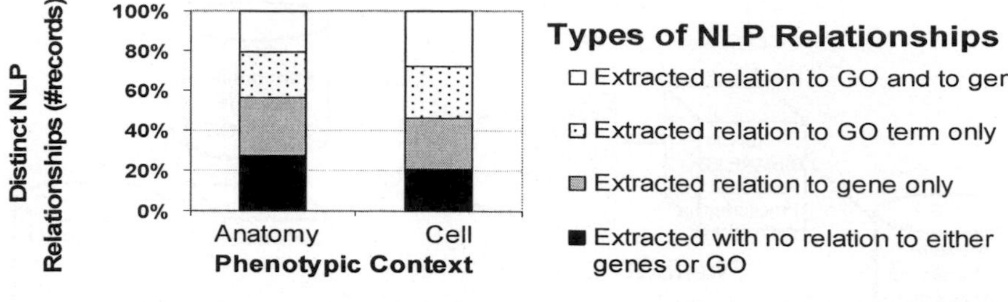

Figure 3. Descriptive Statistics of the NLP Component: Coded Phenotypes according to the Method 1a. The analysis was performed on data captured downstream of the the BioMedLEE NLP component (*1a, Fig. 2*) and upstream (before) the PhenOS component (*2, Fig. 2*) that uses the types of NLP relations provided by the NLP to match the cellular and anatomical phenotypes with specific GO annotations.

ciated with the article, it is more likely that the phenotype may not be related to the pair. Thus, to help analyze the type of relations based on NLP processing that affect performance of PhenoGO, we record the types of relations and matches (i.e. GO-gene-phenotype, gene-phenotype, phenotype-GO, no match). If a GO code is extracted that does not match the database GO code, it may also be because the NLP system obtained a more specific code than the one in GOA. PhenOS is then used to determine whether an ancestor of the extracted GO code matches the code in GOA; in that case this GO code is considered matched. For example, codes are considered to be matching if the extracted code corresponds to 'negative regulation of biological process' and the curated code to 'regulation of biological process'.

2.3 Evaluation

Medline Abstracts, Sampling and Gold Standard. We have focused the experiment on 3,705 PMIDs of the GOA of the Mouse Genome Database (MGI), which contains 2,327 distinct GO terms, 4,269 distinct MGI genes and 12,220 GO-gene pairs. Random samples of PMIDs were selected for creating the *Gold Standard* (**GS**) as described below. For evaluating recall, a sample of 50 PMIDs was randomly selected from the 3,705 MGI PMIDs. In this sample, each occurrence of anatomical and cellular phenotypes as well as their relevance to their respective MGI GO annotations were manually curated by one curator and confirmed by another. Samples for calculating precision were randomly selected from the set of coded results applicable to our study for evaluating, respectively, the cellular and the anatomical phenotypic contexts assigned by PhenoGO. The precision GS comprised a total of 575 curated phenotypes associated with genes and GO terms.

Accuracy Measures. We measured the precision and recall of the assignment of coded anatomies and cell types based on the individual and combined NLP and the

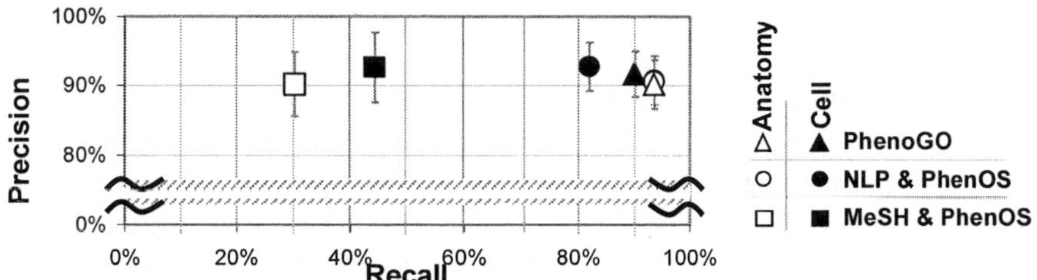

Figure 4. Precision and recall of the assignment of coded phenotypes to gene-GO term pairs by PhenoGO and by its NLP and MeSH components separately (*refer to Fig.2, components 1a & 1b*). Confidence intervals are based on the binomial distribution.

knowledge components of PhenoGO. Recall was calculated as the ratio of the number of distinct [GO-MGI gene-phenotype] triplets that were identified by the mapping method (***Figure 2***) that matched those in the GS, divided by the total number of triplets in the GS, (TP)/(TP+FN). Precision was measured as the same numerator as recall divided by the total number of triplets predicted by the mapping method (NLP&PhenOS, MeSH&PhenOS, PhenoGO), TP/(TP+FP). Thus, in this evaluation, in order to count a true positive score, the PhenoGO system must accurately encode a phenotype and also relate it with its gene-GO term pair.

3 Results

Overall, PhenoGO provided phenotypic context for 96% of the 3,705 PubMedIDs. The NLP coded a larger percentage of phenotypes than the knowledge-based method, and the joint NLP-KB method was significantly better for the cellular annotations demonstrating that the combined NLP-KB method yielded 50% more cellular annotations (***Figure 4***). The NLP process provided several times more annotations than the knowledge-based one, though the difference was more important in anatomies above the cell level than for cellular anatomies (***Figure 4***). As shown in ***Figure 3,*** the majority of the NLP annotations were related to the genes only (e.g. context and gene in same NLP structure with no GO), followed by relationships to GO only; a smaller number were not related to GO or genes (i.e. context not in same structure as GO or gene), and finally an even smaller number were related to both genes and GO terms (example of this relation is shown in ***Figure 1***). By design, the MeSH headings have no mapping to gene or GO before they are received by the PhenOS component (***Figure 2***), thus there is only one type of contextual assignment method for MeSH terms because the phenotypic context is always assumed to be related to the gene and GO pair as specified in GOA. In order to provide a summary of the scales of anatomies mapped by the system, the following are the counts of distinct types of concepts mapped according to their ontologies: MA:345, MP: 305, CO:148, MeSH: 460 (Embryo:32, Organs and body parts:240, Tissues:42, Systems: 22, Cells:124), UMLS: 1,259 (Embryo:97,Organs and body parts: 786, Tissues: 100, Systems:37, Cells:239). The PhenoGO precision and recall for cells and anatomies combined are 91% and 92% respectively (the accuracy measures the combined coding to the ontology and the assignment of the correct gene-GO pair with the phenotype). Details of the results are in ***Figure 4*** showing that the "cell" recall of the PhenoGO system is substantially better than that of the NLP, while the precision remains unchanged.

4 Discussion

PhenoGO performed well in accuracy scores. Precision, which is the most important metric, was over 90% for both the MeSH and the NLP methods. Of note, the NLP component, which had access to only the title and abstract, was not significantly different from the MeSH component which is based on manual curation of the complete paper by experts who focus on main concepts of the article. In addition, the NLP system did not have the expert knowledge of curators, and therefore could not discern whether or not the context was incidentally mentioned or significant to the paper. It is quite interesting that only 1 error in precision was caused by an incorrect association of an anatomical location with a gene-GO pair. Thus, based on our evaluation, use of NLP to augment the GO database with phenotypic information is very promising. Moreover, recall of the NLP component was much higher than that of the MeSH component, possibly because curators do not focus on indexing context. It is also striking that the results for the NLP component are associated with performance in coding and in assigning these codes to the right GOA, which are much more difficult tasks than extraction alone, and thus, the NLP performance in extraction precision is likely to be higher. BioMedLEE was not trained for this particular task, and it is likely that further revisions will enable it to perform better.

An analysis of the BioMedLEE errors was performed, and the most frequent types of errors are shown in ***Table 1***. Not surprisingly, word sense ambiguity was the most frequent cause of error in precision. Often a gene name was also a phenotypic entity, and the incorrect sense was chosen. For example, *Notch* is a gene and also an anatomical part according to the UMLS. Another cause of error in precision was due to use of the synonym lists associated with the ontologies because they often list incomplete terms as synonyms of more complete terms, causing coding and word errors. For example *band* is listed as a synonym of *band form neutrophil* in cell ontology, but it occurred as a different sense in an article. It appears that on

Table 1 - Most frequent types of errors in precision and recall are shown along with examples.

	Reason	Example
Precision	Ambiguity	*Notch* was interpreted as **anatomy** but is **gene** in *defects of notch pathway*
	Ontology	*Band* incorrectly mapped to *band form neutrophil* when it occurred in *50 k-Da bands* since *band* is listed as a synonym of *band form neutrophil* in CO
	Term recognition	*Finger* was interpreted as **anatomy** instead of part of term *zinc finger protein*
	Incorrect relation	*Skeletal defect* was associated with gene *Lfng* in article instead of *delta-like 3*
Recal	Mapping to ontology	*Lymphoid & adipose cell* not mapped to *lymphoid tissue & adipocyte*
	Lexicon	*Epithelia, epididymus* not defined in lexicon
	Term recognition	*Mast cell* not captured as **anatomy** since it is part of term *mast cell tryptase*
	Ontology	*Precursor* was incorrectly listed as synonym of *blood precursor* in UMLS

tologies make assumptions, which may be applicable within a particular domain, but not across broader domains. Other causes of precision errors occurred when a biomolecular term was not recognized, but part of the term was. Thus, *finger* was assigned as context when it occurred in the phrase *zinc finger protein.* The most frequent causes of error in recall were due to failures in coding and to terms that were not recognized because they were missing from the lexicon. Coding errors were typically caused when a synonym of a term was not listed in any of the ontologies. Thus, *lymphoid* (where the word *tissue* was omitted) could not be associated with a code. Lexical omissions typically occurred when a rare variant form of a term occurred in an article.

PhenOS Contextual Assignment Evaluation . We have validated the hypothesis that the curated phenotypes found in a MeSH terms pertained to the whole article, thus to every GO annotation of that article. Indeed, the evaluation measures both the validity of the phenotypic encoding and that of its assigned contextual relationship to a gene-GO term pair. Additionally, the precision of PhenoGO assignments based on NLP are 91% and 93% for the anatomies and the cell types respectively (**Fig. 4**). It is likely that some association relations confer higher quality to the phenotypic context extracted from the abstract by the NLP. For example, when both gene and GO term associations are found related to a phenotype, we predict that the average precision would be higher than that of no associations at all.

There were *some limitations to this study.* Two students who had a background in biology were used to sample the abstracts and results, and create the gold standard. Some of the evaluation required reading the entire documents or the abstracts, both of which are time-consuming tasks, and therefore a limited number of samples were used in the evaluation. An additional limitation is that the study was performed using articles selected from GOA also focused on the mouse. Results of the NLP component may be different for other organisms.

Significance of the Integrated PhenoGO System. An integrated system that combines existing knowledge with NLP coded information has many advantages. First, through GOA, knowledge of the precise gene-GO pair and model organism associated with articles that were annotated is known, providing a fairly accurate way to resolve the identity of an ambiguous gene for contextual assignment. Although the name of an ambiguous gene is associated with more than one identifier, if one of the identifiers matches the one found in the article, it is highly likely to be the correct one. Another very significant advantage is that PhenoGO can be scaled up very quickly and can be deployed to automatically create a database of substantial size. It is scalable in several dimensions. Using the NLP component, it is possible to process huge volumes of journal articles as well as textual database fields where there are no MeSH codes. However, the NLP system must be trained for phenotypic and genotypic information associated with the organism that corresponds to the text. Using the MeSH component it is possible to rapidly determine contextual information across organisms, and thus to augment GOA with context.

Future Work. We are revising PhenoGO to increase performance by improving the BioMedLEE NLP system as well as the PhenOS contextual assignment method and we are mapping to additional types of phenotypes which are beyond this study, such as morphologies and diseases. We are currently using PhenoGO to process every PMID associated with GOA for Mus musculus and Homo sapiens, and we will perform additional evaluations of the results as we extend to every species in GOA. Our ultimate objective is to provide an accurate and regularly update open source PhenoGO database of phenotypic and contextual annotations for the biological and informatics communities.

5 Conclusion

We developed and evaluated an automatic NLP system, PhenoGO, for augmenting gene product and GO associations using MeSH headings, the UMLS hierarchy, and an existing NLP system that maps terms in text to identifiers in multiple biological ontologies. Results demonstrated that the PhenoGO NLP system encodes anatomical and cellular ontologies to GO annotations with high recall and precision. The system is scalable in a number of dimensions: (i) it enables high throughput, (ii) it encodes in multiple ontologies and terminologies (GO, CO, MA, MP, UMLS, MeSH), (iii) it provides different modifiers for the encoded phenotypes, (iv) it uses external knowledge bases when they exist to increase accuracy (e.g. MeSH), (v) it can provide other types of context, such as diseases. In addition, the system can also map to other species and encode textual data from biological databases (results not shown). Of significance, the hypothesis that MeSH anatomical knowledge generally applies to every GO annotation assigned to the same PMID has been confirmed, which is likely to allow for rapid scalability across GOA of every species. In summary, the PhenoGO database is expected, upon completion, to provide valuable high throughput resources for biological in silico experiments as one can investigate in high throughput the differential expression of gene and their GO annotations across different cell types, organs, systems, etc. For example, this tool could be used to investigate the role of genes in the cellular differentiation of complex organisms using GOA with their phenotypic contexts.

Acknowledgments

This study was supported in part by NIH/NLM grants 1K22 LM008308-01(YL) and R01 LM007659(CF). The authors thank Lyudmila Shagina and Jianrong Li for their respective contribution in the development of BioMedLEE and PhenOS.

References

1.Gene Ontology Consortium. *Nat. Genet.* **25**, 25–29 (2000).
2.Camon EB, Barell DG, Dimmer EC,et al. *BMC Bioinform* **6** S1(2005).
3.Raychaudhuri S,Chang J,Sutphin P,Altman RB.*Genome Res.* **12**(1),203-14(2002).
4. Simpson SJ, et al. *Eur J Immunol.* Sep;**25**(9):2618-25 (1995).
5.Camon E, Magrane M, Barrell D, et al. *Nucleic Acids Res.* **1**;32:D262-6 (2004).
6.Bard J, Rhee SY, Ashburner M. *Genome Biol.* **6**(2):R21.(2005).
7.Hayamizu TF, Mangan M, Corradi JP, et al. *Genome Biol.* **6**(3):R29 (2005).
8.Smith CL, Goldsmith CA, Eppig JT. *Genome Biol.* **6**(1):R7 (2005).
9.Lindberg C. *J Am Med Rec Assoc.***61**(5):40-2 (1990).
10.Wheeler DL, Chappey C, Lash AE, et al. *Nucleic Acids Res* **1**;28(1):10-4(2000).
11.Tiffin N, Kelso JF, Powell AR, et al. *Nucleic Acids Res.* **33**(5):1544-52(2005).
12.Krauthammer M, Nenadic G. *J Biomed Inform.* **37**(6):512-26(2004).
13.Cohen AM, Hersh WR. *Brief Bioinform* **6**(1):57-71 (2005)
14.Hirschman L, Yeh A, Blaschke C, Valencia A. *BMC Bioinform* **6** S1(2005)
15.Perez AJ,Perez-Iratxeta C,Bork P, et al.*Bioinformatics* **20**(13),2084-91(2004).
16. Koike A, Niwa J, Takagi T. : *Bioinformatics.* Apr 1;21(7):1227-36(2005).
17.Aronson AR. *Proc AMIA Symp* 17-21 (2001).
18.Nadkarni P, Chen R, Brandt C. *J Am Med Inform Assoc.* **8**(1):80-91(2001).
19.Friedman C, Shagina L, Lussier Y, Hripcsak G. *JAMIA* **11**(5):392-402 (2004).
20.King OD, Lee JC, Dudley AM, et al. *Bioinform* **19** Suppl 1:i183-9 (2003).
21.Rogers FB. Medical subject headings. *Bull Med Libr Assoc.* **51**, 114-6 (1963).
22.Cantor M, Sarkar,Bodenreider O,Lussier YA.*Pac Symp Biocomp.* 103-14(2005).
23.Chen L, Friedman C *Medinfo* **11**(Pt 2):758-62 (2004).
24.Lussier YA, Li J. *Pac Symp Biocomput* 202-13(2004).
25.Sarkar I, Cantor M, Lussier YA. *Pac Symp Biocomput* 439-50(2003).

LARGE-SCALE TESTING OF BIBLIOME INFORMATICS USING PFAM PROTEIN FAMILIES

ANA G. MAGUITMAN[†], ANDREAS RECHTSTEINER[‡,*]
KARIN VERSPOOR[‡], CHARLIE E. STRAUSS[‡], LUIS M. ROCHA[†]

[†]*School of Informatics, Indiana University*
1900 East Tenth Street, Bloomington, IN 47408
E-mail: anmaguit@indiana.edu, rocha@indiana.edu

[‡]*Los Alamos National Laboratory*
PO Box 1663, Los Alamos, NM 87545
E-mail: arechtsteiner@gmail.com, verspoor@lanl.gov, cems@lanl.gov

Literature mining is expected to help not only with automatically sifting through huge biomedical literature and annotation databases, but also with linking bio-chemical entities to appropriate functional hypotheses. However, there has been very limited success in testing literature mining methods due to the lack of large, objectively validated test sets or "gold standards". To improve this situation we created a large-scale test of literature mining methods and resources. We report on a specific implementation of this test: how well can the Pfam protein family classification be replicated from independently mining different literature/annotation resources? We test and compare different keyterm sets as well as different algorithms for issuing protein family predictions. We find that protein families can indeed be automatically predicted from the literature. Using words from PubMed abstracts, of 3663 proteins tested, over 75% were correctly assigned to one of 618 Pfam families. For 90% of proteins the correct Pfam family was among the top 5 ranked families. We found that protein family prediction is far superior with keywords extracted from PubMed abstracts than with GO annotations or MeSH keyterms, suggesting that the text itself (in combination with the vector space model) is superior to GO and MeSH as a literature mining resources, at least for detecting protein family membership. Finally, we show that Shannon's entropy can be exploited to improve prediction by facilitating the integration of the different literature sources tested.

1. Introduction

Biology was until recently essentially a hypothesis driven science in which experiments were carefully designed to answer one or very few specific questions — e.g. test the function of a specific protein in a specific context. In the last decade, fueled by the widespread use of high-throughput technology, we have witnessed

*Address since Oct 2005: Center for Genomics and Bioinformatics, Indiana University, 1001 East 3rd St, Bloomington, IN 47405.

the emergence of a more data-driven paradigm for biological research. Since high-throughput experiments are frequently conducted for the sake of discovery rather than hypothesis testing, and due to the sheer amount of measured variables they entail, it is very difficult to interpret their results. Moreover, since the goal of many experiments is to uncover bio-chemical and functional information about genes and proteins, there is an obvious need to understand the linkages amongst biological entities in literature and databases which allow us to make inferences. Literature mining[18] is expected to help with those inferences; its objective is to automatically sort through huge collections of literature and suggest the most relevant pieces of information for a specific analysis task, e.g. the annotation of proteins[9]. Another application is to uncover similarities of genes according to "publication space", or the more tongue-in-cheek term "bibliome"[8].

Since literature mining hinges on the quality of available sources of literature as well as their linkage to other electronic sources of biological knowledge, it is particularly important to study the quality of the inferences it can provide. Indeed, the *Bibliome* is not just the collection of publications and annotations available; its usefulness ultimately depends on the quality of linking resources that allow us to associate experimental data with publications and annotations. Interestingly, while literature mining is receiving considerable attention in Bioinformatics, it has not been hitherto seriously validated. Towards improving this situation, we present here our large-scale testing and comparison of literature mining algorithms, *paired* with specific bibliome resources.

We present a general method for testing bibliome resources and literature mining algorithms in the context of classification of biological entities. This method formalizes and extends a previous study in which we tested how well is the Pfam protein family classification inferred from PubMed as indexed by the MeSH keyterm vocabulary[16,14]. We expand on these results by testing additional bibliome resources such as GO annotations and text extracted from PubMed abstracts for the same classification problem. We additionally propose a new method based on Shannon's entropy to integrate results from different bibliome resources, and show that it significantly improves protein family predictions.

2. From Text Mining to the Bibliome: Looking for a "Gold Standard"

There exists extensive cross-linkage amongst biomedical databases which can be exploited for bioinformatics analysis. For instance, gene chip identifiers can be linked to protein entries in SWISSPROT which in turn can be linked to PubMed documents. Indeed, in the bibliome, documents are linked to or indexed by various semantic (textual) tags which describe their content; these include Medical Subject Headings (MeSH), Gene Ontology (GO) annotations, PubMed abstract text,

HUGO nomenclature for human genes, GenBank accession numbers for gene sequences, etc. Therefore, in order to fully capture the potential of the bibliome for analysis, integration and dissemination of biological knowledge, in addition to research on text mining and natural language processing, literature mining needs more research on the quality of links amongst the resources that make up the bibliome. Text Mining is particularly applicable to the discovery of relevant information inside text — e.g. discovering a portion of text in a document appropriate to annotate a given protein[9]. But given the highly cross-linked nature of the bibliome, in addition to text mining, we need to approach bibliome informatics from an Information Retrieval (IR) perspective.

Several research groups have been exploiting the cross-linked nature of the bibliome, particularly with semantic annotations such as MeSH and GO, for instance the systems developed by Masys et al[13] and Jenssen et al[11] for identifying sets of keyterms associated with sets of genes. Tools that are similar in spirit are PubMatrix[2], MedMiner[22], MeshMap[21] and others. While these systems are potentially very useful, the quality of their results has not been thoroughly validated. For instance, we have applied Latent Semantic Analysis (LSA) to discover functional themes[15,14] from the literature for microarray experiments dealing with the response to human cytomegalovirus infection. Though the functional themes we discovered automatically matched our previously published manual annotation of the same experiments[4], and even uncovered novel functional themes[15,14], such validation by a few expert biologists is done without access to a "*gold standard*".

By "gold standard" we mean a standardized test data which allows us, unambiguously, to decide if a given inference is correct. Homayouni et al were able to build such gold standard for evaluating the performance of LSA, but only by focusing on a very small set of genes[10]. Unfortunately, for data-driven experiments there is no clear expectation of what functional associations are to be found. Therefore, bibliome tools are typically tested by sampling some of their output and presenting it to experts. The problem is that experts typically disagree or cannot be an expert on all the topics involved. Even more systematic approaches such as Biocreative suffer from variability in experts' opinions[5,9,3], leading to potentially unreliable answers.

3. Large-scale standard for bibliome informatics: Methods and Data

3.1. *A general large-scale bibliome informatics test*

The first requirement for our testing methodology is the existence of a biological classification $\mathcal{C}$, accepted as a true standard, and defined on a large set P of biological entities p (e.g. proteins or genes), where each entity p is associated with a single class $C(p)$. Given that the Bibliome is defined not only by publication and annotation resources, but also by their linkage, we also need a high-quality linking resource L_D between P and the documents of some publication or annotation resource D — where $L_D(p)$ denotes the set of documents of D associated with entity p. Given a $\mathcal{C}$ and L_D pair, our *large-scale bibliome informatics test* (LSBIT) can be applied to any pair, $\langle A, K_D \rangle$, of classification algorithm A and keyterm set K_D extracted from D — where $K_D(p)$ denotes the set of keyterms that index documents $L_D(p)$[a]. The objective of the LSBIT is then to **establish how well a given algorithm A can discover a known classification $\mathcal{C}$ of biological entities P, from a publication resource D using an associated keyterm set K_D and a bibliome linking resource L_D between P and D.**

3.2. *Bibliome Resources*

3.2.1. *Defining $\mathcal{C}$ and L_D*

We chose the Pfam protein sequence classification[20] as $\mathcal{C}$ for our tests. Pfam is a manually curated collection of protein families, currently encompassing several thousands of families. Pfam is an ideal classification for objective evaluation and comparison of Bibliome informatics due to it being based on sequence, which is a physical property of proteins that typically leads to functional similarity. Having settled on Pfam for our classification standard $\mathcal{C}$, our biological entities P are proteins. Therefore, a most appropriate linking resource L_D to test various $\langle A, K_D \rangle$ is the SWISSPROT (now UNIPROT[19]) database, which is a protein sequence database curated by experts. Besides the amino acid sequence of a protein it also lists different types of annotations, cross-references to other databases (including the Pfam family of a protein), as well as references to relevant publications for each protein. Therefore, the LSBIT with $\mathcal{C}$ = Pfam and L_D = SWISSPROT, can be applied to classify proteins p under various pairs $\langle A, K_D \rangle$. The expert nature of Pfam and SWISSPROT allows us to use them as a standard for the classification of proteins.

However, before the LSBIT may be performed, some preprocessing of the set of proteins to be tested is necessary. We extracted all the SWISSPROT protein IDs which contained a single Pfam classification. Multiple Pfam family assignments occur for 15% of all SWISSPROT proteins, possibly because some proteins have more than one classified domain. Because we are interested in constructing a large, unambiguous data set for validating bibliome methods, we removed multi-classification proteins. We do not consider those to be erroneous in any way, but they simply do not serve the purposes of out testing standard, which needs to be unambiguous. After pre-processing (details in[14,16]), we obtained a dataset with

[a]We use keyterm to refer to both keywords and keyphrases depending on available resources.

$|P| = 15,217$ proteins from $\mathcal{C} = 1611$ Pfam families. Each protein p is associated with a unique Pfam family $C(p)$.

3.2.2. *Defining publication/annotation resources D*

Since SWISSPROT lists PubMed IDs, a very natural publication resource is PubMed; let us denote it as D_{PM}. Via SWISSPROT, our linking resource L_D, we retrieve different keyterm sets K_D from PubMed, detailed in the next subsection. Another annotation resource we used was GO, which we denote as D_{GO}, derived from the GOA/UNIPROT dataset provided by the GOA project, run by the European Bioinformatics Institute (EBI). Because we needed to compare and integrate the tests using D_{PM} and D_{GO}, we looked at a reduced set of proteins for which links to both PubMed publications and GO annotations were found, that is $P^r = \{p : L_{D_{PM}}(p) \bigcap L_{D_{GO}}(p) \neq \emptyset\}$. We also restricted our study to Pfam families with at least 3 proteins. This reduced dataset P^r contains 3663 proteins from 618 distinct Pfam families, where 179 of these families contain only 3 proteins and the largest 3 families contain 17 proteins. Mean and median family size is 5.9 and 5 proteins, respectively; standard deviation is 3.3.

3.3. *Keyterm Sets K_D to Test*

We have adapted the IR vector space model[1] to represent proteins as vectors in a keyterm space. Four different keyterm sets were used in our analysis. Three of these sets contain keyterms extracted from PubMed (D_{PM}) publications associated with proteins, while the fourth was based on term annotations in the Gene Ontology (D_{GO}). The first keyterm set $K_{D_{PM}}^{MeSH}$ contains MeSH terms. MeSH (Medical Subject Headings) is a hierarchically organized vocabulary produced by the National Library of Medicine to index MEDLINE/PubMed. $K_{D_{PM}}^{MeSH}$ contains all MeSH terms occurring in the $L_{D_{PM}}(p)$ set of PubMed records associated with all proteins $p \in P^r$.

For the second keyterm set, $K_{D_{PM}}^{Words}$, we used all words (after stop-word filtering) extracted from PubMed abstracts associated with all proteins $p \in P^r$. To build the third keyterm set, $K_{D_{PM}}^{Stems}$, we reduced the words in $K_{D_{PM}}^{Words}$ to their linguistic stems, using a morphological normalization tool, called BioMorpher, which we have used previously[23]. Finally, the fourth keyterm set $K_{D_{GO}}^{Terms}$ contains terms from the $L_{D_{GO}}(p)$ set of GO annotations associated with all proteins $p \in P^r$. Notice that many of the annotations in GOA are electronically inferred (e.g. they are based on hits from sequence similarity searches or are transferred from database records). To avoid circularity in our argument we used the GOA evidence code to filter out term annotations inferred from electronic annotations (IEA), limiting our selection to those annotations assigned due to experimental evidence or published literature.

For each of the keyterm sets, we computed a protein-keyterm co-occurrence matrix where each positive entry denotes that the respective keyterm occurs in a document or annotation linked to the respective protein. The rows of the Matrix define the protein vectors for each protein $p \in P^r$ in the respective keyterm space. Table 1 shows the number of non-zero entries for each matrix and the average number of keyterms per protein in each of the four keyterm sets.

Table 1. A comparison of the four keyterm sets.

	$K_{D_{PM}}^{MeSH}$	$K_{D_{PM}}^{Words}$	$K_{D_{PM}}^{Stems}$	$K_{D_{GO}}^{Terms}$
total protein-keyterm associations	98707	560639	484072	14583
avg. keyterms per protein	27	153	132	4

3.4. *Protein Vectors and Protein Similarity*

The entry for a given protein-keyterm pair in the protein-keyterm co-occurrence matrix is a weight representing the relative importance of the keyterm for that protein. This weight is defined by multiplying a local and a global weight for the protein-keyterm pair. The local weight is the *term frequency* tf_{ik}, defined as the number of documents or annotations cited for protein p_i in SWISSPROT that are also indexed by keyterm k in publication resource D being tested.

The coefficients of the protein vectors are then scaled by a global weight to capture the relative importance of each keyterm in the space. The global weight we applied is related to the *Inverse Document Frequency* (IDF) in IR[7]. We named it *inverse protein family frequency* (IPFF) and defined it as $ipff_k = \log(\frac{N^{PF}}{n_k^{PF}})$ where N^{PF} is the total number of Pfam families in $\mathcal{C}$ and n_k^{PF} is the number of Pfam families that contain a protein with at least a document/annotation indexed by keyterm k. Finally, the protein-keyterm co-occurrence matrix W is defined by $w_{ik} = tf_{ik} \cdot ipff_k$ where row i denotes protein vector i in keyterm dimension/column k. Figure 1 depicts this process.

To measure protein similarity in keyterm space, we used the IR cosine measure[1]: given protein vectors $\mathbf{p}_i$ and $\mathbf{p}_j$ in a n-dimensional keyterm space, the cosine similarity $\sigma_{\cos}$ between them is their normalized dot product:

$$\sigma_{\cos}(\mathbf{p}_i, \mathbf{p}_j) = \frac{\mathbf{p}_i \cdot \mathbf{p}_j}{||\mathbf{p}_i|| ||\mathbf{p}_j||}$$

3.5. *Prediction Algorithms A*

Our first LSBIT experiments, designed to establish how well we can predict the Pfam family of proteins using the bibliome resources described above, tested two classification algorithms closely related to the k-nearest neighbor algorithm[6]. Given a protein keyterm vector $\mathbf{p}_i$ and an angle α, the first algorithm, A_α, assigns a score to each Pfam family j based on the number of proteins of that family

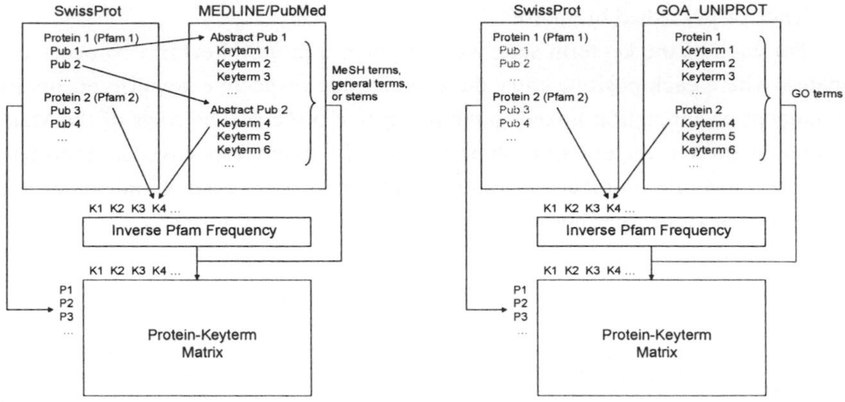

Figure 1. The process of building a protein-keyterm matrix using different linkage information sources: MEDLINE/PubMed (left) and GOA_UNIPROT (right).

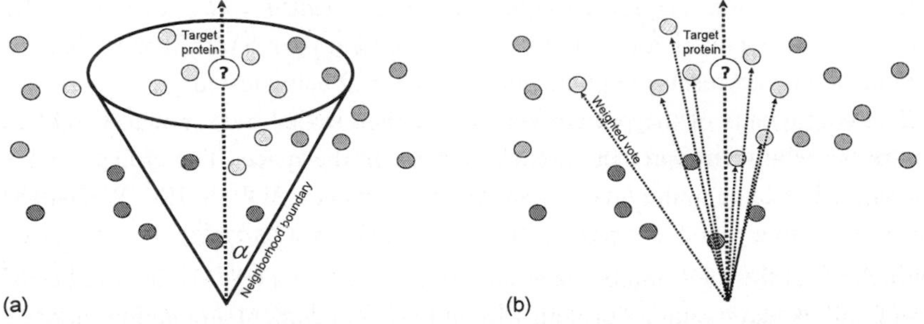

Figure 2. (a) A_α prediction algorithm: target protein neighborhood defined by the hypercone with opening angle α and centered around the target protein vector. (b) A_{WV} prediction algorithm: proteins voting in proportion to their cosine similarity to the target protein.

found in a hypercone defined by the angle α and centered around $\mathbf{p}_i$, as illustrated in figure 2(a). Thus, A_α returns a ranking of Pfam families based on this score:

$$A_\alpha : \mathbf{Pfam}_j(\mathbf{p}_i, \alpha) = |\{\mathbf{p}_k \in \mathit{pfam}_j : \sigma_{\cos}(\mathbf{p}_i, \mathbf{p}_k) \geq \cos(\alpha)\}|$$

The family with most proteins in the neighborhood is ranked first, and so forth. This algorithm is described in detail in[14,16].

A problem with the A_α algorithm is that it depends on an angle α. If α is large, unrelated proteins may be included in the neighborhood; if α is small the neighborhood may contain very few proteins or may be empty, in which case no prediction can be made. A second problem is that it is biased towards ranking

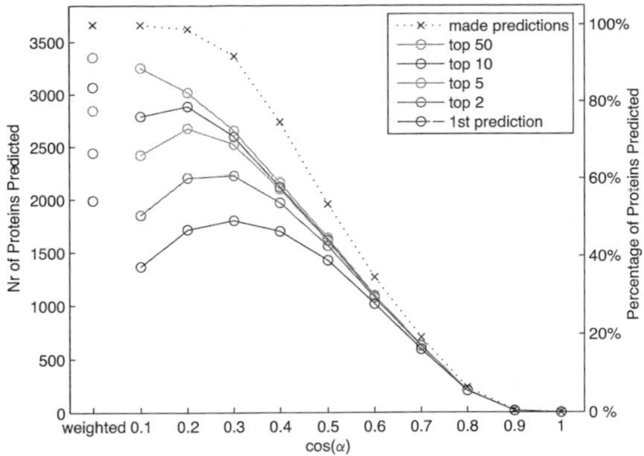

Figure 3. Prediction success using algorithms A_{WV} and A_{α}, and keyterm set K_{DPM}^{MeSH}.

larger families first. We have adapted A_{α} to deal with both these issues. In the new algorithm, A_{WV}, every protein in the space issues a "weighted vote" for its Pfam family (not just those inside a neighborhood hypercone):

$$A_{WV} : \mathbf{Pfam}_j(\mathbf{p}_i) = \frac{\sum_{\mathbf{p}_k \in pfam_j} \sigma_{\cos}(\mathbf{p}_i, \mathbf{p}_k)}{\sqrt{|pfam_j|}}$$

The weight of each protein's vote is given by the cosine of the angle between its vector and the vector of the protein being classified. In order to weaken the bias towards larger families, the family score is normalized with a division by the square root of family size. Figure 2(b) illustrates this process. A_{WV} improves on our first algorithm because it does not require a neighborhood angle to be defined in advance and it always issues a prediction for any protein vector in the space. Additionally, as we will see next, it has a higher prediction success than A_{α}.

4. Results: Testing $\langle A, K_D \rangle$

The two algorithms A_{α} and A_{WV} were tested using the four keyterm sets K_{DPM}^{MeSH}, K_{DPM}^{Words}, K_{DPM}^{Stems} and K_{DGO}^{Terms}. Figure 3 shows the prediction success of our algorithms using K_{DPM}^{MeSH} in terms of *true-positives*, i.e. the number of proteins for which the Pfam family was predicted correctly. The first entry on the x-axis (labelled weighted) corresponds to the weighted-voting algorithm A_{WV}. The remaining entries on the x-axis (labelled 0.1, 0.2, etc.) indicate the cosine of α for the A_{α} algorithm. The y-axis shows the number of proteins correctly predicted out of a total of $3663 \in P^r$. The black, dashed curve shows the number of proteins for which a prediction was made using A_{α} for various angle α. As the cosine threshold increases, the number of predictions made by A_{α} decreases.

A_{WV} outperformed A_α in all our tests, therefore for the other three keyterm sets, we only display results for A_{WV} summarized in Table 2. Noticeably, the three keyterm sets extracted from PubMed records performed better than the one extracted from GO annotations. This might be due to fewer GO than PubMed keyterms per protein (see table 1). Among the three keyterm sets based on PubMed, the two obtained from abstract words significantly outperform the one containing MeSH terms; the stem-based keyterms provided slightly better results than plain words.

Table 2. Prediction success for A_{WV}.

	K_{DPM}^{MeSH}	K_{DPM}^{Words}	K_{DPM}^{Stems}	K_{DGO}^{Terms}
1st prediction	54.35%	75.27%	75.89%	38.08%
top 2	66.72%	84.17%	84.22%	45.65%
top 5	77.70%	88.83%	89.30%	55.53%
top 10	83.76%	91.13%	91.48%	61.86%
top 50	91.54%	94.02%	94.40%	75.59%

5. Integrating Predictions from Different Keyterm Sets

We noticed that the sets of correctly predicted proteins using different keyterm sets do not completely overlap. Therefore, using Shannon's measure of entropy[12] we can select the lower-uncertainty class predictions from the different keyterm sets, leading to a more successful algorithm that efficiently integrates information from those distinct sources.

Let $\rho_K(\mathbf{p}_i, \mathit{pfam}_j, \alpha)$ be the probability of selecting pfam_j as the protein family predicted for protein $\mathbf{p}_i$ using keyterm set K and a neighborhood bounded by angle α. We estimate this probability as follows:

$$\rho_K(\mathbf{p}_i, \mathit{pfam}_j, \alpha) = \frac{|\{\mathbf{p}_k \in \mathit{pfam}_j : \sigma_{\cos}(\mathbf{p}_i, \mathbf{p}_k) \geq \cos(\alpha)\}|}{|\{\mathbf{p}_k : \sigma_{\cos}(\mathbf{p}_i, \mathbf{p}_k) \geq \cos(\alpha)\}|}.$$

Then, we compute the entropy of a prediction for protein i as follows:

$$H_K(\mathbf{p}_i, \alpha) = \begin{cases} \infty & \text{if } |\{\mathbf{p}_k : \sigma_{\cos}(\mathbf{p}_i, \mathbf{p}_k) \geq \cos(\alpha)\}| = 0 \\ -\sum_j \rho_K(\mathbf{p}_i, \mathit{pfam}_j, \alpha) \log \rho_K(\mathbf{p}_i, \mathit{pfam}_j, \alpha) & \text{otherwise.} \end{cases}$$

Finally, we compute the prediction uncertainty of protein i using keyterm set K, $U_K(\mathbf{p}_i)$, as the average entropy on a finite set of angle thresholds T:

$$U_K(\mathbf{p}_i) = \begin{cases} \infty & \text{if } \forall\, \alpha \in T, H_K(\mathbf{p}_i, \alpha) = \infty \\ \langle H_K(\mathbf{p}_i, \alpha) \rangle & : \alpha \in T \wedge H_K(\mathbf{p}_i, \alpha) \neq \infty. \end{cases}$$

Using the uncertainty measure, we implemented and tested a novel algorithm that integrates protein family predictions issued by each keyterm set, by selecting the lower uncertainty predictions. Let $\mathcal{K}$ be a set of keyterm sets. For $K \in \mathcal{K}$, let $\mathbf{Pfam}_j^K(\mathbf{p}_i)$ be the score assigned to protein family j when predicting protein i using keyterm set K. Then, our integration algorithm based on uncertainty, A_U is implemented as follows:

$$A_U : \mathbf{Pfam}_j^{U_\mathcal{K}}(\mathbf{p}_i) = \mathbf{Pfam}_j^K(\mathbf{p}_i) \text{ where } K = \underset{K' \in \mathcal{K}}{\operatorname{argmin}}\, U_{K'}(\mathbf{p}_i).$$

As a baseline for comparison, we implemented a simple prediction algorithm, $A_{\langle K \rangle}$, that also integrates the predictions issued by the four keyterm systems by computing the average score $\langle \mathbf{Pfam}_j^K(\mathbf{p}_i) \rangle$ over all $K \in \mathcal{K}$. Table 3 summarizes the results obtained by these algorithms, highlighting the usefulness of an uncertainty-based method for the top predictions. Indeed, in addition to clearly outperforming $A_{\langle K \rangle}$, A_U outperforms the best results of A_{WV} with a single keyterm set (K_{PM}^{Stems}) (see table 2) for correct first and top 2 predictions.

Table 3. Prediction with combined keyterm sets.

	$A_{\langle K \rangle}$	A_U
1st prediction	70.84%	77.15%
top 2	80.02%	84.77%
top 5	87.50%	88.86%
top 10	91.35%	90.88%
top 50	95.93%	93.80%

6. Discussion and Conclusions

Our experiments show that the Pfam classification of SWISSPROT proteins is quite well inferred, independently, from the publication resources and associated keyterm sets (MeSH, GO, PubMed abstracts), we tested with thc LSBIT. The publication space with associated keyterms largely captures the functional information structure represented by the Pfam classification. Moreover, we have shown that Shannon's measure of entropy can be used to integrate the predictions from various keyterm sets, resulting in an improved protein Pfam prediction algorithm. Algorithm A_{WV} always issues a prediction, which is desirable when we want to maximize the number of true-positives. However, for certain tasks it may be desirable to minimize the number of false-positives, or to use a certainty factor to express how reliable we judge a prediction to be. We are exploring the use of Shannon's measure of entropy to implement this scheme.

An interesting finding for us was that for all tested algorithms, protein family prediction is far superior with keywords extracted from PubMed abstracts than with terms extracted from GO annotations. This suggests that although GO is becoming the standard annotation resource for gene and protein annotation, PubMed abstracts, and even MeSH keyterms, are far superior as resources for literature mining. Given our results, it is fair to conclude that PubMed abstracts and MeSH terms contain more semantic and functional information to classify proteins. In future work, we will investigate what specific information is missing in the GO annotations which causes the lower performance.

Our results also show that the simple vector space model from IR is capable of well representing the semantics entailed in PubMed abstracts for protein family prediction: e.g. for 90% of proteins the correct Pfam family was among the top 5 ranked families (see table 2). In preliminary tests, we have observed that LSA improves the results only when using PubMed abstract words, and not with the other keyword sets. These results suggest that abstract keyterms have more synonymy and polysemy than MeSH and GO, but the details of that analysis are forthcoming. In future work we intend to produce working bibliome informatics tools that build up on the knowledge and algorithms of this study. We will also extend this study with additional algorithms and resources. This includes extending our algorithms by exploiting the Ontology nature of MeSH and GO with similarity measures, testing additional uncertainty-based methods, and methods based on our network analysis methodology[23,17].

Acknowledgements

We are grateful to IU's Research and Technical Services (especially Steve Simms and George Turner) for technical support. The AVIDD Linux Clusters used in our analysis are funded in part by NSF Grant CDA-9601632. This work was also supported by the Department of Energy under contract W-7405-ENG-36 to the University of California. We particularly thank Tom Terwilliger at the Los Alamos National Laboratory for the motivation to conduct this study.

References

1. R. Baeza-Yates and B. Ribiero-Neto. *Modern Information Retrieval*. Pearson Education, 1999.
2. K. G. Becker, D. A. Hosack, G. Dennis, R. A. Lempicki, T. J. Bright, C. Cheadle, and J. Engel. PubMatrix: a tool for multiplex literature mining. *BMC Bioinformatics*, 4(1):61, Dec 2003.
3. E. B. Camon, D. G. Barrell, E. C. Dimmer, V. Lee, M. Magrane, J. Maslen, D. Binns, and R. Apweiler. An evaluation of go annotation retrieval for biocreative and goa. *BMC Bioinformatics*, 6 Suppl 1:S15, 2005.
4. J. Challacombe, A. Rechtsteiner, R. Gottardo, L. M. Rocha, E. P. Brown, T. Shenk, M. Altherr, and T. Brettin. Evaluation of the host transcriptional response to human cytomegalovirus infection. *Physiological Genomics.*, 18(1):51–62, 2004.
5. M. E. Colosimo, A. A. Morgan, A. S. Yeh, J. B. Colombe, and L. Hirschman. Data preparation and interannotator agreement: Biocreative task 1b. *BMC Bioinformatics*, 6 Suppl 1:S12, 2005.

6. R. O. Duda, P. E. Hart, and D. G. Stork. *Pattern Classification*. Wiley, New York, NY, 2nd edition, 2000.
7. S. Dumais. Enhancing performance in latent semantic indexing, 1990.
8. W. Hersh, R. T. Bhupatiraju, and S. Corley. Enhancing access to the Bibliome: the TREC Genomics Track. *Medinfo*, 11(Pt 2):773–777, 2004.
9. L. Hirschman, A. Yeh, C. Blaschke, and A. Valencia. Overview of biocreative: critical assessment of information extraction for biology. *BMC Bioinformatics*, 6 Suppl 1:S1, 2005.
10. R. Homayouni, K. Heinrich, L. Wei, and M. W. Berry. Gene clustering by Latent Semantic Indexing of MEDLINE Abstracts. *Bioinformatics*, 21(1):104–115, 2005.
11. T. K. Jenssen, A. Laegreid, J. Komorowski, and E. Hovig. A literature network of human genes for high-throughput analysis of gene expression. *Nat. Genet.*, 28(1):21–28, 2001.
12. G. J. Klir and M. J. Wierman. *Uncertainty-Based Information : Elements of Generalized Information Theory*. Studies in Fuzziness and Soft Computing. Physica-Verlag, 1999.
13. D. R. Masys, J. B. Welsh, J. Lynn Fink, M. Gribskov, I. Klacansky, and J. Corbeil. Use of keyword hierarchies to interpret gene expression patterns. *Bioinformatics*, 17(4):319–26, 2001.
14. A. Rechtsteiner. *Multivariate Analysis Of Gene Expression Data And Functional Information: Automated Methods For Functional Genomics*. PhD thesis, Portland State University, 2005.
15. A. Rechtsteiner and L. M. Rocha. MeSH key terms for validation and annotation of gene expression clusters. In *Currents in Computational Molecular Biology. RECOMB 2004*, pages 212–213, 2004.
16. A. Rechtsteiner, L. M. Rocha, and C. E. Strauss. Clustering of protein families in literature keyword space. In *Currents in Computational Molecular Biology. RECOMB 2005*, Boston, MA, 2005.
17. L. M. Rocha, T. Simas, A. Rechtsteiner, M. DiGiacomo, and R. Luce. Mylibrary@lanl: Proximity and semi-metric networks for a collaborative and recommender web service. In *The 2005 IEEE/WIC/ACM International Conference on Web Intelligence (WI 2005)*, pages 565–571, Compiegne, France, Sep 2005.
18. H. Shatkay and R. Feldman. Mining the biomedical literature in the genomic era: An overview. *Journal of Computational Biology*, 10(6):821–856, 2003.
19. SIB/EBI. UniProt/Swiss-Prot. http://www.ebi.ac.uk/swissprot/, 2004.
20. E. L. Sonnhammer, S. R. Eddy, and R. Durbin. Pfam: a comprehensive database of protein domain families based on seed alignments. *Proteins*, 28(3):405–420, Jul 1997.
21. P. Srinivasan. MeSHmap: a text mining tool for MEDLINE. *Proc AMIA Symp*, pages 642–646, 2001.
22. L. Tanabe, U. Scherf, L. H. Smith, J. K. Lee, L. Hunter, and J. N. Weinstein. MedMiner: an Internet text-mining tool for biomedical information, with application to gene expression profiling. *Biotechniques*, 27(6):1210–1214, Dec 1999.
23. K. Verspoor, J. Cohn, C. Joslyn, S. Mniszewski, A. Rechtsteiner, L. M. Rocha, and T. Simas. Protein annotation as term categorization in the gene ontology using word proximity networks. *BMC Bioinformatics*, 6 Suppl 1:S20, 2005.

PREDICTING GENE FUNCTIONS FROM TEXT USING A CROSS-SPECIES APPROACH

EMILIA STOICA AND MARTI HEARST
SIMS, UC Berkeley
estoica@sims.berkeley.edu, hearst@sims.berkeley.edu

We propose a cross-species approach for assigning Gene Ontology terms to LocusLink genes based on evidence extracted from biomedical journal articles. We make use of information from orthologous genes to derive and merge two sets of GO codes for a given target gene. For the first set, we restrict GO code assignments to be selected from only those codes which have already been assigned to the target gene's ortholog. Since this approach results in high precision but low recall, for the second set, we allow any GO code to be a candidate, but then eliminate those codes which are illogical to pair with a GO code that is known to be associated with the orthologous gene. Experimental results on three datasets show that the F-measure obtained with this algorithm is consistently higher than the F-measure of other current solutions.

1. Introduction

The complexity of molecular biology is reflected in the large number of experimental results reported in MEDLINE documents, which provide valuable information about the functions of genes and gene products. Extracting these functions from literature (also known as functional annotation), may be a step forward toward understanding diseases and identifying drug targets[4].

Given the large variability in expression of concepts in medical literature, researchers have created a common language for functional annotation, the Gene Ontology (GO)[9]. GO is a controlled vocabulary of over 17,600 terms, also known as GO codes. Each GO code consists of tokens, which are words or punctuation characters. GO codes are organized into three distinct direct acyclic graphs, corresponding to molecular functions (*MF*), biological processes (*BP*) and cellular components/locations (CC) of gene products. More general terms act as parent nodes of the less general ones. For example, the GO code *development* (GO:0007275) is the parent of *embryonic development* (GO:0009790), which in turn is the parent of *somitogenesis* (GO:0001756).

Extracting gene functions from literature is currently done manually, a laborious and time consuming process. Human curators read each document and annotate genes with GO codes if the text contains evidence that supports the annotation[2,5]. Given the enormous number of publications in MEDLINE, manual curation cannot keep pace with the data generation. However, automatic functional annotation is a challenging task, for the following reasons, among others:

(1) When a GO code is assigned to a gene, its GO tokens may not explicitly occur in the text. For example, in document (with PubMed Id) 11401564, GO code *3'-5'-exoribonuclease activity* (GO:0000175) occurs as *3' to 5' exoribonuclease activity*, while in document 11110791 occurs as *3' → 5' exoribonuclease activity*. Similarly, in document 10692450, GO code *negative regulation of cell proliferation* (GO:0008285) occurs as *inhibition of cell proliferation*.

(2) GO tokens do not necessarily appear contiguously in the annotated text. For example, in document 10734056, gene *MIP-1 alpha* is annotated with GO code *G-protein coupled receptor protein signaling pathway* (GO:0007186), based on the following paragraph: *Results indicate that CCR1-mediated responses are regulated ... in the signaling pathway, by receptor phosphorylation at the level of receptor/G protein coupling. ...CCR1 receptor binds MIP-1 alpha with high affinity.*

(3) Algorithms that attempt to assign GO codes to documents based just on the fact that the tokens from the GO codes occur in the text, yield a large number of false positives. Even when the GO tokens occur in text, the curator may not annotate the gene with the GO code because (a) the text does not contain evidence to support the annotation, or (b) the text contains evidence for the annotation, but the curator knows the gene to be involved in a function that is more general or more specific than the GO code that was matched in the text. For example, the Gene Ontology provides guidelines of what the evidence for annotation should be, (e.g., the text should mention *co-purification* or *co-immunoprecipitation* experiments). However, an algorithm that uses this information (e.g., annotates a gene with a GO code only if the text contains words like *co-purification*) does not perform any better than an algorithm that ignores these hints about evidence.

To address these challenges, we propose a cross-species approach for assigning Gene Ontology terms to LocusLink genes, making use of information about orthologous genes. (Orthologous genes are genes from different species that have evolved directly from an ancestral gene.) Our assumption is that since there is an overlap between the genomes of the two species, their orthologous genes may share some functions, and consequently, some GO codes.

We use information from orthologous genes in two ways. First, for a target gene we search in biomedical journal text for only the GO codes previously assigned to its orthologous gene. This yields precise results but at the expense of missing many codes. In the second method, for a given gene we search in biomedical text for any of the 17,600 possible GO codes, but eliminate those codes that are illogical, based on which GO codes are known to co-occur with the GO codes for the ortholog of the gene. This approach is less precise but uncovers more valid codes. We then merge the results of the two processes. Results on three datasets show that our algorithm obtains higher F-measure than previous solutions.

The rest of the paper is organized as follows. Section 2 presents related work. Section 3 describes our solution. Section 4 presents experimental results, and Section 5 concludes and suggests future work.

2. Related Work

Functional annotation from medical documents is a relatively new problem, although there is significant related work for annotating a gene with functions using gene expression time profiles[12,16], sequence-derived protein features[17] and multiple alignments of complete sequences[6]. Many approaches search for uncharacterized sequences across GO-mapped protein databases and assign to them the GO codes of the best hits[13,19,27].

Functional annotation from bioscience articles has been mainly studied by the participants in the BioCreAtIve[15] and the TREC Genomics track[14] competitions. BioCreAtIve addressed the problem of annotating a gene with the exact GO codes and thus has created a defacto benchmark for functional annotation from bioscience literature. TREC made the task easier; rather than exact GO codes, participants had to predict the GO category (molecular function, biological process or cellular component) the GO code belongs to. Below we summarize the methods proposed by the participants in the BioCreAtIve competition.

Chiang and Yu[7,8] observe that there are phrase patterns commonly used in sentences describing gene functions. Examples are *"gene plays an important role in function"*, or *"gene is involved in function"*. To learn the patterns they divide a sentence s into five segments (*prefix, tag1, infix, tag2, suffix*), where *tag1, tag2* are gene products or functions. The *prefix, infix* and *suffix* are divided into tokens and the patterns are learned by seeking out consecutive tokens common to multiple sentences. To predict the overall likelihood that a sentence describes a gene-function relation, they use a Naive Bayes classifier.

Ray and Craven[23] learn a statistical model for each GO code from a training set of four GO annotated databases. In particular, they learn which words are likely to co-occur in the paragraphs containing the tokens of a GO code. They use a multinomial Naive Bayes classifier for every GO category to re-rank the results from pattern matching. Features are words, as well as the distance between the protein and the GO code in the text, and the score of the match.

Couto et al.[10] annotate a gene with a GO code if what they call the *information content* of the GO code, computed as a function of the words that match in text, is larger than its information content computed as a function of all the words in the GO code. Verspoor et al.[26] compute an association strength between words based on how often they co-occur in the paragraphs of a set of documents. Every GO code is expanded with the words having a high association strength with the words in the GO code. GO codes are assigned to genes using a Gene Ontology Categorizer[18] which utilizes the structure of the Gene Ontology to find the best covering nodes.

Ehler and Ruch[11] treat each document as if it was a query to be categorized into GO categories. GO codes are assigned scores based on pattern matching and $TF \times IDF$ weighting and the top GO codes are annotated to the gene. Rice et al.[25] learn a support vector machine classifier for each GO code. Target genes are tested against each classifier and are assigned the highest scoring GO codes.

The literature contains a few other solutions for functional annotation, although these systems did not participate in the BioCreAtIve competition. Raychaudhuri et al.[24] compare three document classification techniques (Maximum Entropy Modeling, Naive Bayes and Nearest Neighbor) for assigning only 21 GO codes to gene products. Koike et al.[21] use shallow parsing and rule-based techniques to semi-automatically enrich GO codes with other terms that appear in the same sentence based on co-occurrence and collocation similarities. Finally, Xie et al.[28] combine both text mining and sequence similarity searches to annotate gene products with GO terms. The results are reported on various datasets, thus it is difficult to compare our solution against them.

3. Algorithms

In this section, we describe our algorithms for annotating genes with GO codes. We make use of information from orthologous genes to derive two sets of GO codes for a given target gene. For the first set (called CSM, **C**ross **S**pecies **M**atch), we restrict GO code assignments to be selected from only those codes which have already been assigned to the target gene's ortholog. Since this approach results in high precision but low recall, for the second set (called CSC, **C**ross **S**pecies **C**orrelation), we allow any GO code to be a candidate, but then eliminate assignments that cannot pair with the gene's ortholog. The final set of annotations is the union of the two sets. Figure 1 shows the block diagram of the annotation process.

In every document, we eliminate stop words and punctuation characters and divide the text into tokens using spaces as delimiters. We analyze text at the

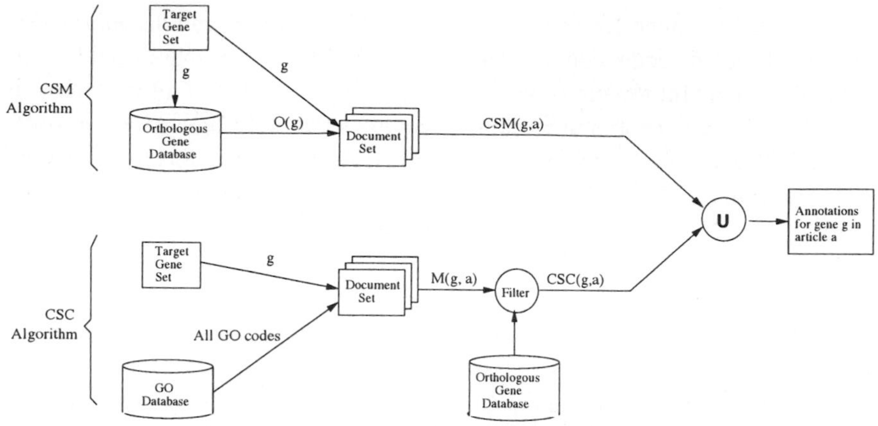

Figure 1. The annotations for gene g computed as the union of two sets CSM and CSC.

sentence level. Similarly, we divide GO codes into tokens. We perform gene name recognition by normalizing and matching different variations of gene names using the algorithm of Bhalotia et al.[1] For every sentence in which a target gene is found, we consider a GO code to be found if the sentence contains a percentage of tokens from the GO code that is larger than a threshold. This threshold is set to 75% for the CSM algorithm, and 100% for the CSC algorithm.

3.1. *CSM: Using the GO codes of Orthologous Genes*

The GO ontology contains 17,600 GO codes (as of July 12, 2004). Our experimental results show that searching in text for all the GO codes results in a large number of false positives and thus low precision. For this reason we aim to limit the set of GO codes that are possible candidates. We achieve this by searching in text for only the GO codes previously annotated to orthologous genes.

As mentioned above, the main assumption behind this algorithm is that for two species that have descended from a common ancestor, the orthologous genes of the two species may have the same functions, and consequently may be annotated with the same GO codes.

For a target gene g, let $O(g)$ represent the set of GO codes that have been assigned to the ortholog of that gene for another species. For a given article a, this algorithm finds all sentences that contain the gene g and then searches only for those GO codes in $O(g)$. We define $CSM(g, a)$ to be the subset of GO codes in $O(g)$ matched in article a for gene g.

It is important to note that many genes are annotated by automated or manual transfer of annotations from other genes with sequences similar to the target genes. Such annotations are marked with the evidence codes *Inferred from electronic annotation (IEA)* and *Inferred from Sequence Similarity (ISS)*. While these annotations are very useful, using them in our case may unrealistically boost our performance. This is because in some cases the annotations of an orthologous gene may have been derived from the annotations of the target gene. To avoid this kind of circular reference, we do not use any annotations of orthologous genes marked with the evidence codes *IEA* and *ISS*.

3.2. *CSC: Using All GO Codes and Eliminating "Illogical" Ones*

Although searching in text for only the GO codes of orthologous genes yields high precision, it limits recall since these codes are only a small subset of those available. To improve recall we use a general observation: if two GO codes tend to occur together in a database, then a gene annotated with one GO code is likely to be annotated with the other one as well. Similarly, if one GO code tends to occur in the orthologous genes' annotations when another does not, then for the target species these two GO codes may not be allowed to both be assigned to the same gene.

The idea is that GO codes co-occur if it makes sense for a gene to support both of their functions; in many cases the underlying biological function will make it illogical for two codes to co-occur. For example, if we find *rRNA transcription* (GO:0009303), *nucleolus* (GO:0005737) and *extracellular* (GO:0005576), then we eliminate *extracellular* because transcription cannot happen outside of the cell.

King et al.[20] use a similar idea to predict how to augment those GO codes that have already been assigned to a gene, once some annotations for the gene are known. Given a database of genes and their GO annotations, they use machine learning algorithms trained on one part of the dataset to predict the annotations for the rest of the database. They do not use cross-species information, nor do they use the correlations to find GO codes in text.

For every pair of GO codes in the orthologous genes database, we compute a χ^2 coefficient[22] using occurrence counts. Let N be the number of GO codes and:

O_{11}: # of times the orthologous gene is annotated with both GO_1 and GO_2
O_{12}: # of times the orthologous gene is annotated with GO_1 but not with GO_2
O_{21}: # of times the orthologous gene is annotated with GO_2 but not with GO_1
O_{22}: # of times the orthologous gene is not annotated with any of GO_1 or GO_2

Then the χ^2 coefficient is

$$\chi^2 = \frac{N * (O_{11} * O_{22} - O_{12} * O_{21}) * (O_{11} * O_{22} - O_{12} * O_{21})}{(O_{11} + O_{12}) * (O_{11} + O_{21}) * (O_{12} + O_{22}) * (O_{21} + O_{22})}$$

```
CSC(g,a) = {};
for every GO_1 in M(g,a)
        count = 0;
        for every GO_2 in O(g)
                if ((χ^2(GO_1, GO_2) > 3.84)&&(GO_1 ≠ GO_2))
                        count ++;
        if (count > p * o)
                add GO_1 to CSC(g,a);
```

Figure 2. *Pseudocode for computing set CSC for gene g in article a.*

For every gene g in article a we search for all 17,600 GO codes. Let $M(g,a)$ be the set of GO codes matched and let o be the size of $O(g)$. Also let $CSC(g,a)$ (**C**ross **S**pecies **C**orrelation) be a set of initially empty annotations for gene g in article a. Figure 2 shows the algorithm for computing the set $CSC(g,a)$.

For every GO code GO_1 in $M(g,a)$, we count how many GO codes GO_2 in $O(g)$ have a χ^2 coefficient larger than 3.84[a]. If the count is larger than o multiplied by a percentage p (0.2 in our experiments[b]) then we consider GO_1 logically related to the GO codes in $O(g)$, and we add it to the set $CSC(g,a)$. Otherwise it is discarded. The final set of annotations for gene g in article a is the union between sets $CSM(g,a)$ and $CSC(g,a)$.

4. Results

In this section we present experimental results. We test our algorithms on the dataset of task 2.2 of BioCreAtIve competition[3], where we compare our results with the performance of the participants in the contest. In addition, we test our algorithms on two other GO annotated databases: EBI human[c] and MGI [d].

4.1. *Results on the BioCreAtIve Dataset*

Task 2.2 of the BioCreAtIve competition provided participants with a set of gene-article pairs and asked them to annotate the genes with the GO codes found in the articles along with the passages supporting the annotations.

[a]For a probability level of 0.005, and one degree of freedom the probability of error threshold for χ^2 is 3.84[22].

[b]Intuitively, we may expect higher percentages to work better. However, since genes may be involved in several unrelated functions, a GO match in text is generally correlated with a small percentage of functions in $O(g)$.

[c]http://www.ebi.ac.uk/Databases.

[d]http://www.informatics.jax.org.

The test set consisted of 138 human genes and 99 full text articles. Human curators judged each annotation. An annotation was marked as "perfect prediction" if the gene name appears in the retrieved passage of text and the passage provides evidence for annotating the gene with the GO code. There was no official evaluation measure but the committee of judges reported, for each system, the total number of predictions, the number of perfect predictions and precision. In total, participants found 237 "perfect predictions". Since the competition organizers did not report numbers for recall, we use the number 237 as the total number of relevant documents for computations of recall.

We conducted this research after the contest had past, so our annotations could not be judged by human curators, which makes it impossible to fully determine how well our performance compares with the other systems. To get around this limitation, we measure our performance using the "perfect predictions" made by the participants. (Note that this may be unfairly penalizing our algorithm as it may be finding relevant documents not found by the other systems.) We consider an annotation we make as correct if it exactly matches a "perfect prediction" made by another system. For example, for gene vhl in PubMed Id 12169961, a "perfect prediction" made by one of the participants annotates the gene with *transcription* GO:0006350 using the following passage in text as evidence: *VHL inhibits transcription elongation, mRNA stability, Sp1-related promoter activity and PKC activity.* For the same gene-article pair we consider our prediction to be correct if we find *transcription* GO::0006350 in exactly the same passage of text.

Since the target genes are human, and since mouse is a species with a genome similar to humans', for each target gene we compute the set $CSM(g, a)$ by searching in the articles for only the GO codes annotated to its mouse orthologous gene (except the GO codes marked with evidence codes *IEA* and *ISS* to avoid circular references). The orthologous databases we used are MGI and the part of SwissProt related to mouse genes[e].

For each human gene, we extract from MGI and SwissPro the GO annotations of the mouse gene with the same name as the target gene or with a name found in the Human-Mouse Orthology maps available from MGI[f]. We were able to find GO codes for about 61% of the human genes. For the genes whose orthologs had no GO annotations, we did not perform any search, so for these genes the sets $CSM(g, a)$ are empty. Next, for each gene g in article a we compute set $CSC(g, a)$ by searching in text for all possible 17,600 GO codes and eliminating illogical annotations using the χ^2 coefficient. Sets CSM and CSC are the union

[e]http://au.expasy.org/sprot/sprot-top.html, as of July 12, 2004.
[f]ftp://ftp.informatics.jax.org/pub/reports/index.html.

of all $CSM(g, a)$ and $CSC(g, a)$ over all genes and articles.

Table 1. Results on BioCreAtIve dataset.

System	Precision	TP (Recall)	F-measure
CSM	0.364	16 (0.068)	0.114
CSC	0.182	44 (0.185)	0.178
CSM + CSC	0.241	51 (0.215)	**0.227**
Ray and Craven[23]	0.213	52 (0.219)	0.216
Chiang and Yu[7,8]	0.327	37 (0.156)	0.211
Ehler and Ruch[11]	0.123	78 (0.329)	0.179
Couto et al.[10]	0.089	58 (0.245)	0.131
Verspoor et al.[26]	0.055	19 (0.080)	0.065
Rice et al.[25]	0.035	16 (0.068)	0.046

Table 1 compares the performance of our algorithms ($CSM, CSC, CSM + CSC$), with the performance of the participants in the competition as presented in Blaschke et al[3]. For each participant we report the best results they obtained in the competition. In the second column, TP stands for the number of true positive predictions made by a system (the number of predictions where both the protein and the GO code found in the passage are correct). Recall is computed as the ratio between TP and the total number of correct predictions (237).

In general, the results show the trade-off between precision and recall, and the systems that did well on precision obtained a recall much lower than the systems that did well on recall (which in turn obtained a lower precision). For example, Chiang and Yu's system has the best precision, 0.327, although the recall 0.156, is much lower than the best recall obtained by Ehler and Ruch's system, 0.329, which in turn had a lower precision, 0.123.

Although high precision is desirable, high recall is also important. For this reason, the F-measure (defined as the the harmonic mean of precision and recall) is considered a better metric for comparing results since a system has to maximize both precision and recall. The best F-measure is obtained by Ray and Craven's system, 0.216 with a precision of 0.213 and a recall of 0.219.

CSM obtains an F-measure of 0.114, although its precision, 0.364 is higher than any precision obtained in the competition. In turn, $CSM + CSC$ obtains an F-measure of 0.227, which is higher than the best F-measure in the competition 0.216, obtained by Ray and Craven's system.

CSC obtains an F-measure of 0.178. This result shows the effect of the CSC heuristic on our task but further analysis would needed to determine how often the co-occurring GO codes truly reflect logical or illogical combinations.

4.2. *Results on the EBI Human and MGI Mouse Datasets*

In this section we further compare the performance of our algorithms by evaluating them on much larger datasets and comparing them with the performance of Chiang and Yu's system[8], which performed well in the BioCreAtIve competition[g].

We present experimental results on two GO-annotated databases: EBI human and MGI Mouse (July 12 2004 versions). On each database, for every gene-document pair, we attempt to predict the manually annotated GO codes. Similarly to Chiang and Yu[8] we restrict our study to the genes we found in abstracts only, although curation is done on the full text.

The EBI human test set consisted of 4,410 genes annotated with 13,626 GO codes in 5,714 abstracts. The MGI test set consisted of 2,188 mouse genes annotated with 6,338 GO codes in 1,947 abstracts. For human genes, the orthologous databases we used are MGI and the part of SwissPro related to mouse genes. For mouse genes the orthologous databases are EBI human and the part of SwissPro related to human genes.

Table 2 shows the results obtained by CSM, $CSM + CSC$ and Chiang and Yu's system on both datasets. Chiang and Yu used the same data for both training and testing[8], which artificially inflates how well it would perform under real test conditions. In our case the test collection represents new data for our algorithm. While Chiang and Yu's algorithm generally achieves higher precision, $CSM + CSC$ obtains a better F-measure. On EBI human, Chiang and Yu obtain an F-measure of 0.105 while $CSM + CSC$ obtains 0.118. On MGI, Chiang and Yu obtain an F-measure of 0.089 while $CSM + CSC$ obtains 0.140.

Our experimental results also show that predicting molecular functions and cellular components may be easier than predicting biological processes. For example, on EBI, $CSM + CSC$ obtains an F-measure of 0.154 (for MFs), 0.124 (for CCs) and only 0.08 (for BPs). A possible explanation for this could be the fact that BPs have longer strings which are more difficult to match in text.

5. Conclusions

We propose a method that annotates genes with GO codes using the information available from other species [h]. In particular, we search in text for only the GO codes annotated to a gene that is an orthologous of the target gene. Since this

[g]These authors have made publicly available the annotations that their system assigns to *all* genes in LocusLink, http://gen.csie.ncku.edu.tw/meke3, as of September 2002. To obtain a fair comparison, our evaluation uses only the genes they annotate and only documents published before September 2002.

[h]Annotations and software available at http://biotext.berkeley.edu.

Table 2. Results on EBI human and MGI datasets.

Test set	System	Precision	Recall	F-score
EBI	CSM	0.289	0.033	0.060
	CSM + CSC	0.163	0.092	**0.118**
	Chiang and Yu	0.318	0.063	0.105
MGI	CSM	0.328	0.049	0.086
	CSM + CSC	0.168	0.121	**0.140**
	Chiang and Yu	0.332	0.051	0.089

approach results in low recall, we also search for all the GO codes in the Gene Ontology, but eliminate illogical annotations using the correlations between GO codes computed on the orthologous genes database. We test our algorithm on three collections: BioCreAtIve, EBI human and MGI. Experimental results show that our algorithm consistently achieves higher F-measure than other solutions.

In the future we plan to explore how to improve the performance of our system; one possibility is to combine or use a voting scheme to decide between the predictions made with our system and the predictions made using a machine learning algorithm like that of Ray and Craven[23]. In addition, we plan to investigate how effective using genes with sequences similar to the target gene (but not orthologous to the gene) is for predicting GO annotations.

Acknowledgements. This research was supported by NSF grant DBI-0317510 as well as a gift from Genentech.

References

1. G. Bhalotia, P.I. Nakov, A.S. Schwartz, and M.A. Hearst. Biotext team report for the trec 2003 genomic track. In *Proceedings of TREC 2003*, pages 612–621, 2003.
2. J.A. Blake, J.E. Richardson, C.J. Bult, J.A. Kadin, J.T. Eppig, and the members of the Mouse Genome Database Group. Mgd: The mouse genome database. In *Nucleic Acids Res*, volume 31, pages 193–195, 2003.
3. C. Blaschke, E. A. Leon, M. Krallinger, and A. Valencia. Evaluation of biocreative assessment of task 2. *BMC Bioinformatics*, 6(S1), 2005.
4. D.L. Brutlag. Genomics and computational molecular biology. *Current Opinion in Microbiology*, 1(3):340–345, 1998.
5. E. Camron, D. Barrell, V. Lee, E. Dimmer, and R. Apweiler. The gene ontology annotation (goa) database - an integrated resource of go annotations to the uniprot knowledgebase. 4(1):5–6, 2004.
6. F. Chalmel, A. Lardenois, J. D. Thompson, J. Muller, J.-A. Sahel, T. Lveillard, and O. Poch. Goanno: Go annotation based on multiple alignment. *Bioinformatics*, 19(11):1417–1422, 2005.
7. J. Chiang and H. Yu. Extracting functional annotations of proteins based on hybrid text mining approaches. In *Proc. of BioCreative Workshop*, 2004.
8. J.-H. Chiang and H.-C. Yu. Meke:discovering the functions of gene products from

biomedical literature via sentence alignment. *Bioinformatics*, 19(11):1417–1422, 2003.

9. The Gene Ontology Consortium. Gene ontology: tool for the unification of biology. *Nature Genet.*, 25(1):25–29, 2000.
10. F. M. Couto, M. J. Silva, and P. Coutinho. Figo: Finding go terms in unstructured text. In *Proc. of BioCreative Workshop*, 2004.
11. F. Ehler and P. Ruch. Preliminary report on the biocreative experiment. In *Proc. of BioCreative Workshop*, 2004.
12. M.B. Eisen, P.T. Spellman, P.O. Brown, and D. Botsein. Cluster analysis and display of genome-wide expression patterns. *Proc. Natl. Acad. Sci*, 95(25):14863–14868, 1998.
13. S. Hennig, D. Groth, and H. Lehrach. Automated gene ontology annotation for annonymous sequence data. *Nucleic Acids Res*, 31(13):3712–3715, 2003.
14. W.R. Hersh, R.T. Bhuptiraju, L. Ross, A.M. Cohen, and D.F. Kraemer. Trec 2004 genomics track overview. In *Proceedings of TREC 2004*, 2004.
15. L. Hirschman, A. Yeh, C. Blaschke, and A. Valencia. Overview of biocreative: critical assessment of information extraction for biology. *BMC Bioinformatics*, 6(S1), 2005.
16. V.R. Iyer, M.B. Eisen, D.T. Ross, G. Schuler, T. Moore, J.C. Lee, J.M. Trent, L.M. Staudt, J. Hudson, and M.S. Boguski M.S. et.al. The transcriptional program in the response of human fibroblasts to serum. *Science*, 283(5398):83–87, 1999.
17. L.J. Jensen, R. Gupta, H.H. Staerfeldt, and S. Brunak. Prediction of human protein function according to gene ontology categories. *Bioinformatics*, 19(5):635–642, 2003.
18. C.A. Joslyn, S.M. Mniszewski, A. Fulmer, and G. Heaton. The gene ontology categorizer. *Bioinformatics*, 4(20):1169–1177, 2004.
19. S. Khan, G. Situ, K. Decker, and C. J. Schmidt. Gofigure: Automated gene ontology annotation. *Bioinformatics*, 19(18):2484–2485, 2003.
20. O.D. King, R.E. Foulger, S.S. Dwight, J.V. White, and F.P. Roth. Predicting gene function from patterns of annotation. *Genome Research*, 13(5):896–904, 2003.
21. A. Koike, Y. Niwa, and T. Takagi. Automatic extraction of gene/protrin biological functions from biomediacal text. *Bioinformatics*, 21(7):1227–1236–1422, 2005.
22. C. Manning and H. Schutze. *Foundations of Statistical Natural Language Processing*. MIT Press, 1999.
23. S. Ray and M. Craven. Learning statistical models for annotating proteins with function information using biomedical text. *BMC Bioinformatics*, 6(S1), 2005.
24. S. Raychaudhuri, J. T. Chang, P. D. Sutphin, and R. Altman. Associating genes with gene ontology codes using a maximum entropy analysis of biomedical literature. *Genome Research*, 12(1):203–214, 2002.
25. S. B. Rice, G. Nenadic, and B. J. Stapley. Protein function asignment using term-based support vector machines. In *Proc. of BioCreative Workshop*, 2004.
26. K. Verspoor, J. Cohn, C. Joslyn, and S. Mniszewski. Protein annotation as term categorization in the gene ontology. In *Proc. of BioCreative Workshop*, 2004.
27. A. Vinayagam, R. Koenig, J. Moormann, F. Schubert, R. Eils, K.H. Glatting, and S. Suhai. Applying support vector machines for gene ontology based gene function prediction. *BMC Bioinformatics*, 5(1), 2004.
28. H. Xie, A. Wasserman, Z. Levine, A. Novik, V. Grebinski, A. Shoshan, and L. Mintz. Large-scale protein annotation through gene ontology. *Genome Res*, 12(5):785–794, 2002.

BOOTSTRAPPING THE RECOGNITION AND ANAPHORIC LINKING OF NAMED ENTITIES IN *DROSOPHILA* ARTICLES

ANDREAS VLACHOS, CAROLINE GASPERIN, IAN LEWIN, TED BRISCOE

Computer Laboratory,
University of Cambridge,
15 JJ Thomson Avenue, CB3 0FD
E-mail: FirstName.LastName@cl.cam.ac.uk

This paper demonstrates how *Drosophila* gene name recognition and anaphoric linking of gene names and their products can be achieved using existing information in FlyBase and the Sequence Ontology. Extending an extant approach to gene name recognition we achieved a F-score of 0.8559, and we report a preliminary experiment using a baseline anaphora resolution algorithm. We also present guidelines for annotation of gene mentions in texts and outline how the resulting system is used to aid FlyBase curation.

1. Introduction

Curated databases are critical in the biomedical sciences as a method of systematizing and making accessible the rapidly expanding scientific literature[1,2]. However, curation is expensive because it requires considerable manual effort on the part of domain experts. In this paper, we describe the development of an adaptive textual information extraction (IE) system using bootstrapping machine learning techniques, designed to function as part of an interactive system to aid curation by supporting thematically-guided navigation of the article being curated in terms of the entities of interest.

Most IE systems for biomedical and other domains have been developed either using supervised machine learning techniques requiring large quantities of annotated data[3] or by manually encoding domain specific rules[4]. Here we describe how we have replicated and extended the approach of

Morgan *et al.*[5] using FlyBase[a] and the Sequence Ontology[6b] to bootstrap an initial unsupervised system with state-of-the-art performance.

The link to extant public-domain resources, such as FlyBase and the Sequence Ontology, both supports the initial automatic adaptation of the system and also provides essential functionality. The association of gene names with FlyBase gene identifiers is a useful extension of classic named entity recognition in the context of FlyBase curation. However, the means by which this link is obtained and by which the initial *Drosophila* gene name recognizer is bootstrapped also relies critically on the availability of such extant resources which, while not developed to support creation of IE systems, contain valuable information which can be exploited to adapt IE technology to the domain. Similarly, the Sequence Ontology encapsulates general genomic knowledge concerning genes, their components, their products, and their products' subclasses and components which can be exploited effectively in the *Drosophila* literature to compute the anaphoric link, whether coreferential or associative, between gene mentions and mentions of proteins, RNA and other gene products.

2. *Drosophila* Gene Name Recognition

In the biomedical domain, there is a paucity of annotated text and none which is focused entirely on the *Drosophila* literature. We extend recent approaches to bootstrapping systems for name recognition, partly by necessity, but also because annotation is expensive and does not constitute a viable long-term approach to the development of IE systems.

2.1. *Reproducing the Morgan et al. experiment*

FlyBase provides a dictionary of all *Drosophila* genes and their synonyms that appear in the extant curated literature together with links to the literature indicating where a specific name is used to refer to a particular gene. Morgan *et al.*[5] exploited this information to create annotated material to train a gene name recognizer. In brief, abstracts were tokenized and then genes names linked to specific abstracts in FlyBase were tagged applying longest-extent pattern matching. The process resulted in a large but noisy corpus which was in turn used to train a hidden Markov model (HMM) recognizer.

---

[a]www.flybase.net

[b]http://song.sourceforge.net/

We replicated this experiment with an enlarged dataset and different software. We built a list of all articles mentioned in the FlyBase bibliography for which the database also recorded at least one gene as having been mentioned within it. We then retrieved all the abstracts of those articles using the NCBI Entrez Programming Utilities[7]. This gave a total of 16609 abstracts (9.5% more than Morgan *et al.*). The abstracts were tokenized using the RASP toolkit[8c]. Then, following Morgan *et al.*, in each abstract we annotated all the gene name mentions licensed by the associated FlyBase gene name list.

The 16609 abstracts contained approximately 7800 distinct gene names representing 5243 distinct genes out of a total of over 44000K names recorded in FlyBase[d]. Many gene names and synonyms do not appear in the training material. As Morgan *et al.* note, there are gene synonyms that are common English words, such as *to* and *by*, resulting in precision errors in the training data. Sometimes genes mentioned in abstracts are not in the respective FlyBase gene lists of those articles (as only some relevant sections of the article are curated), resulting in recall errors.

The recognizer used in our experiments is the open source toolkit LingPipe[e]. The named entity recognition module is a 1st-order HMM model using Witten-Bell smoothing. For each token $t[n]$ and possible label $l[n]$, the following joint probability is computed, conditioned on the previous two tokens and the previous label:

$$P(t[n], l[n] | l[n-1], t[n-1], t[n-2]) \quad (1)$$

Unknown tokens are analyzed using a morphologically-based classifier, which we modified slightly to adapt it to the domain. The approach is highly lexical and conservative compared to others (e.g. Crim *et al.*[9]) which deploy more abstract and general features to achieve greater domain-independence. Lingpipe achieves high precision by only generalizing to unseen names in lexical contexts which are clearly indicative of gene names in the training data.

We tested the performance of the trained recognizer on the test data used in Morgan *et al.*[5]. The data consists of 86 abstracts doubly-annotated

[c] http://www.cogs.susx.ac.uk/lab/nlp/rasp/

[d] Exact figures depend on how much normalization (e.g. homogenizing punctuation, Greek letters, capitalization and whitespace) one applies to the names before counting them.

[e] http://www.alias-i.com/lingpipe/

by a biologist curator (Colosimo) and a computational linguist (Morgan). The performance of LingPipe on each annotation (Recall/Precision/F-score) was 0.8086/0.7485/0.7774 and 0.8423/0.8483/0.8453, respectively. To calculate these figures we used the evaluation script used for the BioNLP2004[f] shared task. Morgan *et al.*, evaluating on the the first set of annotations, reported 0.71/0.78/0.75. Our performance appears better, especially in terms of recall.

2.2. *NER Annotation guidelines*

There is a large difference in performance (0.0679 in F-score) between the two annotations of the dataset due to difficulties in applying the annotation scheme. According to the guidelines used in Morgan et al.[5], gene names are tagged not only when they refer to genes, but also when they are part of mentions of proteins or transcripts, as in *the zygotic Toll protein.* Only *Drosophila* genes are tagged, excluding reporter genes, genes that are not part of the natural *Drosophila* genome, families, particular alleles or protein complexes. However, *Drosophila* genes can be synonymous with foreign genes (e.g. *Hsp90*), family names are often synonymous with specific names (e.g. *CSP*), and foreign and reporter genes are often not flagged as such in text. Additionally, mutant genes, which are not part of the natural genome, are usually referred to using the name of the original gene, leading to inconsistencies in annotation in cases like *dunce mutations* or *eye PKCI700D mutant.*[g]

We developed revised guidelines, partially based on those developed for ACE[10]. We did not exclude foreign genes, reporter genes and families, as they are often of interest to curators and users of FlyBase. Like Morgan *et al.*, gene names (<*gn*>) are tagged not only when they refer to genes but also when they are found in pre-nominal modifier positions. Following ACE, we annotate the surrounding noun phrase (NP). The NP is tagged either as a *gene-mention* (<*gm*>) or as *other-mention* (<*om*>), depending on whether it refers to a gene or not (see 1) and 2) in Figure 1). In cases of alleles, mutants or protein complexes, the gene name is tagged and the remaining tokens of the NP are tagged *other-mention* (see 5) in Figure 1).

---

[f] http://research.nii.ac.jp/ collier/workshops/JNLPBA04st.htm

[g] Overall the biologist Colosimo's annotations are more accurate given the annotation guidelines used. He avoids tagging reporter genes synonymous with specific ones (e.g. *Gal4*), mutants, or gene families (e.g. *Hedgehog Hh*), resulting in fewer genes tagged (909) than by Morgan (989).

(1) <*gm*>the <*gn*>dunce</*gn*> gene</*gm*>
(2) <*om*>the <*gn*>dunce</*gn*> mutations</*om*>
(3) <*gm*>the human <*gn*>IL-2</*gn*> gene</*gm*>
(4) <*om*>the unrearranged <*gn*>TcR delta</*gn*> gene expression</*om*>
(5) <*om*><*gn*>eye</*gn*> PKCI700D mutant</*om*>

Figure 1. Annotated examples

As also reported by Dingare *et al.*[11], the data used in the BioNLP[3] and BioCreative[12] evaluations contained many cases in which modifiers of gene names and nouns modified by gene names were variably annotated. Using the guidelines suggested in this paper, the annotation of such cases becomes clearer. In 3) and 4) in Figure 1, the guidelines are applied to cases with inconsistent annotation reported by Dingare et al.[11]. 2) was inconsistently annotated by Colosimo and Morgan. Annotation of NPs is also relevant to recovery of anaphoric links (see § 3) and aids annotation of gene names within coordinated NPs.

LingPipe's performance using our guidelines to resolve differences between Morgan and Colosimo was Recall/Precision/F-score 0.8081/0.8493/0.8282. Further results below will be reported on both our re-annotation (called "merged") and the gold standard used in Morgan *et al.* (called "morgan").

2.3. *Inspecting errors and improving performance*

We tried to identify the main sources of errors and ameliorate them taking account of the specific HMM model utilized. Our first step was to perform an individual evaluation on seen and unseen tokens. This evaluation didn't take into account multi-token genes, because there were many cases where the one boundary of such multi-token cases was incorrect. Therefore, our system was not penalized for partially recognized genes and received/lost a point for each gene token recognized/missed. This token-wise definition of Recall/Precision is used only when reporting results on seen or unseen tokens. In all other cases, the standard definitions are used.

Evaluating on seen tokens on the "merged" dataset, we achieved 0.8272/0.9022/0.8631 Recall/Precision/F-score, which suggests that there are many gene names that are missed, even though they exist in the training data. For example, the gene *gurken* appears 97 times in the training data, of which 90 times it is tagged correctly as a gene on the basis of FlyBase links. However, the few false negatives in the training data cause LingPipe to fail to tag *gurken* as a gene name during testing. We experimented with a non-conservative version of Morgan *et al.*'s procedure in which all gene names recorded in FlyBase were annotated as such in all of the training data. However, this resulted in many false positives in the training data and overall worse performance.

In general, gene names appearing in the abstracts are mentioned in Fly-Base gene lists. So, we post-processed the training data by reannotating tokens as genes when this token was annotated as a gene in the overall training data more than a certain percentage of the time. By doing this though, we risk changing common English words correctly tagged as ordinary words to genes, since some genes have common English words as synonyms. With the percentage set at 80% we obtained Recall/Precision/F-scores of 0.8567/0.8551/0.8559 on the "merged" dataset and of 0.8614/0.7565/0.8056 on "morgan".

On unseen tokens, compared to Morgan *et al.* our performance is significantly higher (F-scores of 0.619 on "merged" and 0.5365 on "morgan" compared to their 0.33). However, LingPipe is rather conservative in classifying unseen tokens as genes (Recall/Precision was 0.4642/0.9285 on "merged").

As with the seen tokens, we tried to improve recall, as it is important for curation to have a system that is able to recognize unseen gene names. For each token classified, we estimated the entropy of the distribution of Equation 1 computed by LingPipe, which gave us an indication of how (un)certain the classifier was of its decision. We observed that many of the recall errors occurred in cases in which the HMM model classified a token with entropy close to 1, i.e. with high uncertainty. We post-processed the output of the classifier by re-annotating as genes unseen tokens that were classified as ordinary words with entropy higher than a specified threshold. As before, the lower this threshold was set, the higher the recall at the expense of precision. By setting this threshold at 0.6 and evaluating on the "merged" dataset, we improved the performance on unseen tokens to 0.7058 F-score. However, this resulted in more partially recognized genes, which slightly reduced the performance when evaluating using the standard definition of the metrics (from 0.8559 to 0.8545). Also, only 49 out of 16779 tokens in the test set were not seen in the training data. In order to demonstrate the value of this method, we performed an experiment using only 20% of the available training data, which resulted in 1040 unseen tokens in the test set. In this case, using the uncertainty of the classifier in the way described earlier, the performance on unseen tokens rose from 0.4424 to 0.6111 in F-score, while the overall performance using the standard definition of F-score rose from 0.5847 to 0.6487, evaluating on "merged".

2.4. *Reference resolution*

Our recognizer identifies strings that are names of genes. Reference resolution requires determining the FlyBase identifier of a gene name. Frequently, *Drosophila* gene names are not unique identifiers. For ambiguity resolution, a quite effective and simple strategy (around 89% accuracy) is to associate names with the entities that those names most frequently denote using FlyBase's lists of gene names occurring in articles. For orthographic variants, FlyBase's gene synonym lists are a good resource for calculating commonly occurring types of variation (e.g. prefix by *D.* or *Dm.*) which can be applied to previously unseen name strings. Some exploration of variants of these strategies is undertaken in Ma[13].

3. Biomedical anaphora resolution

In FlyBase curation, the "gene" is an organizing concept around which other information is recorded. In order to extract all the information about a gene in a text it is necessary to identify all textual entities (like pronouns, definite descriptions and proper names) that are anaphorically linked to that gene or coreferential with it. These entities may refer to proteins, RNA, alleles, mutants and other gene "products" rather than the gene itself, and may therefore be associative rather than coreferential anaphoric links. Here we report work on resolving anaphoric definite descriptions (DDs; phrases beginning with the definite article *the* e.g. *the faf gene*) and proper nouns (PNs), since in biomedical texts there are fewer cases where pronouns are used.

The first step towards resolving anaphora is selecting the anaphoric expressions to be resolved and their possible antecedents. We first parse the text using RASP, then select all noun phrases (NPs) in the text and filter them to find the ones referring to relevant entities using information from the gene name recognizer and the Sequence Ontology (SO).

3.1. *Biotype information: semantic tagging*

If a NP is headed by a gene name according to the recognizer (i.e. its rightmost element is a gene name), then it refers to a gene. Otherwise, we use information in the SO to search for the *biotype* of the head noun which stands in one of the four following possible relations to a gene: "part-of", "type-of", "subproduct" or "is-a".

SO relates entities by the following relations: derives_from, member_of, part_of, is_a, among others[h]. For instance, we extract the unique path of concepts and relations which leads from gene to protein, shown in Figure 2:

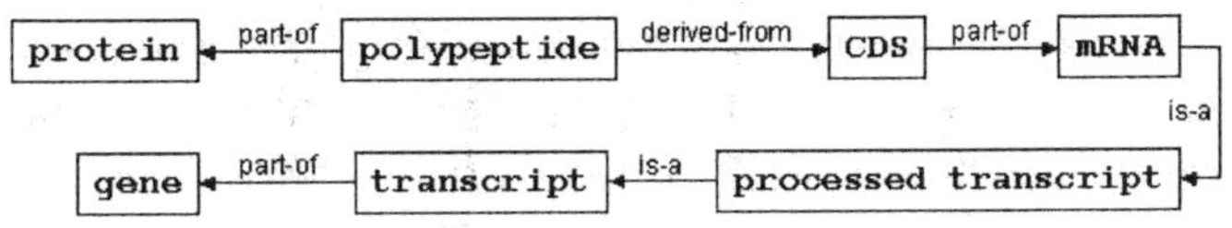

Figure 2. SO path from gene to protein.

Besides the facts directly expressed in this path, we also assume the following:

(1) whatever is-a transcript is also part-of a gene
(2) whatever is part-of a transcript is also part-of a gene
(3) mRNA is part-of a gene
(4) whatever is part-of a mRNA is also part-of a gene
(5) CDS is part-of a gene
(6) polypeptide is a sub-product (derived-from) of a gene
(7) whatever is part-of a polypeptide is also a sub-product of a gene
(8) protein is a sub-product of a gene

We then use these assumptions to add new derivable facts to the original path. For example, an *exon* is a part of a transcript according to SO, therefore, by the 2nd assumption, we add the fact that an *exon* is a part of a gene. We also extract information about gene types that is included in the ontology as an entry called "gene class". Using the derived information, we would tag *the third exon* with "part-of-gene". NPs that remain untagged after this search are tagged as "other-bio" if any head modifier is a gene name. These biotyped NPs are then considered for anaphora resolution.

[h]The member-of relation is considered a type of the part-of relation, so we do not make this distinction and consider both as part-of relations.

3.2. *Anaphora resolution*

Our baseline unsupervised system for anaphora resolution that we present here makes use of lexical, syntactic, semantic and positional information to identify the antecedent of an anaphoric expression. The lexical information consists of the words themselves. The syntactic information consists of NP detection and the distinction between head and premodifiers (extracted from RASP output). The distance (in words) between the anaphoric expression and its possible antecedent is taken into account as positional information. The semantic information comes from the gene recognizer and the SO-based tagging described above. Thus, the system is bootstrapped from a variety of extant resources without any domain-specific tuning.

As anaphoric expressions to be resolved we take all PNs and DDs among the filtered NPs. To link anaphoric expressions to their antecedents we look at three aspects of the corresponding NPs: the head noun, the premodifiers of the head noun, and the biotype.

The pseudo-code to find the antecedent for the DDs and PNs is given below:

- Input: a set A with all the anaphoric expressions (DDs and PNs); a set C with all the possible antecedents (all NPs with biotype information)
- For each anaphoric expression A_i:
  - Let antecedent 1 be the closest preceding NP C_j such that head(C_j)=head(A_i) and biotype(C_j)=biotype(A_i)
  - Let antecedent 2 be the closest preceding NP C_j such that head(C_j)$\neq$ head(A_i) and biotype(C_j)$\neq$ biotype(A_i), but
    head(C_j)=premodifier(A_i), or
    premodifier(C_j)=head(A_i), or
    premodifier(C_j)=premodifier(A_i)
  - Take the closest candidate as antecedent, if 1 and/or 2 are found; if none is found, the DD/PN is treated as non-anaphoric
- Output: The resolved anaphoric expressions in A are linked to their antecedents.

For example, in the passage "Dosage compensation, which ensures that the expression of *X-linked genes(C_j)* is equal in males and females ... the hypertranscription of *the X-chromosomal genes(A_j)* in males", the candidate C_j meets the conditions for antecedent 1 to be linked to the anaphoric expression A_j. In "... the role of *the roX genes(C_k)* in this process ... which

MSL proteins interact with *the roX RNAs(A_k)*", C_k meets the conditions for antecedent 2 to A_k.

3.3. *Experimental Results - Related Work*

We have annotated two articles from PubMed central containing 334 sentences and 7641 tokens in total. 334 anaphoric expressions (90 DDs and 244 PNs) with the relevant biotypes were found and their antecedents were manually annotated when they were functioning anaphorically. When we tested the anaphora resolution algorithm on this annotated data using the manually corrected syntactic and biotype information, the algorithm achieves Recall/Precision/F-score of 0.62/0.64/0.63. However, on the same text using automatic parsing and biotype tagging, performance drops to 0.37/0.43/0.40, primarily because of errors in identifying NPs and extracting their head nouns.

Most previous work on anaphora resolution in (biomedical) texts has used supervised machine learning techniques and different knowledge sources for biotype classification. For instance, Yang *et al.*[14] assigns biotypes using a named entity recognizer trained on the GENIA corpus together with other features as part of a supervised approach; Castano *et al.*[15] uses the UMLS (Unified Medical Language System)[i] to type DDs in MEDLINE abstracts and describes an unsupervised approach. The SO is more focussed on the functional genomics domain and therefore more appropriate for FlyBase curation.

4. Conclusions and Future Work

The two modules described are integrated into an interactive environment for FlyBase curators to help them in the task of literature curation. The environment allows navigation by anaphorically-linked entities and links the current paper with information derived from FlyBase and the SO.

The gene recognizer achieves state-of-the-art performance via bootstrapping but may be further improved by training on full articles with a greater variety of lexical contexts and by the use of additional feature types. Anaphora resolution requires improvement. We plan to use the baseline system to generate noisy training data for a statistical anaphora resolution module. Both components will be incrementally improved using

---

[i] http://www.nlm.nih.gov/research/umls/

active training with curators correcting a small number of highlighted low confidence cases in each presented article.

Acknowledgments

This work is part of the BBSRC-funded FlySlip[j] project. We would like to thank Alexander Morgan for making the annotated test data available to us and for advice on replication of the experiment reported in Morgan *et al.*[5], Chihiro Yamada for his expert help with annotation of *Drosophila* articles, and Bob Carpenter for help with LingPipe. Caroline Gasperin is funded by a CAPES award from the Brazilian government.

References

1. L. Hirschman, J. C. Park, J. Tsujii, L. Wong, and C. H. Wu. Accomplishments and challenges in literature data mining for biology. *Bioinformatics*, 18(12):1553–1561, 2002.
2. H. Liu and C. Friedman. Mining terminological knowledge in large biomedical corpora. In *Pacific Symposium on Biocomputing*, pages 415–426, 2003.
3. J. Kim, T. Ohta, Y. Tsuruoka, Y. Tateisi, and N. Collier, editors. *Proceedings of JNLPBA, Geneva, Switzerland*, August 28–29 2004.
4. R. Gaizauskas, G. Demetriou, P. J. Artymiuk, and P. Willet. Protein structures and information extraction from biological texts: The "PASTA" system. *BioInformatics*, 19(1):135–143, 2003.
5. A. A. Morgan, L. Hirschman, M. Colosimo, A. S. Yeh, and J. B. Colombe. Gene name identification and normalization using a model organism database. *J. of Biomedical Informatics*, 37(6):396–410, 2004.
6. Karen Eilbeck and Suzanna E. Lewis. Sequence ontology annotation guide. *Comparative and Functional Genomics*, 5:642–647, 2004.
7. Eric Sayers and David Wheeler. *Building Customized Data Pipelines Using the Entrez Programming Utilities (eUtils)*. NCBI.
8. E. J. Briscoe and J. Carroll. Robust accurate statistical annotation of general text. In *Proceedings of the 3rd International Conference on Language Resources and Evaluation*, pages 1499–1504, 2002.
9. J. Crim, R. McDonald, and F. Pereira. Automatically annotating documents with normalized gene lists, 2004.
10. Annotation guidelines for entity detection and tracking (EDT).
11. S. Dingare, J. Finkel, M. Nissim, C. Manning, and C. Grover. A system for identifying named entities in biomedical text: How results from two evaluations reflect on both the system and the evaluations. In *The 2004 BioLink meeting at ISMB*, 2004.

---

[j] http://www.cl.cam.ac.uk/users/av308/Project_Index/Project_Index.html

12. Christian Blaschke, Lynette Hirschman, and Alexander Yeh, editors. *Proceedings of the BioCreative Workshop*, Granada, March 2004.
13. Ning Ma. Using author trails to disambiguate entity references. Master's thesis, University of Cambridge, Computer Laboratory, 2005.
14. X. Yang, J. Su, G. Zhou, and C. L. Tan. An NP-cluster based approach to coreference resolution. Geneva, Switzerland, August 2004.
15. J. Castano, J. Zhang, and J. Pustejovsky. Anaphora resolution in biomedical literature. In *Proceedings of the International Symposium on Reference Resolution for NLP*, 2002.

SEMANTIC WEBS FOR LIFE SCIENCES

ROBERT STEVENS

School of Computer Science, University of Manchester
Oxford Road, Manchester, M13 9PL, UK
E-mail: Robert.stevens@manchester.ac.uk

OLIVIER BODENREIDER

U.S. National Library of Medicine
8600 Rockville Pike, MS 43, Bethesda, Maryland, 20894, USA
E-mail: olivier@nlm.nih.gov

YVES A. LUSSIER

Departments of Biomedical Informatics and Medicine
Columbia University, New York, NY 10032, USA
E-mail: yves.lussier@dbmi.columbia.edu

The Semantic Web is a vision for the next generation of the Web [1]. The Web is a huge interlinked information resource, but is largely restricted to human use because the information is represented only in natural language. The goal of the Semantic Web is to make these data – the facts on the Web – amenable to computational processing. To date, the Semantic Web has largely been pushed by technology development, but the life sciences are seen to be a huge area for potential application development. Indeed, a recent workshop on this subject saw over 100 attendees, indicating a great interest in the community[1].

The Semantic Web's broad goal parallels that of many bioinformaticians. There are vast quantities of biological data and associated annotations, or knowledge, now available on the Web. These resources are highly distributed and heterogeneous. This heterogeneity exists at many levels, the most pernicious of which are the semantic heterogeneities in the schema and the values placed in those schema. Semantic Web technologies and the vision itself offer a solution to this long-standing problem in creating an integrated view of bioinformatics.

Lincoln Stein describes this situation as being akin to the rival city states in medieval Italy and talks of the need to create a "bioinformatics nation" [6]. This vision is at one level of heterogeneity – the programmatic access to bioinformatics resources. Stein describes the use of Web Services, a Semantic Web technology, to provide a common form of access to distributed resources with heterogeneous

[1] http://www.w3.org/2004/07/swls-ws.html

platforms and access paradigms. We already see well over a thousand Web Services offering access to bioinformatics resources[2] not seen before in bioinformatics.

Semantic Web technology also offers solutions for the problems of semantic heterogeneity and these technologies have a growing influence. The aim of the Semantic Web is to make facts amenable to machine processing. The Resource Description Framework[3] (RDF) provides a common data model of triples for this purpose. An RDF triple, a subject, predicate (verb) and object, enables any statement to be represented in a simple, flexible, common framework.

Each part of a triple names a resource using either a Uniform Resource Identifier (URI) or a literal. As many resources are transformed to this data model, the common naming scheme will mean that facts can be aggregated, forming a vast graph of descriptions of resources [8]. The Life Science Identifier (LSID) is a form of URI that can be used to uniquely identify and version bioinformatics entries [3]. *Uniprot* is already available in RDF[4] and shows this aggregation happening using LSID. Similarly, *YeastHub* [2] is a system that has transformed many yeast resources into RDF and allows querying, using an RDF query language, to provide access to these aggregated data.

One advantage of the RDF model is the open world assumption. In an open world, only that which is explicitly stated is known – we cannot assume that something does not exist simply because it has not been stated. This means that new statements can be added without fear of breaking the data model, which happens all too easily with existing schema mechanisms such as XML schema [8].

Even with this flexible model, the description of the resources themselves with relationships (predicates) and the objects that provide values for these descriptions are still highly heterogeneous. In RDF, the collection of names formed by the URI provides a vocabulary. True semantic integration requires a common, shared vocabulary. This is the role of ontologies and Semantic Web technology provides languages for this purpose. RDF Schema is an RDF vocabulary for ontologies. It enables classes and their relationships to be defined and used in an RDF graph. The Web Ontology Language[5] (OWL) offers a variety of dialects for building and maintaining ontologies, with strict and precise semantics not offered by RDFS. OWL ontologies can be delivered as RDFS for Semantic Web use.

Again, we see OWL and RDFS being used within the life sciences. The *Gene Ontology*™ [4] is available in RDFS and can be used to annotate, for instance, the RDF version of *Uniprot*. OWL is used in the *BioPAX* ontology [5], which is used to exchange data between pathway resources. OWL is used by the MGED Society to provide an ontology for marking up microarray experiments [7].

There are an increasing number of bioinformatics applications using Semantic Web technologies. There are, however, few real Semantic Webs of Life Sciences, where large quantities of diverse data are aggregated with RDF, described with

---

[2] See, for example, http://www.mygrid.org.uk.
[3] http://www.w3.org/TR/2003/PR-rdf-concepts-20031215/
[4] http://www.isb-sib.ch/~ejain/rdf/
[5] http://www.w3.org/TR/owl-ref/

RDFS vocabularies and then exposed for querying and automatic reasoning. *YeastHub* and *BioDash*, working over *BioPAX* data, come the closest to this vision, as will be seen in the Semantic Webs for Life Sciences session.

This session reflects the early adoption of the Semantic Web by the bioinformatics community. While most papers still focus on foundational issues such as namespaces, ontology creation, mapping and adaptation to Semantic Web formalisms, some contributions present pioneering applications implementing the vision of the Semantic Web.

For instance, two papers address human computer interactions and demonstrate how to use Semantic Web technologies to facilitate the access to otherwise heterogeneous semantics of human interfaces of computerized bioinformatics resources by scientists: *BioDash* from Neuman and Quan provides an integrated web-based dashboard for drug development, and *BioGuide* from Cohen-Bulakia et al. is a user-centric framework to help scientists to choose tools according to their preferences and strategies.

Beyond the interface, organizing heterogeneous information sources can also be approached with Semantic Web technologies. Indeed, as shown by the work of Mukherjea and Sahay, the semantic relationships presented by different applications over the Web can be mined through search engines using Semantic Web technologies and these relations can be elicited explicitly. At a more automated level of communications and automated machine processing, Yip et al. present a Semantic Web approach, *SemBiosphere*, to build a matchmaking system that automatically provides recommendations to users about microarray clustering algorithms by reasoning over the Semantic Web service descriptions of these methods.

Another group of papers focus on ontologies, a necessary technology for Semantic Web development. Zhang et al. verified the consistency of the *Foundational Model of Anatomy* by first transforming its representation in OWL and then using the best "reasoner" to identify unclassifiable classes. Good et al. propose an important and necessary improvement over the development and maintenance of ontologies: a protocol to attain affordability. Indeed, affordability issues remain one of the big challenges for sustainability of the Semantic Web. Kushida et al. describe the design of a new biomedical ontology for annotating biological pathways component. Finally, Kazic reviews the fundamental assumptions of current Semantic Web technologies and proceeds systematically to demonstrate their potential and structural limitations.

As the field matures and as a critical mass of Semantic Web resources (e.g., ontologies, Web Services) becomes available, the number of Semantic Web applications is expected to grow dramatically in the next few years, illustrating the fact that "the combined effect of global naming, universal data structure and open world assumption is that resources exist independently but can be readily linked with little, if any, precoordination." [8].

References

1. Berners-Lee, T., J. Hendler, et al. (2001). "The Semantic Web." Scientific American **284**(5): 34-43.
2. Cheung, K. H., K. Y. Yip, et al. (2005). "YeastHub: a semantic web use case for integrating data in the life sciences domain." Bioinformatics **21**(1): i85-i96.
3. Clark, T. and S. M. Liefeld (2004). "Globally Distributed Object Identification for Biological Knowledgebases." Briefings in Bioinformatics **5**(1): 59-70.
4. Gene Ontology Consortium (2000). "Gene Ontology: Tool for the Unification of Biology." Nature Genetics 2**5**(1): 25-29.
5. Luciano, J. (2005). "PAX of mind for pathway researchers." Drug Discov Today **10**: 937-942.
6. Stein, L. (2002). "Creating a bioinformatics nation." Nature **417**(9): 119-120.
7. Stoeckert, C. J. and H. Parkinson (2003). "The MGED ontology: a framework for describing functional genomics experiments." Comparative and Functional Genomics **4**(1): 127-132.
8. Wang, X., R. Gorlitsky, et al. (2005). "From XML to RDF: how semantic web technologies will change the design of 'omic' standards." Nature Biotechnology **23**(9): 1099-1103.

SELECTING BIOLOGICAL DATA SOURCES AND TOOLS WITH XPR, A PATH LANGUAGE FOR RDF

SARAH COHEN-BOULAKIA[†], CHRISTINE FROIDEVAUX[†] AND EMMANUEL PIETRIGA[‡]

[†]*LRI, CNRS UMR 8023, Université Paris-Sud*
91405 Orsay, CEDEX, France
{cohen, chris}@lri.fr

[‡]*INRIA Futurs, LRI*
91405 Orsay, CEDEX, France
emmanuel.pietriga@inria.fr

As the number, richness and diversity of biological sources grow, scientists are increasingly confronted with the problem of selecting appropriate sources and tools. To address this problem, we have designed BioGuide[1], a user-centric framework that helps scientists choose sources and tools according to their preferences and strategy, by specifying queries through a user-friendly visual interface. In this paper, we provide a complete RDF representation of BioGuide and introduce XPR (eXtensible Path language for RDF), an extension of FSL[2] that is expressive enough to model all BioGuide queries. BioGuide queries modeled as XPR expressions can then be saved, compared, evaluated and exchanged through the Web between users and applications.

1. Introduction

The number and size of new biological data sources together with the number of tools available for analysing this data have increased exponentially in the last few years, to a point where it is unrealistic to expect scientists to be aware of all of them. However, as these sources and tools are often complementary, focused on different objects and reflecting various experts' points of view, scientists should not limit themselves to the sources they already know well, and thus have to face the problem of selecting sources and tools when interpreting their data.

For instance, the European HKIS platform[a] offers a set of analysis scenarios where at each step users may have to ask questions necessitating the consultation of various sources. For this, we have designed the DSS algorithm[3] that builds paths allowing to navigate through data sources. DSS reflects how oncologist partners involved in the project select sources, and takes into account their preferences.

[a]www.hkis-project.com

Two other systems considering paths between sources and exploiting preferences have been developed in the same spirit: Biomediator[4] and Bionavigation[5]. We then wanted to investigate in a systematic way the need for various ways of querying biological sources. A thorough analysis of needs[1], by means of a questionnaire, revealed the importance of (i) expressing *transparent* queries[4] (ii) exploiting *preferences* (e.g. *reliability*) with respect to both sources and tools and (iii) following a specific *strategy*.

In response to these findings, we have designed BioGuide[1], a **user-centric framework** which helps scientists choose sources and tools according to their preferences and strategy. BioGuide has proven itself to be useful to obtain complementary data through the use of alternative ways of finding information, and to deal with divergent data by exploiting preferences. Moreover, BioGuide provides a framework which is general enough to take into account all preferences of current systems (DSS, Biomediator and Bionavigator), to simulate their behaviors by means of strategies and to specify new preferences and strategies. The BioGuide system is available for use from http://www.lri.fr/~cohen/bioguide/bioguide.html.

BioGuide provides a simple visual interface which allows users to specify the biological entities, sources and tools they are interested in. In this paper we want to make it possible for BioGuide users to save their queries (reusability), exchange them (collaboration between experts), compare them (expressiveness) and evaluate them (efficiency). We provide a complete RDF representation of BioGuide, which is well-suited to the uniform representation of biological entities and sources structured as multi-labeled graphs. We then exploit this RDF representation of BioGuide data to model the queries expressed visually by users as queries on RDF models. For this, we introduce XPR (eXtensible Path language for RDF), an extension of FSL[2] that is expressive enough to model all BioGuide queries.

For the sake of readability, we consider in this paper a simpler version of BioGuide where strategies are not described and preferences are simplified.

2. BioGuide: selection of sources and tools

BioGuide aims at assisting users in the specification of their queries. A thorough study of how scientists consider the query process revealed that from a question expressed in natural language, they first identify the underlying biological entities and the relationships between them. For instance, in question *"On which chromosome is the BAC of my CGH array located?"*, the underlying entities are CHROMOSOME and BAC. In BioGuide, the user is supported in his task by a graphical representation of the biological domain, represented through the **entities graph** (Fig.1), in which nodes are biological entities and labeled edges are symmetric relationships. Two kinds of relationships exist: biological relationships (e.g. `causes`, `encodedBy`) and relationships achieved by tools (e.g. `similarSeq`, `mapsWith`).

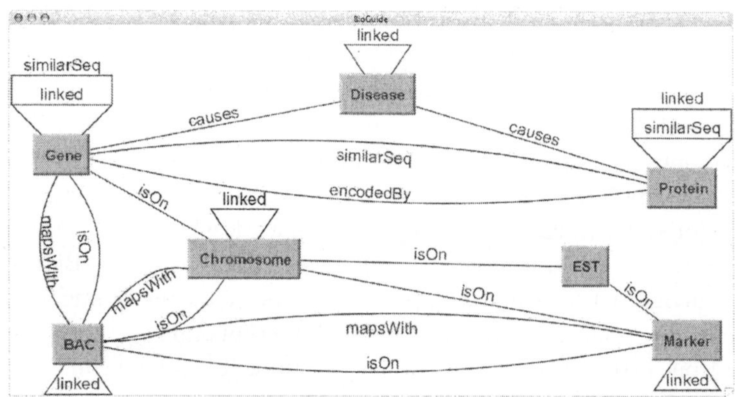

Figure 1. Entities graph fragment (BioGuide's Graphical User Interface)

Scientists can make use of this graph to build BioGuide *initial queries* by selecting entities and, possibly, relationships between these entities. It is also possible to ask the system to not only consider entities given in the query, but also to consider or avoid *additional entities* (*navigating* strategy). This can be done by explicitly referring to them or by specifying the kinds of relationships (e.g. those achieved by tools) used to reach these additional entities.

Furthermore, our study has revealed that users want to know which sources and tools can be accessed[6,7], and punctually need to cite some sources or tools. In BioGuide, they are supported in this task by a graphical representation of sources, provided by the **sources-entities graph** (Fig. 2), in which each node represents an entity in a source and arrows indicate the links between two entities (in the same source or in another). Labels on arrows specify the kind of link: cross-reference (*CrossRef*), internal link (*Internal*)– links between entities in the same source – and tools (e.g. *Blast*).

Using the sources-entities graph, scientists can thus complete their *initial query* by an *extended query* in which they (possibly) specify the sources and tools to access or to avoid. This graph is also used to visualize which

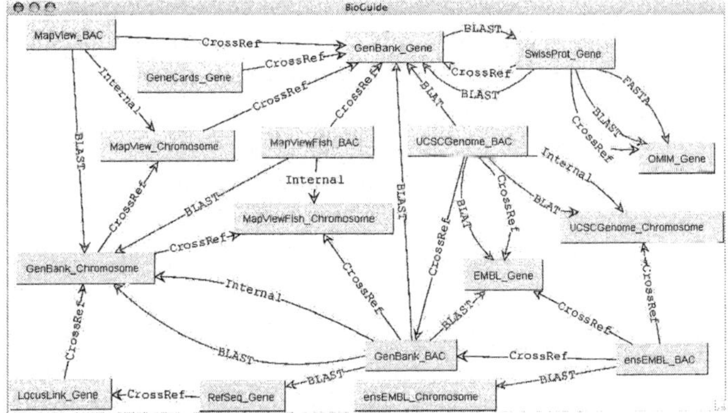

Figure 2. Sources-entities graph fragment relating BAC, CHROMOSOME and GENE

sources-entities contain a given entity and which links achieve a given relationship: the two graphs are *in correspondence* with each other. We now briefly recall BioGuide[1]'s principles and main steps (I to IV).

(I) The *initial user query* Q consists of (i) the entities and relationships underlying the user's query; (ii) the user's choice regarding the navigating strategy criterion and (iii) if necessary, the entities and kinds of relationships to avoid. (II) From Q, the *EPG* (Entity Path Generator) module yields P_e, the set of acyclic paths in the entities graph generated following the choice about the navigating strategy.

In the previous example, following the navigating strategy and avoiding the MARKER entity, the following paths are returned by *EPG*: p_1="BAC isOn CHROMO", p_2="BAC mapsWith CHROMO", p_3="BAC isOn GENE isOn CHROMO" and p_4="BAC mapsWith GENE isOn CHROMO".

In step (III) the *extended user query* Q_{se} consists of (a) P_e, the output of *EPG*, and (b) preferences (e.g. only *reliable* sources, where reliability ratings can be parameterized by each user). (IV) Using Q_{se} and the sources-entities graph, the *SEPT* (Source-Entity Path Translator) module generates the list L_{pse} of paths in the sources-entities graph which meet specified preferences and *correspond* to the paths of P_e.

SEPT first constructs the list of *entity-centered* paths, i.e., paths of cross-references or internal links between sources-entities containing some given entity. Secondly, SEPT relates paths centered on the different entities by considering the links between sources-entities which *achieve* the relationships specified in the paths of entities. All possible paths of sources-entities are generated as alternative ways of finding information. In our previous example, the following two paths (among others) are returned by *SEPT*: (MapView,BAC) $\stackrel{Internal}{\rightarrow}$(MapView,CHROMO) (corresponding to p1) and (GenBank,BAC) $\stackrel{Blast}{\rightarrow}$ (RefSeq,GENE) $\stackrel{CrossRef}{\rightarrow}$ (LocusLink, GENE) $\stackrel{CrossRef}{\rightarrow}$ (GenBank, CHROMOSOME) (corresponding to p_4).

BioGuide thus provides users with alternative *ways of finding data* based on an internal data model that allows them to specify powerful graph queries involving entities, sources and tools through a simple visual interface. More information about the architecture and data model is available[1].

Our goal is now to make it possible for BioGuide users to save their queries, exchange them with other users, compare them and evaluate them on graphs containing different components or based on different settings. BioGuide data structures thus need to be represented within a uniform, open and extensible format and queries expressed within a formal language.

3. A framework for exchanging BioGuide data

Most formats for data interchange between users over the Web and between heterogeneous applications are now based on XML. But as Semantic Web technologies mature, languages such as RDF become more attractive, further increasing interoperability and knowledge exchange capabilities, as well as allowing autonomous software agents to exploit and reason on data that is marked up semantically with RDFS/OWL ontology-based vocabularies.

3.1. *Modeling BioGuide in RDF*

RDF and its companion languages[b] offer a general-purpose framework for representing information about Web resources with domain-specific vocabularies in a minimally constraining yet uniform way. They thus represent a good candidate solution for promoting data integration from Web biological sources. Furthermore, RDF's data model[8] fits naturally with BioGuide's as both are based on directed labeled graphs (as opposed to XML's tree-based data model), making the mapping between BioGuide's data structures and RDF straightforward.

RDF describes Web *resources* identified by their URI (*Uniform Resource Identifier*) in terms of property-value pairs representing relationships between resources and characteristics of these resources. An RDF model is a collection of statements taking the form of (*subject*, *predicate*, *object*) triples. This set of statements can be represented as a directed labeled graph[8]. In

[b]http://www.w3.org/RDF/

the remainder of this paper we take a graph-centric view of RDF models in which *nodes* are resources (depicted as ovals) or literal values such as strings or integers (depicted as rectangles), and *arcs* are properties. Arcs are labeled by a URI identifying the property's type. Property types and resource classes (resources can be stated to be instances of classes with property `rdf:type`) are defined in RDF schemas using RDFS, RDF's vocabulary description language.

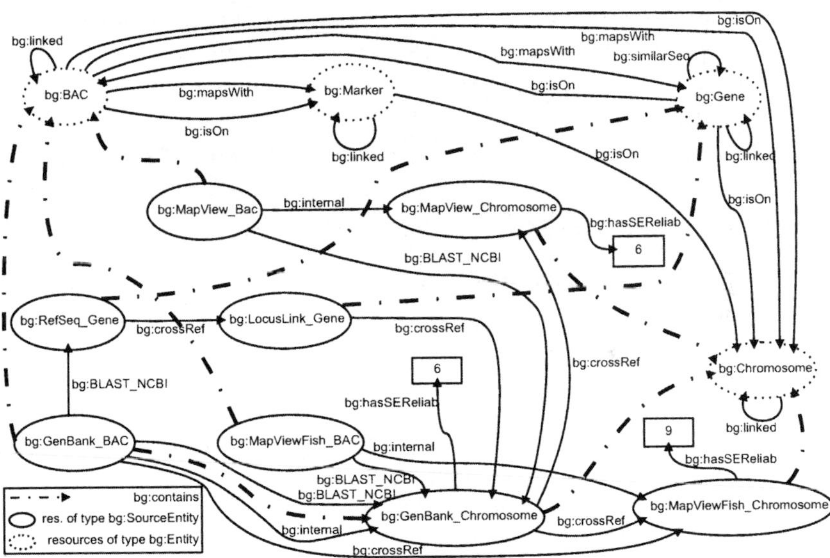

Figure 3. Fragment of BioGuide's RDF model relating various entities (schematic repr.)

Data from both graphs (Fig. 1 and 2) is modeled in a single RDF graph (Fig. 3), as elements of the two graphs are put in correspondence using property `bg:contains` (see e.g. `bg:LocusLink_Gene` and `bg:Gene`). Preferences associated with sources-entities are modeled with properties such as `bg:hasSEReliab` pointing to literal values (e.g. giving access to the level of Reliability of a given Source-Entity). There is an inverse property for each `bg:mapsWith` and `bg:isOn` property; these have been removed from the figure for legibility purposes.

Each RDF vocabulary being identified by a namespace, we introduce the BioGuide namespace whose URI is bound to prefix `bg` in this paper. Usual prefixes are bound to common namespaces: `rdf` and `rdfs` for the RDF and RDFS vocabularies. Figure 4 contains a subpart of the RDF

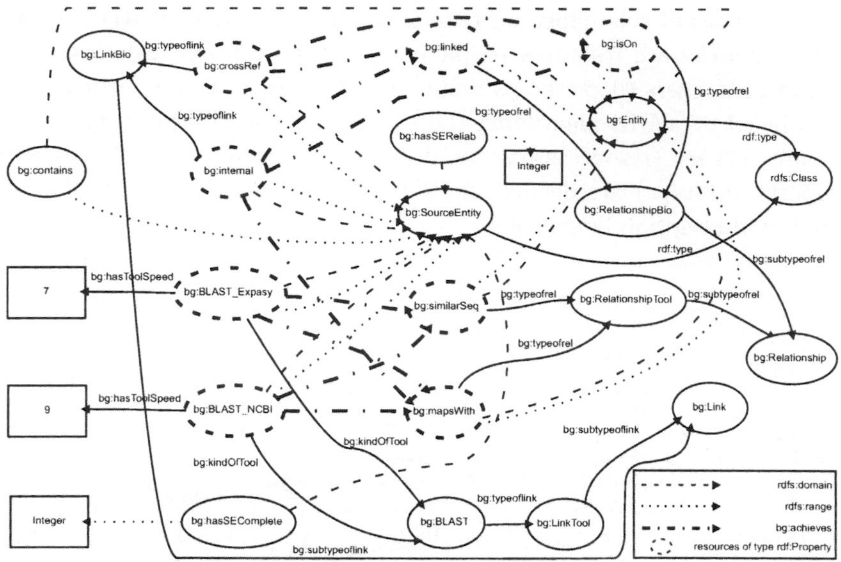

Figure 4. Fragment of BioGuide's RDF schema (schematic representation)

schema describing BioGuide's vocabulary. The schema defines all elements of the BioGuide namespace: properties (e.g `bg:contains`, `bg:internal`) and classes of resources (e.g. `bg:Entity`, `bg:SourceEntity`) that are then used in BioGuide's RDF model (Fig. 3). All properties are defined as resources in the schema graph and each have a *domain* and *range*, indicating which class of resources and typed literal values they can connect. The schema also contains BioGuide-specific, non-RDFS meta-properties, which characterize BioGuide properties in a model-wide manner: `bg:typeOfLink` and `bg:typeOfRel` classify types of links and relationships; `bg:achieves` specifies which relationship(s) a given link achieves; other meta-properties such as `bg:hasToolReliab` are used to indicate preferences values.

3.2. *Exploiting BioGuide's RDF representation*

Our goal is to exploit the RDF representation of BioGuide data to model the queries – expressed visually by users through BioGuide's user interface[1] – as queries on RDF models. BioGuide queries being essentially traversal paths in graphs, we are interested in a language for modeling paths in RDF graphs. Such a language should make it possible to express filtering conditions on nodes and arcs, and provide means of exploiting the meta-information contained in schemas associated with instance data (subclass relationships, characterization of the kinds of links, tool preferences, etc.).

Most of the existing RDF query languages (e.g. RQL[9], SPARQL[10]) offer essential features such as schema/ontology awareness and limited inference capabilities, but are not well-suited to modeling traversal paths in RDF graphs. ρ-Queries[11] consider pairs of RDF resources, identify complex relationships between them, and return those relationships as property sequences. Although such sequences can be seen as paths, these are query results, i.e., instances of paths in the graph; ρ-Queries themselves cannot model queries as traversal paths.

Several proposals for RDF path languages have been made, but almost none of them is fully and formally specified: the proposals are often drafts giving short descriptions of the language constructs and a few example queries. One exception is FSL[2], a fully specified language for modeling traversal paths in RDF graphs, designed to address the specific requirements of a selector language for Fresnel[12] (an RDF vocabulary for modeling RDF data presentation knowledge). Trying to avoid reinventing the wheel, FSL is inspired by XPath[13] (W3C's language for expressing traversal paths in XML trees), reusing many of its concepts and syntactic constructs while adapting them to RDF's graph-based data model. In FSL, RDF models are considered as directed labeled graphs according to RDF Concepts and Abstract Syntax[8]. FSL is therefore fully independent from any serialization of RDF and meets many of our requirements. However, it is designed for use in the context of Fresnel and tries to be as simple as possible (its goal is not to be a full so-called RDFPath language). It is therefore not expressive enough to model all BioGuide queries. In the next section we introduce a proposal for an extension of FSL called XPR (eXtensible Path language for RDF) that addresses this problem.

4. XPR

4.1. *Extending FSL*

An FSL expression represents a path from a node or arc to another node or arc, passing by an arbitrary number of other nodes and arcs. As in XPath, steps are separated by the slash symbol (e.g. `nodeStepA/arcStepB/nodeStepC`). Each step on the path, called a *location step*, follows the XPath location step syntax: `AxisSpecifier::Test[Predicates]` where `AxisSpecifier` is an optional axis declaration specifying the traversal direction in the directed graph, `Test` is a type test taking the form of a URI reference represented as an XML qualified name (QName), or a `*` when the type is left unconstrained, `Predicates` is an optional list of further filtering conditions on the nodes or arcs to be matched by this step. The type test constrains property arcs to be labeled with the URI represented by the QName, and resource nodes to be instances of the class identified by this QName (found in the schema). In other words, type tests specify constraints on the types of properties and classes of resources to be traversed and selected by paths. Constraints on the URI of resources can be expressed as predicates associated with node location steps using function call `uri(.)`. As in XPath, a dot represents the node or arc considered by the current location step. The following example models paths starting at the resource identified by `bg:BAC` (with no constraint on its type), going through arcs labeled by `bg:mapsWith` and ending at any resource of type `bg:Entity` (e.g. `bg:Gene` and `bg:Marker` in Fig. 3).

$$*[uri(.)="bg\!:\!BAC"]/bg\!:\!mapsWith/bg\!:\!Entity$$

More information about FSL, including its grammar, data model and semantics, as well as examples, can be found on the FSL Web page[2].

A limitation of the FSL language is that it does not provide an equivalent of XPath's `descendant` axis (often abbreviated `step1//step2`) that specifies an unconstrained number of elements between `step1` and `step2` (possibly equal to zero). Such a construct adds significant complexity to a path language for RDF and its implementations, mainly because of the possibly cyclic nature of RDF graphs. Considered too costly by the designers of FSL with respect to its added value in the context of Fresnel, the construct is however useful from a more general RDF perspective and is mandatory to express BioGuide queries. XPR extends FSL and allows the use of the `//` notation to specify an unconstrained number of arc and node steps between two explicit location steps[c]. As shown in the expression below, it is possible to express constraints on the nodes and arcs traversed between those two explicit steps: predicates before the semi-colon are evaluated against nodes, predicates after the semi-colon against arcs. Concrete examples of use of this extension are given in section 4.2.

$$step1//_{[nodePredicateExpr;arcPredicateExpr]}step2$$

The second extension to FSL made by XPR is a new function named `rp` (for reify property). Given a property arc, `rp` returns the resource typed

[c]As BioGuide queries implicitly state that paths cannot traverse the same arc twice, the closure mechanism defined by XPR to handle cycles in RDF graphs is not detailed here.

as `rdf:Property` that represents this property in the corresponding RDF schema. For instance, the following **arc** location step:

$$*[rp(.)/bg\!:\!achieves/bg\!:\!isOn]$$

selects all property arcs which are stated to achieve `bg:isOn` in the RDF schema. Applied to the graph in Figure 3, it would select all `bg:crossRef` and `bg:internal` arcs since in the RDF schema of Figure 4 only those two properties are stated to achieve `bg:isOn`.

XPR has been defined as a general purpose extension to FSL which does not make any assumption about the first location step's nature. In the context of BioGuide, we only consider XPR path expressions starting with a node location step, and thus remove any ambiguity about the nature of the first location step. The result of evaluating an XPR expression on a graph is the set of all node/arc sequences in this graph that are instances of the path described by the XPR expression.

4.2. *BioGuide queries as XPR expressions*

The following example illustrates how BioGuide queries can be easily expressed as XPR path expressions, and how such expressions evolve through the various steps of the BioGuide query process. We first specify a query introduced in section 2: *"On which* CHROMOSOMES *are* BACS *located?"*, with the additional constraint that intermediate entities cannot be MARKER, but without restriction on the kind of relationships to avoid.

$$*[uri(.) = "bg\!:\!BAC"]//_{[uri(.)!="bg:Marker";*]}*[uri(.) = "bg\!:\!Chromosome"]$$

From this query, the EPG module generates instantiated paths in the entities graph, such as `bg:BAC/bg:isOn/bg:Chromosome` (Fig. 3). Each of these result paths is then rewritten in terms of query paths involving the *corresponding* sources-entities with additional constraints based on user preferences about the sources and tools to be selected.

As an example, the following expression specifies that source *GenBank* should not be considered for entity BAC, and that only *reliable* sources (e.g. reliability higher than 7) should be considered for entity CHROMOSOME.

$$*[bg\!:\!contains/*[uri(.) = "bg\!:\!BAC"] \text{ and } uri(.) \mathrel{!=} "bg\!:\!GenBank\_BAC"] \quad (1)$$
$$//[*[bg\!:\!contains/*[uri(.) = "bg\!:\!BAC"]] \text{ and } uri(.) \mathrel{!=} "bg.GenBank\_BAC"; \quad (2)$$
$$*[uri(.) = "bg\!:\!crossRef" \text{ or } uri(.) = "bg\!:\!internal"]]$$
$$*[arc(.) \text{ and } rp(.)/bg\!:\!achieves/*[uri(.) = "bg\!:\!isOn"]] \quad (3)$$
$$//[*[bg\!:\!contains/*[uri(.) = "bg\!:\!Chromosome" \text{ and } bg\!:\!hasReliab/text() > 7]]; \quad (4)$$
$$*[uri(.) = "bg\!:\!crossRef" \text{ or } uri(.) = "bg\!:\!internal"]]$$
$$*[node(.) \text{ and } bg\!:\!contains/*[uri(.) = "bg\!:\!Chromosome" \text{ and } bg\!:\!hasReliab/text() > 7]] \quad (5)$$

This single XPR expression has been split for better legibility: lines (1), (3) and (5) are location steps and double lines (2) and (4) are // subpaths with conditions on nodes and arcs respectively on the first and second line. Lines (1) and (5) as well as // subpaths (2)(4) model (possibly empty) *entity-centered* paths, i.e., paths of sources-entities containing a given entity (first BAC, then CHROMOSOME), going through cross-references or internal links, and meeting reliability preferences. Function `rp` is used (3) to relate *entity-centered* paths by selecting only links which achieve relationship `bg:isOn`.

Lastly, the SEPT module generates paths between sources-entities as alternative ways of finding data, such as (Fig. 3):

```
bg:MapviewFish_BAC/bg:internal/MapviewFish_Chromosome
```

User queries, intermediate results and final output of BioGuide are thus expressed in a single unified modeling framework based on XPR.

5. Conclusion

We have defined a complete RDF representation of BioGuide data and introduced XPR, an RDF path language extending FSL[2]. As a general-purpose path language for RDF, we expect XPR to be of interest in many application domains[12,14,15,16]. In the context of BioGuide, XPR makes it possible to express queries involving biological entities, source-entities, kinds of links and preferences in a uniform way, and to simulate the *navigating* strategy. The language is thus used to model the queries expressed visually by scientists through the system's graphical interface (users will never have to deal directly with XPR path expressions), allowing them to save, evaluate, exchange and possibly publish on the Web BioGuide queries and query results.

We are currently studying how BioGuide queries can be compared by building XPR patterns, and how to obtain statistics about them (e.g. find out the most frequently asked source). We are also developing a module to enable the use of BioGuide on top of the well-known SRS[17] platform in order to automatically retrieve instances corresponding to elements of the paths generated by BioGuide.

References

1. Cohen-Boulakia, S., Davidson, D., Froidevaux, C.: A User-centric Framework for Accessing Biological Sources and Tools. *To appear in Proc. DILS, Data Integration for the Life Sciences*, Springer-Verlag, Lecture Notes in Computer Science (LNCS) series. (2005).

2. Pietriga E.: Fresnel Selector Language for RDF (FSL), (2005) *http://www.w3.org/2005/04/fresnel-info/fsl/*
3. Cohen-Boulakia, S., Lair, S., Stransky, N., Graziani, S., Radvanyi, F., Barillot, E., Froidevaux, C.: Selecting biomedical data sources according to user preferences, *Bioinformatics, Proc. ISMB/ECCB04*, **20**, i86-i93 (2004).
4. Shaker, R., Mork, P., Brockenbrough J.S., Donelson L., Tarczy-Hornoch P.: The BioMediator System as a Tool for Integrating Biologic Databases on the Web. *Proc. VLDB Workshop on Information Integration on the Web* (2004).
5. Lacroix, Z., Raschid, L., Vidal, M.: Efficient Techniques to Explore and Rank Paths in Life Science Data Sources, *Proc. Data Integration in the Life Sciences*, 187-202 (2004).
6. Buneman, P., Khanna, S., Tan, W.: Why and Where: A Characterization of Data Provenance, *Proc. Int. Conf. on Database Theory (ICDT)*, 316-330 (2001).
7. Zhao, J., Wroe, C., Goble, C., Stevens, R., Quan, D. and Greenwood, M.: Using Semantic Web Technologies for Representing e-Science Provenance *Proc Semantic Web Conference (ISWC)*, 92-106 (2004).
8. Klyne G., Carroll J.: Resource Description Framework (RDF): Concepts and Abstract Syntax, (2004) *http://www.w3.org/TR/rdf-concepts/*
9. Karvounarakis, G., Alexaki, S., Christophides, V., Plexousakis, D., Scholl, M: RQL: A Declarative Query Language for RDF. *Proc. World Wide Web conference (WWW)* (2002).
10. Prud'hommeaux, E., Seaborne, A.: SPARQL Query Language for RDF, W3C Working Draft, (2005) *http://www.w3.org/TR/rdf-sparql-query/*
11. Anyanwu, K., Seth A.: ρ-Queries: Enabling Querying for Semantic Associations on the Semantic Web. *Proc. World Wide Web Conf. (WWW)* (2003).
12. Bizer C., Lee R., Pietriga E.: Fresnel - A Browser-Independent Presentation Vocabulary for RDF, *Proceedings of the Second International Workshop on Interaction Design and the Semantic Web* (2005)
13. Clark J., DeRose S.: XML Path Language (XPath) Version 1.0, (1999) *http://www.w3.org/TR/xpath*
14. Angeles, R., Gutierrez, C.: Querying RDF Data from a Graph Database Perspective, *To appear in Proc. Europ. Semantic Web Conf. (ESWC)* (2005).
15. Pietriga E.: Styling RDF Graphs with GSS, XML.com, (2003) *http://www.xml.com/pub/a/2003/12/03/gss.html*
16. Haase, P., Broekstra, J., Eberhart, A. and Volz, A.: A comparison of RDF query languages. *Proc. Int. Semantic Web Conference (ISWC)* (2004).
17. Etzold, T., Ulyanov, A. and Argos, P.: SRS: information retrieval system for molecular biology data banks. *Methods Enzymol*, **266**, 114-128 (1996).

FAST, CHEAP AND OUT OF CONTROL: A ZERO CURATION MODEL FOR ONTOLOGY DEVELOPMENT

BENJAMIN M. GOOD, ERIN M. TRANFIELD, POH C. TAN,MARLENE SHEHATA*, GURPREET K. SINGHERA, JOHN GOSSELINK, ELENA B. OKON, MARK D. WILKINSON

*James Hogg iCAPTURE Centre for Cardiovascular and Pulmonary Research, St. Paul's Hospital, University of British Columbia, Vancouver, British Columbia V6Z1Y6, Canada. *University of Ottawa Heart Institute H355, Ottawa, K1Y-4W7, Canada*

During two days at a conference focused on circulatory and respiratory health, 68 volunteers untrained in knowledge engineering participated in an experimental knowledge capture exercise. These volunteers created a shared vocabulary of 661 terms, linking these terms to each other and to a pre-existing upper ontology by adding 245 hyponym relationships and 340 synonym relationships. While ontology-building has proved to be an expensive and labor-intensive process using most existing methodologies, the rudimentary ontology constructed in this study was composed in only two days at a cost of only 3 t-shirts, 4 coffee mugs, and one chocolate moose. The protocol used to create and evaluate this ontology involved a targeted, web-based interface. The design and implementation of this protocol is discussed along with quantitative and qualitative assessments of the constructed ontology.

1. Introduction

Ontologies provide the mechanism through which the "semantic web" promises to enable dramatic improvements in the management and analysis of all forms of data [1]. Already, the importance of these resources to the bio/medical sciences is made clear by the more than 1000 citations[a] of the original paper describing the Gene Ontology (GO) [2]. Because of the broad range of skills and knowledge required to create an ontology, they are generally slow and expensive to build. To illustrate, the cost of developing the GO has been estimated at upwards of $16M (Lewis, S, personal communication). This bottleneck not only slows the initial development of such systems but also makes them difficult to keep up to date as new knowledge comes available.

Conversely, projects such as DMOZ (http://dmoz.org) and BioMOBY[3][4] take a more open approach. Rather than paying curators, DMOZ lets "net citizens" build hierarchies (now utilized by Google among many others) that organize the content of the World Wide Web. BioMOBY, a web services-based interoperability framework, depends on an ontology of biological data objects

---

[a] 1064 Google Scholar citations (http://scholar.google.com) on Sept. 8, 2005

that can be extended by anyone. The successful, open, and ongoing construction of the DMOZ directories and the BioMOBY ontology hints that the power of large communities can be harnessed as a feasible alternative to centralized ontology design and curation.

We describe here a protocol meant to overcome the knowledge-acquisition bottleneck to rapidly and cheaply produce a useful ontology in the bio/medical domain. The key features of the approach are 1) the use of a web-accessible interface to facilitate collaborative ontology development and 2) the deployment of this interface at a targeted scientific conference. This paper describes the protocol and presents the results of a preliminary evaluation conducted at the 2005 Forum for Young Investigators in Circulatory and Respiratory Health (YI forum) (http://www.yiforum.ca/).

1.1. *Experimental context and target application for the YI Ontology*

The YI forum did not (outside of this study) include any research on knowledge capture or artificial intelligence. The topics covered spanned aspects of circulatory and respiratory health ranging from molecular to population-based studies, and analysis of quality of health-service provision. Attendees included molecular biologists, health service administrators, statisticians, cardio/pulmonary surgeons, and clinicians. The target task for the YI Ontology was to provide a coherent framework within which to organize the abstracts submitted to this broadly-based yet specialized conference. This framework would take the form of a simple subsumption hierarchy composed of terms associated with individual abstracts, and/or added by individual experts during the construction process. Such an ontology could be used to facilitate searches over the set of abstracts by providing legitimate, semantically-based groupings.

1.2. *Motivation and novelty of conference-based knowledge capture*

Research in natural language processing and machine learning is yielding significant progress in the automatic extraction of knowledge from unstructured documents and databases [5][6]; however these technologies remain highly error-prone and, to our knowledge, no widely used public ontology in the life sciences has ever been built without explicit, extensive expert curation. Thus, given the costs of curation, it would be preferable to identify methodologies that facilitate extraction of machine-usable knowledge directly from those who possess it. In order to achieve this, several preliminary steps seem necessary:

1. Domain experts need to be identified
2. These experts need to be convinced to share their knowledge.

3. These experts must then be presented with an interface capable of capturing their specific knowledge.

Scientific conferences seem to provide a situation uniquely suited to inexpensive, rapid, specialized knowledge capture because the first two of these requirements are already met by virtue of the setting; experts are identified based on their attendance and, at least in principle, they attend with the intention of sharing knowledge. Clearly, the principle challenge lies in generation of an interface that facilitates extremely rapid knowledge acquisition from expert volunteers.

2. Interface Design

The architecture chosen for this project borrows techniques from a new class of knowledge acquisition systems that attempt to harness the power of the Internet to rapidly create large knowledge bases. Projects in this domain are premised on the assumption that, by distributing the burden of knowledge representation over a large number of people simultaneously, the knowledge acquisition bottleneck can be avoided [7][8][9][10]. Two active projects in this domain are Open Mind Common Sense[10], and Learner2[11][12] . Both of these efforts focus on gathering "common sense" knowledge from the general public with the aim of producing knowledge-based systems with human-like capabilities in domains such as natural language understanding and machine translation.

These large, open, Internet-based projects are premised on the idea that there is little or no opportunity for explicit training of volunteers, and in principle no *strong* motivation to participate. This is similarly true of the conference participants engaged in this study, and thus based on these similarities, the interface developed for this knowledge capture experiment was modeled after the template-based interface of the Learner2 knowledge acquisition platform (http://learner.isi.edu).

Learner2 follows two basic design patterns:

1. Establish a system that allows the knowledge engineer to passively control knowledge base *structure*, while allowing its *content* to be determined entirely by the subject matter experts.
2. Use a web-enabled, template-based interface that allows all volunteers to contribute to the same knowledge base simultaneously and synergistically in real-time.

The "iCAPTURer" knowledge acquisition system presented here applies and adapts these principles to the task of knowledge capture in the conference setting.

2.1. *Specific challenges faced in the conference domain*

The iCAPTURer experiment faced unique challenges by virtue of its expert target-audience. Learner2 is designed to capture "common sense" knowledge, and operates by generating generic, user-agnostic fill-in-the-blank templates. For example, in order to collect statements about objects and their typical uses, a volunteer might be presented with "A [blank] *is typically used* to smash something" and asked to fill in the blank. In order to capture specific, expert knowledge however, it is necessary to adapt the contents of these templates to target each volunteer's specific domain of expertise. The following section details our adaptation of the Learner2 approach to meet this challenge.

3. Methods - Introducing the iCAPTURer

3.1. *Preprocessing*

Prior to the conference, terms and phrases were automatically extracted from each abstract using the TermExtractor tool from the TextToOnto ontology engineering workbench [5]. The TermExtractor was tuned to select multi-word terms using the "C-value" method [13]. This process produced a corpus of terms and phrases linked directly to the abstracts. This corpus provided the first raw material for the construction of the ontology and provided a mechanism to match the contents of the templates to the volunteer's area of expertise.

In addition, the nascent ontology was seeded with a concept hierarchy taken from the Unified Medical Language System Semantic Network (UMLSsn; http://www.nlm.nih.gov/research/umls/). The UMLSsn was selected as the "upper ontology" in order to provide a common semantic framework within which to anchor the knowledge capture process [14].

3.2. *Priming the knowledge acquisition templates - term selection*

Two priming models were employed to ensure that relevant knowledge was captured and that expert volunteers were presented with templates primed with concepts familiar to them. After logging into the system, the volunteer first makes a choice between priming the system with a keyword entered as free text, or priming the system through selection of a specific abstract (preferably their own).

In the abstract-driven model, the term to be evaluated is randomly selected from the pre-processed auto-extracted terms associated with the selected abstract. In this way, the expert is preferentially asked about terms from an abstract that they are presumptively familiar with, though there is nothing stopping them from selecting abstracts at random.

In the keyword-driven model, the system first checks the knowledge base for partial matches to the keyed-in term, and if found, selects one at random. If no matches are found the term is added to the knowledge base and is considered meaningful.

3.3. *Term evaluation*

After the volunteer chooses an abstract or enters a keyword, they are presented with the term-evaluation page. This page presents them with a term and requests them to decide if it is "meaningful", "not-meaningful", or if they do not understand it ("X is a meaningful term or phrase {True, False, I don't know}"). If they are unable to make a judgment on the term, another term is presented and the process repeats. If they indicate that the term is not valid, then the term's "truth value" is decremented in the knowledge base and another term is presented for judgment. Only terms above a set truth value are presented. This allows for rapid pruning of invalid entries from the active knowledge base without any permanent corpus loss. Approximately 50% of the terms extracted using text mining were judged nonsensical, hence this pruning was a critical step in the development of the ontology. If a term is rated as "meaningful", its truth value is raised and the term is considered selected.

3.4. *Relation acquisition*

Once a valid concept is selected, the system directs the volunteer to attach relations to the concept that will determine its position in the ontology. Two types of relation were targeted in this study, synonymy (same as) and hyponymy (is a).

To capture synonyms, a simple fill-in-the-blank template was presented. For example, if the term "muscle" was selected as valid, the volunteer would then be invited to enter synonyms through a template like: *The term or phrase [blank] means the same thing as "muscle".*

A different format was used for capturing the hyponym relation. The hyponym template asks the volunteer to select a parent-term from a pre-existing hierarchical vocabulary (initially seeded with the UMLSsn) rather than letting them type one in freely. This approach was selected with the goal of producing a sensible taxonomic structure. During the knowledge capture process, terms added to this hierarchy became new classes that future terms could be classified under, thus allowing the ontology to grow in depth and complexity.

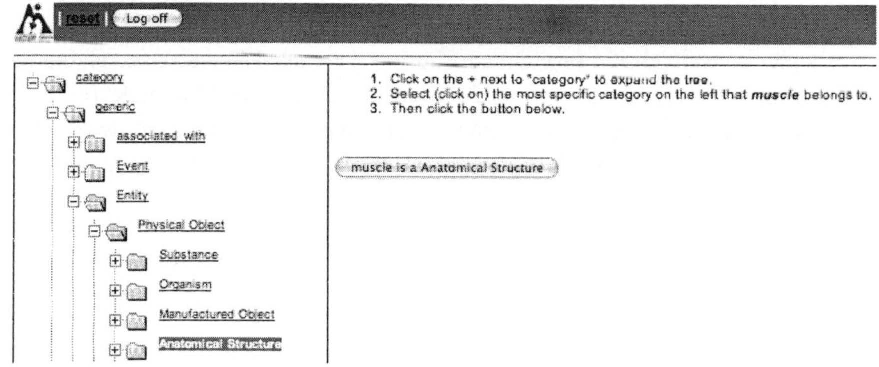

Figure 1: Hyponym collection. "Muscle" is being placed as a child of "Anatomical Structure".

As each task is completed, the volunteer is returned to a task selection screen and the completed task's button is removed. When each of the tasks are completed for the select term, another term is selected and the process repeats.

3.5. *Volunteer recruitment and reward*

To assist in volunteer-recruitment, conference attendees were motivated by a 5 minute introductory speech at the welcome reception, by flyers included in the conference handouts, and by the promise of mystery prizes for the most prolific contributors. Points were awarded to the user for each piece of knowledge added to the system. A simple user management system allowed the users to create accounts, log out, and log back in again while keeping track of their cumulative score throughout all sessions. Anonymous logins were also allowed.

4. Observations

In this preliminary study, qualitative observation of volunteer response to the system was a primary objective. As such, the enthusiastic response the project received from the organizers and the participants in the conference was encouraging, and the willingness of the volunteers to spend significant amounts of time entering their knowledge was unanticipated. From conversations with the participants, it became clear that the competitive aspect of the methodology was often their primary motivation, and this was especially true for the most prolific contributors who indicated a clear "determination to win". Some volunteers also indicated a simple enjoyment in playing this "intellectual game".

Another important observation was that the tree-based interface used to capture the hyponym relation (see figure 1) was not readily understood by the majority of participants. This interface required the user to understand relatively arbitrary symbols and to click multiple times in order to find the correct parent for the term under consideration. In contrast, the interface used in the later qualitative evaluation (discussed in section 6) required just a single click for each evaluation, resulting in no confusion or negative comments and *more than 11,000 collected assertions in just three days* from a similar number and composition of volunteers.

5. Quantitative Results

5.1. *Volunteer contributions*

During the 2 active days of the conference, 68 participants out of approximately 500 attendees contributed to the YI Ontology. Predominantly, volunteers contributed their knowledge during breaks between talks and during poster sessions at a booth with computer terminals set up for the purpose; however several participated from Internet connections in their hotel rooms. The quantity of contributions from the different participants was highly non-uniform, with a single volunteer contributing 12% of the total knowledge added to the system.

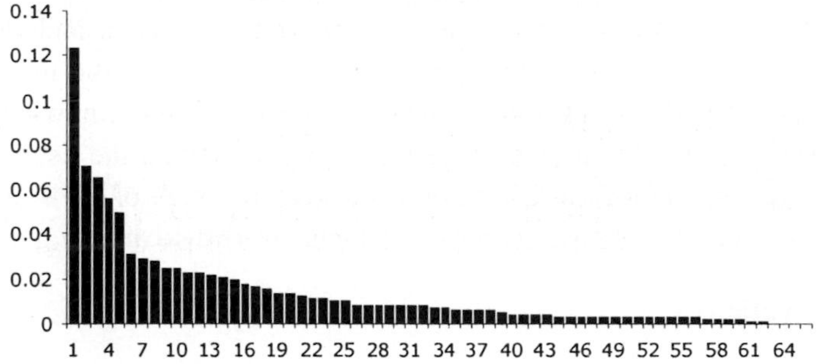

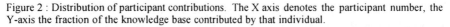

Figure 2 : Distribution of participant contributions. The X axis denotes the participant number, the Y-axis the fraction of the knowledge base contributed by that individual.

5.2. *Composition of the YI Ontology*

The pre-processing text mining step yielded 6371 distinct terms associated with the 213 abstracts processed. These auto-extracted terms were not added to the ontology until they had been judged meaningful by one of the volunteers via the term-evaluation template. 464 auto-extracted terms were evaluated by the conference volunteers. Of these, 232 were judged meaningful and 232 were judged not meaningful. In addition, the 429 terms entered directly by volunteers (in the keyword initialization) were all considered to be meaningful. Thus in total the potential corpus for the ontology consisted of 661 validated terms.

Table 1. Captured Terms

	text-extracted	judged meaningful	judged not meaningful	Added directly	Total meaningful
Count	6371	232	232	429	661

5.3. *Relationships in the YI Ontology*

Of the 661 concepts, 207 were assigned parents in the UMLSsn rooted taxonomy. Of these, 131 concepts came from the auto-extracted set and 76 came from the directly entered set. As terms could be linked to different parents, 38 additional parental relationships were assigned to terms within this set, bringing the total number of hyponym relations assigned up to 245. 219 of the accepted terms were associated with at least one synonym, with many linked to multiple synonyms.

Table 2. Hyponyms

Total number of categories (including the UMLSsn)	469
Total categories added -at the YI forum	207
Added categories created from auto-extracted terms	131
Added categories created from terms added as keywords	76

Table 3. Synonyms

Total distinct targets (number of distinct synonyms entered)	340
Total distinct sources (number of terms annotated with a synonym)	219
Sources from auto-extracted terms	153
Sources from terms added as keywords	66

6. Quality Assessment

The evaluation of the YI Ontology was conducted in similar fashion to the initial knowledge capture experiment. Following the conference, the 68 participants in the conference study and approximately 250 researchers at the James Hogg, iCAPTURE Centre for Cardiovascular and Pulmonary Research were sent an email requesting their participation in the evaluation of the YI ontology. The email invited them to log on to a website and answer some questions in exchange for possible prizes. 65 people responded to the request. Upon logging into the website, the evaluators were presented with templates that presented a term, a hyponym relation, or a synonym relation from the YI

Ontology. They were then asked to make a judgment about the accuracy of the term or relation. For synonyms and hyponyms, they were asked to state whether the relationship was a "universal truth", "true sometimes", "nonsense", or "outside their expertise". For terms, they were asked whether the term was a "sensible concept", "nonsense", or "outside their expertise". After making their selection, another term or relation from the YI ontology that they had not already evaluated was presented and the process repeated.

Again, participants were provided motivation through a contest based on the total number of evaluations that they made (regardless of what the votes were and including equal points for indicating "I don't know"). Participation in the evaluation was excellent, with 5 responders evaluating every term and every relation in the ontology. During the three days of the evaluation, 11,545 votes were received, with 6060 on the terms, 2208 on the hyponyms, and 3277 on the synonyms. 93% of the terms, 54% of the synonyms and 49% of the hyponyms enjoyed more positive than negative votes overall.

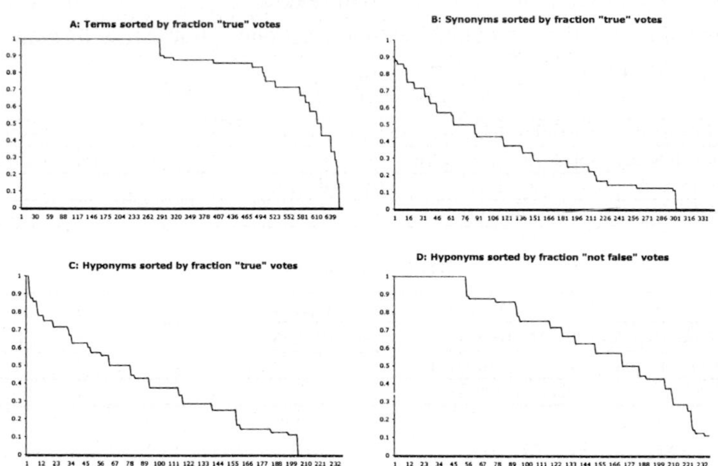

Figure 3 : The positive consensus agreement for captured terms (A), synonyms (B), and hyponyms (C). For A, B and C, the y-axis indicates the fraction of the votes for "universal truth". This value is used to sort the assertions indicated on the X-axis. The y-axis on D indicates the level of positive consensus for the hyponyms if the "true sometimes" votes are counted with the "universal truth" votes indicating a "not-false" category.

Figures 3a, 3b, and 3c display plots of the fraction of "true" votes received for each term, synonym and hyponym in the ontology. These curves illustrate strong positive consensus for the large majority of captured terms, but considerable disagreement regarding the quality of the captured synonyms and hyponyms. To some extent this may have been caused by the exclusion of the "sometimes" category from the term evaluations, but even when the "sometimes" votes are merged with the "true" votes, there are still considerably fewer positive votes for the hyponyms and synonyms and less agreement among the voters. This is illustrated for the hyponyms in Figure 3d.

Table 4. Examples of assertions and associated votes

Assertion	% positive	% sometimes	% negative
Term: "*wild type*"	100	NA	0
Term: "*epinephrine e*"	50	NA	50
Term: "*blablala*"	0	NA	100
Hyponym: "*asthma* is a *disease*"	100	0	0
Hyponym: "*factor xiia* is a *coagulation factor*"	50	50	0
Hyponym: "*stem cells* are a kind of *transmission electron microscopy*"	0	11	89
Synonym: "*positive arrhythmia* is the same as *abnormal pacing of the heart*"	89	11	0
Synonym: "*lps treatment* is the same as *lipopolysaccaharide treatment*"	50	37.5	12.5
Synonym: "*Cd34* is the same as *aneurysm*"	0	14	86

Table 4 gives some examples of the contents of the YI ontology. These examples illustrate that the voting process successfully identified high quality components that should be kept, low quality components that should be discarded, and questionable components in need of refinement. These assessments could be used to improve the overall quality of the ontology through immediate pruning of the obviously erroneous components and by guiding future knowledge capture sessions meant to clarify those components lacking a strong positive or negative consensus.

Summary

Between April 29th and April 30th 2005, 661 terms, 207 hyponym relations, and 340 synonym relations were collected from 68 volunteers at the CIHR National Research Forum for Young Investigators in Circulatory and Respiratory Health. In a subsequent community evaluation, 93% of the terms, 54% of the synonyms and 49% of the hyponyms enjoyed more positive than negative votes overall. The rudimentary ontology constructed from these terms and relationships was composed at a cost of the 4 t-shirts, 3 coffee mugs, and one chocolate moose that were awarded as prizes to thank the volunteers.

Discussion

This work addresses the key bottleneck in the construction of semantic web resources for the life sciences. Ontology construction to date has proven to be extremely, possibly impractically, expensive given the wide number of expert knowledge domains that must be captured in detail. Thus, it is critical that a rapid, accurate, inexpensive, facile, and enjoyable approach to knowledge capture be created and ubiquitously deployed within the life science research community. To achieve this, a paradigm shift in knowledge capture methodologies is required. The open, parallel, decentralized, synergistic protocol presented in this study represents a significant deviation from the centralized, highly curatorial model employed in the development of all of the major bio/medical ontologies produced to date.

The positive consequences of this approach are that 1) knowledge can be captured directly from domain experts with no additional training, 2) a far larger number and diversity of experts can be recruited than would ever be feasible in a centralized effort and 3) because the approach involves no paid curators, the overall cost of ontology development is very low.

The negative aspect of the approach is that the knowledge collected is "dirty", requiring subsequent cleaning to achieve high quality. Future versions of the iCAPTURer software will attempt to improve on the quality of the captured knowledge by integrating the evaluation phase directly with the knowledge capture phase. In this "active learning" approach, the questions will be tuned on-the-fly to direct knowledge capture efforts to areas of uncertainty or contention within the developing ontology and to quickly weed out assertions that are clearly false. The present study describes just one step of such a multi-step process, with obvious opportunities for immediate improvement in the next iteration based on the knowledge gathered during the evaluation.

In comparison to existing methodologies, which tend to separate the biologists from the ontologists, the iCAPTURer approach demonstrates dramatic improvements in terms of cost and speed. If future work confirms that this approach can also produce high quality ontologies, the emergence of a global semantic web for the life sciences may occur much sooner than expected.

Acknowledgments

Funding provided by Genome Canada, Genome British Columbia, Genome Prairie, the Canadian Institute for Health Research, and the Michael Smith Foundation. Thanks to Yolanda Gil and in particular to Timothy Chklovzki for important contributions during the design and conception of the iCAPTURer. Thanks also to the organizers of the YI Forum, in particular Ivan Berkowitz and Bruce McManus. We would also like to thank all of the volunteer knowledge engineers without whom this work would simply not be possible.

References

1. T. Berners-Lee, J. Hendler, and O. Lassila. "The Semantic Web". *Scientific American*, **284**:5 pp 34-43, May (2001)
2. M. Ashburner *et al*, "Gene Ontology: Tool for the Unification of Biology". *Nature Genetics.* **25**:1 pp 25–29 (2000)
3. M.D.Wilkinson, M. Links, "BioMOBY: an open-source biological web services proposal". *Briefings In Bioinformatics* 3:4. pp 331-344 (2002)
4. M.D. Wilkinson, H. Schoof, R. Ernst, D. Haase. "BioMOBY successfully integrates distributed heterogeneous bioinformatics web services. The PlaNet exemplar case", *Plant Physol* **138**, pp 1-13 (2005)
5. A. Maedche, S. Staab, "Ontology learning". In S. Staab and R. Studer, editors, *Handbook on Ontologies*, pp 173-189 (2004)
6. P. Cimiano, A. Hotho, S. Staab, "Clustering Concept Hierarchies from Text", in *Proceedings of 4th International Conference on Language Resources and Evaluation* (2004)
7. T. Chklovski, "Using Analogy to Acquire Commonsense Knowledge from Human Contributors", PhD. thesis, MIT Artificial Intelligence Laboratory technical report AITR-2003-002 (2003)
8. T. Chklovski. "LEARNER: A System for Acquiring Commonsense Knowledge by Analogy", in *Proceedings of Second International Conference on Knowledge Capture.* (2003)
9. M. Richardson, P. Domingos "Building Large Knowledge Bases by Mass Collaboration", In *Proceedings of the International Conference on Knowledge Capture* (2003)
10. P. Singh, T. Lin, E.T. Mueller, G. Lim, T. Perkins, W. L. Zhu, "Open Mind Common Sense: Knowledge Acquisition from the General Public", *Lecture Notes in Computer Science,* **2519**, pp 1223–1237 (2002)
11. T. Chklovski, "Designing Interfaces for Guided Collection of Knowledge about Everyday Objects from Volunteers", in *Proceedings of 2005 Conference on Intelligent User Interfaces* (2005)
12. T. Chklovski, Y. Gil, "Towards Managing Knowledge Collection from Volunteer Contributors", in *Proceedings of 2005 AAAI Spring Symposium on Knowledge Collection from Volunteer Contributors* (2005)
13. K.T. Frantzi, S. Ananiadou, J. Tsujii, "The c-value/nc-value method of automatic recognition for multi-word terms", *Lecture Notes in Computer Science,* **1513**, pp 585-600. (1998)
14. Niles, A. Pease, "Towards a Standard Upper Ontology", in *Proceedings of the international conference on Formal Ontology in Information Systems* (2001)

PUTTING SEMANTICS INTO THE SEMANTIC WEB: HOW WELL CAN IT CAPTURE BIOLOGY?

TONI KAZIC

Dept. of Computer Science
University of Missouri
Columbia, MO 65201
toni@athe.rnet.missouri.edu

Could the Semantic Web work for computations of biological interest in the way it's intended to work for movie reviews and commercial transactions? It would be wonderful if it could, so it's worth looking to see if its infrastructure is adequate to the job. The technologies of the Semantic Web make several crucial assumptions. I examine those assumptions; argue that they create significant problems; and suggest some alternative ways of achieving the Semantic Web's goals for biology.

1. Introduction

Imagine you are interested in purine salvage. You go to KEGG's maps [1] and see that the reactions

$$\text{deoxyadenosine} \rightleftharpoons \text{adenine} \tag{1}$$

$$\text{deoxyinosine} \rightleftharpoons \text{hypoxanthine} \tag{2}$$

are both catalyzed by EC 2.4.2.4. When you click on the link for that EC number, you discover the name of the enzyme is thymidine phosphorylase, and its reaction is

$$\text{thymidine} + \text{phosphate} \rightleftharpoons \text{thymine} + \text{2-deoxy-}\alpha\text{-D-ribose 1-phosphate} \tag{3}$$

Thymidine isn't a purine nucleoside, so how can it catalyze the cleavage of deoxyadenosine and deoxyinosine? Maybe KEGG's chart is mixed up, so you go to to the web site of the Joint Committee on Biochemical Nomenclature, which is the international body in charge of classifying enzymes [2]. Same result — EC 2.4.2.4 is thymidine phosphorylase — but now you notice a comment saying the enzyme can catalyze reactions like those of EC

2.4.2.6, nucleoside deoxyribosyltransferase,

$$\text{2-deoxy-D-ribosyl-base}_1 + \text{base}_2 \rightleftharpoons \text{2-deoxy-D-ribosyl-base}_2 + \text{base}_1 \quad (4)$$

under some circumstances. That's a fundamentally different reaction than that shown for EC 2.4.2.4; one cannot logically substitute a nucleotide base for a phosphate or the ribosyl moiety of nucleosides. If one rewrote the KEGG reactions to fit that of nucleoside deoxyribosyltransferase, *e. g.*,

$$\text{deoxyadenosine} + \text{thymine} \rightleftharpoons \text{thymidine} + \text{adenine}, \quad (5)$$

the result is still not the reaction shown either for EC 2.4.2.4 or on KEGG's map. Adding to the confusion, you notice that two synonyms for the enzyme's name are "blood platelet-derived endothelial cell growth factor" and "gliostatins". Statins stop things and growth factors stimulate growth; endothelial cells are not the same as glial cells; so how can the same enzyme stimulate the growth of one cell and inhibit the growth of another? And why would an enzyme of nucleotide salvage (you're still not sure it's the right enzyme) be involved in cell growth anyway?

Your ability to check, understand, and reconcile apparently contradictory information depends on understanding the semantics of terms such as thymidine, purine, statin, growth factor, glia, and endothelial cell; questioning the apparent contradictions; and synthesizing information from multiple sources. The hope for the Semantic Web is that it would do just these things automatically, accurately, and transparently on the Internet. Simple questions, usually the hardest, would be simply, rapidly, and at least plausibly answered. The net would become a connected engine of knowledge and inference [3].

This powerful and alluring vision has stimulated great excitement. But the utility of the Semantic Web to address biological questions depends as much on the adequacy of its infrastructure as it does on the passion of its advocates. Computations that return biologically incorrect or misleading answers are not helpful, especially if automation returns more "results". To ensure the *scientific* validity of the Semantic Web's computations, it must sufficiently capture and use the semantics of the domain's data and computations: for example, it mustn't confuse reactions of enzymes with reactions to drugs (unpleasant side-effects). Accurate semantics are even more important if the goal of the computation is to return plausible answers, since plausibility depends on persuading someone that the implicit relationships among rather disparate facts are strong enough to form a reasonable hypothesis (*e. g.*, reactions 1 and 3 are the same). Since the Semantic Web's fundamental technologies will be the foundation for any domain-specific extensions, it's important to ask how well their structure fits the semantics of biology [4].

One can think of the Semantic Web and the scientific databases and algorithms it would call as a collection of languages that denote information. To translate among them, their semantics must be adequately specified. Two of the most important requirements for a system to scale are that its operations are automatic and that its methodologies are distributed. Here I examine several of the fundamental assumptions of the Semantic Web's infrastructure to estimate their limits *in the context of computations of biological interest.* With respect to semantics, I consider the structure of RDF and its denotational semantics; and for scalability, the topology and automation of semantic translation. While the only solutions I can suggest are as partial in their ways as the current approaches (see Section 6), I hope to stimulate broader consideration of the technical foundations of the Semantic Web's application to scientific computing. Thus, this paper is in the spirit of BioMOBY's goals of figuring out what's needed for the scalable big picture [5].

2. What are the Assumptions?

- ***A simple syntax is sufficient.*** The non-logical relationships among concepts are ultimately captured in the RDF specification of the terms, usually called an RDF Schema. The assumption is that the <subject> <predicate> <object> syntax is sufficient [6]; I argue in Section 3 that it is not.
- ***An implicit semantics is effective.*** The semantics of a term are given in a natural language comment and in the applications that use the term. Neither of these forms can be computationally inspected to determine the semantics of the term; the Semantic Web relies on humans reading the comments and code to determine the semantics. I argue in Section 4 that this implicit semantics is less effective than an explicit, by which I mean computationally determinable, semantics.
- ***Bilateral mappings, manual translation, and automated inference are just right.*** Because the tasks of definition and implementation are distributed to the community, and DAML+OIL-powered inference engines would translate among different definitions using manually constructed bilateral mappings, the claim is

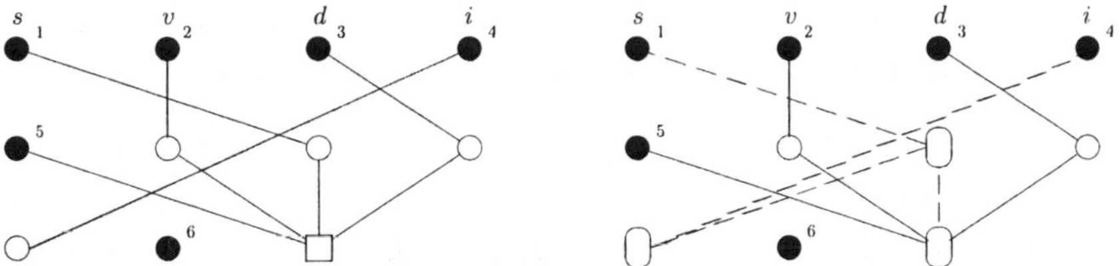

Figure 1. Structural inadequacy of RDF's syntax. The sentences are read from left to right. Each column is a part of speech (labelled s, subject; v, verb; d, direct object; i, indirect object). (left) Successive transformations of a complex sentence to fit RDF's syntax by nesting the subject, direct object, and indirect object of the original sentence together. Introduction of terms implied in the original sentence enables the nesting. Unique terms are denoted by numbered closed circles; open circles denote terms unchanged by syntactic transformation; the square denotes the nested, complex object in the final transformation. Solid lines indicate a term's syntactic transformations and the direction of transformation is from the top down. (right) Differences in term semantics can be concealed by the syntax. The enzyme denoted by term 4 has bound the substrates denoted by term 1, but the syntactic transformations needed to fit the original sentence to RDF give no clues about this change in the semantics of term 4 to another application. The ovals denote terms whose properties are not uniformly inherited. The dashed lines indicate the inheritance relationships and inheritance is from the top down. The terms are: 1, "thymidine and phosphate"; 2, "*convert*"; 3, "*thymine and 2-deoxy-α-D-ribose 1-phosphate"; 4, "*thymidine phosphorylase"; 5, "*reaction"; 6, "catalyzes".

that the Semantic Web will scale. I argue in Section 5 that scalability and precision will be very poor.

3. Is the Structure Sufficient?

RDF captures information in <subject> <predicate> <object> phrases. Thus, "<A> <is an element of> <B>" and "<A> <parses> <B>" fit into this syntax. When the syntax one would naturally use is more complex — say, "thymidine and phosphate are converted into thymine and 2-deoxy-α-D-ribose 1-phosphate by the enzyme thymidine phosphorylase" — one approach is to group the terms through nesting, then use that in a second phrase. This is illustrated in the left panel of Figure 1. Thus, thymidine, phosphate, thymine, and 2-deoxy-α-D-ribose 1-phosphate are ultimately compressed into a phrase that serves as a direct object: "<thymidine phosphorylase> <catalyzes> <*reaction *convert* thymidine and phosphate *thymine and 2-deoxy-α-D-ribose 1-phosphate>" (*s are the omitted prepositions, articles, and auxiliary verbs).

However, the reaction equations and our model sentence have a much richer set of connotations, and these connotations don't readily fit into the syntax. For example, enzymes bind their substrates before catalyzing the reaction, and this binding partitions the populations of molecules into rapidly exchanging, random subpopulations of enzyme-substrate complexes. This information is expressible in natural language. But when compressed into RDF's syntax, the phrasal transformation has separated the interacting molecules (thymidine phosphorylase, thymidine, phosphate), concealing that the semantics of "thymidine phosphorylase" *in the reaction* are different from the semantics of the unbound enzyme (Figure 1, right panel). So another application may be able to unify its "thymidine phosphorylase" with this one, but the semantics of the two instances can differ and the rest of the phrase will not necessarily provide any clues as to the difference. Building a tree of phrases to emulate binding (*e. g.*, "thymidine phosphorylase binds thymidine or phosphate", "the complex of thymidine phosphorylase and thymidine or phosphate binds phosphate or thymidine", *etc.*) forces one to say explicitly something one may not know (*e. g.*, whether the binding is random or sequential, what the order of any sequential binding is, how many substrates are bound per enzyme). By expanding the detail to accommodate the phrasal structure, essential and useful ambiguities have been lost; the task of deciding where to unify an instance of "thymidine phosphorylase" from another application has been complicated; and the ease of description has vanished.

4. Are an Implicit Semantics Effective?

Like other languages, the semantics of the Semantic Web depend on those of its grammar and terms, their "context", and the applications that use them [7]. The terms are named either directly in an RDF document or in a referenced ontology. By "context" of a term, the Semantic Web workers generally mean the URI at which the term is found; if the same term is used in different URIs, its semantics are assumed to be different unless explicitly stated otherwise. But the more common meaning of context is especially important to the Semantic Web, because the *way* a term is used helps bound its semantics. In the Semantic Web, this usage is intended to be by programs. The only way a program can "know" that it is doing something biologically meaningful to the data retrieved by a term from another application is if it can check that the application's definition of the term and its definition of the term are identical and unambiguous.

Terms used in an RDF *may* be defined, in English, in the comments (more often these are partial descriptions of the denoted concepts rather than definitions) [4, 8, 9]. In this situation, the semantics of the language are implicit in three important senses. First, any definitions are in natural language, which remains notoriously difficult for machines to understand. Second, much of the semantics of a construct are carried in the applications — easy for machines to process, but less accessible to humans trying to understand exactly what a program means by a "gene" or "reaction". Third, some of the semantics are intensional, relying on automated reasoning systems such as DAML+OIL, the logical features of OWL, or the predicate calculus to draw (so far relatively simple) inferences.

An implicit, non-machine computable, semantics raises three problems. The first is that people must do the job of reading, understanding, and reifying the connotations of a term or a program before they can implement any resource (*e. g.*, see references 3, 8). How many people are really willing or able to do this, especially when the domain is as specialized as biology? Reactions 3 and 4 are stated to be related, but it requires much more semantics than just knowing that thymidine is a member of the set of 2-deoxy-D-ribosyl-bases to determine how to map these, and there are probably more biologists and chemists who know that information than developers. The second problem is the fact that humans interpret constructs differently; we don't all know what the words mean because we vary so much in the way we use natural languages (for some biological examples, see reference 10; broader cultural examples are found in references 11, 12). Terms, whether natural language words or RDF properties, merely point to a variable set of connotations [13]. Are reactions 1 and 3 the "same"? The answer depends on whether you think KEGG's map has omitted other reactants and your willingness to believe thymidine phosphorylase is broadly specific.

The third problem is that most of the semantics are pushed onto the applications. While reading other people's code can be frustrating, the fundamental problem is that the semantics are far less transparent than they could be. Suppose one wants to test whether thymidine phosphorylase could catalyze reaction 1, and there's a resource that retrieves "related" reactions. One's interpretation of the output will be determined by how that resource defines "related", which will be determined by its code and any underlying databases. Just seeing the result won't divulge the resource's notion of "related", and returning the expected result doesn't test the hypothesis of relationship unless one knows how the test was made. Lack of semantic transparency limits reuse, since each developer must inspect the code of possible components for him or herself.

5. Will It Scale?

Scaling the current model of the Semantic Web requires enough pairs of resources that translate between them, and robust enough inference engines, so that local subnets can be automatically connected. The assumption is that enough translating pairs would spontaneously arise so that a suitably equipped inference engine could compute a directed path between any two resources. The path is directed because one can assume only that the translations will be asymmetric; symmetric systems would be multigraphs, and very welcome too.

Obviously this scaling by $n(n-1)$ is more labor-intensive than if applications refer to a common semantic middle layer, and there have been enough pleas to keep URIs stable and to reuse ontologies to warn the naïve that the machinery could break for trivial reasons. Similarly, people reading and reconciling the system's semantics from ontologies and code is slow. A more fundamental problem is that automatically drawn inferences can explode all too easily. The "signal" – useful or novel inferences — is often lost in the "noise" of the huge collection of trivial inferences (for example, see reference 14). Why wouldn't this happen in the Semantic Web? The usual answer is that declarations that site A "trusts" only site B for certain kinds of information will sufficiently bound the inferences. Perhaps; but then for the Semantic Web to be a web, site C must decide to trust A, rather than site D; and once site E trusts B and D, forming a web, E will have to decide what to do with contradictions, incomplete information, and semantic inconsistencies. (Here I push the Semantic Web beyond its stated plan of not worrying about the accuracy of inferences [3], because automating incorrect scientific inference, *e. g.*, thymidine phosphorylase is not an enzyme of purine salvage because thymidine is not a purine nucleoside, is not a step forward.) As the topology of the Semantic Web changes from disconnected small graphs to larger connected components, the number of paths among the possibly relevant URIs in a component will also increase explosively [15]. One might even run into resource issues, in the sense of available cycles to compute all those inferences and paths.

6. Towards Solutions

This list is not exhaustive, and none of these suggestions will solve all aspects of the problems raised. Indeed, some could interfere with the results of others. But each offers at least one alternative to the problems of the present infrastructure.

Enrich the Allowed Syntax If RDF's syntax isn't robust enough, why not enrich it? Adding more parts of speech, such as adjectives and indirect objects, and permitting its trees to become networks, might well solve the problems of Section 3. Rather than efforts to map more complex relations onto RDF, why not let RDF accommodate these constructs directly, for example by semantic networks [16, 17]? One might begin by watching biologists diagram and explain the relationships among concepts in research papers, which will often be among sentences.

Let Biologists Build Since biologists know the semantics — even if they disagree — one way to develop applications is to let them define the semantics *in a structured way.* Exploiting biological expertise is the fundamental power of the UMLS, the GO, nomenclature committees, and similar efforts, and spreading the effort is central to the Semantic Web's philosophy [17–19]. One possible advantage of a richer syntax is that it might enable faster definition by biologists. But to prevent cacophony, we must either all agree on the semantics of the terms or find a way to translate among them. Agreement is slow, socially difficult, and scientifically inflexible; translation is hard.

Make the Semantics Finely Grained One reason agreement is hard is because we tend to focus on relatively "big" ideas rather than on their component notions. As a crude example, rather than arguing over whether reactions 1 and 4 are similar, one could define different types of similarities, and then allow each person or application to mix and match those types to suit their needs. (The example is crude because in practice there are many much more finely grained notions underneath, such as the tautomerism of the bases.) In an ideal world, the most finely grained ideas would be so axiomatic as to be uncontroversial; and for mixing and matching to be unambiguous, the semantics of the axioms and the symbols denoting them would have to be unique. Two problems would likely arise: ensuring that the axioms, their denotations, and semantics were unique; and deciding how to scope the axioms so that they were truly elementary for the scientific domain in question. For example, whether one maps a molecule's name to SMILES string or a Hamilitonian depends strongly on one's domain.

Very fine granularity exacerbates a problem ontology developers have already experienced: keeping track of the terms so that relevant ones are easily found and all their semantics are disjoint. Solving this tracking problem would let us to avoid meetings and arguments and help suppress synonymy among the terms.

Make the Semantics Computable The obvious solution to implicit

```
<owl:AnnotationProperty rdf:about="&foo;purine"/>

<owl:Class rdf:about="#nucleoside">
  <rdfs:label>nucleoside</rdfs:label>
  <foo:purine>deoxyadenosime</foo:purine>
</owl:Class>
```

Figure 2. A mythical nucleoside ontology *foo*.

semantics is to make them computationally explicit. At its most basic, this would mean a set of signs — not terms in a natural language, but computational data structures — such that *each one's semantics can be algorithmically determined.* Then every application, database, and query are self-describing by referring to the signs; and the semantics of the descriptions can be computed by any other application.

In the Semantic Web, the computations are either ontological on the term (this term has some relationship to another term) or applications acting on the data denoted by the term. For example, some of the key relations in OWL are those of set theory (*e. g.*, *subClassOf*, *sameAs*, *cardinality*, *disjointWith*) and annotation tags (*e. g.*, *AnnotationProperty*, *OntologyProperty*, *isDefinedBy*, *subPropertyOf*) [9,20]. The best candidate for describing the semantics of biology is likely to be the *AnnotationProperty* [21]. For example, one might have a reference ontology *foo* that described nucleosides (see Figure 2). One could compute that deoxyadenosine is a member of the nucleosides using this ontology. But if the ontology's relations were incomplete — for example, that some nucleosides have the structure of 2-deoxy-D-ribosyl-base — then one would need other code to compare the structures of thymidine and adenosine to decide if thymidine is a 2-deoxy-D-ribosyl-base. The *semantics* of the biology are given by the *structural relationships* of the molecules, not the words used to denote particular molecules or classes of molecules. Conversely, the semantics of OWL's literals are given by natural language definitions — that is, they are implicit.

In contrast, I'm suggesting one determine the *definition* of a computational structure denoting a biological idea from that structure and rules governing its formation from very finely-grained axioms. Semantically, this is what compilers do. Several years ago I developed a formal language to do that for biology, called *Glossa*, and we demonstrated this idea works for the semantically most demanding queries we found in a relational database of maize genetics [22,23]. *Glossa*'s capabilities haven't yet been thoroughly tested for significant areas of biology; it suffers from the tracking problem even more acutely than the ontologies do; and it's even less user-friendly than ontologies. So it's premature to believe *Glossa* will be the solution. Description Logics may offer still another route if they can capture the biology sufficiently and escape the limitations of OWL and *Glossa* [24].

Constrain Inferences to Improve Scaling Making semantics computably explicit might decrease scaling by facilitating automatic inferences. Schemes that scoped inferences by semantics ("thymidine phosphorylase accelerates an enzyme reaction, not a drug reaction") obviously depend on being able to compute something about semantics. Or one might permit inferences over the entire web, testing (perhaps at each inferential step) by some set of quality metrics. For example, one might prefer rarer inferences built over longer chains of reasoning; inferences whose components had the fewest number of direct contradictions (or for contrarians, the maximum number); or the most frequent inferences that appear within N reasoning steps. Or perhaps a more explicitly theorem-proving approach is desirable: if a proposition can be proved it is of interest. One can imagine a smorgasbord of such metrics, and it would be fun to compare them. Sequential inference over structured resources is hard enough, but often the most important answers come from unstructured context. For example, it is very easy to find web pages about EC 2.4.2.4, but much harder to see why being both a growth factor and a statin is very provocative, let alone how this occurs. Right now, determining that means reading a large number of retrieved documents. Once again, we're back in the realm of natural language processing.

7. Prospects

So is the Semantic Web the wrong vision for biology? Perhaps not; but there are some fundamental gaps between the infrastructure of the Semantic Web and the needs of distributed computations for biology.

Whatever infrastructure is developed, one problem I've ignored so far — usage — will affect how well the Semantic Web ultimately facilitates scientific computation. At present, only manual inspection can tell if a term retrieves or produces semantically identical types of data within, let alone among, resources. Even when people read the directions (in this case, definitions), consistently implementing them is extremely hard, especially as the volume of data increases. One's view of the meaning of the definitions changes as more instances are worked through; usually more than one person builds a resource, such as a database, and they all have slightly different ideas; and people make mistakes. Ideally, a device would use the definitions to check the usage of terms in each contextual instance, making the Semantic Web self validating. Since many clues to inappropriate usage come from the connotations stored in a wide knowledge of the field, the device would have to somehow compare usage with the definitions and that knowledge — bringing us full circle to why the Semantic Web is an important idea worthy of effort.

Acknowledgments Frank Olken, Robert Stevens, Olivier Bodenreiter, and Daniel McShane encouraged me and gave me an excuse to go out on this particular limb (they are blameless). Jonathan W. King beautifully demonstrated the thymidine phosphorylase example to me. Armani Valvo prompted many useful walks. The reviewers' incisive comments considerably improved the paper. I am grateful to all of you. This work is supported by a grant from the U.S. National Institutes of Health (GM-56529).

References

1. Minoru Kanehisa, Susumu Goto, Hiroyuki Ogata, Hiroko Ishida, Sanae Asanuma, Toshi Nakatani, Saeko Adachi, Kana Matsumoto, Noriko Man, Rumiko Okada, Hidemasa Bono, Kazushige Sato, Toyoko Katrurada, Tomomi Kamiya, Mayuko Egoshi, Wataru Fujibuchi, Hiromi Adachi, Takaaki Nishioka, and Atshuhiro Oka. *From Sequence to Function. An Introduction to the KEGG Project.* Univeristy of Kyoto, Uji, Japan, `http://www.genome.-ad.jp/kegg/kegg1.html`, 1996–present.
2. G. P. Moss. *Biochemical Nomenclature. International Union of Pure and Applied Chemistry and International Union of Biochemistry and Molecular Biology, IUPAC-IUBMB Joint Commission on Biochemical Nomenclature, and Nomenclature Commission of IUBMB Home Page.* Department of Chemistry, Queen Mary and Westfield College, `http://www.chem.qmw.ac.-uk/iubmb/`, 1996.
3. Tim Berners-Lee. *Semantic Web Road Map.* W3C, `http://www.w3.org/-DesignIssues/Semantic.html`, 1998.
4. Patrick Hayes, editor. *RDF Semantics.* W3C, `http://www.w3.org/TR/-2004/REC-rdf-mt-20040210/`, 2004.
5. BioMOBY.org. *Moby.* BioMOBY.org, `http://www.biomoby.org//`, 2005.
6. Dave Beckett, editor. *RDF/XML Syntax Specification (Revised).* W3C, `http://www.w3.org/TR/2004/REC-rdf-syntax-grammar-20040210/`, 2004.
7. Dan Brickley and R. V. Guha, editors. *RDF Vocabulary Description Language 1.0: RDF Schema.* W3C, `http://www.w3c.org/TR/rdf-schema`, 2004.
8. Frank van Harmelen, Peter F. Patel-Schneider, and Ian Horrocks, editors. *Reference Description of the DAML+OIL Ontology Markup Language (March 2001).* W3C, `http://www.daml.org/2001/03/reference.html`,

2001.

9. Mike Dean and Guus Schreiber, editors. *OWL Web Ontology Language Reference.* W3C, `http://www.w3.org/TR/owl-ref/`, 2004.
10. Toni Kazic. Representation, reasoning and the intermediary metabolism of *Escherichia coli.* In Trevor N. Mudge, Veljko Milutinovic, and Lawrence Hunter, editors, *Proceedings of the Twenty-Sixth Annual Hawaii International Conference on System Sciences*, volume 1, pages 853–862, Los Alamitos CA, 1993. IEEE Computer Society Press.
11. Kate Fox. *Watching the English. The Hidden Rules of English Behaviour.* Hodder and Stoughton, London, 2004.
12. Deborah Tannen. *That's Not What I Meant!* Ballantine Books, New York, 1986.
13. Umberto Eco. *Semiotics and the Philosophy of Language.* Indiana University Press, Bloomington IN, 1984.
14. Larry Wos. *Automated Reasoning: Introduction and Applications.* McGraw-Hill Book Company, New York, second edition, 1992.
15. M. Garey and D. Johnson. *Computers and Intractability: A Guide to the Theory of NP-Completeness.* W. H. Freeman and Co., San Francisco, 1979.
16. John F. Sowa. *Conceptual Structures: Information Processing in Mind and Machine.* Addison-Wesley Publishing Co., Reading MA, 1984.
17. Betsy L. Humphreys and Donald A. B. Lindberg. The UMLS project: making the conceptual connection betweeen users and the information they need. *Bull. Med. Libr. Assoc.*, 81:170–177, 1993.
18. Michael Ashburner, C. A. Ball, J. A. Blake, David Botstein, H. Butler, John M. Cherry, A. P. Davis, K. Dolinski, S. S. Dwight, J. T. Eppig, M. A. Harris, D. P. Hill, L. Issel-Tarver, A. Kasarskis, Suzanna Lewis, J. C. Matese, Jane E. Richardson, M. Ringwald, Gerald M. Rubin, and G. Sherlock. Gene Ontology: tool for the unification of biology. *Nature Genet.*, 25:25–29, 2000.
19. International Union of Biochemistry and Molecular Biology. *Enzyme Nomenclature. Recommendations (1992) of the Nomenclature Committee of the International Union of Biochemistry and Molecular Biology.* Academic Press, Inc., London, 1992.
20. Jeremy J. Carroll and Jos De Roo, editors. *OWL Web Ontology Language Test Cases.* W3C, `http://www.w3.org/TR/2004/REC-owl-test-20040210/`, 2004.
21. Peter F. Patel-Schneider and Ian Horrocks, editors. *OWL Web Ontology Language Semantics and Abstract Syntax.* W3C, `http://www.w3.org/TR/-owl-semantics/`, 2004.
22. Toni Kazic. Semiotes — a semantics for sharing. *Bioinformatics*, 16:1129–1144, 2000. Also at `http://www.biocheminfo.org/repository/-semiotes.ps`.
23. Phani Chilukuri and Toni Kazic. *Semantic Interoperability of Heterogeneous Systems.* University of Missouri Bioinformatics Technical Report 2004-01, `http://www.biocheminfo.org/repository/parser.ps`, 2004.
24. Carsten Lutz, editor. *Description Logics.* dl.kr.org, `http://dl.kr.org`, 2005.

EVENT ONTOLOGY: A PATHWAY-CENTRIC ONTOLOGY FOR BIOLOGICAL PROCESSES

TATSUYA KUSHIDA*

Institute for Bioinformatics Research and Development, Japan Science and Technology Agency, AIST Bio-IT Research Building, 2-42 Aomi, Koutou-ku, Tokyo 135-0064, JAPAN

TOSHIHISA TAKAGI †

Graduate School of Frontier Sciences, University of Tokyo, 5-1-5 Kashiwanoha, Kashiwa-shi, Chiba, 277-8562, JAPAN

KEN ICHIRO FUKUDA‡

Computational Biology Research Center, National Institute of Advanced Industrial Science and Technology, AIST Bio-IT Research Building, 2-42 Aomi, Koutou-ku, Tokyo 135-0064, JAPAN

Event ontology is a new biomedical ontology developed to annotate pathway components in a pathway database. It organizes the concepts and terms of sub-pathways, pathways, biological phenomena, experimental conditions, medications, and external stimuli appearing in biological pathways (*e.g.* signal transduction, disease-, metabolic-, molecular interaction-, genetic interaction pathways, *etc.*). Concepts in the Event ontology are extracted manually from scientific literature. Each term has links to external databases such as Gene Ontology, Reactome, KEGG, BioCyc, and PubMed.

1. Introduction

Pathway databases are becoming increasingly important for the elucidation of mechanisms that underlie various types of biological phenomena, including disease. A number of pathway databases achieve these aims. For example, "KEGG"[1] and "BioCyc"[2] manage metabolic pathways in various organisms; "Reactome"[3] and "INOH"[4] attempt to represent various types of biological events such as immune response- and gene expression mechanisms as biological

* kushida-tatsuya@aist.go.jp

† tt@k.u-tokyo.ac.jp

‡ fukuda-cbrc@aist.go.jp

pathways; "aMaze",[5] "Patika",[6] and "Biocarta"[7] manage signal transduction- and disease pathways, and "BIND",[8] "DIP",[9] and "IntAct"[10] are protein–protein interaction databases. "Cytoscape"[11] and "GenMAPP"[12] are tools that utilize pathway data for data analysis. "BioPAX"[13] is a community project in which members engaged in the development of these pathway databases and tools cooperate to establish a data-exchange format for biological pathway data.

While conventional pathway data are highly structured, they use open vocabularies to name a pathway object. As this hampers data integration or mapping between data from different sources, the community needs a methodology to annotate data in a coherent way.

To solve this problem, we propose to use a set of annotation ontologies for pathway data annotation. In the INOH database, pathway diagrams are represented by compound graphs[14] and each pathway component is annotated by several biological ontologies. For example, protein objects are annotated by the MoleculeRole ontology[15] that manages the relations among generic protein names and concrete protein names that appear in the scientific literature. Consequently, INOH provides high-quality pathway data that allow advanced ontological searches, thereby improving accuracy by extending the search range using ontological trees.

Gene Ontology manages biological phenomena divided into "biological process", "molecular function", and "cellular component" categories and is generally used to annotate the functions of genes and gene products.[16] However, it is not designed to cope with the relations between pathways and sub-pathways and between pathways and their related biological phenomena. For example, Gene Ontology contains about 200 terms corresponding to "pathways", but few pathways include their sub-pathways with a part-of relationship [*e.g.*, transforming growth factor beta receptor signaling pathway (GO:0007179), I-kappaB kinase/NF-kappaB cascade (GO:0007249), and JAK-STAT cascade (GO:0007259)].

Furthermore, concepts regarding biological phenomena managed by Gene Ontology are too large and exhaustive for pathway data annotation. Terms, such as "actin filament-based process (GO:0030029)," seldom appear in articles that discuss biological pathways and are not used for pathway annotation.

GO slims are cut-down versions of Gene Ontology; they contain a subset of the terms included in Gene Ontology.[17] However, the terms in GO slims regarding pathway-related biological phenomena are insufficient.

In BioCyc, one of the most popular metabolic pathway databases, each "pathway" consists of "reactions" whose relations are classified hierarchically.[2] However, knowledge regarding signal transduction pathways is not rich enough to cover pathways in general.

Event ontology is designed to satisfy the need for a new pathway-centric biomedical ontology to annotate sub-pathways, pathways (biological processes), and their related biological phenomena in pathway databases. The knowledge was extracted from the scientific literature by manual expert curation. This ontology is expected to provide a structured high-quality controlled vocabulary for pathway data. The use of this ontology, instead of free text, for pathway or sub-pathway annotations facilitates advanced ontological searches and eases the pathway-annotation task of curators.

2. Curation Methods

Terms were collected by curators with a molecular biology background from the text and illustrations of review articles and original papers published in biomedical journals. In the present study, information on biological pathways in mammals (human and non-human) was curated. Event ontology manages the following concepts: (1) sub-pathways, (2) pathways, (3) biological phenomena related to these pathways, (4) experimental conditions under which pathways are investigated, and (5) external stimuli inducing a specific pathway. The organized terms are used to annotate sub-pathways, pathways, biological phenomena, experimental conditions, and external stimuli.

Event ontology was developed using DAG-Edit.[18] Terms are structured as a DAG (directed acyclic graph) and two relations, "is-a" and "part-of", are used. For each term, PubMed IDs are entered as reference information in the General Dbxrefs column in the text-editor panel. Sentences cited from the literature are entered in the Definition column. If there are corresponding terms in Gene Ontology[16], KEGG[1], BioCyc[2], Reactome[3], *etc.*, the IDs are entered in the General Dbxrefs. Synonyms for the terms are entered in the Synonyms column. In addition, if terms related to specific proteins or chemicals are found, their corresponding IDs in the MoleculeRole ontology[15] are entered in the General Dbxrefs column (Figure 1).

In the case of concepts investigated in only a specific organism, its name is recorded after each term name. Terms without a specific organism name indicate generic concepts that are investigated extensively in various organisms. Specific terms (*e.g.*, "Wnt signaling pathway [Mouse]") are entered under the corresponding canonical term (*e.g.*, "Wnt signaling pathway") with an "is-a" relationship.

Event ontology is distributed as an OBO file and can be converted to OWL. Therefore, it is useful for the application of/to semantic web technologies.

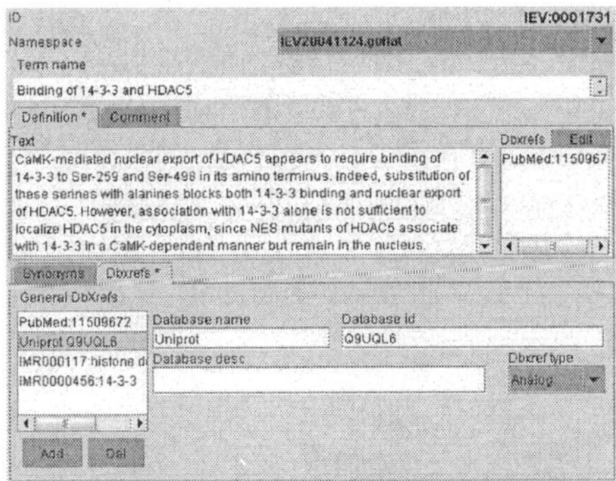

Figure 1. The text editor panel of DAG-Edit

3. Structure of Event ontology

3.1. *Biological event and Environmental event*

The ontology root has two subclasses, "Biological event (IEV:0000003)[§]", which includes sub-pathways (*e.g.*, "Binding of Smad3 and PIASy" and "MAPKKK cascade"), pathways (*e.g.*, Wnt signaling), and biological phenomena (*e.g.*, cell growth), and "Environmental event (IEV:0000185)", which includes experimental conditions (*e.g.*, medium), drugs/medications (*e.g.*, FK506 medication), and external stimuli (*e.g.*, ultraviolet irradiation).

3.2. *Pathways and Sub-pathways*

This is an example of how we manipulated the knowledge resources of Gene Ontology during the construction of Event ontology. We preserved the mapping between Event ontology and Gene Ontology if the latter contained a corresponding term, and we designed the DAG structure of terms by following the Gene Ontology conventions (*e.g.*,. "regulation", "negative regulation"). Furthermore, we collected just enough terms so as not to reduce pathway-curation efficiency.

As Gene Ontology does not include sufficient information regarding the relations between sub-pathways and pathways, we constructed these relations thoroughly in Event ontology. As in Gene Ontology, a sub-pathway is located under the pathway with a "part-of" relationship. For example, "Binding of

[§] "IEV" is the ID prefix of the Event ontology.

antigen and BCR complex (IEV:0001278)" is part-of "B cell receptor signaling (through IKK-NF-kappaB cascade) (IEV:0001298)" (Figure 2).

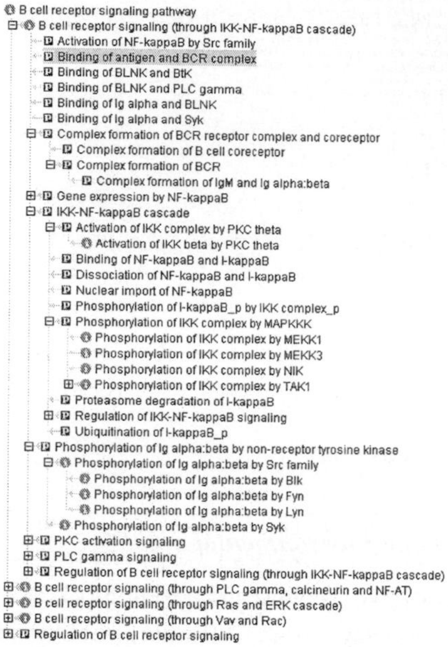

B cell receptor signaling pathway
B cell receptor signaling (through IKK-NF-kappaB cascade)
Activation of NF-kappaB by Src family
Binding of antigen and BCR complex
Binding of BLNK and BtK
Binding of BLNK and PLC gamma
Binding of Ig alpha and BLNK
Binding of Ig alpha and Syk
Complex formation of BCR receptor complex and coreceptor
Complex formation of B cell coreceptor
Complex formation of BCR
Complex formation of IgM and Ig alpha:beta
Gene expression by NF-kappaB
IKK-NF-kappaB cascade
Activation of IKK complex by PKC theta
Activation of IKK beta by PKC theta
Binding of NF-kappaB and I-kappaB
Dissociation of NF-kappaB and I-kappaB
Nuclear import of NF-kappaB
Phosphorylation of I-kappaB_p by IKK complex_p
Phosphorylation of IKK complex by MAPKKK
Phosphorylation of IKK complex by MEKK1
Phosphorylation of IKK complex by MEKK3
Phosphorylation of IKK complex by NIK
Phosphorylation of IKK complex by TAK1
Proteasome degradation of I-kappaB
Regulation of IKK-NF-kappaB signaling
Ubiquitination of I-kappaB_p
Phosphorylation of Ig alpha:beta by non-receptor tyrosine kinase
Phosphorylation of Ig alpha:beta by Src family
Phosphorylation of Ig alpha:beta by Blk
Phosphorylation of Ig alpha:beta by Fyn
Phosphorylation of Ig alpha:beta by Lyn
Phosphorylation of Ig alpha:beta by Syk
PKC activation signaling
PLC gamma signaling
Regulation of B cell receptor signaling (through IKK-NF-kappaB cascade)
B cell receptor signaling (through PLC gamma, calcineurin and NF-AT)
B cell receptor signaling (through Ras and ERK cascade)
B cell receptor signaling (through Vav and Rac)
Regulation of B cell receptor signaling

Figure 2. The relations between B-cell receptor signaling and the sub-pathways. The symbol "i" represents an "is-a" relationship; "P" indicates a "part-of" relationship.

3.3. *Types of sub-pathways*

In Event ontology, a sub-pathway is defined as a reaction in enzymatic reactions, an association or dissociation between molecules, or as a translocation of molecules. A sub-pathway in Event ontology corresponds with a "reaction" and a part-of "pathway" in the BioCyc[2] or Reactome.[3]

The sub-pathways are classified by molecular interaction types. Event ontology has about 30 molecular interaction types such as "Binding", "Phosphorylation", "Nuclear import", and "Deacetylation". "Deacetylation (IEV:0001681)" is-a "Hydrolysis (IEV:0000184)". This is-a relationship is used to define "kind-of" as in the further specialization of a class.

3.4. *Classification of sub-pathways using MoleculeRole ontology*

Each term name of a sub-pathway has a compositional structure and is based on molecular interaction types and the relevant molecule names. For molecular names (protein and chemical names), terms are defined in MoleculeRole Ontology,[15] an ontology of protein names that appear in the scientific literature. It manages the relations among "*function names*" and "*abstract (generic) names*" and "*concrete names*". Thus, each term name of a sub-pathway in Event ontology may consist of a "*molecular function name*" such as "protein serine/threonine kinase", an "*abstract (generic) molecule name*" such as "MAPK", or a "*concrete molecule name*" such as "p38" (*e.g.*, "Phosphorylation of p38 by MKK3/6" (IEV:0000235)).

In Event ontology, "Phosphorylation of MKK4/7 by ASK1 (IEV:0000811)" is located under "Phosphorylation of MKK4/7 by MAPKKK (IEV:0000237)" with an "is-a" relationship. This is-a relationship is used to define specific instances of a class of protein. Thus, the vertical relations among the same molecular interaction type are determined by the relative degree of abstraction of the molecules constituting the sub-pathways.

3.5. *Cellular localization of Sub-pathways*

Furthermore, each sub-pathway term is classified according to its cellular localization. For example, "Phosphorylation in cytosol (IEV:0000025)" is located under "Phosphorylation (IEV:0000005)" with an "is-a" relationship. This is-a relationship is used to define different locations. On the other hand, "Phosphorylation of IKK complex by MAPKKK (IEV:0000250)" is located under "Phosphorylation in cytosol (IEV:0000025)" with an "is-a" relationship. This is-a relationship is used to define participants in processes. Thus, all sub-pathways are classified according to the relations between sub-pathways and pathways, the molecular interaction types, and cellular localization. Usage and the types of terms representing the cellular localization differ according to the molecular interaction types.

In the case of a direct interaction such as "Binding" and a metabolic reaction such as "Phosphorylation", one term, representing the cellular localization where the molecular interaction occurs, is used (*e.g.*, "Phosphorylation in nucleus (IEV: 0000231)"). On the other hand, in the case of a molecular translocation such as "Nuclear export", two terms indicating the start and end-point of the translocation are used [*e.g.*, "Translocation of Bax from mitochondrial membrane to cytosol (IEV:0001692)" and "Translocation of Calcium ion from ER to cytosol through calcium ion channel (IEV:0000274)"].

In Reactome,[3] "Cellular component" is used to represent the cellular localization of "Events" and "EntityCellular compartment" is used to represent the cellular localization of "molecules". The "EntityCellular compartment" is part of the "Cellular component". Furthermore, the "Cellular component" of

Reactome is part of "Cellular compartment" of Gene Ontology. However, we ascertained that the cellular localization information of Events was inconsistent in Reactome. For example, the cellular localization of the reaction "Internalization of the insulin receptor (ReacomeID:74718)" and "Translocation of tBID to mitochondria (RactomeID:139920)" is "cell" and "cytosol", respectively. Therefore, it is not reasonable to represent the cellular localization of the Event (*i.e.*, sub-pathway and pathway) by using a single term from "Cellular compartment" of Gene Ontology.

3.6. *Classification of biological phenomena related to pathways*

Biological phenomena related to biological pathways such as signal transduction were extracted from the literature and then classified into 4 levels of phenomena: "Molecular event (IEV:0000071)", "Cellular event (IEV:0000069)", "Organism event (IEV:0000082)", and "Physiological event (IEV:0001330)". "Cellular event" and "Physiological event" of Event ontology correspond to "cellular process (GO:0009987)" and "physiological process (GO:0007582)" of Gene Ontology, respectively.

3.7. *Relations between pathways and biological phenomena*

Event ontology manages the relations among a biological phenomenon (including disease) and the related pathway(s). In Event ontology, a pathway involved in a specific biological phenomenon is located under the corresponding biological phenomenon with a "part-of" relationship.

For example, the sentence, "Thc BCR induces the signals that are required for survival and proliferation of B cells. These include activation of PI3K-regulated AKT, RAS–RAF–ERK (extracellular signal-regulated kinase) and NF-B pathways"[19] is transcribed in Event ontology as: "B cell proliferation (IEV:0001562)" as a biological phenomenon is located above "B cell receptor signaling (through IKK-NF-kappaB cascade), (IEV:0001298)", "B cell receptor signaling (through Ras and ERK cascade), (IEV:0001295)", and "B cell receptor signaling (through Vav and Rac), (IEV:0001296)" as a pathway.

3.8. *Treatments & Media*

Event ontology manages the concepts of experimental conditions and abiotic external stimuli that induce a pathway and lead to a specific biological phenomenon as an "Environmental event (IEV:0000185)" with two subclasses, "Medium condition (IEV:0000293)" and "Treatment (IEV:0000387)". For example, the "Medium condition" class includes "BME medium

(IEV:0000313)", the "Treatment" class includes "Anticancer drug medication (IEV:0001566)", "Ultraviolet irradiation (IEV:0001568)", "Hydrogen Peroxide treatment (IEV:0001573)", and "Heat shock treatment (IEV:0001565)". The sentence: "Caloric restriction activates cell survival signaling and then induces cell survival"[20] is expressed in Event ontology as "Caloric restriction (IEV:0001678)" is a "Treatment" and a part of "Caloric restriction cell survival signaling (IEV:0001686)" that is a part of "Cell survival (Inhibition of apoptosis) (IEV:0000153)".

4. Application of Event ontology

4.1. *Protein and Gene annotation using Event ontology*

Event ontology manages the relations among sub-pathways, pathways, and biological phenomena. The terms representing sub-pathways contain molecule names controlled by MoleculeRole ontology (*e.g*, "R-smad") and molecular interaction types (*e.g*, "Binding"). Furthermore, the terms have links to external references, such as Gene Ontology,[16] Reactome,[3] KEGG,[1] BioCyc,[2] MoleculeRole ontology,[15] and PubMed.

Thus, to a researcher interested in a gene or protein, querying the molecule name using Event ontology returns information regarding all sub-pathways involving the molecule and provides links to the actual pathway data annotated by Event ontology terms. Furthermore, knowledge regarding related biological phenomena such as apoptosis and immune responses can be acquired with Event ontology from the relations among sub-pathways and pathways, and from relations among pathways and biological phenomena. By annotating proteins or genes using Event ontology, the molecules can be linked to pathway information and pathway data.

In this sense, as Event ontology connects molecules to various levels of biological phenomena via sub-pathways and pathways in the ontological structure, it facilitates the seamless navigation among the concepts of molecules, sub-pathways, pathways, and biological phenomena.

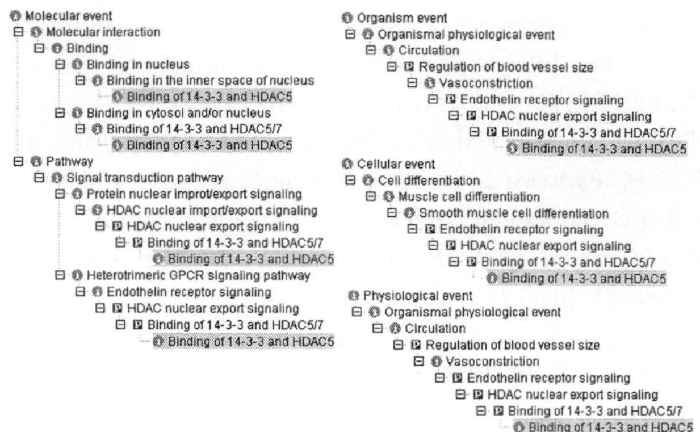

Figure 3. The pathways and biological phenomena involved in the sub-pathway "Binding of 14-3-3 and HDAC5"

4.2. *Example of an ontological search using Event ontology*

To demonstrate a query with Event ontology, we present a search example using DAG-Edit.[18] When we entered the query term "14-3-3" in the query column of DAG-Edit, 14 sub-pathways were found (*e.g.*, "Binding of 14-3-3 and HDAC5, (IEV:0001731)" and "Gene expression of 14-3-3 by p53 (IEV:0001112)"). We selected "Binding of 14-3-3 and HDAC5". The DAG viewer of DAG-Edit showed the cellular localization of "Binding of 14-3-3 and HDAC5" to be the "cytosol" or "within the nucleus" (Figure 3) and indicated that it was a sub-pathway of "Endothelin receptor signaling (IEV:0001693)". The pathway was shown to be involved in "vasoconstriction (IEV:0001822, GO:0042310)" and "muscle cell differentiation (IEV:0001824, GO:0051145)". In addition, PubMedIDs were provided as reference information in Dbxrefs of the DAG-Edit. "Binding of 14-3-3 and HDAC5" was also identified as a sub-pathway of "HDAC nuclear export signaling (IEV:0001866)".

The definition and Dbxrefs of the term "Binding of 14-3-3 and HDAC5, (IEV:0001731)", *e.g.*, PubMedID:11509672 and Uniprot:Q9UQL6 (Figure 1) provide comprehensive information to the user that HDAC5 is a deacetylation enzyme, and the nuclear export of HDAC5 as a result of 14-3-3 and HDAC5 binding leads to the inhibition of deacetylation of histones in nuclear DNA, and that the event controls expression of some gene(s).

5. Conclusions

5.1. *Comparison between Event ontology and other ontologies*

Before developing Event ontology, we investigated whether available ontologies were reasonable to annotate the pathways and sub-pathways in INOH[4] pathway diagrams. We concluded that Gene Ontology,[16] the most widely used biomedical ontology for gene function annotation, was not suitable for pathway data annotation because (1) it does not manage thoroughly the relations among sub-pathways and pathways, (2) it does not manage the relations between pathways and related biological phenomena, and (3) its set terms are too large and exhaustive for annotation of the pathway components.

GO slims (*e.g.*, Generic GO slim) are particularly useful for providing a summary of the results of GO annotations of a genome, microarray data, or cDNA collection when a broad classification of gene product function is required. However, the terms regarding the biological phenomena related to pathways are insufficient in GO slims. For example, Generic GO slim does not have any terms for pathway-related biological phenomena such as "actin filament organization (IEV:0001324, GO:0007015)" or "stress fiber formation (IEV:0000095, GO:0043149)" that are led, for example, by "Integrin signaling pathway (through Rho) (IEV:0000616)" and "TGF beta super family signaling pathway (IEV:0000090)".

Event ontology is a pathway-centric complement to the GO biological process ontology. We carefully followed the Gene Ontology conventions to structure our terms. Since there is a certain amount of complementarity, we are planning to submit our terms to the Gene Ontology consortium.

In Reactome,[3] each "reaction" or "event" is not annotated with controlled vocabularies. For example, terms such as "translocation", "transport", and "internalization" are used to represent the translocation of molecules. The definitions of these vocabularies and relations are not recorded in Reactome [*e.g.*, "Translocation of BIM to mitochondria [Homo sapiens] (ReactomeID:139919)", "Notch 2 precursor transport to golgi [Homo sapiens] (ReactomeID:157077)", and "Internalisation of the insulin receptor [Homo sapiens] (ReactomeID:74718)"]. This complicates retrieval by the system of all translocation-related reactions.

In BioCyc,[2] each "pathway" consists of "reactions" and the relations are classified hierarchically with the hierarchy mainly covering metabolic pathways. We are planning to add the entire metabolic pathway classification of BioCyc under the "Metabolic pathway (IEV:0000818)" term in Event ontology.

Event ontology attempts to manage concepts of various types of biological pathways such as signal transduction, metabolic-, molecular interaction-, genetic interaction-, and disease pathways, and it yields thorough information concerning the relations among sub-pathways and pathways and among pathways and related biological phenomena. Event ontology is a new biomedical ontology to annotate various types of biological pathway components.

5.2. *Statistics*

Table 1 shows the latest statistics of Event ontology. Currently, the number of total terms, at 2289, is about 1/8 of Gene Ontology (19,081 in June 2005).[16] We selected the terms in Event ontology carefully to achieve effective pathway annotation. Event ontology is an ongoing project and is updated on a regular basis.

Event ontology and MoleculeRole ontology in OBO format can be downloaded from our project web site (http://www.inoh.org/).

Table 1. The latest statistics of Event ontology (June 30, 2005). *: "Others" includes class names to classify concepts.

Pathways	*e.g.*, Wnt signaling	121
Sub-pathways	*e.g.*, Binding of Wnt and Frizzled	1794
Biological phenomena	*e.g.*, Apoptosis	168
Environmental events	*e.g.*, FK506 medication	127
Others*	*e.g.*, Binding in cytosol	79
Total		2289

5.3. *Future work*

Event ontology has a compositional structure. Mapping terms from other ontologies such as MoleculeRole ontology and Cell location ontology will enrich the ontology and the computation of implicit relations from these term relations represents an interesting research topic. We are part of the BioPAX working group and strongly believe Event ontology should incorporate existing communitiy-based standards, *e.g.,* BioPAX and OBO. The current Event ontology is limited to provide a structured controlled vocabulary for (pathway) data annotations. Decomposing the current "is-a" relation into more explicit relations such as "kind_of", "located_in" and "has_participant" relations[21] is the subject of ongoing work in our laboratory.

Acknowledgments

This work was supported by BIRD of Japan Science and Technology Agency.

References

1. M. Kanehisa and S. Goto, *Nucleic Acids Res.* **28**, 27-30 (2000).
2. CJ. Krieger, P. Zhang1, LA. Mueller, A. Wang, S.e Paley, M. Arnaud, J. Pick, S. Y. R. and PD. Karp, *Nucleic Acids Res.* **32**, 438-442 (2004).
3. G. Joshi-Tope, M. Gillespie, I. Vastrik, P. D'Eustachio, E. Schmidt, B. de Bono, B. Jassal, GR. Gopinath, GR. Wu, L. Matthews, S. Lewis, E. Birney and L. Stein, *Nucleic Acids Res.* **33**, D428-432 (2005).
4. INOH database website: http://www.inoh.org/
5. C. Lemer, E. Antezana, F. Couche, F. Fays, X. Santolaria, R. Janky, Y. Deville1, J. Richelle and SJ. Wodak, *Nucleic Acids Res*, **32**, 443–448 (2004).
6. E. Demir, O. Babur, U. Dogrusoz, A. Gursoy, A. Ayaz, G. Gulesir, G. Nisanci and R and Cetin-Atalay, *Bioinformatics* **20**, 349–356 (2004).
7. BioCarta website: http://www.biocarta.com/
8. BIND website: http://bind.ca/
9. DIP website: http://dip.doe-mbi.ucla.edu/
10. IntAct website: http://www.ebi.ac.uk/intact/
11. P. Shannon, A. Markiel1, O. Ozier, NS. Baliga, JT. Wang, D. Ramage, N. Amin, B. Schwikowski and T. Ideker, *Genome Research* **13**, 2498-2504 (2003).
12. KD. Dahlquist, N. Salomonis, K. Vranizan, SC. Lawlor, BR. Conklin, *Nat Genet.* **31**, 19-20 (2002).
13. BioPAX website: http://www.biopax.org/
14. K. Fukuda and T. Takagi, *Bioinformatics* **17**, 829-837 (2001).
15. S. Yamamoto, T. Asanuma, T. Takagi and K. Fukuda, *Comparative and Functional Genomics* **5**, 528-536 (2004).
16. The Gene Ontology Consortium, *Genome Res.* **11**, 1425-1433 (2001).
17. GO Slims website: http://www.geneontology.org/GO.slims.shtml
18. DAG-Edit website: http://sourceforge.net/projects/geneontology
19. H. Niiro H and EA. Clark, *Nature Reviews Immunology* **2**, 945-956 (2002).
20. HY. Cohen, C. Miller, KJ. Bitterman, NR. Wall, B. Hekking, B. Kessler, KT.Howitz, M. Gorospe, R. de Cabo, DA. Sinclair, *Science* **305**, 390-392 (2004)
21. B. Smith, W. Ceusters, B. Klagges, J. Kohler, A. Kumar, J. Lomax, C. Mungall, F. Neuhaus, AL. Rector and C. Rosse, *Genome Biology* **6**, R46 (2005)

DISCOVERING BIOMEDICAL RELATIONS UTILIZING THE WORLD-WIDE WEB

SOUGATA MUKHERJEA

IBM India Research Lab,
Hauz Khas, New Delhi, India
E-mail: smukherj@in.ibm.com

SAURAV SAHAY

College of Computing, Georgia Institute of Technology,
Atlanta, Ga, USA
E-mail: ssahay@cc.gatech.edu

To crate a Semantic Web for Life Sciences discovering relations between biomedical entities is essential. Journals and conference proceedings represent the dominant mechanisms of reporting newly discovered biomedical interactions. The unstructured nature of such publications makes it difficult to utilize data mining or knowledge discovery techniques to automatically incorporate knowledge from these publications into the ontologies. On the other hand, since biomedical information is growing explosively, it is difficult to have human curators manually extract all the information from literature. In this paper we present techniques to automatically discover biomedical relations from the World-wide Web. For this purpose we retrieve relevant information from Web Search engines using various lexico-syntactic patterns as queries. Experiments are presented to show the usefulness of our techniques.

1. Introduction

A Semantic Web for Life Sciences storing information about all the biomedical concepts as well as relations between them will enable researchers and autonomous agents to efficiently retrieve information as well as discover unknown and hidden knowledge. However, the current situation of the Semantic Web is one of a vicious cycle in which there is not much of a Semantic Web due to the lack of semantic markup of data, and there is such a lack because there is no easy way to semantically annotate biological knowledge.

One of the goals of Semantic Web research is to incorporate most of the knowledge of a domain in an ontology that can be shared by many applications. Various ontologies and knowledge bases have been developed for Life Sciences including *Unified Medical Language System (UMLS)*[1] and Gene Ontology[2]. These ontologies organize information of various biological concepts, each with their attributes, and describe simple relationships like *is-a* and *part-of* between concepts. However, these ontologies are not up-to-date and may not have the information about newly discovered biomedical entities. Moreover, they generally do not incorporate complex relationships between biomedical entities. For example, although UMLS contains details about many diseases, viruses and bacteria, it does not incorporate relations between diseases and the causes of the diseases. Therefore, representing these ontologies and knowledge sources in Semantic Web languages like OWL will not be sufficient to create a Semantic Web for Life Sciences.

Journals and conference proceedings represent the dominant mechanisms of reporting biomedical results. The unstructured nature of such publications makes it difficult to utilize automated techniques to extract knowledge from these sources. Therefore information about new biomedical entities or relations between them need to be added to the ontologies manually. However, because of the very large amounts of data being generated, it is difficult to have human curators extract all these information and keep the ontologies up-to-date.

A large section of the research literature is available online and is therefore searchable by Web Search engines like Google. Although databases like PubMed are not readily accessible to the Google crawler, many of the PubMed abstracts has been crawled by Google (by following the links to these abstracts specified in other Web pages). Moreover, publications available from researchers' homepages or from conference Websites as well as biomedical information sources other than research publications can be also accessed by Google.

This paper presents a technique to automatically discover biomedical relations from the World-wide Web. We first query Web search engines with hand-crafted lexico-syntactic patterns to retrieve relevant information. The knowledge extracted from the search results can be used to augment the ontologies and knowledge bases and create a Semantic Web for Life Sciences. Different types of relations between biomedical entities can be discovered by this technique. For example, given a biomedical term and a class, one can determine whether the entity belongs to the class. Our technique is efficient and does not require any Web page download.

The paper is organized as follows. The next section cites related work.

Section 3 explains how our technique can be used to classify biomedical terms. Section 4 discusses how we can automatically discover any arbitrary relation between biomedical entities. Finally Section 5 is the conclusion.

2. Related Work

2.1. *Biomedical Information Extraction*

Our objective is to automatically discover biomedical information. Automatic extraction of useful information from online biomedical literature is a challenging problem because these documents are expressed in a natural language form.

The first task is to recognize and classify the biological entities in scientific text. Biological term extraction systems can be broadly divided into two types: those with a rule base and those with a learning method. In [3] protein names are identified in biological papers using hand-coded rules. On the other hand, in [4] supervised learning methods based on Hidden Markov Models are used. We have developed the BioAnnotator system[5] which uses rules and dictionary lookup for identifying and classifying biological terms.

After the biological entities are recognized, the next task is to identify the relations between these entities. To determine the relations between biological entities (for example protein-protein interactions), one approach is to use templates that match specific linguistic structures[6]. Natural Language processing techniques that use parsers of increasing sophistication have also been utilized. For example in [7], a bi-directional incremental parsing technique based on combinatory categorical grammar is used. Recently, research has gone beyond treatment of single sentences to look at relations that span multiple sentences through the use of co-reference[8].

Since it is very difficult to extract information from unstructured text, in this paper we introduce a completely different technique of identifying biomedical knowledge which utilizes a Web search engine.

2.2. *Knowledge Extraction from the World-wide Web*

Marti Hearst had suggested that hyponyms could be acquired from Large Text Corpora[9]. For example, consider the sentence *"The bow lute, such as the Bambara ndang, is plucked"*. Even if we have not encountered the terms *bow lute* and *Bambara ndang*, we can infer from the sentence that *Bambara ndang* is a kind of *bow lute*. Thus lexico-syntactic patterns can be utilized to discover information from a large Text corpus.

This technique has been successfully utilized to discover knowledge from the World-wide Web, the largest Text corpus. Oren Etzioni introduced the metaphor of an *Information Food Chain* where Search engines are herbivores "grazing" on the Web and intelligent agents are *"information carnivores"* that consume output from various herbivores[10].

Several systems have been built based on this principle. Instead of gathering information from the Web directly, these systems utilize Web search engines which have already crawled and indexed the information. For example, Know-it-all[11] was able to extract thousands of facts automatically using Web search engines. Similarly, PANKOW[12] could automatically discover names of countries, cities and rivers. We believe that ours is the first system that utilizes this technique to discover biomedical knowledge. Moreover, we have extended the technique to identify relations between entities.

3. Classifying Biomedical Terms

Biological knowledge sources like UMLS can be utilized to create a Semantic Web for Life Sciences by representing them using languages like RDF[13] and RDFS[14]. The biological concepts in UMLS can be represented as RDF resources and the Semantic Network classes can be represented as RDFS classes in the Semantic Web. The *RDF:type* property will link a concept to the classes it belongs to. However UMLS is not comprehensive and does not contain information about all biological terms present in the research literature. In this section we discuss a technique for determining the biomedical class of an unknown biological term so that it can be included in the Semantic Web.

3.1. *Methodology*

Marti Hearst had introduced several patterns that indicate the *"is-a"* relation in English text[9]. More patterns have been identified by others[12]. Examples of such patterns together with instances from the biomedical domain are[a]:

- *NP_term* is a *NP_class*
  ... *malaria* is a *disease*

[a]Here *NP_term* indicates the noun phrase for the term and *NP_class* indicates the noun phrase of the class

- *NP_class* such as *NP_term*
  ... *genes* such as *p53*
- *NP_term* or other *NP_class*
  ... *amylase* or other *proteins*
- *NP_class* including *NP_term*
  ... *vitamins* including *riboflavin*
- the *NP_class NP_term*
  ... the *peptide somatostatin*

If a biological term belongs to a particular class, there would be a large number of the above patterns in the World-wide Web. Thus there will be several occurrences of the phrase *"malaria is a disease"* and the phrase *"diseases including malaria"* in the Web. On the other hand there will be very few occurrences of phrases such as *"the hormone malaria"* or *"hormones such as malaria"*.

```
isa(t,c) {
  let PATTERNS be the set of patterns to determine IS-A relationships
  count = 0
  for each pattern in PATTERNS {
     queryString = pattern with NP_term replaced by t
                   and NP_class replaced by c
     resultCount = GoogleSearchResultCount(''queryString'')
     count += resultCount
  }

  if (count <= THRESHOLD)
     return false
  else return true
}
```

Figure 1. Pseudo code to classify a biological term

Based on these observations, we can determine whether a term t belongs to a class c using the procedure *isA(t,c)* as shown in Figure 1. For each pattern that indicates the *isA* relationship, we determine the number of such phrases in the WWW. We utilize the *Google APIs*[15] for searching the Web. If the total number of patterns is greater than a predefined constant (*THRESHOLD*), we consider the term to belong to the particular class.

Note that only the search result count is sufficient for our purpose; we do not need to download any Web pages. Therefore the technique is quite efficient.

3.2. *Experiments*

Although the objective of our technique is to classify terms not in the ontologies, it is not possible to evaluate the effectiveness of the classification of unknown biological terms without the help of domain experts. Therefore, we have evaluated our technique with terms that have already been classified by UMLS. We randomly selected 100 UMLS terms belonging to 10 classes that have many instances in the biomedical literature including *gene, protein, lipid, vitamin*, etc. We utilized our technique to determine whether the term belongs to some of these 10 classes.

We calculated the following statistics from our experiments:

- **True Positive (TP)**: If a term t belongs to a class c and *isa(t,c)* returns *true*.
- **True Negative (TN)**: If a term t does not belong to a class c and *isa(t,c)* returns *false*.
- **False Positive (FP)**: If a term t does not belong to a class c but *isa(t,c)* returns *true*.
- **False Negative (FN)**: If a term t belongs to a class c but *isa(t,c)* returns *false*.
- **Precision** $P = \frac{TP}{TP+FP}$
- **Recall** $R = \frac{TP}{TP+FN}$
- **F-measure** $F = \frac{2*P*R}{P+R}$

Note that we designed our experiment so that there were an equal number of positive and negative examples. Thus $TP + FN = TN + FP$. Table 1 shows the results of our experiments.

Table 1. Precision and Recall of the Classifier

Threshold	Precision	Recall	F-score
0	0.615	0.798	0.695
25	0.875	0.702	0.779
50	0.877	0.596	0.71

The best results were obtained at a *THRESHOLD* of 25 when our technique could classify biological terms with a precision of 87.5% and recall of 70.2%. At a lower threshold there were many false positives which reduced the precision. At a higher threshold there were many false negatives reducing the recall.

False Negatives occur when we find very few matching patterns for a term that belongs to a particular class. (Since at threshold 0 recall is not 100%, it indicates that in some cases not even a single matching pattern could be found). This mostly occurs for uncommon terms like *dipalmitoyl-lecithin* (a lipid). Moreover, some terms have many synonyms. If a synonym is not common, we may not be able to classify it. Thus we could not classify *riboflavine* as a vitamin but *riboflavin* could be correctly classified.

False Positives occur mostly because sometimes an IS-A pattern may occur in a different context. For example, the sentence "*diseases caused by viruses such as aids*" matches our IS-A pattern "*NP_class* such as *NP_term*" which indicates that aids is a virus. Patterns like "*the aids virus*" are also common.

4. Discovering Relations between Biomedical Terms

In UMLS Semantic Network the 135 biomedical classes are linked by a set of 54 semantic relationships (like *prevents*, *causes*). However there are no relationships between the biomedical concepts themselves. To develop a comprehensive Semantic Web, discovering relations between the biomedical concepts is essential. In this section we will discuss how the World-wide Web can be utilized to discover such relationships.

It should be noted that classification of biomedical terms is the determination of IS-A relation between the term and a Biomedical class. However, identifying any arbitrary relation between two biomedical entities is much more challenging. It would be very difficult to determine patterns that are true for any relation between biological terms. On the other hand, if we try to determine some particular types of relations, specifying the patterns is much easier.

Let us assume that our objective is to discover causal relationship between a disease and a biological entity. Given a disease d and a biomedical entity e, we can query Google with phrases like *"e causes d"* or *"d is caused by e"* and count the number of results that are retrieved. However, there are thousands of entities (viruses, bacteria, parasites, etc.) that can cause a disease. Querying Google for each of them is not efficient. It would be more useful if given a disease we can discover the likely causes of the disease.

```
relationIdentifier(t,patterns) {
  initialize a Hash Map resultEntities
  for each pattern in patterns {
    queryString = pattern with NP_term replaced by t
    results = GoogleSearchResultSnippets(''queryString'')
    for each result in results {
      bioAnnotatedResult = BioAnnotate(result)
      relAnnotatedResult = RelationAnnotate(bioAnnotatedResult)
      entity = relationEntity(relAnnotatedResult,t)
      resultEntities{entity}++
    }
  }

  return resultEntities
}
```

Figure 2. Pseudo code to determine the entity that has the relations specified in *patterns* with term *t*

We have implemented a generic framework for discovering relationships between biomedical entities. Patterns that indicate each of these relations have been identified. Figure 2 shows the pseudocode to determine the entity that takes part in relations specified by *patterns* with term *t*. For example, if we want to discover causal relationship between a disease and a biological entity, *patterns* may consist of phrases like *"NP_term causes"* or *"is caused by NP_term"*. In this case just the number of results retrieved by Google for the queries is not sufficient. However, downloading the result pages will make the process very slow. Therefore, we utilize the *result snippets* (the small section of the result pages that contain the query string that is returned with a Google search).

We determine the entity that is related to term *t* from these result snippets. For this purpose we first use BioAnnotator[5] to determine the biomedical entities in the strings. After that a Relation Annotator discovers the relations between the biomedical entities. It uses templates for patterns which specify relationships in sentences. For example some common templates are:

- *Subject Verb_Group Object* (For example, *"HIV causes AIDS"*)
- *Object Passive_Verb_Group Subject* (For example, *"AIDS is caused by HIV"*)

- *Noun (Nominal form of verb) Object Subject* (For example, *"causing of AIDS by HIV"*)

If a template is matched it is assumed that a relation of the matching verb group (or nominal form) has been identified. Note that if there are noun phrases or adjectives between the biological entities and the verb groups in the sentences they are considered as qualifiers for the biological entities.

The combination of BioAnnotator and Relation Annotator creates an annotated string from which the entity taking part in the relation with the term *t* can be easily identified. For example given the result snippet *"AIDS is caused by HIV"*, BioAnnotator will recognize *AIDS* and *HIV*, a Part-of-speech tagger is used to recognize *"is caused by"* as the Verb Group and the Relation Annotator recognizes *HIV* as the entity that is in causal relationship with *AIDS*. On the other hand for the more complex result snippet *"Metabolic bone disease is caused by the lack of Vitamin D3"*, the Relation Annotator recognizes *"Vitamin D3"* as the entity that is in causal relationship with *"Metabolic bone disease"* with the qualifier *"the lack of"*.

Different authors will express the same semantics in different ways. Therefore there will be variations in the snippets that are retrieved by Google. For example, one snippet may state that *AIDS* is caused by *HIV* while another may state that the disease is caused by *Human Immunodeficiency Virus*. However, BioAnnotator will map them to the same biological entity using ontologies (UMLS). Therefore, Relation Annotator will identify the same biological entity from the two snippets. However, this may not be true for all snippets. For example, if one snippet states that *Metabolic bone disease* is caused by *"the lack of Vitamin D3"* and another states that it is caused by *"Calcium deficiency"*, our annotators will not be able to match the two entities. Therefore, as shown in Figure 2, a hash map that has the entities that have the specified relation with the given concept along with the number of occurrences for each of them are returned from the *relationIdentifier* procedure.

4.1. *Experiments*

We have utilized our technique to identify various types of relationships between biomedical entities. However, a formal evaluation of our technique is difficult because there are no test data sets that can be used for the evaluation. For determining the efficiency of our technique, we determined five types of relations. Besides Semantic Network properties *causes*, *diagnoses*, *consists of* and *affects* we also extracted *binds* relations for several entities of UMLS class *Amino Acids, Peptides or Proteins.* Table 2 shows several biomedical relations determined by our technique. Thus we could identify the cause of *Thyphoid* (*Bacterium Salmonella Typhi*) as well as entities that affect *Statin* (*Lipitor, Gemfibrozil, Niaspan*).

Table 2. Some Biomedical Relations determined by our technique

PROPERTY	UMLS ENTITY	RELATION ENTITY
causes	Typhoid	Bacterium Salmonella Typhi
diagnoses	Cyst	Ultrasonography
consists of	Butane	Liquefied Petroleum Gas
affects	Statin	Lipitor, Gemfibrozil, Niaspan
binds	Rhodopsin	Lys296, Transducin

For each property, we determined relations for several entities of some particular UMLS class which has that property. To test the system impartially we have included common as well as rare concepts in our experiments. In the absence of domain experts, we did a literature survey to determine whether the relations identified by our system are correct. We calculated the following statistics for each property from our experiments:

- **N**: Total number of biomedical entities for which we tried to identify relations.
- **F**: The number of entities for which at least one relation was identified by our system.
- **C**: The number of entities for which at least one relation that was identified by our system is correct.
- **Precision (P)** $P = \frac{C}{F}$
- **Recall (R)** $R = \frac{F}{N}$

Table 3 displays the results for each property and the corresponding UMLS class. The results show the promise of our technique, with the precision and recall values exceeding 70% for all properties. The quality of the Relation Identifier is affected by various factors:

- The recall is affected by Google's inability to identify complex class associations such as chemicals, genes, proteins and their relationships. For example, Google is unable to retrieve any results on our queries such as *"binds Auxin Response Factor 1"* or *"Nephroptosis is diagnosed by"*.

- Sometimes the snippet returned by Google may not be able to identify the cause. For example, one snippet retrieved was *"Primary Hypertension is caused by abnormalities of "* with the relevant cause of the disease stripped off.
- Both the precision and recall is affected by the limitations of the Relation Annotator. For example, the Relation Annotator can neither handle complex sentences nor relations expressed in multiple sentences.

Table 3. Precision and Recall of the Relation Identifier

Property	Class	Precision	Recall
causes	Disease	0.82	0.85
diagnoses	Anatomical Abnormality	1.0	0.9
consists of	Organic Chemical	0.75	0.72
affects	Gene	0.8	0.76
binds	Amino Acid, Peptide, or Protein	0.83	0.75

5. Conclusion

This paper introduced a new technique to automatically and efficiently discover biomedical relations. It utilizes the World-wide Web, undoubtedly one of the most comprehensive sources of biomedical knowledge. Since Web Search engines have crawled and indexed most of the information, we query these engines with several lexico-syntactic patterns to retrieve relevant information. This information can be used to classify biomedical terms or discover relations between biomedical entities. Our experiments show the promise of our techniques.

At present we are improving our Relation Annotator system for identifying relations between biomedical entities. We are also extending our system to MedLine to improve the recall. Our ultimate objective is to utilize the discovered relations between biological concepts to develop a Semantic Web for Life Sciences which would store the "meaning" of biological concepts as well as relations between these concepts. This will enable researchers to perform a single semantic search to retrieve all the relevant information about a biological concept.

References

1. UMLS. http://umlsks.nlm.nih.gov.
2. Gene Ontology. http://www.geneontology.org/.
3. K. Fukuda, T. Tsunoda, A. Tamura, and T. Takagi. Toward Information Extraction: Identifying Protein Names from Biological Papers. In *the Proceedings of the Pacific Symposium on Biocomputing*, pages 707–718, Hawaii, 1998.
4. N. Collier, C. Nobata, and J. Tsujii. Extracting the names of Genes and Gene products with a Hidden Markov Model. In *the Proceedings of the 18th International Conference on Computational Linguistics*, pages 201–207, Saarbrucken, Germany, 2000.
5. L.V. Subramaniam, S. Mukherjea, P. Kankar, B. Srivastava, V. Batra, P. Kamesam, and R. Kothari. Information Extraction from Biomedical Literature: Methodology, Evaluation and an Application. In *the Proceedings of the ACM Conference on Information and Knowledge Management*, New Orleans, Lousiana, 2003.
6. L. Wong. PIES: A Protein Interaction Extraction System. In *the Proceedings of the Pacific Symposium on Biocomputing*, pages 520–531, Hawaii, 2001.
7. J.C. Park, H.S. Kim, and J.J. Kim. Bidirectional Incremental Parsing for Automatic Pathway Identification with Combinatory Categorial Grammar. In *the Proceedings of the Pacific Symposium on Biocomputing*, pages 396–407, Hawaii, 2001.
8. J. Pustejovski, J. Castano, J. Zhang, M. Kotecki, and B. Cochran. Robust Relational Parsing over Biomedical Literature: Extracting Inhibit relations. In *the Proceedings of the Pacific Symposium on Biocomputing*, Hawaii, 2002.
9. M. Hearst. Automatic Acquisition of Hyponyms from Large Text Corpora. In *Proceedings of the Fourteenth International Conference on Computational Linguistics*, Nantes, France, July 1992.
10. O. Etzioni. Moving up the Information Food Chain: Softbots as Information Carnivores. In *Proceedings of the Thirteenth National Conference on Artificial Intelligence*, Portland, Oregon, August 1996.
11. O. Etzioni, M. Cafarella, D. Downey, S. Kok, A. Popescu, T. Shaked, S. Soderland, D.S. Weld, and A. Yates. Web-Scale Information Extraction in KnowItAll. In *Proceedings of the Thirteenth International World-Wide Web Conference*, New York, NY, May 2004.
12. P. Cimiano, S. Handschuh, and S. Staab. Towards the Self-Annotating Web. In *Proceedings of the Thirteenth International World-Wide Web Conference*, New York, NY, May 2004.
13. Resource Description Format. http://www.w3.org/1999/02/22-rdf-syntax-ns.
14. Resource Description Format Schema. http://www.w3.org/2000/01/rdf-schema.
15. Google APIs. http://www.google.com/apis/.

BIODASH: A SEMANTIC WEB DASHBOARD FOR DRUG DEVELOPMENT

ERIC K. NEUMANN†
W3C, MIT
Cambridge, MA 02139 USA

DENNIS QUAN
IBM T. J. Watson Research Center, 1 Rogers Street
Cambridge, MA 02142 USA

A researcher's current scientific understanding is assembled from multiple sources of facts and knowledge, along with beliefs and hypotheses of their interpretations. A comprehensive and structured aggregation of all the relevant components is to-date not possible using standard database technologies, nor is it obvious how to include beliefs, such as models and hypotheses into such a bundle. When such information is required as the basis for important decision-making (e.g., in drug discovery), scientists often resort to using commercial presentation applications. This is sub-optimal for the effective use of knowledge, and alternatives that support the inclusion of meaning are urgently needed. This paper describes a prototype Semantic Web application, BioDash[1], which attempts to aggregate heterogeneous yet related facts and statements (using an RDF model) into an intuitive, visually descriptive and interactive display.

1. Introduction

1.1. Today's Research Informatics Problems

Scientific research relies on researchers sharing heterogeneous knowledge, experimental data, and interpretations in meaningful ways that go beyond transmitting data fragments. Although computational methods and data-exchange protocols are common to modern scientific practice, they are a small part of the overall process. Critical interpretation of experimentally derived information and the consolidation of knowledge that include alternative views and hypotheses are essential to the scientific process of debate and rebuttal.

The need for improved information systems is being recognized throughout the pharmaceutical and biotech industry. In a recent report on drug development [1], the U.S. Food and Drug Administration (FDA) stated the need for "a knowledge base built not just on ideas from biomedical research, but on reliable insights into the pathway to patients." Much still needs to be done to meet these

† contact: eneumann@alum.mit.edu
1 http://www.w3.org/2005/04/swls/BioDash/Demo

goals, and current web and enterprise architectures cannot satisfy the functionality specified.

Currently, knowledge is captured in either rigid, hard-to-define databases, or in applications (PowerPoint and Excel) designed for human viewing. The former misses the inclusion of explicit scientific meaning, while the latter are difficult to query, making it hard to find or reuse knowledge (*knowledge cul-de-sacs*). In addition, most analysis and visualization applications are not interoperable via open standards, and cannot "see" connections across data sets that could be presented to users. Finally, items such as context and hypotheses are not well encoded, severely limiting data interpretation.

The essential problem in data management is not how to store large amounts of data but how best to distill insights from them (through analysis), and associate these interpretations with the data. In so doing, knowledge would be organized for higher-level reasoning and decision-making. The Semantic Web (SW) (www.w3.org/2001/sw/), as proposed by Tim Berners-Lee, is supposed to allow meaning (i.e., semantics) to be associated with information on the Web through a universal mechanism that machines can process as well [2, 3]. It is based on two key standards: Resource Description Framework (RDF) for describing objects and the relations between them; and the Web Ontology Language (OWL, based on RDF) for specifying the supported ontologies (semantic systems of concepts and relations). OWL ontologies (one or more per RDF document) are used to define the logical types of objects and how they can relate to one another within an RDF document.

We define six areas where SW technologies could offer critical support to the life sciences: (1) database conversions and wrappers; (2) unique identifiers that are supported by the SW URI model; (3) coordination and management of terminologies and ontologies; (4) tools and viewers conversant in RDF-OWL; (5) knowledge encoding: theories, hypotheses, models; (6) semantics accounts and channels: store and share annotations based on SW. This research addresses the first four and suggests directions for the latter two.

1.2. *BioDash: A Life Science Scenario of How Things Should Be*

A real-world illustration of the need to organize and utilize complex, distributed forms of information is seen in drug discovery. The likely success of a new drug is indicated through the combined analysis of target classes, high-throughput (HT) screening, ADME, toxicity, efficacy, animal testing, and efficient clinical trials. These can determine whether a new drug is successfully launched or terminated.

This paper describes our work on BioDash, a prototype of an "information dashboard" for drug discovery, which involves a scenario based on the drug target Glycogen Synthase Kinase 3 beta (GSK3b) [4], in which multiple forms of knowledge (genomic, biopathway, disease, chemical, and SNP data) residing in disparate repositories are brought together through SW technologies to support the discovery process.

We will begin by summarizing the basic notions of the Semantic Web – universal identifiers, the "quantitization" of knowledge, and Semantic Lenses – from the perspective of life science data integration. Afterwards, we proceed to describe how the BioDash user experience takes advantage of Semantic Web integration technologies. Ultimately, this integration is only worthwhile if it can enable users to gain insights they would otherwise be unable to if the data were kept separate. We give several examples of how BioDash permits the user to "experience" data integration across topics. Finally, we end with a discussion of additional applications that could benefit from the technology.

2. The Semantic Web Data Model within BioHaystack

2.1. *Building on the Web Model*

The World Wide Web succeeded in large part because it allows users to retrieve information from an ever-broadening range of sources through a single tool: the Web browser. In the days before the Web, users had to jump tediously from one system to another to perform complex retrieval tasks. At present, there is a lot less system-hopping thanks to hyperlinks However, there are still barriers to making effective use of that information. For example, applications do not successfully negotiate data from other applications due to differences in data formats.

Ironically, one application that is often caught "hoarding" data is our old friend, the Web browser. The data needed for various tasks are found on public domain Web pages buried in tables, bullet listings, or even prose. The characteristics that make Web pages easily consumable for humans, i.e., context-specific page layouts and inspired uses of formatting, are the very things that inhibit machine processing, which depends on data being laid out in a consistent, predetermined, "boring" fashion. Differences in data formats have made collating data from multiple web pages hard for humans and impossible for machines. While the Web has standardized the way humans retrieve information, until now it has done little to standardize data representations.

As data become easier to consume by applications, new visualization capabilities will be possible, and browsers will evolve to take advantage of them. BioHaystack, on which BioDash is built, is a prototype of a life science "Semantic Web Browser" that specifically supports information formatted for the Semantic Web. It is able to handle RDF and OWL documents, by aggregating, filtering, and rendering RDF data files into viewable and interactive displays. BioHaystack also allows one to create new RDF information and store the new contents, and will be discussed in further detail in Sections 2.5 and 3.1.

2.2. *A more universal data exchange format*

The Web was premised on the idea that human-readable content must be written in a common format (HTML) and made available to Web browsers from content servers through the HTTP protocol. The Semantic Web requires that data published to the Web should utilize the RDF format, making it easier for applications other than the data's origin to read and incorporate them.

The key idea behind RDF is that by introducing some syntactical simplifications on XML, a number of important capabilities are enabled: (1) personal or domain-specific annotations, classifications, and other forms of knowledge can be added to any application's data without interfering with its normal function; (2) information retrieval is made easier, because RDF-enabled Web browsers and search engines can index and extract classification metadata from any RDF file; (3) arbitrary RDF data files, containing pieces of knowledge from multiple applications, can be easily merged to form a larger whole (information integration); (4) automated, rules-based processing is possible using off-the-shelf RDF inference engines.

2.3. *LSID: A More Universal Naming Scheme*

SW also requires objects that are described by RDF data files, such as gene sequences, research papers, or 3D structures, to be referred to by universal names in accordance with the Universal Resource Identifier (URI) standard, an extension of the original URL system (e.g., http://www.w3.org/). A universal naming scheme simplifies the processing of data from a variety of sources, because the application does not need to have specific, hard-coded support for each naming scheme. This allows cross-referencing between data sources to be done implicitly using URI's.

One such effort currently underway is the Life Sciences Identifier (LSID) project [5]. In our demonstration we use LSID's as external references for OMIM records as well as for Uniprot proteins in both the target data as well as the WNT pathway data:

urn:lsid:uniprot.org:uniprot:P49841

This LSID names the protein record in Uniprot that is referred to as P49841. It consists of parts separated by colons: A prefix "urn:lsid:", The authority name; The authority-specific data namespace; and the namespace-specific object identifier ("P49841").

2.4. *Statements: the Quantum Unit of RDF*

The second constraint is that RDF data files are decomposable into fundamental units of information called statements (or triples). A statement has three parts: a subject, a predicate, and an object[2]. Here are some examples of statements:

- GSK3b is-type Protein
- GSK3b has-name "Glycogen Synthase Kinase 3 beta"
- GSK3b interacts-with betaCatenin

Each of these elements are specified by their URIs (the object could also be a string values), eliminating ambiguity for machine processing. The statement from the above set could be recorded as follows:

<urn:lsid:uniprot.org:uniprot:GSK3b>
<http://www.w3.org/1999/02/22-rdf-syntax-ns#type>
<urn:lsid:uniprot.org:uniprot:Protein>

Using the LSID resolver at http://lsid.biopathways.org, Uniprot and NCBI records are returned to BioDash as RDF statements. By breaking data files into quantum units, applications that "see" RDF statements they do not "understand" can safely ignore these (a consequence of logical monotonicity). Formats that are not based on a quantum unit, such as standard XML formats, have blurred boundaries between units of information, so the complete structure must be understood in advance. Consequently, RDF data can be combined by simple concatenation, with metadata included and additional properties appended, without any need to re-program applications that already read these files.

2.5. *Semantic Lenses*

SW requires browsers that not only collect and render the semantic documents visually, but also aggregate select information referenced by documents and data objects. Automated rules can then be applied to filter relations or create new ones so that only the relevant parts of information aggregations are shown and users are not overwhelmed by the data. The browser component that makes this possible is an intelligent information filter and viewer called a "Semantic Lens" that is created to isolate specific meaning within an arbitrary chunk of information. In BioHaystack, lenses are defined using the *Adenine* language, which supports a full set of UI components including lenses and parts[3], as well as query objects for finding and extracting

---

[2] Here "object" is used in the grammatical sense, as in object of a verb or a preposition, instead of in the sense of "entity". To avoid confusion, the SW community refers to entities as "resources".

[3] See http://haystack.csail.mit.edu/documentation/ui.pdf

additional RDF information. XML documents can be converted into RDF using XSLT scripts that can be called through Adenine. Since Adenine is declarative, there is often no need for any additional Java, C, or perl coding. BioDash is defined by the set of lenses it utilizes for Topic Views, Pathway Views, and SNP View.

```
hs:member ${
        rdf:type    data:RDFQueryAspect ;
        data:sourceExistential ?s ;
        data:targetExistential  ?t ;
        rdfs:label  "" ;
        data:existentials @( ?s ?t ?type ) ;
        data:statement ${
                data:subject        ?type ;
                data:predicate      biopax:LEFT ;
                data:object         ?s
        } ;
        data:statement ${
                data:subject        ?type ;
                data:predicate      biopax:RIGHT ;
                data:object         ?t
        }
}
```

Figure 1. Semantic Lens written in Adenine for rendering BioPAX pathways

Lenses are the SW equivalent of cascading style-sheets used by HTML browsers to enhance the HTML being viewed. However, since they handle logic and are active, semantic lenses can be applied on the back-end as well. For example, consider the problem of displaying a pathway encoded with the BioPAX standard (www.biopax.org [6]). As with any machine-readable format, BioPAX allows myriad details of the pathway to be encoded. BioPAX also defines the notion of a reaction to have a left hand side and a right hand side. Most tools today—Web browsers included—will present all of this detail on a single screen, making it difficult to decipher basic properties of the pathway such as which proteins are interacting with each other. For the purposes of an overview, it is often much more useful to filter out everything but this basic level of detail. Figure 1 shows the definition of such a filter. It directs BioHaystack to draw an arrow between the LEFT and RIGHT properties of a reaction. By defining families of lenses, completely different views of the same data can be constructed for multiple concerned parties.

3. Results: The User Experience

3.1. *The BioHaystack Semantic Web Browser*

BioDash is built on the BioHaystack Semantic Web Browser [7], a Java application that enables users to navigate, visualize, annotate, and organize data in highly-customizable ways (www.w3.org/2005/04/swls/BioDash/Demo). Similar to a traditional Web Browser, a Semantic Web Browser provides a graphical interface to data available both locally and on the network. A user can begin his or her browsing session by entering a URI into the "Go to" box. The "pages" that are shown are graphical displays containing hyperlinks to URI's (but not in HTML); by clicking on a hyperlink, a user is taken to a "new object page". The toolbar also provides the familiar "Back", "Forward", "Refresh", and "Home" buttons.

Compared to standard Web Browsers, BioHaystack provides users with improved flexibility in how they view information. Rather than viewing data through layouts predetermined by a Web site designer, BioHaystack allows users to choose the view that is most appropriate to the task at hand. Different views for the same data can be provisioned for bioinformaticists, chemists, pathologists, or other roles. Additionally, the browser itself is capable of data integration, allowing users to incorporate data from custom data sources, such as local files or secondary data stores.

3.2. *Building a Drug Target Model*

Central to most drug development strategies is the mapping of gene/protein target information to bioactive compounds. Targets are typically organized into classes, where a handful of target classes map to 80% of approved drugs. The objective is to identify new classes of targets, which are usually identified and validated based on the significance they play in a disease- a key piece of knowledge for all drug R&D. Additionally, information of "anti-targets" and secondary targets are of interest as well, since these can be used to improve compound selectivity. Such information comes from either empirical HT screening assays (compounds to isolated proteins), or from *in silico* modeling of molecular interactions between ligands and proteins.

BioDash organizes information about targets, investigated compounds, therapeutic areas, and other relevant data into *therapeutic topics*, defined by the LS-Ont bridge ontology (http://www.w3.org/2005/04/swl/BioDash/ls-ont.rdf). The BioDash topic view, seen in Figure 2, incorporates a series of views or lenses (see semantic lenses below) that give an overview of the status of the topic. The *Target Overview* lens shows the chemical entities being considered that target (arrows) GSK3b. The *Primary* (project team disease focus) and *Alternative* (potential future applications) *Disease* lenses render information about the diseases in which the target has been implicated; here we use descriptions from OMIM. The *Group Members* lens gives a listing of the people involved in the effort, as well as their roles and emails. Finally, we have included published antagonists of GSK3b in the RDF demonstration set, including their chemical structures and properties.

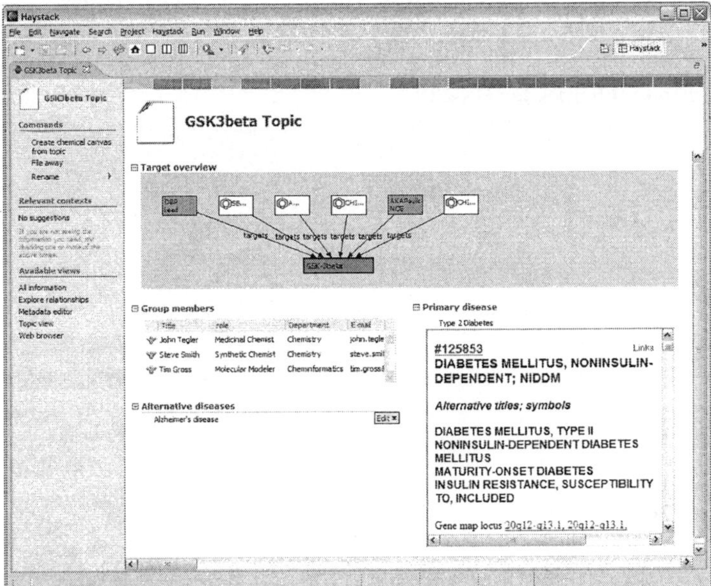

Figure 2. The Topic View containing the Target Overview, Primary Disease, and Group members lenses

3.3. *Finding Multiple Intervention Points*

A powerful way to understand the interaction of drugs with biological systems is to see the relation between compounds and the molecular pathways that they are presumed to affect. Such a perspective can be especially insightful when searching for optimal intervention points that modulate a key process with reduced chances for adverse effects. Examining the data this way often highlights differences in tissue specificity, downstream effects, and regulation type. In addition, by considering molecular processes, multi-target therapies can be developed whereby drug combinations can more effectively modulate a process.

To support this mode of investigation, we have incorporated a *Pathway View* into BioDash that can render pathways encoded using the BioPAX representation. Figure 3 shows a screenshot of BioDash rendering the WNT pathway, in which GSK3b plays a role. The Pathway View shows information that is quite distinct from the Topic View. First, information for the Pathway View comes primarily from public pathway databases (e.g., BioCyC), whereas the Topic View is populated with internal topic-tracking status data. Second, the information rendered by the two views is represented with completely different ontologies. Additionally, the Pathway View is rendered as a full-screen graph, while the Topic View is a segmented display with both graphical and tabular diagrams. Finally, while the two displays both depict GSK3b, the various data sets use different names for GSK3b.

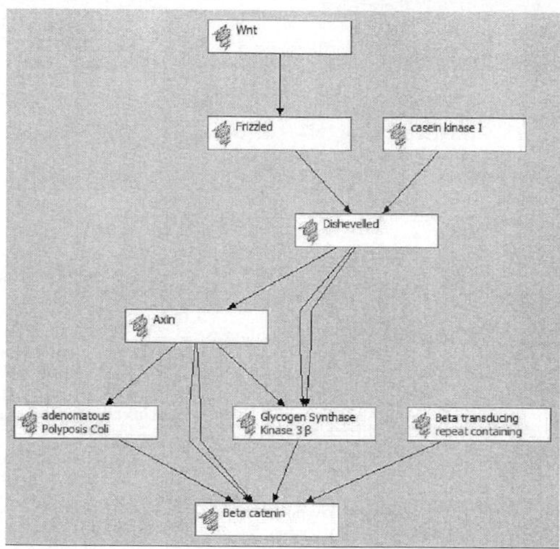

Figure 3. A BioPAX-encoded WNT pathway, rendered using a semantic lens.

Despite these differences, it is still possible to aggregate information from the two views. The Pathway View is designed to accommodate other forms of information, using proteins marked with Uniprot IDs as "pivot points" (see LSID above). In our scenario, the pivot point between the GSK3b Pathway View and the WNT Pathway View is GSK3b itself. If the user drags the red GSK3b icon from the Topic View onto the Pathway View, BioDash merges the two diagrams together (see Figure 4). The significance of this merge is twofold: first, contrary to the commonly held belief that ontologies require significant development effort to interoperate, hardly any coordination was required between the drug topic ontology and the BioPAX ontology; only a common Uniprot identifier was needed. Second, the merge exposes information (using a rule) that was not present in either of the two diagrams alone (but present in the data set): the fact that one of the chemical entities under consideration also targets casein kinase I, another player in the WNT pathway.

3.4. *Sensitivity to Polymorphisms*

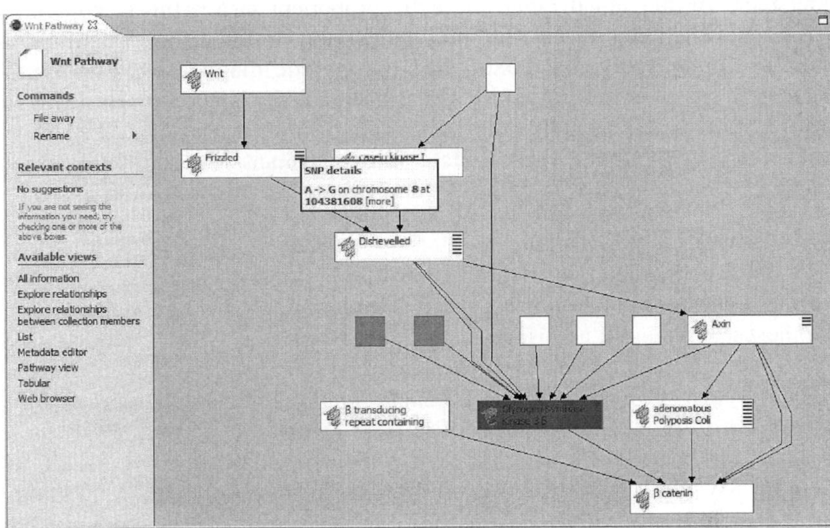

Figure 4. Non-synonymous SNP information (shown as purple dashes) aggregated onto proteins in the WNT pathway along with the compounds that target GSK3 beta all represented together as RDF.

With the advent of *personalized medicine*, pharmacogenomics will play an increasing role in assessing the safety and utility of drugs as determined by the variations of an individual's genetic background. The genotype each person inherits from their parents, tends to follow the distributions of their ancestral sub-population. Genomic variation information today can be obtained for most gene loci from single nucleotide polymorphic (SNP) databases such as dbSNP (http://www.ncbi.nlm.nih.gov/entrez/query.fcgi?db=snp). Consequently, families of polymorphisms can be aggregated onto sets of genes (proteins) that all belong to a common pathway. When overlaid onto pathways, the SNP variations along with their known population distributions can be used to predict what branches of a biological process might be more susceptible to genotypic variations, possibly affecting drug responses or causing side effects for different individuals. This information is queried from dbSNP and the returned XML results can be converted into RDF using XSLT and displayed graphically as purple bars shown on the right hand side of the protein tiles in the Pathway View. By clicking on a bar, a SNP summary pop-up appears that allows more information to be retrieved. Since the polymorphic-pathway is semantically defined as RDF, additional computational reasoning can be performed.

We specifically queried for non-synonymous polymorphisms (i.e., protein sequence changes) for each gene, since these have the highest probability of affecting the function of a pathway component directly, enabling one to understand the functional range of each component in a pathway context. At such a time when clinical data on individuals becomes available, extending this model to handle individual genotypic (via genetic diagnostics) plus clinical evidence to assess which polymorphisms influence therapeutic responses would be straightforward. Thus polymorphic mapping onto pathways could serve as a scaffold for aggregating clinical data into a semantic structure to analyze complex interactions between pathway components and their polymorphisms.

4. Conclusion

In the SW paradigm, we begin to consider biological, chemical, and clinical information as part of a viewable and computable web of related facts and hypotheses, not simply as disassociated data fragments. Many traditional data models were defined at a time when data were submitted in chunks. However, databases such as Entrez [8] and Reactome [9] have much more intrinsic connectivity to related information of diverse forms, though they represent the semantics implicitly. If the semantics were explicitly defined using RDF/OWL, emerging SW applications could make full use of their information. Some data sources including UniProt (www.isb-sib.ch/~ejain/rdf/) have already been converted to RDF. Even so, SW tools such as BioDash can already take advantage of structured life science resources by converting XML files into RDF or mapping databases to RDF using wrappers. In addition, most life science data objects and documents can be uniquely tracked with URI's, either through LSID's or URL's appended with identifiers.

In this project we demonstrate that relevant facts can be collected from multiple sources, combined semantically, and viewed using a SW browser. Semantic Web Browsers will be necessary since the full complement of RDF-based information is too complex for humans to take in all at once. In Drug Discovery, processes are segmented from each other and information from one set needs to be provided to subsequent steps (e.g., selected targets for defining HT screening), using the knowledge perspectives local to each step. It is also possible to postulate hypotheses as RDF statements, and share these points of view as part of the topic. Furthermore, such additions could be distributed using RSS (based on RDF) newsfeed technology. Some open issues still require consideration: standard do not exist yet for semantic lenses; models for aggregation or knowledge sharing are lacking; and memory limitations on the client-side may suggest that large aggregations be performed on back-end servers. Nonetheless, BioDash offers a practical test-bed for asking these questions in different contexts over a broad range of research areas.

The use of aggregators and Semantic Web Browsers can be applied to other areas requiring embedded semantics: medical language systems [10], health care management [11], chemistry [12], cancer research [13], clinical trial management [14], and analytical workflows (myGRID) [15]. Shifting emphasis to knowledge representations allows aggregation and reasoning between all information sets, and can support the managing of information across different communities. Semantic Lenses offer an intelligent and powerful means to organize interlinked information specific to a user's needs, supporting the construction and use of collective knowledge.

Acknowledgement

We would like to thank Melissa Cline, Joanne Luciano, Eric Prud'hommeaux, Susie Stephens, and John Wilbanks for their contributions to this project.

References

1. FDA Report, U.S. FDA, March 2004.
2. T. Berners-Lee, J. Hendler, O. Lassila. Sci. Am. 284:34-43 (May 2001).
3. E. Neumann, Science STKE 2005, pe22 (2005).
4. Cohen P, Goedert M. Nat Rev Drug Discov. 2004 Jun;3(6):479-87. Review. PMID: 15173837
5. T. Clark, S. Martin, T. Liefeld, Brief Bioinform. March. 5, 59–70 (2004).
6. J. Luciano, Drug Discovery Today, Vol 10, No. 13, 938-942 (2005).
7. D. Quan, D. Karger, Proceedings of the 13th International Conference on World Wide Web, pg 255-265 Association Computing Machinery Press.
8. D. L. Wheeler, *et al.,* Nucleic Acids Res. 33, D39–D45 (2005)
9. G. Joshi-Tope, M. Gillespie, I. Vastrik, P. D'Eustachio, E. Schmidt, B. de Bono, B. Jassal, G. R. Gopinath, G. R. Wu, L Matthews, S. Lewis, E. Birney, L. Stein, Nucleic Acids Res. 33, D428–D32 (2005).
10. V. Kashyap, American Medical Informatics Association Annu. Symp. Proc. 2003, 351–355 (2001).
11. G. Goebel, K. L. Leitner, K. Pfeiffer, Medinfo 2004, 1618 (2004).
12. P. Murray-Rust, H. S. Rzepa, S. M. Tyrrell, Y. Zhang, Org. Biomol. Chem. 2 (22), 3192–3203 (2004).
13. S. De Coronado, M. W. Haber, N. Sioutos, M. S. Tuttle, L. W. Wright, Medinfo 2004, 33–37 (2004).

M. N. Kamel Boulos, A. V. Roudsari, E. R. Carson, Med. Inform. Internet Med. Sep. 27, 127–137 (2002).

14. R. D. Stevens, H. J. Tipney, C. J. Wroe, T. M. Oinn, M. Senger, P. W. Lord, C. A. Goble, A. Brass, M. Tassabehji, Bioinformatics 4 (suppl 1.), I303–I310 (2004).

SEMBIOSPHERE: A SEMANTIC WEB APPROACH TO RECOMMENDING MICROARRAY CLUSTERING SERVICES

KEVIN Y. YIP[1], PEISHEN QI[1], MARTIN SCHULTZ[1], DAVID W. CHEUNG[5] AND KEI-HOI CHEUNG[1,2,3,4]

[1]*Computer Science,* [2]*Center for Medical Informatics,* [3]*Anesthesiology,* [4]*Genetics, Yale University, New Haven, Connecticut, USA,*
[5]*Computer Science, University of Hong Kong, Hong Kong*

Clustering is a popular method for analyzing microarray data. Given the large number of clustering algorithms being available, it is difficult to identify the most suitable ones for a particular task. It is also difficult to locate, download, install and run the algorithms. This paper describes a matchmaking system, SemBiosphere, which solves both problems. It recommends clustering algorithms based on some minimal user requirement inputs and the data properties. An ontology was developed in OWL, an expressive ontological language, for describing what the algorithms are and how they perform, in addition to how they can be invoked. This allows machines to "understand" the algorithms and make the recommendations. The algorithm can be implemented by different groups and in different languages, and run on different platforms at geographically distributed sites. Through the use of XML-based web services, they can all be invoked in the same standard way. The current clustering services were transformed from the non-semantic web services of the Biosphere system, which includes a variety of algorithms that have been applied to microarray gene expression data analysis. New algorithms can be incorporated into the system without too much effort. The SemBiosphere system and the complete clustering ontology can be accessed at `http://yeasthub2.gersteinlab.org/sembiosphere/`.

1. Introduction

As modern life sciences research involves high throughput bio-technologies (e.g., sequencing, DNA microarray, and mass spectrometry), a vast quantity of data is being generated, which needs to be stored, analyzed, visualized, and interpreted. As a result, a plethora of analyzing tools have been developed and made accessible via the Internet. Subsequently, it has become difficult to locate the tools that are relevant to the research questions at hand. The current web technologies rely heavily on the use of keyword-based searches in locating web resources, which suffers from the problem of sensitivity and specificity. In addition, even if the relevant software tools are found, users may still experience problems in downloading, installing, and running the programs. It is also difficult for users to keep track of the updates, such as bug fixes and the addition of new features.

1.1. *Microarray Cluster Analysis*

For microarray data, clustering is a popular data analysis method. A large collection of clustering algorithms have been developed, published, and made available in different forms through many web sites. Some of them are available as downloadable software that can be installed on the client computer. For example, Eisen's cluster program[18] can be downloaded from the Eisen Lab website and run on a Windows PC. Some algorithms are available as web server applications that allow users to submit their microarray data to the server on which the cluster analysis is performed (e.g. EPCLUST [3]). More recently, some cluster analysis algorithms have been implemented as web services using SOAP[10], which allow direct machine invocation over the Internet. Biosphere[17] is a representative example. The advantage of publishing the algorithms as web services is that all programs are invoked through the same XML-based interface, regardless of what languages they are written in, and what platforms they are running on.

The availability of a large number of algorithms provides more options to choose from, but at the same time also makes it difficult for one to determine the most suitable algorithm for the current task. This problem is especially true for many microarray experimentalists who may not know much about the technical details of the algorithms. Although reports[16] have been published on the evaluation and comparison of different algorithms, it relies on the human users to study many such reports carefully to figure out which algorithms perform better under different circumstances.

To address this problem, we employ the latest semantic web technologies to transform the basic clustering web services of Biosphere into semantic web services. The semantic layer describes what the algorithms are and how they perform in addition to how they can be invoked. It is thus possible for machines to recommend which algorithms to use based on the user requirement inputs and data properties. In the following we describe SemBiosphere, a matchmaking system that provides such recommendations.

2. SemBiosphere

Figure 1 shows the overall system architecture of SemBiosphere. First, there are algorithm providers who expose their available algorithms as web services. As discussed, the providers have the flexibility to choose the programming language and the running platform, as long as they follow the XML standards to describe their programs. The descriptions involve two parts: a lower part based on

WSDL[11] for specifying the input/output types, and an upper part for specifying semantic descriptions. We developed an ontology for clustering algorithms using OWL[8], and described the algorithms using the ontology. The two types of descriptions are combined by the latest OWL-S language[7], and are stored in a central RDF[9] repository for querying. The recommendation system is a web application that can be accessed using web browsers. It accepts both the data and the matching requirements from users through an HTML interface. It then performs filtering and ranking to produce a sorted list of algorithms according to the matching scores. The users may pick any number of them to use. The system provides a form for inputting the parameters. It then executes the algorithms, and sends an email notification to the users when the results are ready.

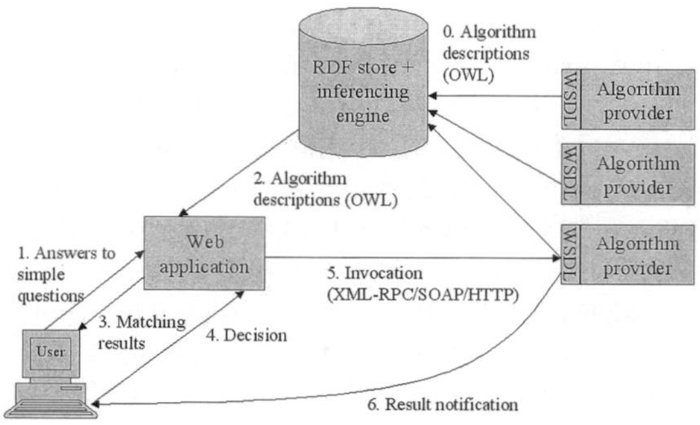

Figure 1. SemBiosphere system architecture.

Compared with other popular microarray software and systems such as Bioconductor [1] and EPCLUST [3], the SemBiosphere system is more extensible as it can integrate programs written in different languages and by different groups dynamically. The potentially larger variety of algorithms allows experimentalists to try out newer methods that may suit their specific needs more. Compared with some keyword-based searching systems, the semantic web approach allows for much more focused and structured queries and the possibility to answer questions based on logical inference rather than text associations.

There are some large-scale projects aiming at facilitating complex *in silico* experiments that adopt semantic web technologies, including myGrid [5] and BioMOBY [2]. The current study is related to, but not covered by, these projects. It is an interesting future work to see how the ideas of the current project can be applied to their general frameworks.

Below we describe in more detail some of the core components of the system.

2.1. *Ontology for Describing Clustering Algorithms*

Figure 2 shows the class hierarchy of our clustering algorithm ontology. It covers three aspects of clustering algorithms: algorithm types, the kinds of data that can be handled, and time complexities.

Algorithms are classified into common categories such as hierarchical, partitional and density-based. The ontology is extensible so that more categories can be added. The categorization is useful for making some quick decisions on which clustering algorithms to use. For example, if a user wants to have a dendrogram as part of the clustering output (such as those produced by TreeView[18]), hierarchical algorithms are by default more suitable.

Each algorithm is allowed to belong to multiple classes. For example, the algorithm ORCLUS[14] involves both a hierarchical phase and a partitional phase in each iteration. Some classes also have subclasses. For example, hierarchical clustering algorithms are further subdivided into agglomerative and divisive algorithms. Besides these main categories, some categories were also defined for special classes of clustering algorithms, such as projected clustering[22] and biclustering algorithms[15]. These categories are orthogonal to the main categories. Each algorithm belongs to at least one main category and any number of these special categories. For example, PROCLUS[13] is both a partitional clustering algorithm, and a projected clustering algorithm. These special categories are useful for matching some special requirements. For example, when the number of experiments (columns) in a dataset is large in compared to the number of genes (rows), algorithms that find clusters in subspaces (e.g. projected clustering algorithms) may perform better[22].

Another important classification of the algorithms is the kind of data attributes that they can handle. For example, some microarray data are discretized, such that the attributes are categorical. We selected a relevant subset of the attribute types in Kaufman and Rousseeuw[21] to define a hierarchy of attribute types. Some Biosphere implementations of the algorithms are modified versions that can handle some attributes that the original algorithms cannot handle. The extra capability is provided by defining special similarity functions. For example, if an algorithm relies on the distance between different expression values in the clustering process, by default they cannot handle categorical attributes in which the distance between different values is not defined. However, if a similarity function can also define the distance between different categorical values, then the algorithm can also work on categorical attributes. In view of this, we defined different similarity functions, each associated with a list of attribute types that it can handle. Each clustering algorithm then picks the similarity functions that it can use, from which the types of attributes that it

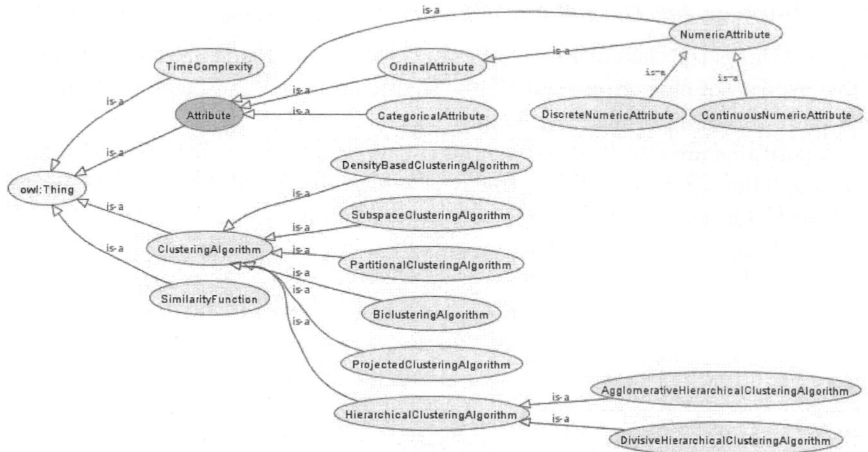

Figure 2. Clustering algorithm ontology.

can handle are defined.

Finally we defined the time complexity of each algorithm. It turned out to be very difficult to define such a category, as time complexities are expressed in functions, and may involve many variables specific to each algorithm. To make things simple, we defined several coarse complexity classes such as constant time, $O(n)$ time and $O(nlogn)$ time, and requires each algorithm to pick the tightest ones with respect to the number of experiments and genes in a dataset. This is certainly not an ideal way to specify the time complexities. Also, time complexity alone docs not directly imply the actual running time of the algorithms. However, the current categories are already very useful in some basic matching. For example, if a dataset contains more than 10000 genes and the user is only willing to wait for seconds, an algorithm with cubic time complexity with respect to the number of genes is unlikely to be a good choice.

2.2. *Semantic Clustering Web Services*

We use the KPrototype algorithm[20] as an example to explain what have to be specified for a semantic web service. Figure 3 shows its descriptions in OWL-S, which contain the following parts:

- The *service* part organizes all other parts of a web service description. The "KPrototypeClusteringService" is described by "KPrototypeClusteringProfile", presents "KPrototypeClusteringProcess" and supports

"KPrototypeClusteringGrounding". The service can be thought of as an API declaration for an entry point that a service provider wants to make accessible.

- The *profile* part tells us "what the service does"; that is, it gives the type of information needed by a service-seeking agent to determine whether the service meets its needs. It also provides the contact information of the service provider. Here, we defined "KPrototypeClusteringProfile" as an instance of "ClusteringService", which is linked to our clustering algorithm ontology by its "algorithm" property. An agent searching for clustering services can first determine the appropriate algorithms it desires by examining the ontology, and then get to the services via the "algorithm" relationships.
- The *process model* tells "how the service works". Our clustering service processes are composed of two alternative atomic processes: *getAlgorithmMetaData* and *execute*. *getAlgorithmMetaData* requires no inputs and returns the metadata description, such as the type and parameters, of the algorithm. *execute* requires three inputs: dataset URL, user email and the set of parameter values. It responds with a task ID and a security token for the retrieval of clustering results when the service is successfully executed.
- *Service grounding* specifies the details of how an agent can access a service. Here we use SOAP RPC as our communication protocol and ground our services on WSDL to specify the port used in contacting the services. Each atomic process is grounded to a WSDL operation and its inputs and outputs to the message parts of that operation. When an atomic process is invoked, the corresponding operation is called in the remote server. Inputs and outputs are transformed between semantic OWL documents and SOAP messages using the XSLT[12] stylesheet embedded in the grounding part of the OWL-S file.

2.3. *Matchmaker*

The matchmaking subsystem takes as input a microarray dataset and some other user requirements for a clustering task (Figure 4). In order not to request users to know too many technical details about the algorithms, the requirements are entered through answering several very simple questions. The system performs the matching and returns a list of algorithms in descending order of their matching scores (Figure 5). Matching details are provided at the bottom of the page for explaining how the matching was performed. From the list, the users can choose one or more algorithms to cluster the data. The request will be sent to

```
- <rdf:RDF xmlns:rdf="http://www.w3.org/1999/02/22-rdf-syntax-ns#" xmlns:xsd="http://www.w3.org/2001/XMLSchema"
    xmlns:soapenv="http://schemas.xmlsoap.org/soap/envelope/" xmlns:xsi="http://www.w3.org/2001/XMLSchema-instance"
    xmlns:soapenc="http://schemas.xmlsoap.org/soap/encoding/" xmlns:sembiosphere-service="http://mcdb750.med.yale.edu/biopax/sembiosphere-
    service.owl#" xmlns:sembiosphere="http://mcdb750.med.yale.edu/biopax/sembiosphere.owl#" xmlns:rdfs="http://www.w3.org/2000/01/rdf-schema#">
  - <sembiosphere:ClusteringAlgorithm rdf:ID="hk.hku.csis.biosphere.algorithm.KPrototype">
      <rdf:type rdf:resource="http://mcdb750.med.yale.edu/biopax/sembiosphere.owl#PartitionalClusteringAlgorithm" />
      <rdfs:comment rdf:datatype="http://www.w3.org/2001/XMLSchema#string">This programme is an implementation of the k-prototype clustering
        algorithm ("Extensions to the k-means Algorithm for Clustering Large Data Sets with Categorical Values", Z. Huang, Data Mining and Knowledge
        Discovery, 1998). The initial partition is determined by randomly drawing k sample points as the cluster seeds. This implementation also accepts
        other similarity score definitions besides the default squared Euclidean score.</rdfs:comment>
      <sembiosphere:objectGrouping rdf:datatype="http://www.w3.org/2001/XMLSchema#string">disjoint</sembiosphere:objectGrouping>
      <sembiosphere:handleOutliers rdf:datatype="http://www.w3.org/2001/XMLSchema#boolean">false</sembiosphere:handleOutliers>
    - <sembiosphere-service:parameter xmlns:sembiosphere="urn:webservices.sembiosphere.cs.yale.edu">
      + <sembiosphere-service:ClusteringAlgorithmParameter rdf:ID="config">
      </sembiosphere-service:parameter>
    - <sembiosphere-service:parameter xmlns:sembiosphere="urn:webservices.sembiosphere.cs.yale.edu">
      + <sembiosphere-service:ClusteringAlgorithmParameter rdf:ID="recOrderCol">
      </sembiosphere-service:parameter>
    - <sembiosphere-service:parameter xmlns:sembiosphere="urn:webservices.sembiosphere.cs.yale.edu">
      + <sembiosphere-service:ClusteringAlgorithmParameter rdf:ID="recOrder">
      </sembiosphere-service:parameter>
    - <sembiosphere-service:parameter xmlns:sembiosphere="urn:webservices.sembiosphere.cs.yale.edu">
      + <sembiosphere-service:ClusteringAlgorithmParameter rdf:ID="tcr">
      </sembiosphere-service:parameter>
    - <sembiosphere-service:parameter xmlns:sembiosphere="urn:webservices.sembiosphere.cs.yale.edu">
      + <sembiosphere-service:ClusteringAlgorithmParameter rdf:ID="k">
      </sembiosphere-service:parameter>
    - <sembiosphere-service:parameter xmlns:sembiosphere="urn:webservices.sembiosphere.cs.yale.edu">
      + <sembiosphere-service:ClusteringAlgorithmParameter rdf:ID="sim">
      </sembiosphere-service:parameter>
    - <sembiosphere-service:parameter xmlns:sembiosphere="urn:webservices.sembiosphere.cs.yale.edu">
      + <sembiosphere-service:ClusteringAlgorithmParameter rdf:ID="dataReader">
      </sembiosphere-service:parameter>
    - <sembiosphere-service:parameter xmlns:sembiosphere="urn:webservices.sembiosphere.cs.yale.edu">
      + <sembiosphere-service:ClusteringAlgorithmParameter rdf:ID="gamma">
      </sembiosphere-service:parameter>
    - <sembiosphere-service:parameter xmlns:sembiosphere="urn:webservices.sembiosphere.cs.yale.edu">
      + <sembiosphere-service:ClusteringAlgorithmParameter rdf:ID="us">
      </sembiosphere-service:parameter>
    - <sembiosphere-service:parameter xmlns:sembiosphere="urn:webservices.sembiosphere.cs.yale.edu">
      + <sembiosphere-service:ClusteringAlgorithmParameter rdf:ID="data">
      </sembiosphere-service:parameter>
    + <sembiosphere-service:parameter xmlns:sembiosphere="urn:webservices.sembiosphere.cs.yale.edu">
    </sembiosphere:ClusteringAlgorithm>
</rdf:RDF>
```

Figure 3. OWL-S representation of the KPrototype clustering web service.

the chosen clustering web services, and the users will be notified by email when the results are ready, which can be viewed on a result page with a standard colored graph for visualizing the clusters (Figure 6).

The current matching system adopts a rule-based approach. Matching is performed based on two types of predefined rules: filtering rules and scoring rules. Filtering rules define which algorithms cannot be used in the current clustering task. For example, if the dataset contains a certain kind of attributes that an algorithm cannot handle, the algorithm will be filtered. In general, a set of preconditions can be specified for each algorithm so that it can be used for a clustering task only if all the preconditions are satisfied.

The algorithms that remain are evaluated by the scoring rules. Each rule takes into account an evaluation criterion and gives a score for each algorithm. For example, one rule evaluates the speed performance of the algorithms in terms

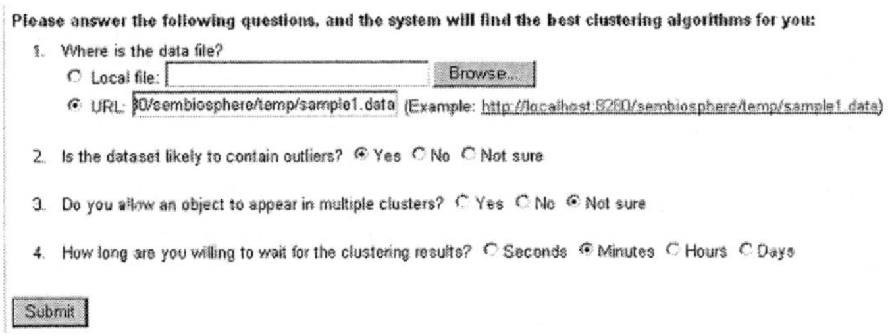
Please answer the following questions, and the system will find the best clustering algorithms for you:
1. Where is the data file?
Local file:
Browse...
URL: 80/sembiosphere/temp/sample1.data (Example: http://localhost:8280/sembiosphere/temp/sample1.data)
2. Is the dataset likely to contain outliers? Yes No Not sure
3. Do you allow an object to appear in multiple clusters? Yes No Not sure
4. How long are you willing to wait for the clustering results? Seconds Minutes Hours Days
Submit

Figure 4. User input.

Recommended algorithms:

1. PROCLUS (score = 4)
2. CAST (score = 2)
3. BAHC (score = 0)
4. HARP (score = 0)
5. KPrototype (score = -2)
6. CLARANS (score = -4)
7. DIANA (score = -6)

Dataset information:

- URL: http://localhost:8280/sembiosphere/temp/sample1.data
- Number of rows: 2884 (large)
- Number of columns: 17 (small)
  - 0 categorical
  - 17 numeric

Matching details:

Algorithm: PROCLUS (score = 4)

- The dataset does not contain categorical attributes. Whether the algorithm accepts any similarity functions that can handle categorical attributes is not important. **(score +0)**
- The dataset contains numeric attributes. The algorithm can be used since it accepts some similarity functions that can handle numeric attributes. **(score +0)**

Figure 5. Matched algorithms and match description.

of the number of genes. Basically, algorithms with higher time complexities will receive lower scores. However, the importance of the rule depends on the size of the dataset as well as the time that the user is willing to wait. If the dataset is large, an algorithm with $O(n^2)$ time complexity may run significantly faster than one with $O(n^3)$ time complexity. Yet if the dataset is small, the difference may be negligible and the scores for the two algorithms will not have a large difference. Similarly, if the user wants the clustering result be ready within seconds, the execution time is very important and the rule will give a large score difference between algorithms with different time complexities. But if the user is willing to wait for days, the rule will give similar scores to the algorithms.

The final score for an algorithm is a weighted sum of the individual scores. The weights depend on the dataset, the algorithm, and also the user require-

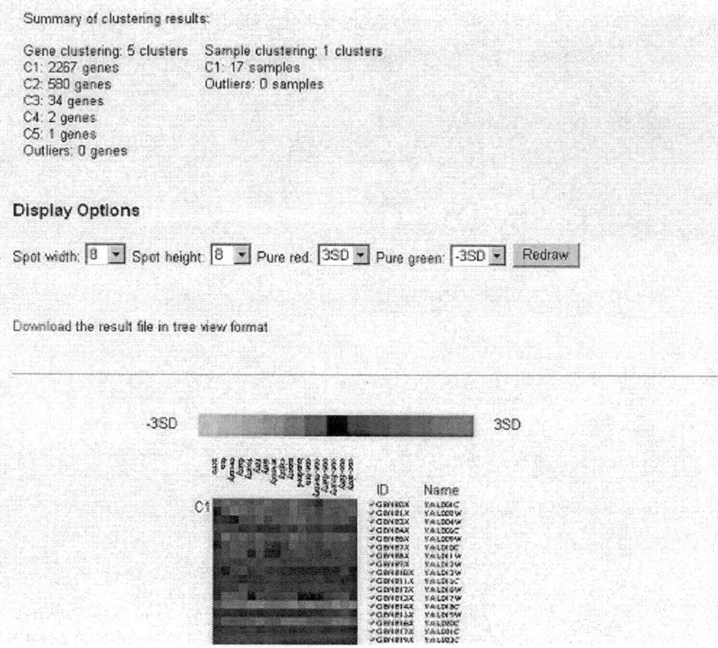

Figure 6. Visualization of clustering results.

ments. Currently the weights are set by fixed rules, but it is also possible to learn the personalized weights by collecting user feedbacks after actually running the algorithms. A difficulty is that for a reasonably large set of rules, the number of training examples required is huge. The learning may therefore need to be based on the feedbacks of all the users, but with a special emphasis on the part of the particular user. We leave this learning capability as a future research direction.

3. Discussion

While we considered only clustering algorithms in this study, many concepts are brought from the more general data analysis or machine learning domain, such as the hierarchy of attribute types. It is a good idea to have the ontology general enough to be able to merge with other machine learning ontology. However, over-generalization may make things complicated while not gaining much for the current goal. Also, if the logic inferencing capability is to be heavily used, a complicated ontology is more prone to errors and may incur a large computational overhead. Our suggestion is to keep the ontology minimal and flexible. For example, we did not attempt to include all the possible clustering algorithm types in the ontology, but we kept in mind that in future more types might be defined, so the codes were written in ways such that minimal changes are required when new types are added. This flexibility is very important in a fast growing field such as bioinformatics, since new concepts evolve rapidly. For instance, the concept of finding the corresponding subspace of each cluster has become popular only in recent years due to the production of extremely high dimensional datasets from sources such as microarray experiments. Rather than being static, the ontology needs to grow dynamically with time.

The above discussion is also related to another question: by whom the algorithm descriptions should be defined or changed? One possibility is to have the classes, properties and the relationships involved defined in a commonly agreed ontology. Each group then defines the property values of their own algorithms accordingly. The problem is that since the values directly affect the chance that the corresponding services being chosen, there has to be an objective and fair way to determine the values, which is not easy if this is done by each group individually. On the other hand, it may not be possible to have the values completely determined by a central committee since the performance of the services may be dependent on some information owned only by the service provider, such as the CPU load of their servers. Some monitoring and trust mechanisms are required in such situations. Feedbacks from the user could also help fine-tune the matching mechanism to penalize imprecise descriptions.

The current ontology is mainly based on the computational issues of the algorithms. Since the system is targeted for biologists, there has to be some means to translate their high-level, domain-specific requirements into these low-level technicalities. The current implementation performs this by providing a simple user interface with several non-technically deep questions. For the system to be widely used by biologists, it has to incorporate more biological knowledge. Our proposal is to keep a low-level ontology for computational issues, but at the same time add a high-level ontology for specifying domain-specific requirements. It would then be possible to define more formal rules for the matching between algorithm characteristics and user requirements. While there are existing bio-ontologies related to microarray experiments (e.g. MGED [4]), they are mainly for describing data properties and experiments. Such ontologies are necessary, but not sufficient for specifying the user requirements. The high-level ontology also needs to include ways for describing analysis tasks. Current efforts in this area (e.g. PMML [6]) are not tailored for biological sciences. A biologists-friendly version would be needed for capturing the specific goals and common practices of the field.

We have also observed some fundamental issues at the web service layer. One of them is the need to transfer large amount of data through the network. In our case, the dataset needs to be transmitted between the client and both the recommendation system and the clustering web services. Caching can alleviate the problem if different algorithms are applied to the same dataset, but cannot completely solve the problem. One solution is to run the algorithms at a machine close to the data, so that large data moves are localized to a small region, preferably a local area network. To achieve this, either the services have to be mirrored at different geographically areas, or there has to be some ways to "download" the service to run locally. The concept is similar to Java applets that are downloaded to run at client browsers, but here the concept is generalized in that the programs may also be run on an intermediate powerful server. In order to make this approach possible, the service programs have to be divided into two halves so that only the downloaded part works directly on the data, and only the server part connects directly to server-side resources such as backend databases. In our case, this approach seems feasible since the clustering algorithms are Java classes that can run on any machines with a Java virtual machine, and they do not need to access any server side resource directly. At the same time, server side activities such as authentication and task queuing do not need to deal with the data directly. Whether the approach is generally applicable in other situations is yet to be determined.

4. Conclusion

In this paper, we described how the semantic web approach could be used to address the problem of choosing among the many clustering algorithms available for analyzing microarray data. We demonstrated the importance of semantic web service descriptions (through the use of ontology), and how the semantic web technologies can be used to build a matchmaking system that can make recommendations to users on which clustering web services to use according to their requirements. While our results are preliminary and there are yet issues to be addressed, we believe this kind of web-services-based distributed and collaborative systems will become the trend of the near future. This project can be viewed as a small showcase for exploring the impact of semantic web in modern life sciences research.

5. Acknowledgement

We would like to thank Mark Gerstein, Andrew Smith and other members of the Gerstein Lab for their comments on a preliminary version of the system. This work was supported in part by NIH grant K25 HG02378 from the National Human Genome Research Institute, and NSF grant BDI-0135442.

References

1. Bioconductor. http://www.bioconductor.org/.
2. BioMOBY. http://www.mygrid.org.uk/.
3. EPCLUST. http://ep.ebi.ac.uk/EP/EPCLUST/.
4. The MGED ontology. http://mged.sourceforge.net/ontologies/MGEDontology.php.
5. myGrid. http://www.mygrid.org.uk/.
6. The predictive model markup language (PMML). http://www.dmg.org/pmml-v3-0.html.
7. W3C OWL-S: Semantic markup for web services. http://www.w3.org/Submission/OWL-S/.
8. W3C OWL web ontology language overview. http://www.w3.org/TR/owl-features/.
9. W3C RDF primer. http://www.w3.org/TR/rdf-primer/.
10. W3C SOAP version 1.2 part 0: Primer. http://www.w3.org/TR/2003/REC-soap12-part0-20030624/.
11. W3C web services description language (WSDL) version 2.0 part 0: Primer. http://www.w3.org/TR/2004/WD-wsdl20-primer-20041221/.
12. W3C XSL transformation (XSLT) version 1.0. http://www.w3.org/TR/xslt.
13. C. C. Aggarwal, C. Procopiuc, J. L. Wolf, P. S. Yu, and J. S. Park. Fast algorithms for projected clustering. In *ACM SIGMOD International Conference on Management of Data*, 1999.
14. C. C. Aggarwal and P. S. Yu. Finding generalized projected clusters in high dimensional spaces. In *ACM SIGMOD International Conference on Management of Data*, 2000.
15. Y. Cheng and G. M. Church. Biclustering of expression data. In *Proceedings of the 8th International Conference on Intelligent Systems for Molecular Biology*, 2000.
16. S. Datta and S. Datta. Comparisons and validation of statistical clustering techniques for microarray gene expression data. *BioInformatics*, 19(4):459–466, 2003.
17. R. de Knikker, Y. Guo, J. long Li, A. K. H. Kwan, K. Y. Yip, D. W. Cheung, and K.-H. Cheung. A web services choreography scenario for interoperating bioinformatics applications. *BMC Bioinformatics*, 5(25), 2004.
18. M. B. Eisen, P. T. Spellman, P. O. Brown, and D. Botstein. Cluster analysis and display of genome-wide expression patterns. *Proc. Natl. Acad. Sci. USA*, 95:14863–14868, 1998.
19. J. A. Hartigan and M. A. Wong. A K-means clustering algorithm. *Applied Statistics*, 28, 1979.
20. Z. Huang. Clustering large data sets with mixed numeric and categorical values. In *The First Pacific-Asia Conference on Knowledge Discovery and Data Mining*, 1997.
21. L. Kaufman and P. J. Rousseeuw. *Finding Groups in Data: An Introduction to Cluster Analysis*. Wiley Inter-Science, 1990.
22. K. Y. Yip, D. W. Cheung, M. K. Ng, and K.-H. Cheung. Identifying projected clusters from gene expression profiles. *Journal of Biomedical Informatics (JBI)*, 37(5):345–357, 2004.

EXPERIENCE IN REASONING WITH THE FOUNDATIONAL MODEL OF ANATOMY IN OWL DL

SONGMAO ZHANG [1], OLIVIER BODENREIDER [2], CHRISTINE GOLBREICH [3]

[1] *Institute of Mathematics, Academy of Mathematics and System Sciences, Chinese Academy of Sciences, Beijing, China {smzhang@math.ac.cn}*

[2] *U.S. National Library of Medicine, National Institutes of Health, Bethesda, MD, USA {olivier@nlm.nih.gov}*

[3] *LIM, University Rennes 1, Rennes, France {christine.golbreich@univ-rennes1.fr}*

The objective of this study is to compare description logics (DLs) and frames for representing large-scale biomedical ontologies and reasoning with them. The ontology under investigation is the Foundational Model of Anatomy (FMA). We converted it from its frame-based representation in Protégé into OWL DL. The OWL reasoner Racer helped identify unsatisfiable classes in the FMA. Support for consistency checking is clearly an advantage of using DLs rather than frames. The interest of reclassification was limited, due to the difficulty of defining necessary and sufficient conditions for anatomical entities. The sheer size and complexity of the FMA was also an issue.

1 Introduction

As virtually all other biomedical ontologies relate to them, reference ontologies for core domains such as anatomical entities and small molecules form the backbone of the Semantic Web for Life Sciences. One such ontology is the Foundational Model of Anatomy (FMA). However, existing reference ontologies sometimes need to be adapted to Semantic Web technologies before they can actually contribute to the Semantic Web. Converting ontologies into the formalisms supported by the Semantic Web can also benefit these ontologies as such formalisms enable consistency checking and reasoning support. This study explores the benefits of converting the FMA, a large reference ontology, to the Web Ontology Language OWL.

Biomedical terminologies and ontologies are increasingly taking advantage of Description Logic (DL)-based formalisms in representing knowledge. GALEN [1] and SNOMED Clinical Terms® (SNOMED CT)[2] were both developed in a native DL formalism. Other terminologies have been converted into DL formalism, including the UMLS® Metathesaurus® [1-3] and Semantic Network [4], the Medical Subject Headings (MeSH) [5], the Gene Ontology™ [6] and the National Cancer Institute Thesaurus [7].

[1] http://www.opengalen.org/

[2] http://www.snomed.org/snomedct_txt.html

However, many ontologies developed in the frame paradigm – often with the ontology editor Protégé (e.g., the Foundational Model of Anatomy) – cannot benefit from the reasoning support provided by description logics and cannot directly contribute to the Semantic Web.

While developed out of frame-based structures, description logics provide more precise specification of domain knowledge and enable powerful reasoning support. The most popular description logic formalism is currently the Web Ontology Language (OWL) [8, 9]. Serving as the logical basis for the Semantic Web, OWL is used to formalize a domain, assert properties about individuals and reason about classes and individuals. OWL comes in three flavors (OWL Lite, OWL DL and OWL Full), corresponding to different levels of expressivity (i.e., what knowledge can be represented with the language) and decidability (i.e., whether reasoning support is assured). OWL Full is maximally expressive but undecidable; in contrast, OWL Lite is efficient but has limited expressivity. Based on description rather than predicate logic, OWL DL offers a trade-off between expressivity and decidability. All versions of OWL use the Semantic Web technology RDF (Resource Description Framework) for their syntax.

In previous work, we proposed a method for converting the Foundational Model of Anatomy (FMA) from its original frame-based representation to OWL DL [10, 11]. In addition to the conversion process, this study focuses on the reasoning support enabled by OWL DL for the FMA. Dameron et al. have explored the conversion of the FMA to OWL Full rather than OWL DL [12]. Their goal is to stay as close as possible to the Protégé representation constructs, which is not possible with OWL DL (e.g., representing metaclasses). Beck et al. also transformed the FMA into a description logic-based representation (but not OWL), with special emphasis on the representation of partitive relations ("Structure-Entirety-Part triplets") [13].

This study is composed of two parts: conversion and reasoning. Section 3 presents the conversion rules we established to automatically convert the FMA from its frame-based representation in Protégé into OWL DL, followed by results (section 4) and optimization issues (section 5). A reasoner (Racer) is used to reason over the OWL version of the FMA converted from Protégé. After a brief overview (section 6), we study satisfiability (section 7) and reclassification (section 8). The benefits and limitations of using description logic to model the FMA are discussed in section 9. Examples of FMA classes in OWL DL could not be included in this manuscript, but are available as supplementary material at: `mor.nlm.nih.gov/pubs/supp/2006-psb-sz/`. They are referenced by "Supp x" markers.

2 The Foundational Model of Anatomy

The Foundational Model of Anatomy[3] (FMA) is an evolving ontology that has been under development at the University of Washington since 1994 [14, 15]. The FMA is implemented in Protégé[4], a frame-based ontology editing and knowledge acquisition environment developed at Stanford University [16]. The objective of the FMA is to conceptualize the physical objects and spaces that constitute the human body. 70,169 classes cover the entire range of macroscopic, microscopic, and subcellular canonical anatomy. Additionally, 187 slots are specified and used. Seven of them correspond to partitive relationships (*e.g.*, *CONSTITUTIONAL_PART_OF* and *2D_PAR_ OF*). In canonical anatomy, all partitive relationships have inverses (*e.g.*, *CONSTITUTIONAL_PART* and *2D_PART*, respectively). 80 slots represent associative relationships between classes, of which 42 have inverses (*e.g.*, *BRANCH* / *BRANCH_OF* and *CONTAINS* / *CONTAINED_IN*); *CONTINUOUS_WITH* is its own inverse; 37 slots do not have inverses (*e.g.*, *FASCICULAR_ARCHITECTURE* and *HAS_WALL*). In addition to slots linking classes, there are 61 slots in FMA describing atomic properties of classes (*e.g.*, the slot *HAS_MASS* accepts a Boolean value: TRUE or FALSE). Finally, 32 slots in the FMA link classes to instances[5] (e.g., *LOCATION* and *PREFERRED_NAME*).

In order to reduce the number of classes under investigation while keeping most of the complexity of the FMA, we ignored the classes differing from their parents solely by laterality (e.g., *Left ligament of wrist* vs. *Ligament of wrist*). The remaining subset comprises 39,337 classes. A CLIPS representation of the FMA was generated in Protégé, provided by the FMA developers. The features in the CLIPS representation of FMA are generally the same as in the Protégé environment. However, slots typed as Boolean in the Protégé environment are represented as type SYMBOL in CLIPS (with allowed-values of TRUE and FALSE). The version of the FMA used in this study is dated of July 2004.

3 Conversion rules

Ontologies developed in Protégé[6] are composed of classes and instances, the classes being organized in a taxonomy. Slots and facets are another important component of frame-based systems: slots specify relationships between classes and describe class properties; facets express constraints on slots. OWL ontologies contain classes, proper-

---

[3] http://fma.biostr.washington.edu/

[4] http://protege.stanford.edu/

[5] Instances in FMA correspond to special types of slot values, not to the realization of anatomical concepts as it is generally understood. (See Supp 4 and Supp 9 for examples)

[6] Throughout this paper, Protégé refers to the "core Protégé", i.e., the frame-based editor, ignoring its popular OWL plugin.

ties and individuals. Classes are specified by necessary conditions and/or defined by necessary and sufficient conditions.

We designed conversion rules and implemented them in order to convert the FMA into OWL DL automatically. In practice, our tools convert the original CLIPS file into an OWL file. The conversion can be summarized as follows. Classes in Protégé become classes in OWL DL[7]. Slots in Protégé become properties in OWL DL (including annotation properties). Finally, necessary and sufficient conditions are defined for OWL DL classes.

3.1 *Converting slots of the FMA in Protégé into properties in OWL DL*

All slots used in the FMA are represented in a top-level slot class. Each of these slots is converted into a property in OWL DL. Slots have a type specification (*e.g.*, INTEGER and SYMBOL) and constraints about the allowable values (*i.e.*, in allowed-parents / allowed-classes / allowed-values), which are used to delimit the type and range of property in OWL DL, as shown in Table 1. Additionally, the number of values allowed in a slot (single-slot or multi-slot specification) corresponds to the cardinality (at most one or multiple) of the corresponding property. Slots with single-slot specification are converted into functional properties in OWL DL. Finally, slots having inverses (inverse-slot specification) are converted as to stand in a owl:inverseOf relation in OWL DL; when a slot is its own inverse, the corresponding property becomes symmetric in OWL DL.

Slot of the FMA in CLIPS	Property in OWL DL
Typed INTEGER, FLOAT or STRING	owl:DatatypeProperty with range being XML Schema datatypes integer, float and string, respectively
Typed SYMBOL with allowed-values TRUE and FALSE	owl:DatatypeProperty with range being XML Schema datatype Boolean
Typed SYMBOL with allowed-values that are neither TRUE nor FALSE	owl:ObjectProperty with range being an enumerated class of all individuals in allowed-values
Typed SYMBOL with allowed-parents	owl:ObjectProperty with range being owl:unionOf all classes in allowed-parents
Typed INSTANCE with allowed-classes	owl:ObjectProperty with range being owl:unionOf all classes in allowed-classes

Table 1 – Rules for converting slots of the FMA into properties in OWL DL

In addition to the overall top-level definition, slots can be introduced in class descriptions in CLIPS, representing that the class is allowed to have the slot. We use such specification to delimit the domain of property in OWL DL. If slot *S* is introduced in

[7] OWL classes are either named or unnamed. Throughout this paper, unless we explicitly specify "unnamed", "class" refers to named classes.

class X, then X becomes an element of the domain of the property S. As one slot can be introduced into multiple classes, the domain of S is the union of all these classes. (see Supp 1- Supp 4 for examples)

In order to convert slots of type SYMBOL with allowed values other than TRUE or FALSE into properties having an enumerated class as their range, one individual has to be generated in OWL DL for each of the allowed values of these slots (see Supp 5).

3.2 *Converting classes of the FMA in Protégé into classes in OWL DL*

Every class of the FMA is represented both as a metaclass and an instance of another metaclass in Protégé, "as a technical solution for enabling the selective inheritance of attributes" [16]. The *metaclass definition* of a class, inherited by its subclasses, specifies its name, its direct superclass(es), and the slots introduced in this class. Therefore, allowable slots for a class include the slots introduced in this class and those inherited from its superclasses. In contrast, the *instance definition* of the class, not inheritable, specifies the metaclass of which this class is an instance (*i.e.*, metaclass instantiation), and all the values for the slots in this class. When converted into a class in OWL DL, the metaclass and instance definitions of the class in Protégé are merged, as shown in Table 2. (See Supp 6-Supp 8 for examples).

	Class of the FMA in CLIPS	Class in OWL DL
Metaclass definition	Every taxonomic relation to direct superclass	rdfs:subClassOf axiom to a named class representing the direct superclass
	Every slot introduced with allowed-parents (or allowed-classes)	property restriction with owl:allValuesFrom constraint on owl:unionOf all classes in allowed-parents
	Every slot introduced with allowed-values and a concrete value specification	property restriction with owl:hasValue constraint on the value
Instance definition	Metaclass instantiation	rdfs:subClassOf axiom to a named class representing the metaclass that this class is an instance of
	Every slot value where the slot is converted into a datatype property	property restriction with owl:hasValue constraint on the value
	Every slot value where the slot is converted into an object property ranging over an enumerated class	rdfs:subClassOf axiom to the property restriction with owl:hasValue constraint on the value
	Every slot value where the slot is converted into an object property ranging over a named class or disjunction of named classes	property restriction with owl:someValuesFrom constraint on the value

Table 2 – Rules for converting classes of the FMA into classes in OWL DL

Attributed slots are used to represent the properties of relations. For example, because the partitive relation between *Wall of esophagus* and *Esophagus* is not shared with other anatomical structure, *unshared* is an attribute of this ATTRIBUTED_PART slot. Attributed slots and their values are converted into subClassOf axioms to the property restrictions with owl:someValuesFrom constraints on the nested classes generated for the values (see Supp 9 for an example).

3.3 *Defining classes in OWL DL by necessary and sufficient conditions*

In modeling classes, OWL distinguishes between two types of conditions: *necessary and sufficient conditions* (owl:equivalenceClass) which define classes *and necessary conditions* (owl:subClassOf). Slot values generally correspond to necessary conditions. However, there is no correspondence in Protégé for necessary and sufficient conditions in OWL. One trivial solution consists of simply describing the classes with necessary conditions rather than defining them with necessary and sufficient conditions. In this case, only limited reasoning support can be expected, as reasoners such as Racer rely in part on defined classes. Alternatively, we had to select – somewhat arbitrarily – the properties that would define FMA classes. Intuitively *and as a first approximation*, we considered anatomical structures to be "the sum of their parts" and selected one of the mereological views, the slot *CONSTITUTIONAL PART* – with all its values – as the source for necessary and sufficient conditions for classes in OWL.[8] Other slots and combination thereof could also be selected, leading to different reasoning results in OWL. (See Supp 10 for an example).

In addition to necessary and sufficient conditions, defined classes can also have necessary conditions, called *global axioms* in this case (coming from slots other than those selected for necessary and sufficient conditions). However, global axioms are known to dramatically increase the reasoning complexity in Racer and were therefore purposely removed from defined classes.

3.4 *Designating annotation properties in OWL DL*

Similarly to the necessary and sufficient conditions of classes, annotation properties in OWL have no direct correspondence in Protégé. Slots for identifiers and names of anatomical structures (e.g., *UWDAID*, *PREFERRED_NAME* and *SYNONYMS*) typically become annotation properties in OWL DL. Such slots must be identified manually. Their values are converted into data literals in OWL DL. (See Supp 11 and Supp 12 for examples).

---

[8] A class is defined to be equivalent to a conjunction of its direct superclasses, the metaclass of which this class is an instance, someValuesFrom restriction on S over U_1, ..., and someValuesFrom restriction on S over U_n

4 Results of the conversion

After the conversion of the 39,337 classes and 187 slots from FMA in Protégé (ignoring laterality distinctions), FMAinOWL contains 39,337 classes, 187 properties and 85 individuals. Among the properties, 20 correspond to annotations (including 3 from attributed slots), 19 to datatypes and 148 to object properties (including 29 from attributed slots). 115,203 subClassOf axioms are generated, including 39,331 from taxonomy and 3,406 from metaclass instantiation. Additionally, 2,310 nested classes are generated for the values of attributed slots, and 9,092 subClassOf axioms are contained in these nested classes. 559 classes are defined through equivalentClass axioms after using slot *CONSTITUTIONAL_PART* as source of necessary and sufficient conditions. With these defined classes, the total number of subClassOf axioms in FMAinOWL has decreased to 107,238, including 38,772 from taxonomy, and 3,378 from metaclass instantiation.

5 Optimizing the conversion

Optimization techniques have been explored to downsize FMAinOWL, for the purpose of enabling the OWL reasoning and to make it more efficient. Unlike removing global axioms as presented earlier, the optimization does not change the logical definitions or reasoning results of FMAinOWL.

Optimizing domains. As stated earlier, the domain of a property in OWL DL is the disjunction of all classes where the corresponding slot is introduced. Some properties contain a large number of classes in their domains (*e.g.*, 1,618 for location), leading to inefficient OWL reasoning. Classes that are descendants of other classes in the domain can be removed from the domain without changing the definition or application of the property. The optimization results in downsizing the domain of 40 of the 187 properties. For example, only 2 classes remain in the domain of location after optimization.

Optimizing subClassOf axioms. As stated earlier, classes receive subClassOf axioms to named classes in OWL DL from two sources: taxonomic relations and metaclass instantiation. For example, class *X* is represented as "X is-a Y" in metaclass definition and "[X] of Z" in instance definition. Optimization techniques prevent the generation of 28 reflexive subClassOf axioms (from "[X] of X"), 24,307 duplicate subClassOf axioms (from "X is-a Y" and "[X] of Y") and 11,430 transitively redundant axioms (from "X is-a Y" and "[X] of Z" where Y is a descendant of Z). Overall, only 9% of 39,337 classes end up having subClassOf axioms from both taxonomy and metaclass instantiation.

6 Reasoning over FMAinOWL with Racer

Besides making it available in a popular formalism, the principal motivation for converting the FMA into OWL is to benefit from reasoning support. Because it is mapped to description logic, OWL DL makes use of existing reasoners such as Racer [17]. Reasoning support allows users to check the consistency of the onlology and the hierarchical organization of the classes (classification). Unlike consistency checking, classification requires classes to be defined with necessary and sufficient conditions, not only described with necessary conditions.

The sheer size and complexity of FMAinOWL, even after limiting the number of classes and optimizing the conversion, caused Racer to fail to reason over the whole file. Extracting a subset (e.g., for the cardiovascular system) would alleviate this problem but is likely to hide issues specific to other subsets. Instead, we elected to reason over the whole domain. As suggested by the developers of Racer, we tested only a limited number of properties at a given time (*e.g.*, no properties, only Boolean typed properties, two inverse object properties). In practice, we generated smaller versions of the FMAinOWL file, containing all classes but limited to the properties to be tested. Version 1.7 of Racer was used in this study on a Microsoft Windows platform.

7 Checking consistency: class satisfiability

We checked the consistency of the ontology based on Boolean properties and on the domain and range of properties. Importantly, not only do the descendants of unsatisfiable classes become unsatisfiable themselves, but this is also the case of all classes which have an unsatisfiable class as value for some property.

7.1 *Consistency based on Boolean datatype properties*

In the FMA in Protégé, Boolean slots are used to record differentiae between high-level anatomical categories. For example, material physical anatomical entities have mass (hasValue (has_mass true)), while non-material physical anatomical entities do not (hasValue (has_mass false)). Classes specified as descendants of both Material_physical_anatomical_entity and Non-material_physical_anatomical_entity were identified as unsatisfiable by Racer.

113 such classes were identified by Racer in FMAinOWL. Inconsistencies were traced back to inconsistent descriptions in the FMA (39 cases) and to the conversion process (74 cases). Examples of **inconsistent descriptions in the FMA** include the class Zone_of_cell. This class inherits hasValue (has_mass true) from its ancestor Material_physical_anatomical_entity and has value false in its own slot has_mass. Note that because Zone_of_cell is unsatisfiable, all its descendants also become unsatisfiable. During the **conversion process**, we showed that both taxonomic relations and metaclass instantiation are converted into subClassOf axioms (section 3.2). Merging the two definitions may result in conflicting values for a given Boolean property. For example, the class Compartment_subdivision is a descendant of Material_physical_anatomical_entity and an instance of the metaclass Anatomical_space, itself a descendant of Non-material_physical_anatomical_entity. Again, this

class inherits both true and false for the property has_mass and is therefore unsatisfiable (as are, in turn, its descendants).

7.2 *Consistency based on the domain and range of object properties*

Racer checks the consistency between the domain and range defined for a given object property *P* (see 3.1) and the restriction(s) involving this property in the definition of a class *C*. Consistency checking based on domain and range in OWL is different from type checking in programming languages. Here, consistency implies that the intersection between the domain (or range) of *P* and the value of *P* in *C* is not empty. The class *C* is declared unsatisfiable by Racer if this condition is not met. For example, the property *D2D_PART* has range Non-Material_physical_anatomical_entity. In the class Surface_of_wrist, the property *D2D_PART* has value Anatomic_snuff_box, a descendant of Material_physical_-anatomical_entity. The value of *D2D_PART* in Surface_of_wrist is disjoint from the range of *D2D_PART*. The class Surface_of_wrist is thus identified as unsatisfiable by Racer. Overall, this error is the only one revealed by this type of consistency checking in the FMA.

8 Reclassification

The whole taxonomy of the FMA is built manually by the domain experts under FMA-specific modeling principles [15]. In contrast, Racer automatically recreates the class hierarchies based on the definition of the classes. Discrepancies between the original taxonomy and Racer's hierarchy, i.e., reclassified classes, typically correspond to inconsistent descriptions in the FMA or issues in the conversion process. As for unsatisfiability, reclassification may have far-reaching effects due to propagation.

8.1 *Reasoning on necessary and sufficient conditions*

Based on necessary and sufficient conditions (i.e., the property *CONSTITUTIONAL_PART* in this experiment), 286 classes were reclassified by Racer, bringing to light the following issues: sibling classes having the same constitutional parts become equivalent; a class and its direct superclass having the same constitutional parts become equivalent; a class and its direct superclass become equivalent when the class and one of its indirect superclasses have the same constitutional parts; and a class becomes a subclass of its sibling. An analysis of some of the classes reclassified confirms that the property *CONSTITUTIONAL_PART* – as currently defined in the FMA – is not a reliable source of necessary and sufficient conditions. For example, the class Atrioventricular_valve and its two direct subclasses Mitral_valve and Tricuspid_valve are all identified as equivalent because Mitral_valve and Tricuspid_valve have the same constitutional parts as their indirect superclass Cardiac_valve.

8.2 *Reasoning on transitive properties*

Partitive relationships among (canonical) anatomical entities are generally transitive. Unlike in Protégé, the transitivity of properties is supported in OWL DL. The property *CONSTITUTIONAL_PART*, for example, was defined as transitive, which helped identify additional issues in the class definitions in FMAinOWL, most of which being related to selecting *CONSTITUTIONAL_PART* as the source for necessary and sufficient conditions. One such issue can be summarized as follows. The constitutional parts of Prostate include, by transitivity, cells such as Luminal_cell_of_prostatic_acinus, which have the same values for *CONSTITUTIONAL_PART* as Cell. This causes Prostate to be reclassified – along with 137 other classes – as a direct subclass of Cell. Our point here is to not to argue whether prostate is a kind of cell or not, but rather to emphasize the power of reasoning in identifying modeling or conversion insufficiencies. Of course, adding constraints to the definition of Cell would prevent such infelicitous reclassification [11].

9 Discussion

Reasoning support as a quality assurance tool. Large ontologies are notoriously difficult to develop and maintain in a consistent state, especially when they are developed with little or no support for consistency checking. Frame-based ontology environments such as Protégé do accommodate plugins allowing users to perform consistency checks, but offer little built-in support for consistency checking. "DL-izing" the FMA makes it amenable to reasoning support and can therefore be used for quality assurance purposes. In our experience with the FMA, consistency checking helped detect modeling errors otherwise difficult to identify (e.g., low-level classes inheriting from two disjoint high-level classes). The benefit in terms of reclassification is more subtle, due to the difficulty of defining necessary and sufficient conditions for anatomical entities.

More work is needed to determine the place of DL-based techniques in the validation and verification of ontologies. While our experience seems consistent with recent work on ontology "debugging" [18], such techniques are certainly complementary to visual (e.g., Jambalaya plugin) or other validation approaches (e.g., [19, 20]).

Size matters. The FMA is one of the largest ontologies developed with Protégé so far, and probably the largest to be converted from Protégé to OWL DL. In comparison to the 70,169 classes and 187 properties of the FMA, the NCI Thesaurus contains "only" about 34,000 classes, 100 properties, and 9,000 conditions of classes [21]. Moreover, no necessary and sufficient conditions are defined, nor are any owl:hasValue or owl:allValuesFrom restrictions specified in the NCI Thesaurus. The sheer size and complexity of the FMA represented an issue not for the conversion, but for the reasoning. In fact, Racer could not digest the entire FMA, even after removing 43% of its classes and optimizing the representation. In order to enable consistency checking, properties had to be tested individually or in small groups rather than all together. Reasoning over large ontologies remains technically challenging.

Necessary and sufficient conditions. The biggest challenge in this experiment is certainly to define the classes in OWL DL automatically by selecting the appropriate necessary and sufficient conditions. Other attempts to convert existing ontologies or terminologies into OWL generally did not address necessary and sufficient conditions (e.g., [21]) or deferred this issue to the applications using these ontologies (e.g., [12]). We attempted to define anatomical entities by combining the following properties into necessary and sufficient conditions: taxonomic relations, metaclass instantiation and constitutional parts. This simple method can be automatically implemented as part of the conversion process, but is insufficient in many respects. Defining anatomical entities solely on the basis of their constitutional parts is not correct, in part because no such constitutional parts are defined for most FMA classes. The absence of precisely defined classes was a serious limitation for reasoning support, especially reclassification. A closer collaboration with the authors of the FMA should lead to better class definitions for anatomical entities. Analogously, our conversion strategy generally consisted in preserving most of the features of the frame representation. However, a better understanding of the original modeling choices in Protégé for the FMA would certainly result in a more accurate representation in OWL. Alternative representations should be tested and evaluated.

Acknowledgments

This research was supported in part by the Intramural Research Program of the National Institutes of Health (NIH), National Library of Medicine (NLM).This work was done while Songmao Zhang and Christine Golbreich were visiting scholars at the Lister Hill National Center for Biomedical Communications, NLM, NIH, Thanks for their support to the developers of Racer: Volker Haarslev and Ralf Möller. Our thanks also go to Cornelius Rosse, José Mejino and Todd Detwiler for the FMA.

References

1. Cornet R, Abu-Hanna A. Usability of expressive description logics--a case study in UMLS. Proc AMIA Symp 2002:180-4
2. Hahn U, Schulz S. Towards a broad-coverage biomedical ontology based on description logics. Pac Symp Biocomput 2003:577-88
3. Pisanelli DM, Gangemi A, Steve G. An ontological analysis of the UMLS Methathesaurus. Proc AMIA Symp 1998:810-4
4. Kashyap V, Borgida A. Representing the UMLS Semantic Network using OWL: (Or "What's in a Semantic Web link?"). In: Fensel D, Sycara K, Mylopoulos J, editors. The SemanticWeb - ISWC 2003. Heidelberg: Springer-Verlag; 2003. p. 1-16

5. Soualmia L, Golbreich C, Darmoni S. Representing the MeSH in OWL: Towards a semi-automatic migration. Proceedings of the KR 2004 Workshop on Formal Biomedical Knowledge Representation 2004:81-87 http://CEUR-WS.org/Vol-102/soualmia.pdf.
6. Wroe CJ, Stevens R, Goble CA, Ashburner M. A methodology to migrate the Gene Ontology to a description logic environment using DAML+OIL. Pac Symp Biocomput 2003:624-35
7. Golbeck J, Fragoso G, Hartel F, Hendler J, Oberthaler J, Parsia B. The National Cancer Institute's thesaurus and ontology. Journal of Web Semantics 2003;1(1) http://www.websemanticsjournal.org/ps/pub/2004-6.
8. Antoniou G, van Harmelen F. Web Ontology Language: OWL. In: Staab S, Studer R, editors. Handbook on Ontologies: Springer-Verlag; 2004. p. 67-92
9. Smith MK, Welty C, McGuinness DL. OWL Web Ontology Language Guide, W3C Recommendation, 10 February 2004; 2004 http://www.w3.org/TR/2004/REC-owl-guide-20040210/.
10. Golbreich C, Zhang S, Bodenreider O. Migrating the FMA from Protégé to OWL. Proceedings of the Eighth International Protégé Conference 2005:108-111
11. Golbreich C, Zhang S, Bodenreider O. The Foundational Model of Anatomy in OWL: Experience and perspectives. Proceedings of the workshop "OWL Experiences and Directions", November 11-12, 2005, Galway, Ireland 2005:(electronic proceedings: http://CEUR-WS.org)
12. Dameron O, Rubin D, Musen A. Challenges in converting frame-based ontology into OWL: the Foundational Model of Anatomy case-study. Proc AMIA Symp 2005:(in press)
13. Beck R, Schulz S. Logic-based remodeling of the Digital Anatomist Foundational Model. AMIA Annu Symp Proc 2003:71-5
14. Rosse C, Mejino JL, Modayur BR, Jakobovits R, Hinshaw KP, Brinkley JF. Motivation and organizational principles for anatomical knowledge representation: the digital anatomist symbolic knowledge base. J Am Med Inform Assoc 1998;5(1):17-40
15. Rosse C, Mejino JL, Jr. A reference ontology for biomedical informatics: the Foundational Model of Anatomy. J Biomed Inform 2003;36(6):478-500
16. Noy NF, Musen MA, Mejino JLV, Rosse C. Pushing the envelope: challenges in a frame-based representation of human anatomy. Data & Knowledge Engineering 2004;48(3):335-359
17. Haarslev V, Möller R. Racer: An OWL reasoning agent for the Semantic Web. Proceedings of the International Workshop on Applications, Products and Services of Web-based Support Systems 2003:91-95
18. Schlobach S. Debugging and semantic clarification by pinpointing. In: The Semantic Web: Research and Applications. Proceedings of the Second European Semantic Web Conference: Springer-Verlag; 2005. p. 226-240
19. Noy NF, Musen MA. PROMPT: algorithm and tool for automated ontology merging and alignment. Proc of AAAI 2000:450-455
20. Zhang S, Bodenreider O. Law and order: Assessing and enforcing compliance with ontological modeling principles. Computers in Biology and Medicine 2005:(in press)
21. de Coronado S, Haber MW, Sioutos N, Tuttle MS, Wright LW. NCI Thesaurus: using science-based terminology to integrate cancer research results. Medinfo 2004;11(Pt 1):33-7

THE CHALLENGE OF PROTEOMIC DATA, FROM MOLECULAR SIGNALS TO BIOLOGICAL NETWORKS AND DISEASE

BOBBIE-JO WEBB-ROBERTSON, WILLIAM CANNON, JOSHUA ADKINS, DEBORAH GRACIO

Pacific Northwest National Laboratory
Richland, WA 99352

Mass spectrometry (MS) based proteomics is a rapidly advancing field that has great promise for both understanding biological systems as well as advancing the identification and treatment of disease. Breakthroughs in science and medicine due to proteomics, however, are coupled with our ability to overcome significant challenges in the field. These challenges are multi-scalar, spanning the range from the statistics of molecules and molecular signals, to the phenomenological characterization of disease. The papers presented in this section are a representative snapshot of these challenges that span scale and scientific disciplines.

The multi-scalar challenges are hinted at in figure 1, which depicts a typical MS-based proteomics analysis that is performed in many laboratories. Proteins are first extracted from cells and then may be cut at defined locations in the sequence by adding enzymes called peptidases to the protein extract. The solution of peptides is then partially separated by the use of liquid chromatography. The partially separated peptides are shown in the chromatogram in the top panel of figure 1. Each peak in the chromatogram consists of multiple peptides. A peak is then analyzed by the mass spectrometer attached at the end of the chromatography system. The co-eluting peptides are then introduced into the gas phase and, as shown in the middle panel of figure 1, separated by their mass-to-charge ratios in the mass spectrometer. Ideally, the peptides have been completely separated from each other at this stage. The analysis may stop at this stage, or peaks from the initial mass spectrum may be isolated, and those peptides can be subject to a second round of analysis where the isolated peptides are vibrationally excited by collision with an inert gas. The peptides then fragment at labile bonds and a subsequent mass spectrum is obtained of the fragments of the peptide, shown in the bottom panel of figure 1. Because the peptides tend to fragment into recognizable patterns, the identity of the peptide can frequently be determined from this mass spectrum.

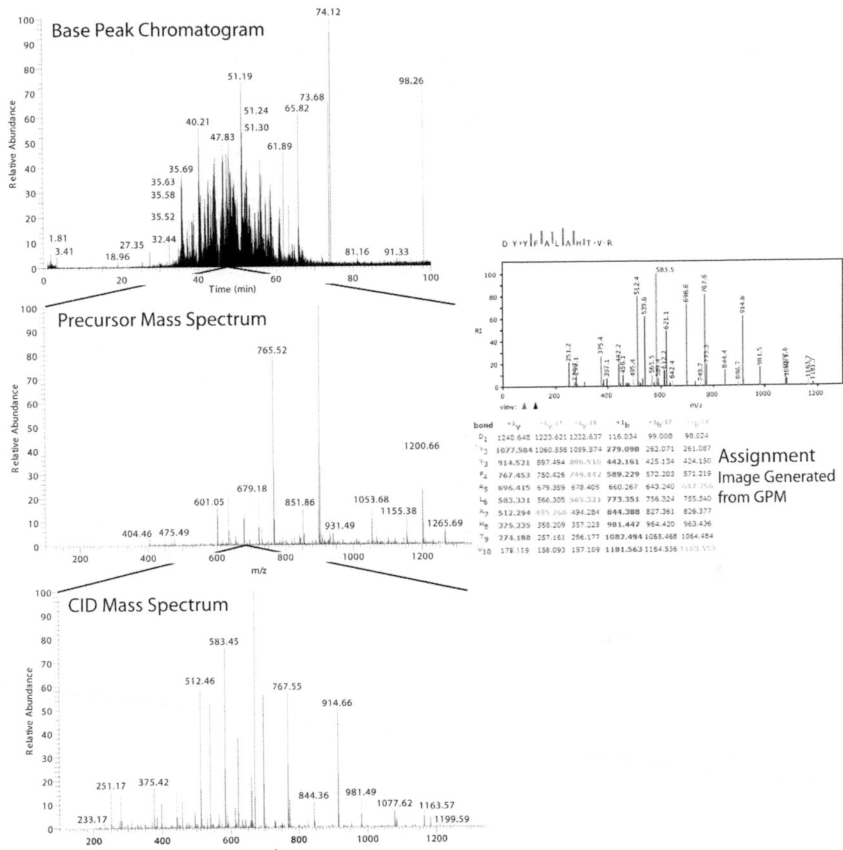

Figure 1. (Top). Peptides are partially separated from each other using liquid chromatography, resulting in a chromatogram in which each peak consists of one or more peptides. (Middle) The peptides co-eluting in a peak from the chromatography system are further separated in the mass spectrometer by their mass-to-charge ratio. The peak intensities reflect the number of molecules that have been isolated. A peptide in the MS spectrum can be isolated and selected for further analysis in which the peptide is collisionally-activated by an inert gas. (Bottom) The peptide may then fragment apart and the resulting mass-to-charge and abundances of the fragments are measured by a subsequent round of mass spectrometry

Currently, many of the instruments use proprietary software for transforming raw spectra into sets of peak locations and intensities. This results in difficulties when gauging the accuracy of the resulting peak intensities and locations; ultimately this limits the interpretations that lead to biological results improvements can be made in the existing methods. Many aspects of this multi-scalar problem can be studied independently, but many more are dependent on the approaches taken earlier in data processing. In this regard, surveying the depth and scope of computational proteomics is key to understanding the information content of each proteomic experiment. We start more detailed discussions with an example to a critical step in computation proteomics in which **Lange *et al.***[b] use open source wavelet approach to peak picking, the process of transforming the raw spectra into the familiar impulse plots used to display intensity and mass-to-charge location information.

Comparative Proteomics

Following quantification from raw mass spectra, a wide number of approaches may be employed to interpret them. In comparative proteomics, a first step is the normalization of the data to enable comparison between experiments and determination of differential abundance levels of the proteins in a specific proteome. In this area, **Wang, *et al.***[b] introduce a rank-inspired probability method for normalizing ion peaks between samples. The ultimate goal is to increase the reliability of the quantitation process so that differential expression of proteins can be measured under various growth conditions.

Another challenge in comparative proteomics is aligning the spectra so that the correct peaks can be compared and the ratio of the normalized intensities determined. A problem that has been encountered with shotgun proteomics experiments in applying this method, however, is that the alignment of the spectra in both elution time and mass-to-charge values is difficult due to the number data points. A typical shotgun experiment in which the cell lysate is digested with trypsin and then separated using liquid chromotagraphy may generate 70,000 data points or more. An alternative method is to label the samples using isotopes that are normally in low abundance in nature. Samples can then be mixed and the alignment step essentially is obviated because isotopic labeling has only a small affect on the elution properties of the peptides and proteins [1]. Further improvements may be made by reducing the complexity of the peptide mixture and isotopically labeling peptides using isotope-encoded affinity tags (ICAT) [2]. Recent advances in isotope labeling technology have enabled the comparison of differential protein expression for up to four samples[3]. **Michailidis and Andrews**[b] present a statistical framework for the analysis of variance of multiple isotopically-labeled proteins and formal hypothesis testing as to whether the proteins are differentially expressed.

---

[b] Session Publication.

Network Inference

If the differential expression can be done in a reliable and informative manner as **Michailidis and Andrews**[b] present, then it would be conceptually possible, with proper experimental design, to infer networks of protein regulation from the mass spectrometry results. One of the persistent questions regarding the use of graphical models to infer networks from RNA or protein expression data is how to optimally design experiments in order to achieve the greatest resolution of the network. Typically this argument is cast in terms of whether one should study cell populations under different treatment or environmental conditions, or whether one should use time series data. The former are attractive because each data series represents an independent observation, while in the latter the observation of expression of each protein is correlated in time and thus is not an independent observation. On the other hand, time-series experiments that focus on a specific sub-network are easier to design. This trade off is considered in detail by **Page and Ong**[b].

Regulatory networks are not the only networks of interest. Rapid advances are currently being made in the determination of protein interaction networks [4-9]. A typical approach in these assays is to use affinity purification techniques to pull-down a preselected *bait* protein and the *prey* proteins that interact, directly or indirectly, with the bait protein. This information can be used to construct a protein interaction network in which all discovered interactions are laid out. However, a protein-protein interaction network is not as informative as actually determining the protein complexes or machines that carry out the biological function in the cell. Determining these complexes from the pull-down data is computationally challenging because multiple-complexes may be present in any given pull-down data set. **Chu *et al.***[b] propose a solution to this problem that combines a kernel method to identify potential complexes, a latent feature model to address the number of complexes, and Bayesian statistics that can ultimately be used to bring in informative prior knowledge

Peptide Identification

Network analyses and any other analyses that seek to determine biological knowledge from proteomics experiments rely on the initial correct identification of peptides. The automation of peptide identifications using computers took a step forward in the late1980s and early 1990s in work by several groups that laid a foundation for the next several years[10-12]. In 1994-1995, the *SEQUEST* method was developed and published [13]. *SEQUEST* was the first example of high-throughput processing of proteomic data and has since become one of the standards of the field. Although the code was developed in a relatively short time, the wide spread use of the tool is a testimonial to its utility.

Now that proteomics is being widely used in industry and research labs, however, there is a pressing need to solve the many of the peptide identification challenges that remain. Currently, only 25% or fewer of peptide spectra are identified with a peptide. There are many reasons for this. First, many fragmentation pathways are poorly understood and the current set of patterns that are searched for is limited. Second, most peptide identification tools only consider parent ions that have charges of +3 or less. Incomplete digestion of peptides by peptidases, such as trypsin, is likely to result in an abundance of higher charge state ions leading to a decrease of spectra that are currently identifiable. In addition, tools such as *SEQUEST* search genome sequence databases for peptides that are likely to result in a spectrum similar to the experimental spectrum under consideration.

In *SEQUEST* and most peptide identification tools, a fragmentation pattern is used in one way or another to determine this similarity. Typically, the peptide sequence is used as a template to generate the pattern, and the pattern or *model spectrum* is compared to the actual spectrum. The accurate development of these patterns is the topic of the work by **Arnold, *et al.***[b], in which they employ a neural network to learn peptide fragmentation patterns from a training database of peptide spectra. This work is significant because it is believed that non-classical fragmentation patterns are largely missed in the identification process. **Wang *et al.***[b] take a different approach in which they use a minimal fragmentation model in a simple scoring function and then analyze the score and properties of the candidate peptides using a support vector machine (SVM). This approach extends the use of SVMs in peptide identification [14, 15] to not only choose spectra that have been correctly matched to a peptide, but to also choose the best candidate peptide from a sequence database for a given spectrum.

Alternate Spectra Assignment Algorithms

Although more and more genome sequences are becoming available, by far the majority of genomes have not been sequenced nor are they likely to be sequenced in the next 5-10 years. The alternative to searching a sequence database is to determine the peptide sequence *de novo* from the spectrum[11, 16, 17], or to use optimization to evolve a peptide sequence to match a spectrum [18]. *De novo* methods are attractive because the idealized problem, that of essentially spelling out a peptide from a set of mass peaks, intuitively corresponds to a problem that can be solved using graph theory. The devil, as usual, lies in the details. Real spectra are noisy and missing peaks from key

---

[b] Session Publication.

fragments are the rule rather than the exception. Graph theory analyses often result in a series of graphs, not all of which are compatible. **Liu, *et al.*** extend recent advances in this area by the application of tree decomposition to the problem that allows for compatible graphs to be found more readily and in faster time.

Post-Translation Modifications

Of the 75% or so of spectra that go unassigned with a peptide in a general database search, many of these are thought to be post-translationally modified peptides. The identification of these peptides is extremely important because the post-translational modifications (PTM) are one of the key mechanisms by which cells respond to external stimuli, resulting in the up and down regulation of genes. **Yan *et al.***[b] describe a point-process model that has the advantage used in dynamic programming models [19] in which the mass offsets for the PTMs are determined automatically, and is deployed in a fast cross-correlation framework.

Final Thoughts

Ultimately, a major motivation for investments into the development of proteomics is to develop advanced methods of disease diagnosis, understanding of disease processes, and remedies. Early detection of disease is important because the clinical outcome is much more favorable, in general, if the disease can be treated in an early stage. As a result, there is much interest in improvements at every level that can yield MS-detectable biomarkers that signal the presence of the disease long before more overt symptoms occur that signal advanced stages of disease. An example of this need and approach is **Pratapa *et al.***[b] present a hierarchical data analysis scheme for the identification of protein biomarkers that are indicative of lung cancer. Using data from mass spectrometric analyses of diseased and normal tissues, they compare a SVM classification with that of a Bayesian sparse logistic regression. They find known biomarkers as well as identify several more candidate biomarkers that may prove to be clinically useful.

The incredible diversity of problems and solutions is well sampled by the efforts of this session authors. Mass spectrometry-based proteomics is likely to offer a central role well into the future for understanding protein function and complex biological systems.

[b] Session Publication.

Acknowledgments

The session organizers would like to thank the authors of the 30 submissions to this session and express our regret that only a handful of the excellent papers can be presented. We would also like to express deep gratitude to the anonymous referees who together volunteered uncountable hours to provide the key input to make this session successful.

References

1. L. Pasa-Tolic et al., *J Am Chem Soc.* **121**(34): 7949-7959 (1999).
2. S.P. Gygi et al., *Nat Biotechnol.* **17**(10): 994-999 (1999).
3. P.L. Ross et al., *Mol Cell Proteomics.* **3**(12) : 1154-1169 (2004).
4. J.S. Bader et al., *Nature.* **22**(1): 78-85 (2004).
5. A.C. Gavin et al., *Nature.* **415**(6868): 141-147 (2002).
6. L Giot et al., *Science.* **302**(5651) : 1727 :1736 (2003).
7. Y Ho et al., *Nature.* **415**(6868) : 180-183 (2002).
8. A.J. Link et al., *Nature Biotech.* **17**(7): 676-682 (1999).
9. G. Butland et al., *Nature.* **443**(7025) : 531-537 (2005).
10. K. Biemann, Methods in Enzymology, ed. J.A. McCloskey. Vol. 193. 1990, San Diego: Academic Press, Inc.
11. C. Bartels, *Biomed Env Mass Spectrom.* **19**: 363-368 (1990).
12. M. Mann, C.K. Meng and J.B. Fenn, *Anal Chem.* **61**(15): 1702-1708 (1989).
13. K. Eng, A.L. McCormack and J.R. Yates III, *J Am Chem Soc.* **5**: 976-989 (1994).
14. D.C. Anderson et al., *Proteome Res.* **2**(2) : 137-146 (2003).
15. Cannon et al., *Proteome Res.* Web release September 10, 2005.
16. V. Dancik et al. *J Comput Biol.* **6**(3/4): 327-342 (1999).
17. J.A. Taylor and R.S. Johnson, *Rapid Comm in Mass Spectrom.* **11**: 1067-1075 (1997)
18. A. Heredia-Langner et al., *Bioinformatics.* **20**(14): 2296-2304 (2004).
19. P.A. Pevner et al., *Genome Res* **11**(2): 290-299 (2001)

A MACHINE LEARNING APPROACH TO PREDICTING PEPTIDE FRAGMENTATION SPECTRA

RANDY J. ARNOLD,[1] NARMADA JAYASANKAR,[2] DIVYA AGGARWAL,[2] HAIXU TANG,[2,3] AND PREDRAG RADIVOJAC[2*]

1) Department of Chemistry, Indiana University, Bloomington, IN 47405
2) School of Informatics, Indiana University, Bloomington, IN 47408
3) Center for Genomics and Bioinformatics, Indiana University, Bloomington, IN 47405
**Corresponding author*

Accurate peptide identification from tandem mass spectrometry experiments is the cornerstone of proteomics. Although various approaches for matching database sequences with experimental spectra have been developed to date (e.g. Sequest, Mascot) the sensitivity and specificity of peptide identification have not yet reached their full potential. This is in part due to the tradeoffs between robustness and accuracy of the existing methods with respect to the non-uniform nature of peptide fragmentation and bond cleavages induced by different mass spectrometers. Accordingly, it is expected that new approaches to *de novo* predicting peptide fragmentation spectra will enable more accurate peptide identification. To address this problem, here we used a data-driven approach to learn peptide fragmentation rules in mass spectrometry, in the form of posterior probabilities, for various fragment-ion types of doubly and triply charged precursor ions. We show that the accuracy of our neural-network based methodology is useful for subsequent peptide database searches and that the most useful rules of fragmentation significantly differ across ion and precursor types.

1 Introduction

Recent advances in separations and mass spectrometry have enabled a surge in the comprehensive analysis of cellular proteins, commonly referred to as proteomics.[1,2] The critical development in this area is the ability to identify a peptide, or in some cases entire proteins, from the fragment ions generated by tandem mass spectrometry.[3] Various dissociation methods have been introduced, including commonly used gas phase collision-induced dissociation (CID),[4] surface-induced dissociation,[5] photodissociation,[6,7] electron-capture dissociation,[8] and electron transfer dissociation.[9] The resulting tandem mass spectra are compared with *in silico*, i.e. computer generated, spectra derived from peptides in the available protein database.[10] The commonly used protein identification tools often use ad-hoc rules[11] or unified probabilistic models[12,13] to estimate the likelihood that a given experimental spectrum was generated from each sequence contained in the database. The final assignments and confidence levels are then based on both the scores and database content.[14,15]

In practice, the peptide fragmentation into various ions may differ for several reasons. For many peptides, cleavages of amide bonds dominate the fragmentation and produce a series of b- and y-ions. For other peptides, the enhanced fragmentation at some types of amino acid residues may dominate. Also, the charge carried by the precursor ion affects electron distribution along the cleaved peptide backbone.

Finally, fragmentation method and the energy level used in the experiments are also known to largely change the global behavior of peptide fragmentation.

The development of chemical theory of peptide fragmentation, e.g. the "mobile proton" model,[16-18] enabled the *de novo* prediction of fragmentation spectra from peptide sequences. Using a kinetic model, Zhang made the first successful attempt at predicting the low-energy CID spectra of singly and doubly charged peptides.[19] He recently introduced a simplified model that can accommodate peptides with three or more charges as well as sequences of increased lengths.[20] However, it is not clear how this approach could be extended to the other types of mass spectrometry instruments.

An ability to obtain large amounts of peptide fragmentation data relatively cheaply sprung the development of data-driven approaches and machine learning techniques. Elias et al.[21] were first to successfully utilize a set of well annotated fragmentation spectra acquired from an electrospray ion-trap mass spectrometer in an attempt to infer the probabilistic rules of fragmentation. As a proof of concept, they learned a decision tree for the b- and y-ion fragmentation of the doubly charged precursors and used their model to significantly improve on Sequest scores of tryptic peptides. In addition, Elias et al. confirmed previously known rules of peptide fragmentation and presented a large set of new ones.

In this paper, we extend this approach to the triply charged precursors in addition to other, harder-to-predict, ion types (b–H_2O, b–NH_3, b–H_2O–NH_3, etc.). We note that it is not a trivial extension for two reasons. First, from the standpoint of protein identification, triply charged (+3) peptides seem to be much more difficult to identify than doubly charged (+2) peptides. For example, in the dataset used herein (Section 2.1), there were roughly four times as many doubly charged peptides (16,056) as triply charged peptides (4,130) that could be reliably identified by Mascot. This indicates that the current peptide identification tools may be better suited to the +2 charged peptides even though there may be a general preference to form +2, rather than +3 precursor ions. The distributions of fragment ions observed in +2 vs. +3 precursor ions are different. For instance, many +2 b- and y- fragment ions can be observed in the fragmentation spectra of the +3 precursors. For the +2 precursors, +1 b- and y-ions dominate the fragmentation spectra, whereas few, if any, of the +2 b- and y-ions can be observed. We illustrate these differences between the +2 and +3 precursors in Figure 1, where the same peptide was used to produce peptide fragmentation spectra. Second, the fragmentation mechanism of +3 ions is less understood than that of the doubly charged ions.[22] As a result, the new rules of fragmentation could be important since it is not as easy to develop a *de novo* prediction method for the +3 precursors as for the +2 precursors.[19]

The results of our study indicate that, for most of the ions, it is possible to predict the peptide fragmentation spectra with a useful accuracy. Furthermore, the obtained predictions can be used in a straightforward way to improve a simple correlation-based scoring function for peptide identification.

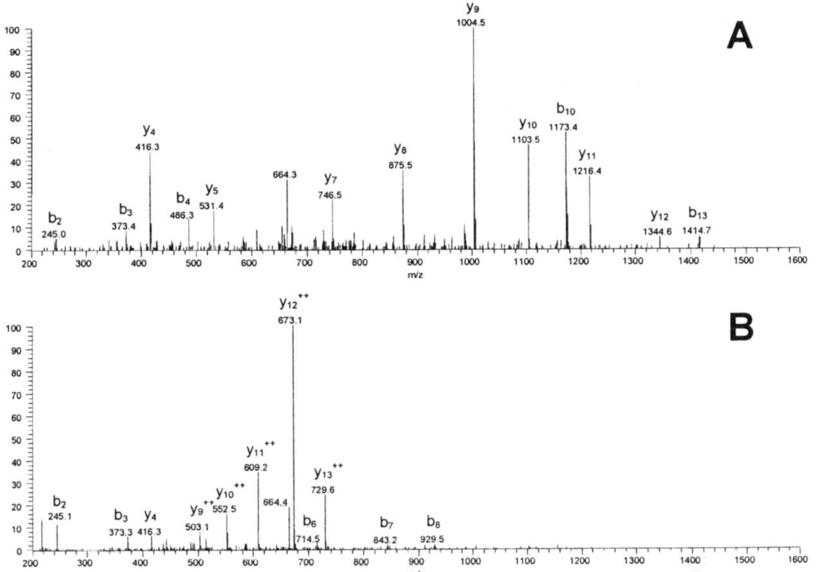

Figure 1. MS/MS spectra of the A) +2 and B) +3 precursor ions of peptide MLQLVEESKDAGIR acquired in consecutive scans of an LC-MS/MS experiment using an ion trap mass analyzer. Selected precursor *m/z* values are A) 795.2 and B) 530.69.

2 Methods

As previously mentioned, the major objective of this study was to use automated techniques in order to learn peptide fragmentation rules in the form of posterior probabilities and then utilize the trained model for peptide identification. The original problem of predicting spectral peak intensities was converted into a simpler and easy-to-interpret classification problem, in which the peak intensity was first normalized and then binned into two groups based on a threshold. Formally, given a precursor sequence S and its charge $q_S \in \{+2, +3\}$, we aimed to estimate the following set of probabilities: $P(I(i) \geq t \mid S, q_S)$, where $I(i)$ is the peak intensity of any fragment ion $i \in \{precursor\text{–}H_2O, b\text{–}H_2O, b\text{–}NH_3, b\text{–}H_2O\text{–}NH_3, y\text{–}H_2O, y\text{–}NH_3, y\text{–}H_2O\text{–}NH_3, b^2, y^2\}$ and t is an appropriately chosen threshold. In this study, by default, t is equal to 1% of the total intensity of the spectrum.

2.1 Datasets

Two groups of samples were prepared from isolated rat brains. The first group was produced by homogenizing hippocampus tissues and separating the lysate into four different fractions by differential sedimentation. The second group was produced by separately lysing tissue from six different brain regions (amygdala, caudate putamen, frontal cortex, hippocampus, hypothalamus, and nucleus accumbens). All samples were digested separately with proteomics grade (modified) trypsin in the presence of an acid-labile surfactant. Tryptic peptides were separated by nano-flow reversed-phase liquid chromatography and electrosprayed directly into a ThermoFinnigan (San Jose, CA) LCQ Deca XP ion-trap mass spectrometer which recorded mass spectra and data-dependent tandem mass spectra of the peptide ions. By using dynamic exclusion, the mass spectrometer was limited to acquiring only one tandem mass spectrum for a given parent m/z over a 60-second window. Tandem mass spectra were filtered based on a total spectrum signal of 1 million counts for the first group of samples and 300 million counts for the second group of samples. All spectra were searched against protein sequences for *R. norvegicus* in the Swiss-Prot database[23] using Mascot[12] for peptide identification. Searches were performed with variable modifications of protein N-terminal acetylation and methionine oxidation selected and a maximum of one missed cleavage site. Mascot result files were parsed using a Protein Results Parser program written in-house to create a single training set with all peptides having Mascot scores of 40 or higher.

We normalized each spectrum to sum to one and divided all precursor peptides into doubly and triply charged. Peak intensities were estimated for the following ion types: *precursor*–H_2O, b, b–H_2O, b-NH_3, b–H_2O–NH_3, y, y–H_2O, y-NH_3, y–H_2O–NH_3, for the doubly charged precursors, while b^2 and y^2 ions were also considered for the triply charged ions. The set of precursor peptides of a given charge was filtered to prevent multiple copies. If two or more precursor peptides were identical, the one with the highest Mascot score over all fragment ions was retained. The dataset contained a significant number of identical precursor sequences and was reduced by factors of 9.3 and 6.3 for the doubly and triply charged peptides, respectively. The total counts of fragment ions corresponding to the set of unique precursor sequences are shown in Table 1.

A preliminary peptide identification using the new scoring based on the predicted fragment spectra was performed using the whole set of available proteins from *R. norvegicus*, containing 35,085 proteins (from the NCBI web site).

2.2 Data representation

To enable learning, each sequence fragment S was encoded into a fixed-length vector representation. More specifically, sequence $S = s_1s_2,\ldots s_n$ was represented by a vector of binary and real-valued features. Assuming the cleavage occurred between positions k and $k + 1$ in $S = s_1s_2,\ldots s_n$, the following features were constructed for all b- and y-ions: (i) amino acid compositions of the prefix subsequence $s_1s_2,\ldots s_k$ and the suffix subsequence $s_{k+1}s_{k+2},\ldots s_n$; (ii) lengths of both fragments, k and $n - k$; (iii) first neighbor prefix/suffix amino acids, s_k and s_{k+1}, and second neighbor prefix/suffix amino acids s_{k-1} and s_{k+2}; (iv) N- and C-terminal residues, s_1 and s_n; (v) parent mass m; (vi) ion masses, m_{prefix} and m_{suffix}; and (vii) N-terminal acetylation. We also incorporated a number of features introduced by Elias et al.: gas phase basicity, helicity, hydrophobicity, and isoelectric point, both average and for the residues s_k and s_{k+1}.[21] Individual amino acids were encoded using a binary data representation,[24] expanded by adding oxidized methionine residue, while the compositional attributes were real-valued. To encode *precursor*–H_2O ions, we ignored the features related to the cleavage site. Overall, b- and y-ions were represented by 202 features, while the *precursor*–H_2O ion encoding contained 76.

Table 1. The total count of ions corresponding to the unique precursor sequences. An ion was considered present (positive cases) when its peak intensity exceeded 1% of the total spectral intensity. Otherwise, the ion was considered absent (negative cases).

Ion	*Doubly charged precursors*			*Triply charged precursors*		
	Positives	*Negatives*	*Total*	*Positives*	*Negatives*	*Total*
precursor – H_2O	239	1484	1723	64	590	654
b	5210	16916	22126	950	12000	12950
$b - H_2O$	1700	20426	22126	206	12744	12950
$b - NH_3$	678	21448	22126	117	12833	12950
$b - H_2O - NH_3$	249	21877	22126	121	12829	12950
b^2	-	-	-	1343	11607	12950
y	9323	12802	22126	1639	11311	12950
$y - H_2O$	431	21695	22126	132	12818	12950
$y - NH_3$	286	21840	22126	101	12849	12950
$y - H_2O - NH_3$	145	21981	22126	107	12843	12950
y^2	-	-	-	1953	10997	12950

2.3 Model selection and training

Predictors of ion intensities were built as ensembles of two-layer feed-forward neural networks, which, if provided with enough data, are known to be universal approximators of bounded functions.[25] A particularly useful property of these models is that the expected number of data points necessary for successful training is linear with the number of weights and that the training is relatively fast. Each model in an ensemble contained $h \in \{1, 2, 4, 8, 16, 32\}$ hidden neurons and one output neuron, all with sigmoidal activation function, and was trained using the resilient propagation algorithm.[26]

Since the threshold t for the quantization of the peak intensities was set to 1% of the total peak intensity, each resulting dataset was high-dimensional and class-imbalanced (even at 1% cutoff there were much fewer positives than negatives). Thus, we randomly under-sampled the majority class to the size of the positive class to train each network from a class-balanced dataset. However, to effectively use whole dataset, a different selection of negatives was made for each network in the ensemble.

Prior to network training, a t-test feature selection filter was employed to filter out unpromising features. The threshold for feature retention, t_f, was varied from the following set of values: {0.001, 0.01, 0.1, 1}. Clearly, in the case of $t_f = 1$ all features were retained. Finally, to remove correlated features we applied the principal component analysis and retained 95% of the variance. Feature selection thresholds and the number of hidden neurons were selected using a separate validation set for each individual model (20% of the training set), therefore producing only the final set of estimated accuracies. Each ensemble contained 30 neural networks.

2.4 Performance evaluation of the fragmentation ion peak prediction

A model was trained for each ion type separately using cross-validation. The non-overlapping folds were chosen at the level of precursor sequences since one +2 precursor ion of length n can produce a combination of the b-ions (b_1, b_2, etc.), y-ions (y_1, y_2, etc.) and their variants with neutral losses which could create information leak if distributed independently over training, validation, and test sets. Triply charged ions could also produce a combination of +2 b-ions (b_1^2, b_2^2, etc.) or +2 y-ions (y_1^2, y_2^2, etc.). Finally, the number of positive examples in each fold was balanced in order to achieve stable and realistic estimate of classification accuracy.

We measured sensitivity (sn) and specificity (sp) for each classifier. Sensitivity is defined as the percentage of positive examples, i.e. peaks over 1% of total intensity, correctly predicted, while specificity is the percentage of negative examples correctly predicted. We also estimated a class-balanced accuracy $acc = (sn + sp)/2$ and the area under the ROC curve (AUC), both of which are essentially unaffected by the disparity in class sizes. The area under the curve was calculated using the trapezoid rule.

2.5 Peptide identification based on the predicted spectra

The predicted fragment spectrum was constructed by combining the outputs of individual predictors for each ion type. Since individual predictors are trained on the class-balanced datasets, it is necessary to adjust their outputs according to the observed prior probabilities of positives and negatives for each ion type.[27] Given the class-balanced training set, an adjusted output score o_{adj} of the predictor can be calculated as

$$o_{adj} = \frac{p \cdot o}{1 - p + o \cdot (2p - 1)},$$

where $o \in [0, 1]$ is the output of a class-balanced trained model and p is the class prior for the positive examples. Different class priors were used for each fragment ion type.

We use the simple correlation coefficient to score the matching between an experimental spectrum $\{m_i^e, I_i^e\}$ and a predicted spectrum $\{m_j^p, I_j^p\}$, defined as

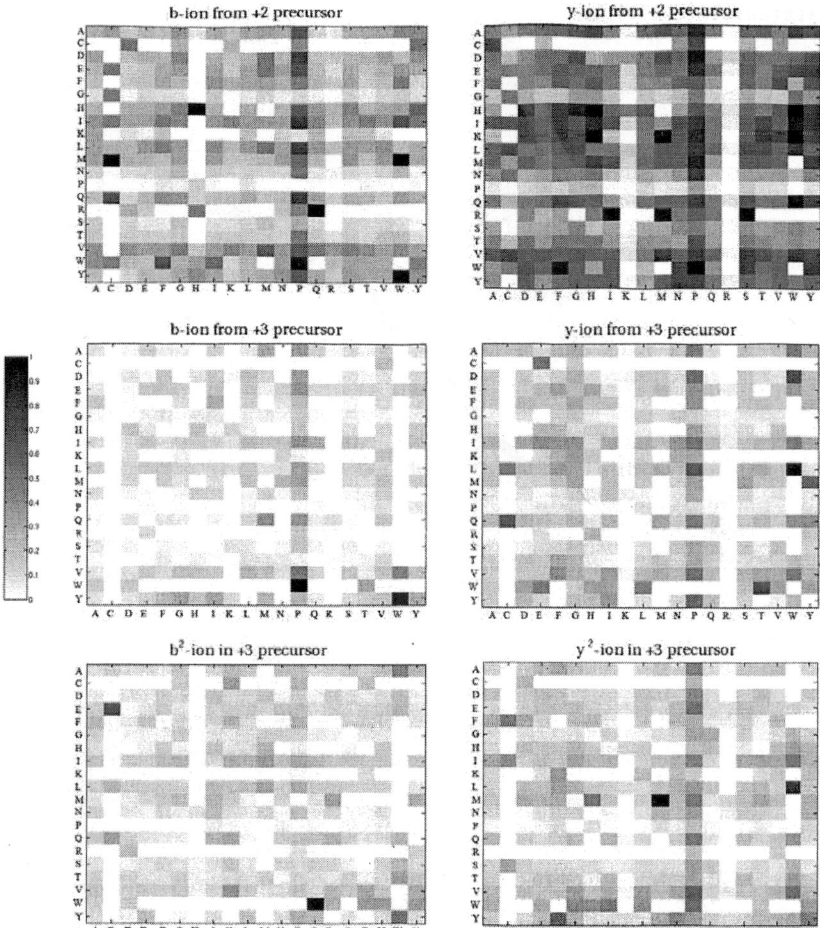

Figure 2. The amino acid preferences for peptide fragmentation. The frequencies of observing ion types (b-, y- or b^2, y^2) were plotted in grey scaling from 0 (white) to 1 (black). The rows indicate amino acid on the left-hand side, while the columns indicate amino acids on the right-hand side of the cleavage site.

$$S = \frac{\sum_{m_i^e \approx m_j^p} I_i^e \cdot I_j^p}{\sqrt{\sum_i (I_i^e)^2 \cdot \sum_j (I_j^p)^2}},$$

where m and I represent the mass and the intensity of each ion, respectively, and $m_i \approx m_j$ means that the difference between these two mass values is smaller than the tolerance t of the mass spectrometer ($t = 0.5$). For the comparison purposes, we also predict spectra using ad-hoc rules for each peptide (Section 3.3).

3 Experiments and Results

3.1 Analysis of amino acid preferences at the cleavage sites

Figure 2 illustrates the amino acid preferences for peptide fragmentation as influenced by the amino acids on both sides of the cleavage site. Several expected trends are observed across all ion types, specifically the preference for proline on the C-terminal side and the preference against proline on the N-terminal side of the cleavage site. The plots for the +2 precursors compare favorably with that shown previously[28] for cleavage intensity ratios for +2 peptides with partially mobile protons, even though the data shown here represent all proton mobility types (mobile, partially mobile, and non-mobile proton). Other interesting trends can be observed, such as the apparent preference for tryptophan (W) in the C-terminal position for +2 and +3 precursors, preference against glycine (G) and serine (S) in the N-terminal position for +2 precursors, and preference for cleavage between amino acid pairs LW (y-ions), WP and YW (b-ions), MM (y^2-ions), and WQ (b^2-ions) for +3 precursors. These trends are not fully understood at this time and may suggest previously unknown enhanced fragmentation sites. Note that Figure 2 does not account for the number of observed amino acid pairs at the cleavage site.

3.2 Evaluation of the peptide fragmentation prediction

Using the methodology presented in Sections 2.2-2.4, we trained classification models for twenty different ion types and evaluated performance of each model. In terms of varying learning parameters, we observed an increase in accuracy by 2-3 percentage points when ensembles of 30 models were used instead of a single model. In addition, the improvement of the non-linear models over the linear (a network with a single hidden neuron) was greater than 5 percentage points in some cases. We also note that in the case of b and y-ions, the number of selected hidden neurons reached its maximum ($h = 32$) indicating that an increase in dataset size and higher expressiveness of the classifier would likely cause an additional improvement in overall performance. The detailed evaluation of the classification accuracy appears in Table 2, while two sample predictions of triply charged precursors are shown in Figure 3.

The performance of our models was also evaluated against a decision tree model proposed by Elias et al.[21] We encoded the full set of features provided by the authors and trained a classifier for the *b* and *y* ions of the +2 precursors. We obtained classification accuracy of 73.5% (sn = 74.8%; sp = 72.1%) for the *b*-ion and 80.4% (sn = 82.3%; sp = 78.6%) for the *y*-ion, the two ion types studies by Elias et al. The differences between our ensemble models and decisions trees are statistically significant, with p-values below 0.01 in both cases (binomial distribution was used to calculate p-values). The C4.5 decision tree software was used with the default parameters.[29] More detailed evaluations can be found at our research home page.

Table 2. Classification accuracy [%] of the predictors on the doubly and triply charged ions; *sn* – true positive rate, *sp* – true negative rate, *acc* = (*sn* + *sp*)/2, *AUC* – area under the ROC curve.

Ion	*Doubly charged precursors*			*Triply charged precursors*		
	sn	*sp*	*acc/AUC*	*sn*	*sp*	*acc/AUC*
precursor – H_2O	72.0	60.8	66.4/70.7	81.3	68.5	74.9/79.7
b	80.4	75.4	77.9/85.8	80.6	71.9	76.3/84.6
$b - H_2O$	76.8	76.3	76.5/84.6	76.2	60.2	68.2/76.8
$b - NH_3$	75.8	76.0	75.9/82.8	76.9	65.0	70.9/78.6
$b - H_2O - NH_3$	69.1	64.6	66.8/73.1	81.8	51.9	66.9/68.1
b^2	-	-	-	88.4	75.8	82.1/88.5
y	84.7	79.3	82.0/89.5	88.9	79.1	84.0/91.4
$y - H_2O$	66.4	66.2	66.3/72.2	82.6	56.5	69.6/73.0
$y - NH_3$	70.3	70.8	70.6/79.0	81.2	59.8	70.5/77.8
$y - H_2O - NH_3$	60.7	51.1	55.9/56.5	83.2	54.3	68.7/69.6
y^2	-	-	-	87.9	72.6	80.2/86.8

3.3 Peptide identification experiments

The quality of peptide identification was estimated by comparing the correlation scores between the experimental spectrum and the computer generated spectra for the true precursor sequence and 500 spurious tryptic peptides selected from the rat proteome. We estimated (i) the average difference between the score for the true peptide and the score for the best scoring random peptide, and (ii) the average rank of the true peptide in the context of 500 candidates. The mass of the true peptide m was used to get candidate tryptic peptides in the rat proteome whose masses approximately matched the mass of the experimentally measured peptide (within $m-1$ and $m+3$). The candidate peptides were allowed to have up to one missed cleavage.

The evaluation of peptide identification is presented in Table 3. Our new scoring scheme was compared to an ad-hoc (also referred to as simple) scoring in which all possible b and y fragment ions are assigned intensities of 1, b–H_2O and y–H_2O are assigned peak intensities of 0.5, b–NH_3 and y–NH_3 are assigned peak intensities of 0.3, while b-H_2O–NH_3, y–H_2O–NH_3, b^2 and y^2 are assigned intensities of 0.1. Other fragment ions were assigned peak intensities of 0, as well as all fragment ions whose mass was greater than 2,000 due to the upper mass limit of the mass spectrometer. A negative score indicates that, on average, highest-scoring random peptides may be selected over the true peptides with higher confidence than the true peptides are selected over highest-scoring random peptides. These experiments provide evidence that our approach to spectrum prediction may provide highly promising peptide identification.

The spectra were classified into 8 categories according the chemical property of the peptides, namely +2 or +3 precursor, mobile (number of precursor charges is greater than total of H, K, and R) or non-mobile proton, and presence or absence of proline. While improvement over simple scoring is observed for all 8 categories, the

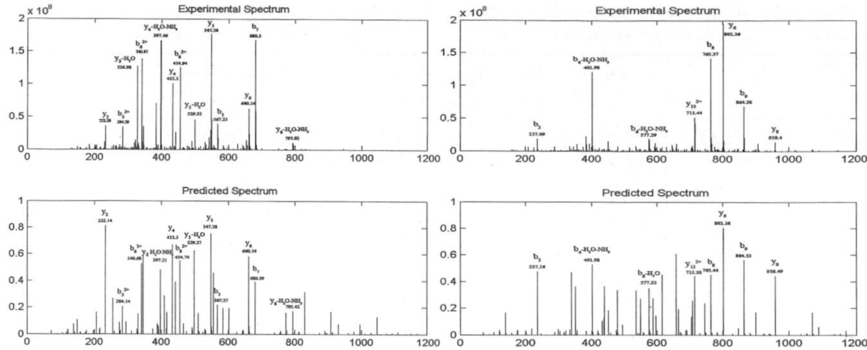

Figure 3. Experimental (upper panels) vs. predicted (lower panels) spectra for two triply charged precursor ions. The left panels correspond to sequence HRDTGILDSIGRZ, while the right panels correspond to the sequence HVLSGTLGCPEHTYR. Note that the first sequence corresponds to the non-mobile proton w/o proline case, while the second sequence corresponds to the non-mobile proton with proline case.

greatest improvement appears to be for triply-charged precursors, especially with a mobile proton and proline.

Table 3. The average difference in scores (*diff*) between the true peptide and the highest scoring random peptide and an average ranking of the true peptide (*rank*). To obtain scores and rankings we used 500 random peptides from the rat proteome having approximately the same precursor mass as the true peptide. The scores are separated for the cases of doubly vs. triply charged precursors, mobile vs. non-mobile proton, and presence vs. absence of proline (Pro) in the precursor sequence. Each field in the table was averaged using a set of 25 randomly taken precursor sequences (with 68% confidence intervals), identical for both scoring schemes.

Scoring scheme		*Doubly charged precursors*				*Triply charged precursors*			
		Mobile proton		*Non-mobile proton*		*Mobile proton*		*Non-mobile proton*	
		w/o Pro	*w Pro*	*w/o Pro*	*w Pro*	*w/o Pro*	*w Pro*	*w/o Pro*	*w Pro*
diff	*New*	.32±.03	.26±.04	.30±.03	.24±.04	.13±.03	.14±.04	.22±.03	.25±.04
	Simple	.22±.02	.14±.03	.23±.02	.15±.02	−.01±.02	−.03±.02	.08±.02	.09±.03
rank	*New*	1.1±0.1	1.4±0.2	1.1±0.1	1.5±0.2	1.8±0.7	1.5±0.2	1.4±0.4	1.2±0.2
	Simple	1.1±0.1	1.4±0.2	1.0±0.1	1.3±0.2	9.0±1.8	19.0±4.5	2.3±1.0	6.1±2.2

4 Discussion

Machine learning approaches have been extensively applied to proteomics research. They, however, mostly focused on either the preprocessing of the spectrum or the post-processing the peptide identification results of conventional tools. For examples, binary classifiers[30] and artificial neural networks[31] were introduced to evaluate the quality of MS/MS spectra before they were used for peptide identification; support vector machines were also used to classify the positive protein identification based on Sequest output.[32] Very little work, however, has addressed the potential of applying machine learning to the peptide identification itself. We extend the previous work by Elias et al.[21] to the more challenging problem of predicting the full fragment spectra of peptides for both doubly- and triply-charged precursors. Our preliminary tests on the scoring of peptide identification showed encouraging results towards a new scoring scheme for peptide identification with better performance.

All classifiers used in this study were trained on balanced samples. Certainly, balanced training provides good insight into the class separability since it is not related to the relative fraction between the positive and negative datasets. On the other hand, application of such a predictor may cause significant overprediction if applied to the representative imbalanced dataset. This problem, however, was easily resolved by adjusting the outputs of the predictor, depending on the class priors. Our choice of neural networks compared to other machine learning techniques (SVMs or decision trees) were based on our experience with using the models and the fact that only minor modifications will be required in order to learn peak intensities in a regression-based approach.

The empirical rules we derived from machine learning approaches in this paper will be also useful for understanding the fragmentation mechanism of triply charged ions. The enhanced cleavage on the N-terminal side of proline is consistent with previous observations for doubly charged ions.[32] The data also suggest that tryptophan can enhance cleavage of the bond on its N-terminal side. Other subtle effects such as the rules regarding mobile versus non-mobile protons for +3 precursors will be the focus of future investigations.

We stress that the method used in this paper is not restricted to any mass spectrometry instrument. As long as a large set of annotated spectra are available, our method can be applied to any proteomics platform. We intend to apply this method to other commonly used MS/MS instruments, e.g. Q-TOF or MALDI/TOF/TOF, as well as more specialized modes of fragmentation, e.g. photodissociation.

The prediction of peptide fragmentation spectra may have other potential applications in protein analysis. It has been shown that the fragmentation patterns may correlate with protein local structures.[33] Including the features of protein local structures in the prediction of peptide fragmentation spectra may result in a potential new approach to protein structure determination.

Acknowledgements

This study was partially funded by the Indiana University Office of the Vice President for Research through a Faculty Research Support grant awarded to RJA, HT and PR.

References

1. Yates JR, 3rd. *Annu Rev Biophys Biomol Struct.* **33** 297 (2004).
2. Russell SA, Old W, Resing KA, Hunter L. *Int Rev Neurobiol.* **61** 127 (2004).

3. Resing KA, Ahn NG. *FEBS Lett.* **579** 885 (2005).
4. Biemann K. *Biomed Environ Mass Spectrom.* **16** 99 (1988).
5. McCormack AL, Jones JL, Wysocki VH. *J Am Soc Mass Spectrom.* 3 859 (1992).
6. Barbacci DC, Russell DH. *J Am Soc Mass Spectrom.* **10** 1038 (1999).
7. Thompson MS, Cui W, Reilly JP. *Angew Chem Int Ed Engl.* **43** 4791 (2004).
8. Zubarev RA, Kelleher NL, McLafferty FW. *J Am Chem Soc.* **120** 3265 (1998).
9. Syka JE, Coon JJ, Schroeder MJ, Shabanowitz J, Hunt DF. *Proc Natl Acad Sci USA.* **101** 9528 (2004).
10. Sadygov RG, Cociorva D, Yates JR, 3rd. *Nat Methods.* **1** 195 (2004).
11. Yates JR, 3rd, Eng JK, McCormack AL, Schieltz D. *Anal Chem.* **67** 1426 (1995).
12. Perkins DN, Pappin DJ, Creasy DM, Cottrell JS. *Electrophoresis.* **20** 3551 (1999).
13. Zhang N, Aebersold R, Schwikowski B. *Proteomics.* **2** 1406 (2002).
14. Moore RE, Young MK, Lee TD. *J Am Soc Mass Spectrom.* **13** 378 (2002).
15. MacCoss MJ, Wu CC, Yates JR, 3rd. *Anal Chem.* **74** 5593 (2002).
16. Biemann K, Martin SA. *Mass Spectrom Rev.* **6** 1 (1987).
17. McCormack AL, Somogyi A, Dongre AR, Wysocki VH. *Anal Chem.* **65** 2859 (1993).
18. Wysocki VH, Tsaprailis G, Smith LL, Breci LA. *J Mass Spectrom.* **35** 1399 (2000).
19. Zhang Z. *Anal Chem.* **76** 3908 (2004).
20. Zhang Z. *Anal Chem* (2005) Electronic publication ahead of print.
21. Elias JE, Gibbons FD, King OD, Roth FP, Gygi SP. *Nat Biotechnol.* **22** 214 (2004).
22. Tabb DL, Smith LL, Breci LA, Wysocki VH, Lin D, Yates JR, 3rd. *Anal Chem.* **75** 1155 (2003).
23. Bairoch A, Apweiler R. *Nucleic Acids Res.* 28 45 (2000).
24. Qian N, Sejnowski TJ. *J Mol Biol.* **202** 865 (1988).
25. Cybenko G. *MCSS, Math Control Signals Syst.* **2** 303 (1989).
26. Riedmiller M, Braun H. *Proc IEEE Internat'l Conf on Neural Networks.* **1** 586 (1993).
27. Saerens M, Latinne P, Decaestecker C. *Neural Comput.* **14** 21 (2002).
28. Kapp EA, Schutz F, Reid GE, Eddes JS, Moritz RL, O'Hair RA, Speed TP, Simpson RJ. *Anal Chem.* **75** 6251 (2003).
29. Quinlan J. *C4.5: programs for machine learning.* San Mateo, CA: Morgan Kaufmann; 1992.
30. Bern M, Goldberg D, McDonald WH, Yates JR, 3rd. *Bioinformatics.* **20** 149 (2004).
31. Baczek T, Bucinski A, Ivanov AR, Kaliszan R. *Anal Chem.* **76** 1726 (2004).
32. Anderson DC, Li W, Payan DG, Noble WS. *J Proteome Res.* **2** 137 (2003).

IDENTIFYING PROTEIN COMPLEXES IN HIGH-THROUGHPUT PROTEIN INTERACTION SCREENS USING AN INFINITE LATENT FEATURE MODEL*

WEI CHU & ZOUBIN GHAHRAMANI

Gatsby Computational Neuroscience Unit,
University College London,
London, WC1N 3AR, UK
E-mail: chuwei,zoubin@gatsby.ucl.ac.uk

ROLAND KRAUSE

Max-Planck-Institute for Molecular Genetics
D-10117 Berlin, Germany
E-mail: roland.krause@molgen.mpg.de

DAVID L. WILD

Keck Graduate Institute of Applied Life Sciences
Claremont, CA 91171, USA
E-mail: wild@kgi.edu

We propose a Bayesian approach to identify protein complexes and their constituents from high-throughput protein-protein interaction screens. An infinite latent feature model that allows for multi-complex membership by individual proteins is coupled with a graph diffusion kernel that evaluates the likelihood of two proteins belonging to the same complex. Gibbs sampling is then used to infer a catalog of protein complexes from the interaction screen data. An advantage of this model is that it places no prior constraints on the number of complexes and automatically infers the number of significant complexes from the data. Validation results using affinity purification/mass spectrometry experimental data from yeast RNA-processing complexes indicate that our method is capable of partitioning the data in a biologically meaningful way.
A supplementary web site containing larger versions of the figures is available at http://public.kgi.edu/~wild/PSB06/index.html.

*This work was supported by the National Institutes of Health Grant Number 1 P01 GM63208. A part of this work was carried out at Institute for Pure and Applied Mathematics (IPAM) of UCLA. We are grateful to Thomas L. Griffiths for the Matlab script of the Gibbs sampler.

1. Introduction

The analysis of protein-protein interactions forms an essential part of the "systems biology" enterprise. Many cellular functions are performed by multi-protein complexes and the identification and analysis of protein complex membership reveals insights into both the topological properties and functional organization of protein networks. Recently, high-throughput techniques have been developed to investigate physical binding between the constituents of protein complexes on a proteome-wide scale. The yeast two-hybrid assay (Y2H), a means of assessing whether two single proteins interact, has been adapted to systematically test pairwise protein interactions on a large scale[1,2], whereas affinity purification techniques using mass spectrometry (APMS)[3] provide a particularly effective approach to identifying protein complexes that contain more than two components. These techniques have been used to perform large scale protein-protein interaction screens in the yeast *Saccharomyces cerevisiae*[4,5,6] and the bacterium *Escherichia coli*[7].

In the APMS techniques, as described by Kumar and Snyder[3], individual proteins are tagged and used as "baits" to form physiological complexes with other proteins in the cells. Then, using the tag, each bait protein is purified, retrieving the proteins to which it binds, which may *sometimes* constitute the entire complex. The proteins extracted with the bait protein are identified using standard mass spectrometry methods. The raw results of these experiments are often referred to as "purifications" and may differ substantially from what is thought to exist in the cell and what is annotated in databases of protein complexes. Identification of actual protein complexes from these "purifications" often involves manual post-processing based on the existence of overlaps between the purifications[4]. Attempts to automate complex identification have involved the use of binary protein-protein interaction graphs[8,9,10], unsupervised clustering based on special similarity measures[13,6] and graph-theoretic approaches[14]. However, these approaches are bedeviled by a number of problems, such as fact that the exact number of complexes is initially unknown; the presence of potential contaminant proteins (which may themselves form complexes); the fact that the experiments do not always retrieve whole complexes, but only sub-complexes; and the presence of shared components, which need to be assigned to more than one complex.

In this paper, we propose a probabilistic algorithm to identify protein complex membership using the data from affinity purification/mass spec-

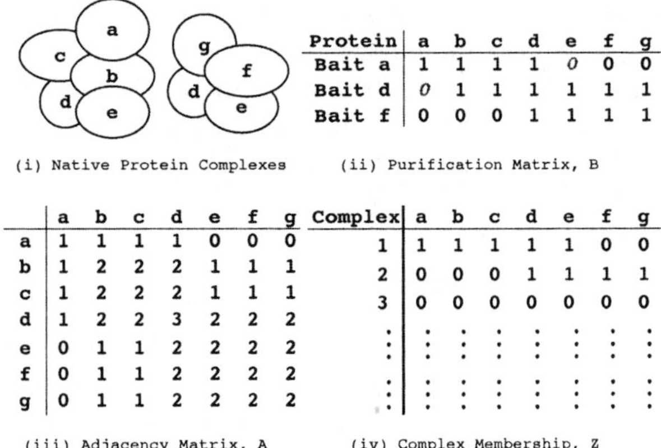

(i) Native Protein Complexes

Protein	a	b	c	d	e	f	g
Bait a	1	1	1	1	*0*	0	0
Bait d	*0*	1	1	1	1	1	1
Bait f	0	0	0	1	1	1	1

(ii) Purification Matrix, B

	a	b	c	d	e	f	g
a	1	1	1	1	0	0	0
b	1	2	2	2	1	1	1
c	1	2	2	2	1	1	1
d	1	2	2	3	2	2	2
e	0	1	1	2	2	2	2
f	0	1	1	2	2	2	2
g	0	1	1	2	2	2	2

(iii) Adjacency Matrix, A

Complex	a	b	c	d	e	f	g
1	1	1	1	1	1	0	0
2	0	0	0	1	1	1	1
3	0	0	0	0	0	0	0
⋮	⋮	⋮	⋮	⋮	⋮	⋮	⋮
⋮	⋮	⋮	⋮	⋮	⋮	⋮	⋮

(iv) Complex Membership, Z

Figure 1. Representations of APMS results. (i) shows the native composition of two protein complexes. (ii) represents the purification results for bait proteins, in which 1's denote positive and 0's denote negative, respectively. (iii) presents the corresponding adjacency matrix. (iv) specifies the complex membership of these proteins, where each entry is a binary random variable which indicates the members of the corresponding protein complex.

trometry (APMS) experiments. The membership between pairs of proteins is represented by a graph diffusion kernel, and the complexes and their constituents are identified by an infinite latent feature model. This model allows for multi-complex membership by individual proteins, which is fundamental for procuring an accurate protein complex catalog. The approach is novel as it identifies the complexes directly from experimental purifications. Our method provides insights into the organization of protein complexes into a core and peripherally located, possibly transiently binding components[11,12].

2. Data Representation

An example of the APMS method is shown in Figure 1. The purification results are usually recorded in the form of a binary matrix as shown in Figure 1(ii). Note that the APMS technology is neither perfectly sensitive nor specific, resulting in the failure to detect certain components (false negatives – FN) and the identification of proteins which are not members of a complex (false positives – FP). FN observations are represented by italic 0's in the example shown in Figure 1(ii). Let $\mathbf{B}$ denote the purification matrix with size $S \times N$, where S is the number of bait proteins and N is the number of proteins found in purifications. The corresponding adjacency matrix $\mathbf{A}$ is defined by $\mathbf{A} = \mathbf{B}^T\mathbf{B}$, which is a symmetric $N \times N$ matrix. The

ij-th element, $\mathbf{A}_{ij}$, is the number of purifications in which both protein i and protein j appear. The similarity between protein pairs can be measured by a graph diffusion kernel[15] based on $\mathbf{A}$. In this work, we focus on the von Neumann diffusion kernel[16] as the closeness measure. Kernel methods have also been applied to the inference of biological networks from other data sources[17,18].

3. The von Neumann Diffusion Kernel

The element $\mathbf{A}_{ij}$ in the adjacency matrix can be thought of as the number of distinct "paths" between protein i and protein j discovered by the APMS experiment. For example in Figure 1, there are two paths between protein d and protein e and no path directly connecting protein a and protein e. However, we could also reach protein e indirectly from protein a via the paths through the neighbors of protein a, e.g. a-b-e a-c-e and a-d-e. The number of distinct paths with length 2 between a pair of proteins can be directly counted by the matrix product $\mathbf{AA}$. More generally, the number of paths from protein i to protein j of length ℓ on the graph can be directly counted as the ij-th element of the matrix $\mathbf{A}^\ell$. The closeness between a pair of proteins can be measured by the number of distinct paths with different length. The von Neumann diffusion kernel[16] is the limit of the sum of the geometric series, defined as

$$\mathbf{K} := \sum_{\ell=1}^{\infty} \gamma^{\ell-1} \mathbf{A}^\ell = \mathbf{A}\left(1 - \gamma \mathbf{A}\right)^{-1}. \tag{1}$$

where γ is the diffusion factor to ensure the longer range connections decay exponentially.[a] The normalized kernel is an appropriate measure of similarity, which is defined as

$$\mathbf{D}_{ij} = \frac{\mathbf{K}_{ij}}{\sqrt{\mathbf{K}_{ii}\mathbf{K}_{jj}}}. \tag{2}$$

Note that the matrix elements are between 0 and 1. $\mathbf{D}_{ij} = 0$ implies protein i is isolated from protein j. On the contrary, $\mathbf{D}_{ij}$ approaches 1 if the protein pair is tightly connected. The elements of the normalized von Neumann kernel (2) provide a probabilistic measure on the pairwise membership of two proteins in the same complex. This interpretation makes $\mathbf{D}_{ij}$ suitable for use as a likelihood in a probabilistic model, which we now describe.

---

[a]The von Neumann kernel (1) is positive definite only if $0 < \gamma < \rho^{-1}$ where ρ is the spectral radius of $\mathbf{A}$. γ could be learnt from the data. In this work, we set $\gamma = (1+\kappa)^{-1}\rho^{-1}$ where κ is the proportion of non-zero elements in the adjacency matrix $\mathbf{A}$.

4. Protein Complex Membership

Protein complex membership can be represented as a binary matrix, denoted as $\mathbf{Z}$ (see Figure 1(iv)). Each column of the matrix $\mathbf{Z}$ is denoted by $\mathbf{z}_i$, known as the feature vector of the protein. The length of $\mathbf{z}_i$ is variable, as the number of protein complexes is actually unknown. The membership of the i-th protein in complex c is indicated by a binary random variable z_{ci}. Note that each protein may belong to multiple complexes. The learning task is to infer a catalog of protein complexes and their constituents from the APMS experimental data.

4.1. *An Infinite Latent Feature Model*

Griffiths and Ghahramani[19] have proposed a probability distribution over binary matrices with a fixed number of columns (proteins) and an infinite number of rows (complexes), which is particularly suitable for use as a prior in probabilistic models that represent proteins with multiple complex membership. In the following we describe this infinite latent feature model[19] in the context of protein complex membership identification. Since the exact number of complexes is initially unknown, we start with a finite model that assumes C complexes, and then take the limit as $C \to \infty$ to obtain the prior distribution over the binary matrix $\mathbf{Z}$. As in other non-parametric models, taking this limit ensures that the model is flexible enough to capture any number of complexes.

We assume that each protein belongs to a complex c with probability π_c, and then given the set $\pi = \{\pi_1, \pi_2, \ldots, \pi_C\}$ the probability of matrix $\mathbf{Z}$ is a product of binomial distributions

$$\mathcal{P}(\mathbf{Z}|\pi) = \prod_{c=1}^{C} \prod_{i=1}^{N} \mathcal{P}(z_{ci}|\pi_c) = \prod_{c=1}^{C} \pi_c^{n_c}(1-\pi_c)^{N-n_c}, \tag{3}$$

where $n_c = \sum_{i=1}^{N} z_{ci}$ is the number of constituent proteins belonging to the complex c. As suggested by Griffiths and Ghahramani[19], the beta distribution is chosen to be beta$(\frac{\alpha}{C}, 1)$ where α is a model parameter.[b] In the probabilistic model we have defined, each z_{ci} is independent of all other memberships and the π_c's are also independent of each other. Given this prior on π, we can simplify this model by integrating over all possible

[b]We set $\alpha = 1$ in this work which represents our prior belief that each protein is expected to belong to one complex but probably not many more.

settings for π, and then compute the conditional distribution for any z_{ci} as follows

$$\mathcal{P}(z_{ci}|\mathbf{Z}_{-i,c}) = \frac{n_{-i,c} + \frac{\alpha}{C}}{N + \frac{\alpha}{C}}, \tag{4}$$

where $\mathbf{Z}_{-i,c}$ denotes the entries of $\mathbf{Z}$ except z_{ci}, and $n_{-i,c}$ is the number of proteins belonging to the complex c, not including the protein i.

The infinite model can be obtained from the finite model by taking the limit of (4) as $C \to \infty$. The conditional distribution of z_{ci} in the infinite model is then

$$\mathcal{P}(z_{ci}|\mathbf{Z}_{-i,c}) = \frac{n_{-i,c}}{N}, \tag{5}$$

for any c such that $n_{-i,c} > 0$. As for the c's with $n_{-i,c} = 0$, it can be shown that the number of new complexes associated with this protein, denoted as ν_i, has a Poisson distribution with the parameter $\frac{\alpha}{N}$ as follows,

$$\mathcal{P}(\nu_i|\mathbf{Z}_{-i,c}) = \left(\frac{\alpha}{N}\right)^{\nu_i} \frac{\exp(-\frac{\alpha}{N})}{\nu_i!}. \tag{6}$$

Details of all these properties of the infinite latent feature model can be found in Ref. 19.

4.2. *Likelihood Evaluation*

Given a particular protein complex membership matrix $\mathbf{Z}$, the pairwise membership can be determined by examining whether $\mathbf{z}_i^T\mathbf{z}_j > 0$ or $\mathbf{z}_i^T\mathbf{z}_j = 0$, which categorizes the protein pairs respectively into two classes, members of the same complex or not. The likelihood can be evaluated by the von Neumann diffusion kernel (2) directly, as $\mathbf{D}_{ij}$ exactly measures the probability of the protein pair being members of a protein complex. Therefore, the likelihood can be evaluated as follows,

$$\mathcal{P}(\mathbf{D}|\mathbf{Z}) = \prod_{\{ij:\mathbf{z}_i^T\mathbf{z}_j>0\}} (\mathbf{D}_{ij})^{\mathbf{z}_i^T\mathbf{z}_j} \prod_{\{ij:\mathbf{z}_i^T\mathbf{z}_j=0\}} (1-\mathbf{D}_{ij}), \tag{7}$$

where $\{ij\}$ denotes any distinct pair and $\mathbf{D}$ denotes the normalized von Neumann kernel matrix obtained from the APMS experiments.

4.3. *Membership Inference*

Based on Bayes' theorem, the posterior distribution of the protein complex membership $\mathbf{Z}$ can be given by $\mathcal{P}(\mathbf{Z}|\mathbf{D}) \propto \mathcal{P}(\mathbf{D}|\mathbf{Z})\mathcal{P}(\mathbf{Z})$, where $\mathcal{P}(\mathbf{D}|\mathbf{Z})$ is defined as in (7), and $\mathcal{P}(\mathbf{Z})$ is defined by the infinite latent feature model.

We have defined a posterior distribution for the protein complex membership that does not assume a fixed number of protein complexes and allows for multiple membership. In the following, we describe a Gibbs sampler to carry out inference in the infinite latent feature model. The critical quantity required in the Gibbs sampling is the conditional distribution

$$\mathcal{P}(z_{ci}|\mathbf{Z}_{-i,c},\mathbf{D}) \propto \mathcal{P}(\mathbf{D}|\mathbf{Z})\mathcal{P}(z_{ci}|\mathbf{Z}_{-i,c}), \tag{8}$$

where the likelihood $\mathcal{P}(\mathbf{D}|\mathbf{Z})$ is defined as in (7), and $\mathcal{P}(z_{ci}|\mathbf{Z}_{-i,c})$ is defined as in (5) for any c with $n_{-i,c} > 0$. While for the complexes with $n_{-i,c} = 0$, the conditional distribution over the number of new complexes taken by the protein can be computed as follows

$$\mathcal{P}(\nu_i|\mathbf{Z}_{-i,c},\mathbf{D}) \propto \mathcal{P}(\mathbf{D}|\mathbf{Z})\mathcal{P}(\nu_i|\mathbf{Z}_{-i,c}), \tag{9}$$

where $\mathcal{P}(\nu_i|\mathbf{Z}_{-i,c})$ is a Poisson distribution defined as in (6). Note that the membership of new complexes does not change the pairwise membership at all. So the likelihood $\mathcal{P}(\mathbf{D}|\mathbf{Z})$ stays equal for any value of ν_i in our model. The overall algorithm can be summarized as follows,

(1) *Initialize* $\mathbf{Z}$ *randomly*, usually start with one complex.
(2) *For* $t = 1$ *to* T
   (a) For each i and each c with $n_{-i,c} > 0$, sample z_{ci} in the distribution (8).
   (b) For each i, sample the number of new complexes in the Poisson distribution (6).
   (c) Save the sample $\mathbf{Z}$.
(3) *Exit*

In this work, we collected 1000 samples after burning in the first 1000 samples as an approximate estimate of the posterior distribution of $\mathbf{Z}$. The computational overhead of this algorithm is approximately $\mathcal{O}(TCN^2)$, where T denotes the number of samples we collect, C denotes the number of significant complexes in the data and N denotes the number of hit proteins. The algorithm has been implemented in Matlab. On a Linux Athlon 1800 desktop, it took about 89.7 seconds to process the RNA Polymerase complex data set described below.

5. Results

To validate our approach, we have applied the above algorithm to two experimental data sets: the purifications corresponding to the proteins contained

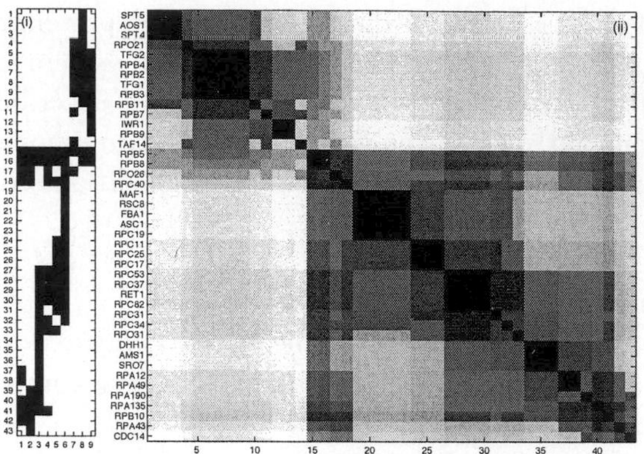

Figure 2. The RNA Polymerase complexes. (i) presents the purification results using 9 bait proteins. (ii) presents the corresponding normalized von Neumann kernel matrix, where the gray scale indicates the probability of pairwise membership defined as in (2). The proteins are sorted according to the inferred complex membership

in the RNA Polymerase complexes from the whole proteome screen of Gavin et. al[4], and the yeast RNA-Processing Complexes data of Krogan et. al[6]. The RNA Polymerase complexes are a particularly suitable and tractable test case, since they share five components and their three-dimensional structure has been determined by X-ray crystallography[20], providing a "gold standard". The data set used comprised 9 purifications (baits) and 43 proteins (hits) extracted from the data of Gavin et al.[4] as shown in Figure 2(i). Figure 2(ii) shows the normalized von Neumann diffusion kernel (2) for this data. The final protein complex assignments are shown in Figure 3. Proteins correctly assigned to the three RNA Polymerase complexes according to the MIPS protein complex database[21] are marked with crosses in Figure 3. TFG1/2 and SPT5 appear as members of the RNAP II complex under the APMS conditions used and thus are included in the prediction. RPB5 and 8 are clearly seen to be shared amongst all 3 complexes, whilst RPO26 and RPC40 are shared between RNAP I and III. Due to the nature of the experimental data, one cannot assume a perfect match, particularly as only a few baits of the whole complex were actually purified.

A subset of the more extensive yeast RNA-processing complex data of Krogan et al. comprising the "reliably identified" complexes[6] consisted of a data set of 71 purifications (baits) and 240 proteins (hits). We restricted the analysis to the data sample used for hierarchical clustering in the study of Krogan et al. (MALDI data). Inspection of the normalized von Neumann diffusion kernel for this data (Figure 4, bottom right) indicated that a subset of this data (24 baits and 49 proteins) formed clear and unam-

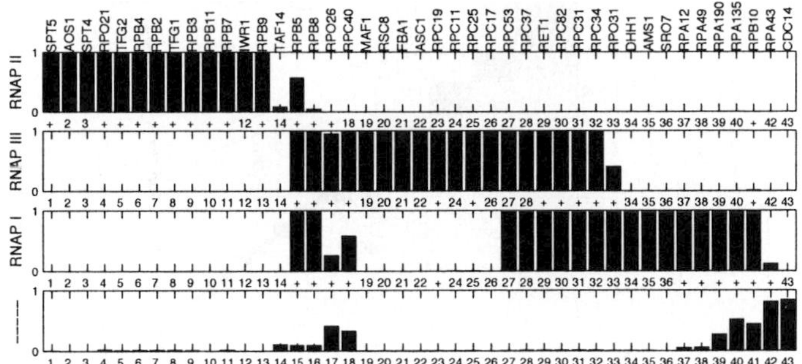

Figure 3. The four largest complexes identified by our algorithm. The bars indicate the probability of membership of the proteins. The top 3 complexes correspond to RNAP II, RNAP III and RNAP I respectively. The cross indices indicate the members of the three RNA polymerase complexes according to the MIPS protein complex database.

biguous clusters amongst themselves, and so were not processed further. This subset of the data comprised 9 clusters with membership ≥ 2, the largest of which was the 19S regulatory subunit of the proteasome comprising RPN1,2,3,5,6,7,9,10,11,12 and RPT1,3,6. Some known proteins of the proteasome are missing because they were not part of the experimental data set. The remaining data set, comprising 47 baits and 191 proteins were analyzed using the method described above, resulting in 20 clusters being identified. The assignments into the 20 identified clusters are shown in Figure 5. Most of the complexes correspond to those identified by Krogan et al.[6] such as the RNA polymerase complexes, the SSU processome, the exosome and U6 specific snRNP core. In the case of the SSU processome, we also identify a distinct UTP-A complex which includes UTP4,5,8,9,15, NAN1 and POL5, and a distinct UTP22/RRP7 complex. Interestingly, this appears as distinct from the casein kinase II complex although it co-purifies with casein kinase II subunits. Our method was also capable of identifying the TFG1/2 complex as a separate entity to the RNAP II complex even though some of its elements were purified using TFG1 as a bait. These, and other examples indicate that our method is capable of partitioning the data in a biologically meaningful way.

6. Discussion and Conclusions

We have demonstrated that the algorithm produces biological meaningful results and yields insights into how the clusters were generated, which is important for the interpretation of the results. In particular, the assignment of a protein to more than one complex and the choice of the number

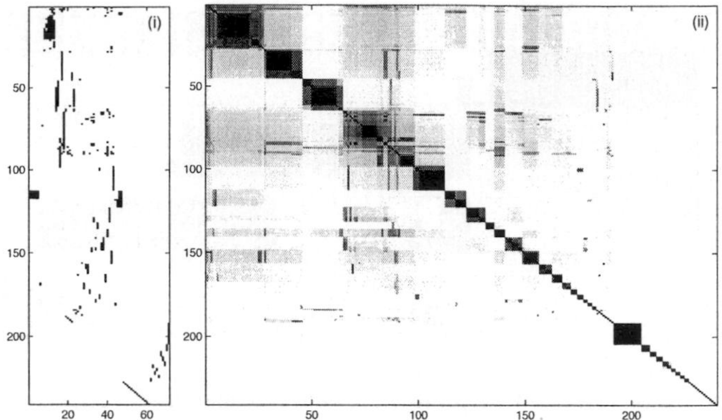

Figure 4. Yeast RNA-processing complex data of Krogan et al. (i) presents the purification results using 71 bait proteins. (ii) presents the corresponding normalized von Neumann kernel matrix of 240 hit proteins where the gray scale indicates the probability of pairwise membership defined as in (2). The proteins are sorted according to the inferred complex membership. A larger version of the figure can be found on the supplementary web site. The names of the first 191 proteins are also indexed in Figure 5.

of complexes can be performed without heuristic assumptions, a major improvement over previous methods.

Obviously, the method relies on experimental data and assignment of some artifacts such as ribosomal contaminants to complexes cannot be avoided. It should also be noted that there is no consensus amongst experts on how to identify artifacts. The interpretation must be applied in a similar fashion across the whole network of protein-protein interactions.

It would be necessary to introduce reference data sets for more comprehensive comparisons with standardized methods for identifying protein complexes. The current data sets are sparse and there is little independent confirmation for most complexes. Our method works well for the data sets used here and further improvements should be obtained when the individual complexes are better sampled.

References

1. P. Uetz, L. Giot, G. Cagney, et al., A comprehensive analysis of protein-protein interactions in *Saccharomyces cerevisiae*, *Nature*, **403**, 623-627, (2000).
2. T. Ito, T. Chiba, R. Ozawa, M. Yoshida, M. Hattori and Y. Sakaki, A comprehensive two-hybrid analysis to explore the yeast protein interactome, *Proc Natl Acad Sci USA*, **98(8)**, 4569-4574 (2001).
3. A. Kumar and M. Snyder, Protein complexes take the bait, *Nature*, **415**, 123-124, (2002).

4. A.C. Gavin, M. Bosche, R. Krause, et al., Functional organization of the yeast proteome by systematic analysis of protein complexes, *Nature*, **415**, 141-147, (2002).
5. Y. Ho, A. Gruhler, A. Heilbut, et al., Systematic identification of protein complexes in *Saccharomyces cerevisiae* by mass spectrometry, *Nature*, **415**, 180-183, (2002).
6. N.J. Krogan, W.T. Peng, G. Cagney, et al., High-definition macromolecular composition of yeast RNA-processing complexes, *Molecular Cell*, **13**, 225-239, (2004).
7. G. Butland, J. M. Peregrin-Alvarez, J. Li, et al., Interaction network containing conserved and essential protein complexes in *Escherichia Coli*, *Nature*, **433**, 531-537, (2005).
8. G.D. Bader and C.W. Hogue, Analyzing yeast protein-protein interaction data obtained from different sources, *Nat Biotechnol*, **20(10)**, 991-997, (2002).
9. G.D. Bader and C.W. Hogue, An automated method for finding molecular complexes in large protein interaction networks, *BMC Bioinformatics*, **4(1)**, (2003).
10. V. Spirin and L.A. Mirny, Protein complexes and functional modules in molecular networks, *Proc Natl Acad Sci USA*, **100(21)**, 12123-12128, (2003).
11. J. Hollunder, A. Beyer and T. Wilhelm, Identification and characterization of protein subcomplexes in yeast, *Proteomics*, **5(8)**, 2082-9, (2005).
12. Z. Dezso, Z. Oltvai and A.L. Barabasi, Bioinformatics analysis of experimentally determined protein complexes in the Yeast *Saccharomyces cerevisiae*, *Genome Res*, **13**, 2450-2454, (2003).
13. R. Krause, C. von Mering and P. Bork, A comprehensive set of protein complexes in yeast: Mining large scale protein-protein interaction screens, *Bioinformatics*, **19(15)**, 1901-1908, (2003).
14. D. Scholtens and R. Gentleman, Making Sense of High-throughput Protein-protein Interaction Data, *Statistical Applications in Genetics and Molecular Biology* **3**, Article 39, (2004).
15. R. Kondor and J. Lafferty, Diffusion kernels on graphs and other discrete structures, *ICML*, 315-322, (2002).
16. J. Shawe-Taylor and N. Cristianini, *Kernel Methods for Pattern Analysis*, (2004).
17. A. Ben-Hur and W.S. Noble, Kernel methods for predicting protein-protein interactions, *Bioinformatics*, **21**, Suppl. 1, i38-i46, (2005).
18. K. Tsuda and W.S. Noble, Learning kernels from biological networks by maximizing entropy, *Bioinformatics*, **20**, Suppl. 1, i326-i333, (2004).
19. T. L. Griffiths and Z. Ghahramani, Infinite latent feature models and the Indian buffet process *Technical Report*, **GCNU TR 2005-001**, University College London, (2005).
20. P. Cramer, D.A. Bushnell, J. Fu, et al., Architecture of RNA polymerase II and implications for the transcription mechnism, *Science*, **288**, 640-649, (2000).
21. H.W. Mewes, D. Frishman, U. Guldener, et al., MIPS: a database for genomes and protein sequences, *Nucleic Acids Res*, **30**, 31-34, (2002).

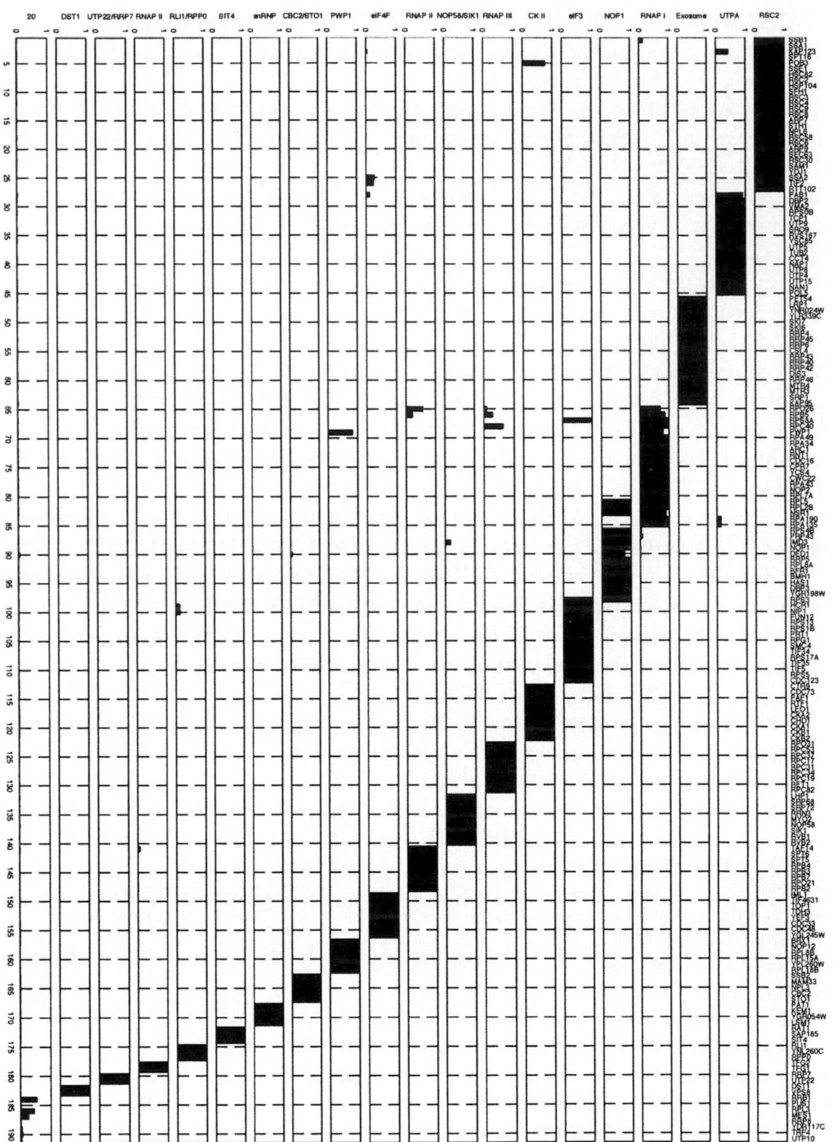

Figure 5. The assignments of the 191 proteins into 20 inferred complexes. The bars indicate the probability of membership of these proteins. Complexes identified (left to right) are: DST1; UTP22/RRP7; RNA Polymerase II; RLI1/RPP0; SIT4; U6 specific snRNP core; CBC2/STO1; PWP1/BRX1/NOP12; mRNA cap-binding/EIF4F; RNA Polymerase II; SRP3/NOP58/SIK1; RNA Polymerase III; Casein Kinase II; eIF3; NOP1; RNA Polymerase I; Exosome; SSU processome (UTPA); RSC2.

HIGH-ACCURACY PEAK PICKING OF PROTEOMICS DATA USING WAVELET TECHNIQUES*

EVA LANGE[†], CLEMENS GRÖPL, KNUT REINERT
Institute of Computer Science, Free University of Berlin
Takustr. 9, 14195 Berlin, Germany
E-mail: lange@inf.fu-berlin.de

OLIVER KOHLBACHER
Center for Bioinformatics, Eberhard Karls University Tübingen,
Sand 14, 72076 Tübingen, Germany
E-mail: oliver.kohlbacher@uni-tuebingen.de

ANDREAS HILDEBRANDT
Center for Bioinformatics, Saarland University,
P.O. 15 11 50, 66041 Saarbrücken, Germany
E-mail: anhi@bioinf.uni-sb.de

A new peak picking algorithm for the analysis of mass spectrometric (MS) data is presented. It is independent of the underlying machine or ionization method, and is able to resolve highly convoluted and asymmetric signals. The method uses the multiscale nature of spectrometric data by first detecting the mass peaks in the wavelet-transformed signal before a given asymmetric peak function is fitted to the raw data. In an optional third stage, the resulting fit can be further improved using techniques from nonlinear optimization. In contrast to currently established techniques (e.g. SNAP, Apex) our algorithm is able to separate overlapping peaks of multiply charged peptides in ESI-MS data of low resolution. Its improved accuracy with respect to peak positions makes it a valuable preprocessing method for MS-based identification and quantification experiments. The method has been validated on a number of different annotated test cases, where it compares favorably in both runtime and accuracy with currently established techniques. An implementation of the algorithm is freely available in our open source framework OpenMS.

*This work is supported by the German federal ministry of education and research, (grant no. 0312705a 'Berlin Center for Genome Based Bioinformatics').
[†]corresponding author

1. Introduction

Mass spectrometry is one of the central technologies for quantitative proteomics as well as for protein identification. The conversion of the "raw" ion count data acquired by the machine into peak lists for further processing is usually called *peak picking.* This is often done by vendor software bundled with the machine. However, it is often desirable to have more control over this process than one has with the limited intervention allowed by the vendor programs. Any algorithm for peak picking has the following main objectives: First, the peak should have a mass to charge ratio that is as accurate as possible, that means as near as possible to the true mass to charge ratio of the measured compound. This is especially important for identification algorithms. Second, the algorithm should run in real time, that means processing the data should never exceed the time of acquiring it. Among the main difficulties in peak picking are: i) there is often considerable asymmetry in the peaks which confounds a correct mass to charge computation; ii) the convolution of isotopic peaks makes it hard to distinguish individual peaks (this depends on the charge state and the resolution of the instrument). Recently, several approaches to the peak picking problem in proteomics data have been proposed [1,2,3,4]. For example Strittmatter and coworkers [3], use a fit of a Gaussian mixture to model the observed asymmetry in peak shapes. In connection with a calibration method for TOF machines they achieve a considerable improvement in mass accuracy for non convoluted ESI-TOF data. Kempka et al [4] elaborate on this mixture modelling and test also other mixtures like a Lorentzian and a Gaussian curve. They compare their results to the ones obtained by commercial peak picking algorithms (SNAP) and conclude that they perform better for most peaks. For small and considerably skewed peaks the improvement in accuracy is up to fivefold. The results were obtained on highly resolved MALDI-TOF data without convoluted peaks, since these algorithms require baseline or close to baseline separation of isotopic patterns.

In this paper we describe an algorithm that addresses the above mentioned goals. It computes accurately the mass over charge ratio not only for well-resolved, but also for convoluted data using an asymmetric peak shape. It achieves this in real time and does not make assumptions about the underlying machine or ionization method (MALDI or ESI), which makes the algorithm robust for different experimental settings. This is achieved by addressing the problem from a signal theoretic point of view, which tells us that spectral data like MS measurements are of an inherently multiscale nature. Different effects, typically localized in different frequency ranges, add up to result in the final signal. In the following, we will assume that the experimentally obtained signal s can be decomposed into three such contributions: a high-frequency noise term n, a low-frequency baseline or background term b, and the information i we are interested in, often referred to as the analytical signal [5], where i occupies a frequency range in between noise and baseline.

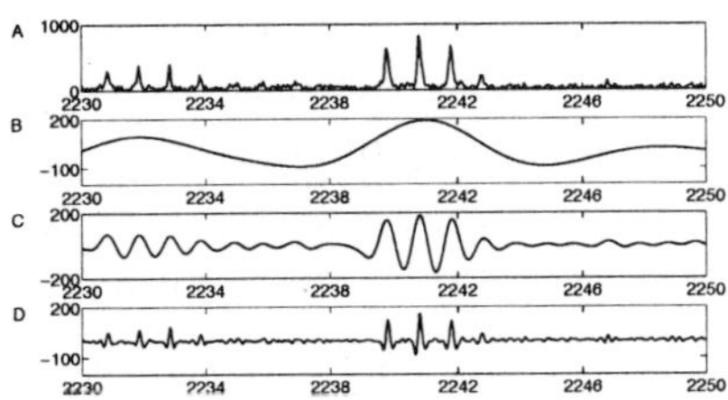

Figure 1. In plot A, B, C, and D the x-axis represents the mass interval between 2230Da and 2250Da, whereas the y-axis shows the intensity. A: Part of a MALDI mass spectrum. Plots B, C, and D show the continuous wavelet transform of the spectrum using a Marr wavelet with different dilation values a (B: $a = 3$, C: $a = 0.3$, D: $a = 0.06$).

The algorithm presented here directly exploits the multiscale nature of the measured spectrum. This becomes possible with the help of a Continuous Wavelet Transform (CWT) – a mathematical tool particularly suited for the processing of data on different scales, where it clearly outperforms classical signal processing techniques like the Fourier Transform, since it preserves information about the localization of different frequencies in the signal in a near-optimal manner [6]. Using the CWT, we can split the signal into different frequency ranges or length scales that can be regarded independently of each other. This is demonstrated in Fig. 1, where we have plotted the transformed signal of a typical region of a mass spectrum on different scales. Apparently, looking at the signal at the correct scale – in our case, a rough estimate of the typical peak width – effectively suppresses both baseline and noise, keeping only the contribution due to the analytical signal. This decomposition allows us to determine each feature of a peak in the domain from which it can be computed best, i.e., either from the frequency range of the analytical signal i, the full signal s, or from a combination of both. Our algorithm is a two-step technique that first determines the positions of putative peaks in the Wavelet-transformed signal and then fits an analytically given peak function to the data in that region. In an optional third stage, the resulting fit can be further improved using techniques from nonlinear optimization. The method has been validated on a number of different annotated test cases, where it compares favorably in both runtime and accuracy with currently established techniques. The algorithm has been implemented in C++. This implementation is freely available in our open-source framework OpenMS[7].

In Section 2 we describe the data sets we used and explain our algorithm in more detail. In Section 3 we demonstrate that our algorithm leads to accurate predictions of the mass over charge position and deconvolutes overlapping peaks more accurately than the vendor software. Finally we discuss further developments in Section 4.

2. Methods

2.1. *Sample preparation and data generation*

Data set A was obtained from a peptide mix (peptide standards mix #P2693 from Sigma Aldrich) of nine known peptides (bradykinin (F), bradykinin fragment 1-5 (B), Substance P (H), [Arg^8]-vasopressin (E), luteinizing hormone releasing hormone bombesin (G), leucin enkephalin (A), methionine enkephalin (C), oxytocin (D)). Sample concentration was 0.25 ng/μl, injection volume 1.0 μl. HPLC separation was performed on a capillary column (monolithic polystyrene/-divinylbenzene phase, 60 mm x 0.3 mm) with 0.05% trifluoroacetic acid (TFA) in water (eluent A) and 0.05% TFA in acetonitrile (eluent B). Separation was achieved at a flow of 2.0 μl/min at 50°C with an isocratic gradient of 0–25% eluent B over 7.5 min. Eluting peptides were detected in a quadrupole ion trap mass spectrometer (Esquire HCT from Bruker, Bremen, Germany) equipped with an electrospray ion source in full scan mode (m/z 500-1500).

Data set B The MALDI-TOF mass spectrum of a tryptic digest of bovine serum albumin (BSA, Aldrich) was acquired from a preparation of an amount corresponding to 50 fmol of the digested protein. In brief, cystines were reduced by incubation with dithiotreitol (DTT) followed by carbamidomethylation using iodoacetamide, prior to proteolysis. The sample was prepared for MALDI using the matrix-affinity sample preparation method with alpha-cyano-4-hydroxycinnamic acid as the matrix [8]. Analysis of positively charged ions in the m/z range 500-5000 was performed on an Ultraflex II LIFT mass spectrometer (Bruker Daltonics, Bremen) operated in the reflectron mode and using Panorama(TM) delayed ion extraction.

A near-neighbour calibration was performed using a peptide standard mixture.

2.2. *The general scheme of our algorithm*

A peak picking technique suitable for high-throughput proteomics applications necessarily needs to combine high accuracy with computational efficiency to provide the results in real time. In our case, a great gain in performance without any negative impact on the accuracy can be achieved by decomposing the mass spectra into smaller parts, so called *boxes*, with a typical length of 10 Da. When splitting a spectrum in this way, special care has to be taken to preserve the signal content. In particular, the split point must not belong to any of the peaks we want to find. After this decomposition of the signal, we compute the wavelet transform of the data in each box. Starting from the maximum position in the wavelet transform, every peak centroid, its height, and its area can be estimated in the raw data. Using these parameters, we are able to represent the raw data peaks using an asymmetric sech^2 and asymmetric Lorentzian function. At this stage of the algorithm, the fitted analytical description is typically in very good correspondence with the experimental signal. To further improve the quality of the fit, the correlation of the resulting peaks with the experimental data can be increased in a subsequent, optional optimization step. This is of particular importance in two cases: first, if neighboring peaks overlap strongly enough that they cannot be fitted well individually, and second, if the resolution of the experimental data is low.
In pseudocode, the general scheme of the algorithm can be formulated as follows:

```
for all mass spectra s in experiment do
  box_list = splitSpectrum(s)
  for all boxes b in box_list do
    peak_list = []
    w_b = continuousWaveletTransformation(b)
    while getNextMaximumPosition(w_b, b, x_0) do
      (x_l, x_r) = searchForPeakEndpoints(b, x_0)
      c = estimateCentroid(x_l, x_r)
      h = intensity(x_0)
      (A_l, A_r) = integrateAreas(x_l, x_r)
      f = fitPeakShape(A_l, A_r, c, h)
      push(f, peak_list)
```

removeRawDataPeak(x_l, x_r, b)
w_b = continuousWaveletTransformation(b)
end while
optimizePeakParameter($peak\_list, b$)
end for
end for

In the following we elaborate on the individual steps.

Decomposing the mass spectrum Our decomposition algorithm ensures that all raw data points belonging to one peak lie in exactly one box. To this end, we split a mass spectrum at a raw data point x if and only if its intensity y is smaller than a given noise threshold, otherwise we search for a minimum in x's neighborhood using the moving average method. If there is no minimum inside a certain search radius we can be sure that x does not belong to a peak with sensible width, and thus cut at x.

Computing the maximum in the Continuous Wavelet Transform The Continuous Wavelet Transform $W_{s(b)}$ of the signal s at position b is defined as

$$W_{\psi}s_a(b) = \frac{1}{\sqrt{a}} \int_{-\infty}^{\infty} s(x)\psi\left(\frac{x-b}{a}\right) dx,$$

with the wavelet ψ and the corresponding dilation parameter a. For ψ we choose the so-called Marr wavelet

$$\psi(x) = (1-x^2)\exp\left(\frac{x^2}{2}\right) = \frac{d^2}{dx^2}\exp\left(-\frac{x^2}{2}\right)$$

since it is known that at least for symmetric peaks, the maximum position in the CWT coincides with the maximum position in the data [9] and is a good first estimate even for asymmetric peaks. For a peak picking application, the scale of the wavelet (and thus its width) should correspond to the typical width of the peaks. On such a scale, ψ is described by only a few data points and thus the convolution of wavelet and signal can be computed very efficiently with pre-tabulated values of ψ.

Searching for the peak endpoints Defining the "ends" of a peak shape becomes difficult when effects like noise or overlapping of peaks have to be considered. In this case, we cannot expect that the peak's intensity drops below a given threshold before the next peak's area of influence is reached. To solve this problem, we start at the maximum position and proceed to the left and right until either a minimum is reached, or the value drops below a pre-defined noise threshold. A minimum might either be caused by the rising flank of a neighboring peak, or could be a mere noise effect. To decide between these two cases, we consider again the CWT in the neighborhood, where noise effects are typically smoothed out and peaks can be clearly discerned.

Estimating the peak's centroid For the estimation of the peak's centroid c we compute the intensity-weighted average of the cap of the peak, which is defined as the consecutive set of points next to the maximum with intensity above a certain percentage of the peak's height.

Fitting asymmetric Lorentzian and sech^2 functions In the literature, several different analytical expressions have been proposed for the representation of mass spectrometrical peaks. Since to our knowledge no universally accepted peak shape exists, our algorithm can fit the data to different peak functions. In the current implementation, we use asymmetric Lorentzian ($\mathfrak{L}$) or sech^2 ($\mathfrak{S}$) functions, which are defined by

$$\mathfrak{L}_{h,\lambda(x),c}(x) = \frac{h}{1+\lambda^2(x)\cdot(x-c)^2}, \tag{1}$$

and

$$\mathfrak{S}_{h,\lambda(x),c}(x) = \frac{h}{\cosh(\lambda(x)(x-c))^2} \tag{2}$$

where

$$\lambda(x) = \begin{cases} \lambda_l, & x \leq c \\ \lambda_r, & x > c \end{cases} \tag{3}$$

but other peak shapes like double Gaussian profiles [3,4] can be easily included. A peak can be fitted to the raw data in several ways. In our implementation, we have chosen to use the peak's previously determined centroid and the area under the experimental signal. Fitting the area of the peak automatically introduces a smoothing effect, yields very good approximations to the original peak shape, and is extremely efficient, since the peak's width can be computed from its area in constant time for the functions considered here. Since the peaks are modelled as asymmetric functions, we integrate from the left endpoint x_l up to the peak centroid c to obtain the left peak area A_l. Analogously, we compute the right peak area A_r between c and the right peak endpoint x_r. From these values, we can finally analytically compute the asymmetric Lorentzian or sech^2 function with centroid position c and height h which has the same area A_l as the raw peak between x_l and c, and A_r between c and x_r, respectively.

Optimizing the peak parameters The peaks computed so far typically yield a reasonable approximation of the true signal, especially for well resolved, clearly separated peaks. To further improve accuracy, we perform an additional (optional) optimization step. This turned out to be particularly useful for poorly resolved data with strongly overlapping, convoluted peak patterns. Let us assume that we have found m peak functions f_i in a box b, which contains n raw data points (x_i, y_i), $i = 1..n$. In the previous stage, each of the peaks has been fitted independently of the others, but for a true separation, we need to fit the sum of all peaks to the experimental signal. This can be achieved using standard techniques from nonlinear optimization, like the Levenberg-Marquardt algorithm [10]. Each function f_i is described by an initial set of four parameters $p_i = (c_i, h_i, \lambda_{l_i}, \lambda_{r_i})$ with height h_i, centroid c_i, and left/right widths $\lambda_{l_i}/\lambda_{r_i}$. The m parameter sets p_i define the parameter vector $p = (p_1, ...p_m)^T$. As loss function, we employ the absolute difference $l_i(p) = |\sum_{j=1}^{m} f_j(x_i) - y_i|$ between estimated and experimental signal. The nonlinear least squares problem then consists in finding the parameter vector p which minimizes the total loss function

$$\Phi(p) = \frac{1}{2}\sum_{i=1}^{n} l_i(p)^2 = \frac{1}{2}||L(p)||^2 \quad (4)$$

The initial approximation provided by the first stage of the algorithm is usually a good starting point, enhancing the convergence properties. In addition, this allows us to penalize strong deviation from the initial solution, resulting in significantly enhanced robustness.

3. Results

Assessing the quality of a peak picking scheme is a non-trivial problem for which no straight-forward and general approach exists. Obviously, such an algorithm should compute the peak's centroid, height, and area as accurately as possible while featuring a high sensitivity and specificity. To determine the accuracy of, e.g., a peak's centroid, the correct mass value is needed, and thus peak picking algorithms are typically tested against a spectrum of known composition, e.g., a standard peptide mixture or the tryptic digest of a certain protein. Comparing the features of the peaks found in the spectrum with the theoretical values then gives a measure of the algorithm's capabilities, typically expressed in the average absolute and relative deviation (measured in ppm). Unfortunately, these results are heavily affected by the quality of the experimental data, and additional issues like calibration. Consequently, peak picking algorithms are typically tested against particularly well-resolved spectra, and internal calibration methods are employed. This usually results in high mass measurement accuracy, but the quality of the peak picking algorithms can not be judged independently of the quality of the calibration scheme. From a user's perspective, on the other hand, obtaining similarly well resolved spectra is often infeasible, and internal calibration is not always an option. We have thus decided to demonstrate the capabilities of our approach mainly on LC-MS data with low resolution, containing severely overlapping isotope patterns. Obviously, this complicates comparison of the resulting mass accuracy to published results for alternative peak picking schemes that were tested on well resolved data subjected to sophisticated calibration. We have therefore decided to use the vendor supplied Bruker DataAnalysis 3.2 software on the same spectra to provide a fair means of comparison.

To assess the performance of our peak picking scheme on a set of LC-MS runs on the peptide mixture (dataset A), we determined how often each peptide was found in the expected retention time interval, whether the corresponding isotope patterns (given by at least three consecutive peaks) were discovered and separated, and computed the resulting relative errors of the monoisotopic peak's centroid compared to the theoretical monoisotopic mass. The same analysis was performed with the Bruker software, using the Apex algorithm recommended for ion trap data. The resolution of the data set is critically low with a Δm value of 0.2, implying that each peak is represented by as little as 3–6 data points, and instead of a sophisticated calibration, we only allowed for a constant mass offset to keep the number of fit parameters as small as possible. Using recommended signal-to-noise settings in the Bruker software turned out to miss a large number of the isotopic patterns due to the poor quality of the data. We therefore decided to perform two tests against the Bruker software, one with the recommended setting, and one with a significantly reduced signal-to-noise threshold and peak bound, leading to a total number of peaks comparable to our method. The results of these tests are shown in Table 1. For each peptide, this table contains the theoretical monoisotopic mass, the average relative error of the monoisotopic position, and the number of scans in which the peptide was correctly identified.

Table 1. Evaluation of dataset A. In the table, I denotes the results of the method presented here, II_a the Apex algorithm with reduced thresholds, and II_b Apex with default settings.

			rel. err. [ppm]			#*occ.*		
	z	m_{theo} [Da]	I	II_a	II_b	I	II_a	II_b
A	1	556.2693	31	35	39	22	35	19
B	1	573.3071	16	24	16	29	57	29
C	1	574.2257	44	60	21	19	44	15
D	1	1007.4365	25	-	94	8	0	5
E	1	1084.4379	18	-	12	3	0	2
F	2	1061.5614	56	64	64	3	2	2
G	2	1183.5730	15	-	-	7	0	0
H	2	1349.736	28	-	13	8	0	1
I	2	1620.8151	37	-	-	13	0	0

Considering the resolution of the raw data, and the lack of sophisticated internal calibration, the mass accuracy that was obtained in these experiments is remarkable. Particularly important is the behaviour on highly convoluted charge two isotopic patterns: as can be seen from the number of correctly identified and separated patterns shown in Table 1, the algorithm presented in this work successfully deconvolutes significantly more of these patterns than the established approaches. The high quality of this separation typically obtained after the optimization stage of our algorithm is shown in Figure 2. In addition, it should be mentioned that the algorithm runs in real time. On the LC-MS spectra of about 100 Mb of data, the peak picking stage took several seconds on a typical PC, while the following optimization run lasted for about 1 to 5 minutes, depending on the number of iterations performed. The applicability of the proposed scheme is not restricted to low-resolution data, nor to ESI data. To demonstrate this, we performed a peak picking on a well-resolved, but difficult MALDI-MS spectrum of a tryptic digest of bovine serum albumin (data set B). This time we performed a Mascot[11] peptide mass fingerprinting query with the peaks determined by our implementation and by the Bruker software. In both cases, the bovine serum albumine was identified with a very high significance, where the results obtained with the vendor software led to a sequence coverage of 44% and our peak picking scheme achieved between 52% and 67%, depending on the applied signal-to-noise threshold. It should be noted that for these results, no internal calibration was performed on the spectrum in order to prevent distortion of the results by possible overfitting due to the calibration procedure. Consequently, the resulting mass accuracy for the peptides identified by Mascot is low with about 95 ppm for Bruker and about 80 to 93 ppm for our method. A simple linear calibration using four monoisotopic masses turned out to reduce the mass error significantly to about 20 to 30 ppm for the same sequence coverages mentioned above.

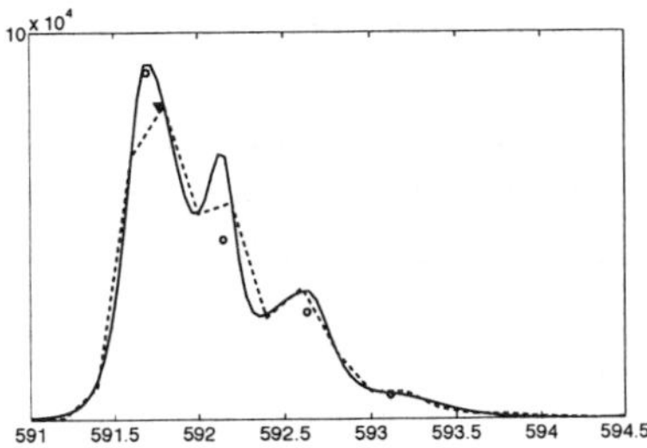

Figure 2. Charge two isotopic pattern of LHRH Decapeptid (solid line: sum of the fitted asymmetric peak shapes, dashed line: linearly interpolated raw data, circles: peak centroids with corresponding peak heights (OpenMS), triangle: peak centroid with corresponding peak height (Bruker Apex)). The relative error of the centroids of the first four peaks as determined by our method are given by 21 ppm, 1.2 ppm, 35 ppm, and 16 ppm.

4. Discussion

We have presented a wavelet-based peak picking technique suited for the application to the different kinds of mass spectrometric data arising in computational proteomics. In contrast to many established approaches to this problem, the algorithm presented here has been particularly designed to work well even on data of low resolution with strongly overlapping peaks. This is especially apparent when deconvoluting for example charge two isotopic patterns with poor separation, as those arising in the LC-MS datasets discussed above. Here, the good performance of our algorithm can be attributed to two of its unique features: the ability to determine the end points of a peak even if it overlaps heavily with another one, which is due to the use of the Wavelet transform as discussed in Section 2, and the optional nonlinear optimization following the peak picking stage. Applied to a high-quality MALDI-TOF spectrum of a tryptic digest, our algorithm yields a high degree of sequence coverage when used as input for a Mascot fingerprinting query. In all applications, it compares very favorably with the algorithms supplied by the vendor of the mass spectrometers. A free open source implementation is available in the OpenMS C++ framework.

Acknowledgements

We would like to thank Bruker Daltonics, Germany for providing access to the CDAL library for direct access to its raw data formats. In particular, we would like to thank Dr. Jens Decker for his support and helpful discussions. In addition we would like to thank Prof. Christian Huber, Saarland University, Saarbrücken for providing data set A, Dr. Johan Gobom, Max-Planck-Institute for Molecular Genetics, Berlin for providing data set B, and Michael Kerber who was involved in the implementation of an early version of the algorithm.

References

1. Breen, E., Hopwood, F., Williams, K., Wilkins, M. Electrophoreses **21** (2000) 2243–2251
2. Yasui, Y., McLerran, D., Adam, B., Winget, M., Thornquist, M., Feng, Z. Biomed. Biotechnol. **2003** (2003) 242–248
3. Strittmatter, E.F., Rodriguez, N., Smith, R.D. Analytical Chemistry **75** (2003) 460–468
4. Kempka, M., Sjödahl, J., Rocraade, J. Rapid Communications **18** (2004) 1208 1212
5. Tan, H., Brown, S. Journal of Chemometrics **16** (2002) 228–240
6. Louis, A., Maass, D.: Wavelets: Theory and Applications. John Wiley & Sons (1997)
7. Kohlbacher, O., Reinert, K.: OpenMS – an open source framework for shotgun proteomics in C++. http://sourceforge.net/projects/open-ms (2005)
8. Gobom, J., Schuerenberg, M., Mueller, M., Theiss, D., Lehrach, H., Nordhoff, E. Analytical Chemistry **73** (20013) 434–438
9. Wu, S., Nie, L., Wang, J., Lin, X., Zhen, L., Rui, L. Journal of Electroanalytical Chemistry **508** (2001) 11–27
10. Press, W., Teykolsky, S., W.T., V., Flannery, B.: Numerical Recipes in C++: The art of scientific computing. Cambridge University Press (2002)
11. Perkins, D., Pappin, D., Creasy, D., Cottrell, J.: Probability-based protein identification by searching sequence databases using mass spectrometry data. Electrophoresis **20** (1999) 3551–3567

FAST *DE NOVO* PEPTIDE SEQUENCING AND SPECTRAL ALIGNMENT VIA TREE DECOMPOSITION

CHUNMEI LIU[1,*], YINGLEI SONG[1], BO YAN[2], YING XU[2], LIMING CAI[1,*]

[1] *Department of Computer Science and* [2] *Department of Biochemistry and Molecular Biology*

University of Georgia, Athens GA 30602, USA

De novo sequencing and spectral alignment are computationally important for the prediction of new protein peptides via tandem mass spectrometry (MS/MS). Both approaches are established upon the problem of finding the longest antisymmetric path on formulated graphs. The problem is of high computational complexity and the prediction accuracy is compromised when given spectra involve noisy data, missing mass peaks, or post translational modifications (PTMs) and mutations. This paper introduces a graphical mechanism to describe relationships among mass peaks that, through graph tree decomposition, yields linear and quadratic time algorithms for optimal *de novo* sequencing and spectral alignment respectively. Our test results show that, in addition to high efficiency, the new algorithms can achieve desired prediction accuracy on spectra containing noisy peaks and PTMs while allowing the presence of both b-ions and y-ions.

1. Introduction

Tandem mass spectrometry (MS/MS) has been extensively used in proteomics to identify and analyze proteins[2,3,9]. In this method, molecules of a protein can be cleaved into short peptide sequences by enzymes. Amino acids in these peptides are then determined and combined to obtain the sequence of the protein. To sequence a peptide, sequences with the same amino acids are fragmented into charged prefix and suffix subsequences (ions) and their mass/charge ratios can be measured by a mass spectrometer. In a theoretical MS/MS spectrum, there are usually two types of ions present: b-ions associated with N-terminals and y-ions with C-terminals. Ideally, fragmentation may occur at any position along the peptide backbone and we thus expect to be capable of inferring the amino acids a peptide contains from its MS/MS spectrum and the masses of single amino acids.

*Corresponding authors: {chunmei, cai}@cs.uga.edu

However, difficulty may arise when we intend to identify the ion types for mass peaks. In addition, experimental spectra are usually incomplete and contain noisy peaks. Therefore, the *de novo* sequencing of a peptide solely from its spectrum remains a challenging task[5,6].

A number of algorithms have been developed for the *de novo* sequencing problem. An early developed algorithm[12] generates all amino acid sequences and the corresponding theoretical spectra to be compared with the experimental spectrum. Since, to find out the best match an exponential number of spectra may need to be generated, the algorithm is not efficient. Prefix pruning approaches have been developed to speed up the search by restricting it to sequences whose prefixes match the spectrum well[13,17,18]. However, heuristic pruning may adversely affect the sequencing accuracy while the computation time may remain expensive. Recently, based on the notion of spectrum graph[6], the *de novo* sequencing problem has been reduced to finding the longest (or maximum scored) antisymmetric path in directed graphs[2,6,7,8,15]. However, a straightforward path-finding algorithm may yield undesired paths containing multiple vertices associated with complementary ions. This issue was resolved later with a linear time dynamic programming algorithm[5] that ensures the path found to be antisymmetric. However, it requires quadratic time to discover one modified amino acid and more time to deal with additional noisy peaks.

Comparing and evaluating the similarity between two spectra are often used in database search for peptide identification[10]. Traditional methods for computing the similarity identify the shared mass peaks between two spectra and use the count as a measure of the similarity. More recently, spectral alignment was proposed as a new method for evaluating spectral similarity; it proves useful for identifying related spectra in the presence of post translational modifications (PTMs) and mutations[9]. In particular, based on finding the longest (k-shift) path in alignment graph, a spectral alignment algorithm can align two spectra of n peaks in time $O(n^2k)$, where k is the maximum number of peak shifts resulting from PTMs[9]. However, the algorithm considers only b-ions or y-ions but not both. To consider both ion types, a dynamic programming algorithm in the same spirit as that for *de novo* sequencing[5] is possible. But it would require a computation time that is polynomial of a much higher degree.

In this paper, we introduce a graphical mechanism to describe related mass peaks in spectra. In particular, the peaks associated with complementary ions are linked with non-directed edges, yielding extended spectrum graphs and extended alignment graphs. Such graphs demonstrate small tree width t (usually $t \leq 6$) for real mass spectra, so a very efficient algorithm for finding the longest antisymmetric path can be devised based on the tree decompositions of these graphs. In particular, the resulting new algorithms for *de novo* sequencing and spectral alignment run in time $O(6^t n)$ and $O(6^t n^2)$ respectively. Based on the notion of tree decomposition, the antisymmetry of complementary vertices can be efficiently ensured on the found path; both ion types can be simultaneously considered by our algorithms. In addition, using the graphical mechanism, vertices for peaks with similar masses can be easily related and considered simultaneously in the tree decomposition-based dynamic programming algorithm. This allows vertices for noisy peaks to be eliminated from the found path.

We have implemented the algorithms and tested their performance on both simulated spectra and real experimental ones with noisy peaks. Our algorithm is able to identify the correct peptide sequences from all the tested spectra with noisy peaks in a few seconds. In particular, the algorithm achieves more than 96% accuracy on spectra in which the number of noisy peaks is the same as that of others. In addition, we used the algorithm to identify PTMs of amino acids based on spectra generated *in silico*. Test results for spectral alignment demonstrated that the algorithm can identify all PTMs accurately in a few seconds.

2. Models and Algorithms

2.1. *Problem Description*

Since theoretically, any ion has its complementary ion contained in the same spectrum[16], we assume the MS/MS spectrum S of a peptide P be a set of mass peaks $\{x_1, x_2, \cdots, x_{2k}\}$, where $x_i > x_j$ for $i > j$. For any mass peak x_i in S, there exists a mass peak x_{2k+1-i} *complementary* to x_i and the sum of their mass values is the total mass M of P. One of x_i and x_{2k+1-i} is a b-ion and the other is a y-ion. A *spectrum graph* $G = (V_s, E_s)$ can be constructed from the mass peaks in S. Specifically, vertex $v_i \in V_s$ represents x_i and, in addition to the mass peaks in S, vertices *source* v_0 and *sink* v_{2k+1} are included in G with virtual mass values 0 and M respectively. Directed edge $(v_i, v_j) \in E_s$ if the mass value difference $x_j - x_i$ is the mass of a single amino acid. Sequencing a peptide from its spectrum thus corresponds to finding the longest antisymmetric directed path from the source v_0 to the sink v_{2k+1}. A path is *antisymmetric* if it includes at most one of the complementary vertices v_i and v_{2k+1-i}, for all $i = 1, \ldots, k$. An *extended spectrum graph* can be obtained from a spectrum graph by

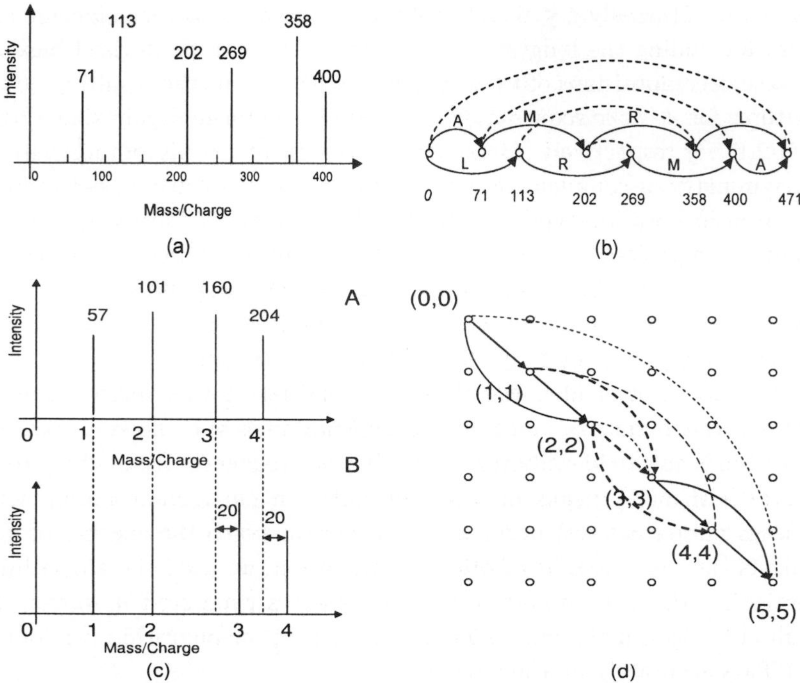

Figure 1. (a) The mass peaks in a tandem mass spectrum with total mass 471. (b) The corresponding extended spectrum graph, where dashed undirected edges connect complementary vertices. (c) The mass peaks in two tandem mass spectra A and B that are to be aligned. The total mass of spectrum A is 261. Mass peaks 3 and 4 in spectrum B have a shift of 20 in their mass values compared to those in spectrum A. (d) The alignment graph constructed based on the mass peaks in A and B. Solid and dashed directed edges represent real and virtual connections respectively; dashed non-directed edges connect complementary vertices; only edges along the diagonal vertices are drawn in the figure.

connecting all pairs of complementary vertices with non-directed edges. Figure 1(a)(b) provide an example for a spectrum and its corresponding extended spectrum graph.

An *alignment graph* $H = (V_a, E_a)$ can be constructed based on two spectra to be aligned. We assume the set of mass peaks for spectra A and B are $S_A = \{x_1, x_2, \cdots, x_{2k_1}\}$ and $S_B = \{y_1, y_2, \cdots, y_{2k_2}\}$ respectively. The set of vertices $V_a = (S_A \times S_B) \cup \{(x_0, y_0), (x_{2k_1+1}, y_{2k_2+1})\}$, where x_0, x_{2k_1+1} and y_0, y_{2k_2+1} are virtual mass peaks with zero and total peptide masses in spectra A and B respectively. (x_i, y_j) is connected to (x_k, y_l) with a *real directed edge* if $x_k > x_i$ and $x_k - x_i = y_l - y_j$. In addition to real directed edges, an *extended alignment graph* may also contain *virtual directed edges* and *non-directed edges.* (x_i, y_j) is connected to (x_k, y_l) with a virtual directed edge if $x_k > x_i$, $y_l > y_j$ and $|(y_l - y_j) - (x_k - x_i)| \leq \Delta_m$, where Δ_m is the maximum mass peak shift due to modified amino acids. Two vertices (x_i, y_j) and (x_k, y_l) are *complementary* if x_i and x_k or y_j and y_l are complementary mass peaks. Complementary vertices are connected with non-directed edges. The *source* and the *sink* in the graph are vertices (x_0, y_0) and (x_{2k_1+1}, y_{2k_2+1}) respectively. The similarity between spectrum A and B can thus be evaluated by finding in H the longest directed antisymmetric path that connects the source and the sink. Figure 1(c)(d) provides an example of two spectra and their extended alignment graph.

Weights on directed edges in the graph are masses. In practice, directed edges in an extended spectrum or alignment graph can be scored based on other experimental parameters. For example, Dancík *et al.*[6] proposed a stochastic edge scoring scheme where each mass peak in the spectrum is generated with a certain probability; the score of a directed edge can be evaluated based on the probabilities of its ends. The sequencing result with the maximum likelihood corresponds to the maximum scored antisymmetric path connecting the source and the sink in the graph.

2.2. *Tree Decomposition and Tree Width*

Definition 2.1. [11] *Let $G = (V, E)$ be a graph, where V is the set of vertices in G, E denotes the set of edges in G (E may contain both directed and non-directed edges). Pair (T, X) is a tree decomposition of graph G if it satisfies the following conditions:*

(1) $T = (I, F)$ defines a tree, the sets of vertices and edges in T are I and F respectively,
(2) $X = \{X_i | i \in I, X_i \subseteq V\}$, and $\forall u \in V$, $\exists i \in I$ such that $u \in X_i$,
(3) $\forall (u, v) \in E$, $\exists i \in I$ such that $u \in X_i$ and $v \in X_i$,
(4) $\forall i, j, k \in I$, if k is on the path that connects i and j in tree T, then $X_i \cap X_j \subseteq X_k$.

The tree width of the tree decomposition (T, X) is defined as $\max_{i \in I} |X_i| - 1$. The tree width of the graph G is the minimum tree width over all possible tree decompositions of G.

Figure 2(a)(b) shows that tree decomposition provides an alternative view over a graph where vertices are grouped into tree nodes according to

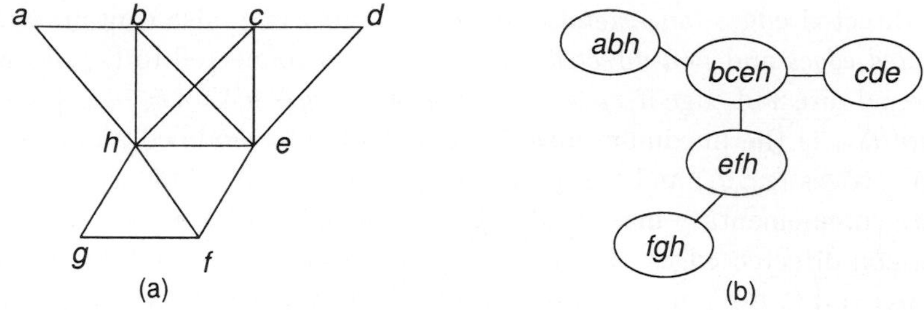

Figure 2. (a) An example of a graph. (b) A tree decomposition for the graph in (a).

their topological relationships (represented by edges, arcs, etc). Our experiments on simulated and real spectra show that the tree width for extended spectrum graphs and extended alignment graphs are generally around 5. The property of having a small tree width makes it possible for us to develop very efficient algorithms for both problems based on the technique of tree decomposition, since partial optimal solutions on subgraphs induced by subtrees can be efficiently extended and combined with exhaustive enumeration restricted to vertices in a single tree node[1].

2.3. *The Path-finding Algorithm*

The algorithm selects a tree node that contains both the source and the sink as the root of a tree decomposition and maintains a dynamic programming table for each tree node. The algorithm follows a bottom-up fashion to fill the tables for all the tree nodes. The table in the root thus stores the length of the longest antisymmetric path connecting the source and the sink. The algorithm then follows a recursive tracing back procedure to find all the vertices in the path.

For a tree node with t vertices, the dynamic programming table contains $2t+1$ columns, of which the first t columns store the *selection* of each vertex in the node to form a subpath. In addition, $t-1$ columns are used to store the *connection state* between each pair of consecutive selected vertices in the tree node. Two additional columns V and L store the *valid bit* and the largest length of the partial path associated with the combination of selections and connection states in the same table entry respectively.

The selection value of a vertex in a tree node is 1 if it is selected to be in the partial optimal path and 0 otherwise. The value of a connection state could be one of the integers in set $\{0, 1, \cdots, l\}$, where l is the number of children of the tree node. The connection state for a pair of consecutive

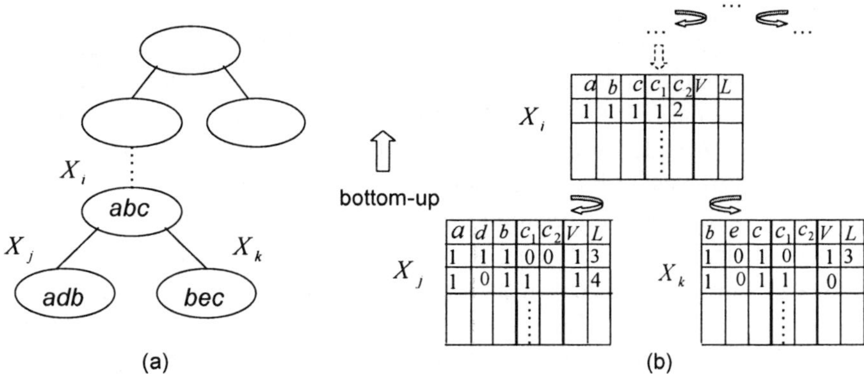

Figure 3. A tree decomposition and its corresponding dynamic programming tables. The algorithm follows a bottom-up fashion starting with leave tree nodes. When computing the dynamic programming tables for an internal node X_i, the tables of its child nodes X_j and X_k need to be queried to compute the validity (V) and the largest path length (L) of a given entry in the table for X_i.

selected vertices in the tree node is 0 if they are contiguous in the path and is i ($i > 0$) if the vertices between the pair of vertices are covered by the subtree rooted at the ith child. The number of possible combinations of selections and connection states can thus be up to $(2(l+1))^t$. However, since we can remove tree nodes with more than two children by generating extra tree nodes, the table for a tree node with t vertices may contain up to 6^t entries. The valid bit for a given entry is set to be 1 if there exists a partial antisymmetric path that follows the combination of selections and connection states in the entry. To determine the relative order of selected vertices in a partial path, the algorithm topologically sorts the vertices in each tree node.

To determine an entry in the table for a leaf node, the algorithm exhaustively enumerates and directly computes the validity and largest path length for every possible combination of selections and connection states for vertices in the node. For an internal node, the algorithm refers to the tables of its children to determine the validity and longest path length for each of its table entry. In particular, for a given entry, the algorithm obtains its selections of vertices and the corresponding connection states and then queries the table contained in each of the child nodes. All valid table entries whose selections of vertices and connection states do not contradict the given entry are queried and the one with the largest path length is selected as the *descendent entry* in the child. The algorithm sets an entry to be invalid if its selection of vertices violates the antisymmetric property or one of the child nodes contains no descendent entries. The largest path length for the entry is then computed by summing up the number of edges covered by the node itself and the largest path length for the descendent entry found in each of its child nodes.

Figure 3 provides an example of computing the table entries for an internal node X_i. Without loss of generality, we assume X_i has two child nodes X_j and X_k, and $X_i = \{a, b, c\}$, $X_j = \{a, d, b\}$ and $X_k = \{b, e, c\}$. To determine the V and L for the entry $\{(1, 1, 1), (1, 2)\}$ in the table for X_i, the algorithm needs to query both of the tables for X_j and X_k since the entry suggests that the vertices on the path between a and b are covered by the subtree rooted at X_j and those between b and c are covered by that rooted at X_k. To query the table for X_j, the algorithm only checks valid entries that select both a and b since $X_i \cap X_j = \{a, b\}$, thus the leading two entries in the table for X_j are checked by the algorithm. Similarly, since $X_i \cap X_k = \{b, c\}$, the algorithm only checks valid entries that select both b and c in the table for X_k.

In the last stage of the computation, the algorithm queries the table in the root node and considers those valid entries that select both the source and the sink and finds the one with the longest path length. The algorithm then follows an up-bottom tracing back procedure to recover the nodes selected to be present on the path. The path found by the algorithm is guaranteed to satisfy the antisymmetric property since, based on the definition of tree decomposition, any pair of complementary vertices is covered by at least one tree node. The computation time needed by the algorithm is $O(6^t N)$, where t is the tree width of the tree decomposition and N is the number of vertices in the graph. The algorithm uses a greedy graph reduction technique to obtain a tree decomposition for a graph[4].

3. Experimental Results

We implemented the path-finding algorithm for both *de novo* sequencing and spectral alignment problems. The programs were tested on simulated and real MS/MS spectra. For *de novo* sequencing, we evaluated the performance of the program on simulated spectra that contain different amount of noise, and then analyzed real experimental MS/MS spectra. For spectral alignment, we generated simulated spectra for peptides with PTMs and identified modified amino acids.

3.1. *De Novo Sequencing*

To evaluate the performance of the program on spectra with different amount of noise, we obtained simulated tandem mass spectra for 100,000 fully tryptic digested peptides of proteins in the Yeast genome. We then filtered out peptides of less than 5 and more than 24 amino acids. In addition to the mass peaks that result from the fragmentation of peptides, we incorporated noisy mass peaks into these simulated spectra and applied the program to obtain the peptide sequences from these noisy spectra. To simulate the noise generally present in real experimental spectra, noisy mass peaks were generated in groups and the differences of mass values for mass peaks in the same group were selected to be those of single amino acids or their combinations. Table 1 shows the performance of the program on spectra with different amount of noise. As we have expected, the tree widths of the spectrum graphs increase when more noisy peaks are inserted into the spectra. The program thus needs more computation time for analyzing a spectrum. In addition, a slight drop in sequencing accuracy is observed when an ideal spectrum is changed into a noisy one.

Table 1. The accuracy of the program on spectra with different amount of noise. N/S is the ratio of the number of noisy peaks to that of others in a spectrum. AC is the percentage of amino acids that are correctly identified by the program; PT(< 5), PT($= 5$) and PT(> 5) are percentages of spectrum graphs whose tree widths are less than 5, equal to 5 and greater than 5 respectively; CT is the average amount of time the program needs to analyze a spectrum.

N/S	AC (%)	PT(< 5) (%)	PT($= 5$) (%)	PT(> 5) (%)	CT(sec)
0.00	98.60	52.45	44.93	2.62	1.54
0.20	98.27	42.48	41.38	16.15	7.24
0.50	98.29	37.69	34.84	27.47	12.37
0.80	97.98	32.98	37.13	29.87	13.10
1.00	96.95	27.64	39.47	32.89	15.46

To evaluate the performance of the program on real experimental spectra, we downloaded 14 tandem mass spectra for peptides in *E. Coli* proteins from the Open Proteomics Database (OPD) and collected 3 experimental FT-ICR data from two different peptide sources. Before we applied the program to a spectrum, the mass peaks in the spectrum were preprocessed. Isotopic mass peaks and mass peaks with intensities less than 0.1 of the maximum intensity value were removed. In addition, a complementary ion was added back for each ion in the spectrum if it was missing. Table 2 shows sequencing results obtained with the program for each spectrum.

The program identified most amino acids correctly with few sequencing errors. However, the reason for these errors is clear. The program is unable to identify I from L and K from Q since the mass of I is equal to that of L, and the mass difference between K and Q is too small to be recognized by the program.

Table 2. The performance of the program on real experimental spectra. TW is the tree width of the spectrum graph; CT is the computation time of the program. Peptides in the first eighteen rows are from OPD. The first six are from EF-Tu protein, followed by seven from glutamate synthase, enolase, sodium-calcium antiporter, HU-2, ferric transport, and S5; the last two are from thioredoxin. The rest three rows are from experimental FT-ICR with the first two from horse myoglobin and the last one from BSA.

Real Sequence	Obtained Sequence	TW	CT(sec)
RAFDQIDNAPEEKA	RAFDQIDNAPEEQA	6	12.11
RPQFYFRT	RPQFYFRT	4	0.42
KVGEEVEIVGIKE	QVGEEVEIVGIKE	6	5.16
KMVVTLIHPIAMDDGLRF	KMVVTILHPIAMDDGIRF	5	7.30
RAGENVGVLLRG	RAGENVGVLLRG	6	13.09
KMVVTLIHPIAMDDGLRF	QMVVTIIHPIAMDDGLRF	5	6.85
KVVRTAIHALARMQHRG	KVVRTAIHAIARMQHRG	6	9.19
KFNQIGSLTETLAAIKM	KFNQIGSLTETLAAIQM	6	10.27
RKFATQYMNLFGIKQ	RKFATQYMNLFGIKK	6	10.15
KTQLIDVIAEKA	QTQIIDVIAEKA	5	5.09
KPVYSNGQAVKD	KPVYSNGQAVQD	5	2.53
KLNIDQNPGTAPKY	KINIDQNPGTAPQY	6	5.80
KNQTLALVSSRP	QNQTLALVSSRP	6	4.85
RVKSQAIEGLVKA	RVKSQAIEGLVQA	6	4.10
HGTVVLTALGGILK	HGTVVLTAIGGILQ	4	0.22
VEADIAGHGQEVLIR	VEADIAGHGQEVLLR	6	10.34
DAFLGSFLYEYSR	DAFLGSFLYEYSR	5	2.18

3.2. *Spectral Alignment*

As an application of spectral alignment, we used the program to identify modified amino acids on peptide sequences with PTMs. We generated pairs of spectra *in silico* for peptides and their modified sequences and perform a spectral alignment between each pair of spectra. The mass modifications can be identified from the longest antisymmetric path found by the program. We introduced two additional parameters, k and Δ, where k is the maximum number of modifications allowed in the peptide and Δ is the maximum amount of mass modification that may occur on a single amino acid. Based on the parameters k and Δ, the number of non-directed edges in the alignment graph can be significantly reduced. In particular, for spectra A and B with sets of mass peaks $S_A = \{x_1, x_2, \cdots, x_{2k_1}\}$ and $S_B = \{y_1, y_2, \cdots, y_{2k_2}\}$ respectively, (x_i, y_j) and $(x_{2k_1+1-i}, y_{j'})$ are connected with an non-directed edge if and only if $|y_{j'} - y_{2k_2+1-j}| \leq k\Delta$. In addition, Δ is used as the Δ_m defined in section 2.1 for adding virtual directed edges to the graph. In our experiment, the values of k and Δ were set to be 3 and 20.0 since, in practice, most of the peptides contain up to 3 modified amino acids. Table 3 shows the results we have obtained on identifying modified amino acids on pairs of spectra we generated *in silico*. The table shows that the tree width of an alignment graph ranges from 4 to 6 and the program is able to identify the modified amino acids accurately in a few seconds.

Table 3. The performance of the program on identifying modified amino acids using spectral alignment. DM is the number of modified amino acids identified by the program; TW is the tree width of the alignment graph; CT is the computation time in seconds. Modified amino acids are superscripted with asterisks.

Peptide	Modified Peptide	DM	TW	CT(sec)
RAIKNLL	RAIK*NLL	1	4	0.04
FKMKRTQVFWKV	FK*MKRTQVFWK*V	2	6	2.43
MALPFQLLRQLGVA	M*ALPFQLLRQLGVA	1	4	0.12
AKYEGGL	AK*YEGGL	1	4	0.07
DFLIKRGV	DFLIK*RGV	1	5	0.73
PKDMILLFATTTTKF	PK*DMILLFATTTTK*F	2	6	2.31
LWEVKDRTAHS	LWEVK*DRTAHS	1	6	3.50
IGALKDKITMS	IGALK*DKITM*S	2	5	0.70
MAIVMGRLEVKAIS	MAIVMGRLEVK*AIS	1	4	0.12
FVPGQKNGIKGDLS	FVPGQK*NGIK*GDLS	2	4	0.04

4. Conclusions

We have extended the notions of MS/MS spectrum graphs and alignment graphs to include relationships among mass peaks such as complementarity and modification. Based on the notion of tree decomposition, such graphs have been exploited for the development of fast optimal algorithms for *de novo* peptide sequencing and spectral alignment. In addition to the efficiency, our work can accurately sequence peptides from noisy spectra and identify post translational modifications of amino acids while allowing the presence of both types of ions. In addition, we expect this approach can be extended to accurately infer partial sequence "tags" from a MS/MS spectrum [14], which can speed up the database search significantly.

Acknowledgement

BY and YX's work is supported in part by the US Department of Energy's Genomes to Life Program under project "Carbon Sequestration in Synechococcus sp: From Molecular Machines to Hierarchical Modeling" (http://www.genomes-to-life.org), by National Science Foundation (#NSF/DBI-0354771 #NSF/ITR-IIS-0407204) and by a "Distinguished Cancer Scholar" grant from Georgia Cancer Coalition.

References

1. S. Arnborg and A. Proskurowski, *Discrete Applied Math.*, 23: 11-24, 1989.
2. C. Bartels, *Biomed. Environ. Mass Spectrom.*, 19: 363-368, 1990.
3. K. Biemann and H.A. Scoble, *Science*, 237:992-998, 1987.
4. H. L. Bodlaender and A. M. C. A. Koster, *Proc. of the 6th Workshop on Alg. Eng. and Exp.*, 70-94, 2004.
5. T. Chen, M. Y. Kao, M. Tepel, J. Rush, and G. M. Church, *Journal of Computational Biology*, 8(3): 325-337, 2001.
6. V. Dancík, T. A. Addona, K. R. Clauser, J. E. Vath, and P. A. Pevzner, *Journal of Computational Biology*, 6(3/4): 327-342, 1999.
7. J. Ferńandez de Cossío, J. Gonzales, and V. Besada, *CABIOS* 11(4): 427-434, 1995.
8. W. M. Hines, A. M. Falick, A. L. Burlingame, and B. W. Gibson, *J. Am. Soc. Mass. Spectrom.*, 3: 326-336, 1992.
9. P. A. Pevzner, V. Dancík, and C. L. Tang, *Proceedings of The Fourth Annual International Conference on Computational Molecular Biology*: 231-236, 2000.
10. P. A. Pevzner, A. Mulyukov, V. Dancik, and C. Tang, *Genome Research*, 11:290-299, 2001.
11. N. Robertson and P. D. Seymour, *Journal of Algorithms*, 7: 309-322, 1986.
12. T. Sakurai, T. Matsuo, H. Matsuda, and I. Katakuse, *Biomed. Mass Spectrom.*, 11(8): 396-399, 1984.
13. M. M. Siegel and N. Bauman, *Biomed. Environ. Mass Spectrom.*, 15: 333-343, 1988.
14. D. Tabb, A. Saraf, and J. R. Yates, *Anal. Chem.*, 75: 6415-6421, 2003.
15. J. A. Taylor and R. S. Johnson, *Rapid Commun. Mass Spectrom.*, 11: 1067-1075, 1997.
16. B. Yan, C. Pan, V. N. Olman, R. L. Hettich, and Y. Xu, *Bioinformatics*, 21(5): 563-574, 2005.
17. J. R. Yates, P. R. Griffin, L. E. Hood, J. X. Zhou, *Techniques in Protein Chemistry II*, 477-485, Academic Press, 1991.
18. D. Zidarov, P. Thibault, M. J. Evans, and M. J. Bentrand, *Biomed. Environ. Mass Spectrom.*, 19: 13-16, 1990.

EXPERIMENTAL DESIGN OF TIME SERIES DATA FOR LEARNING FROM DYNAMIC BAYESIAN NETWORKS

DAVID PAGE AND IRENE M. ONG

Department of Biostatistics & Medical Informatics
University of Wisconsin
Madison, WI 53706 USA
E-mail: page@biostat.wisc.edu, ong@cs.wisc.edu

Bayesian networks (BNs) and dynamic Bayesian networks (DBNs) are becoming more widely used as a way to learn various types of networks, including cellular signaling networks[16,14], from high-throughput data. Due to the high cost of performing experiments, we are interested in developing an experimental design for time series data generation. Specifically, we are interested in determining properties of time series data that make them more efficient for DBN modeling. We present a theoretical analysis on the ability of DBNs without hidden variables to learn from proteomic time series data. The analysis reveals, among other lessons, that under a reasonable set of assumptions a fixed budget is better spent on collecting many short time series data than on a few long time series data.

1. Introduction

Time series data, and dynamic Bayesian networks (DBNs) to model such data, have become more popular as an approach to learning gene regulatory networks[3,12,6,13]. As experimental techniques for collecting proteomic data has improved, there has been a shift to learning protein networks from various types of proteomic data[16,20]. However, the cost of generating experimental proteomic data is still high, hence, we are interested in developing an experimental design for time series data generation that is more efficient for dynamic Bayesian network learning.

In a typical experiment, mass spectrometry (MS) is used to measure protein abundance *at several specific time points* after a particular stimulus to an organism or cell sample. The peaks in the MS data can be discretized to indicate the presence or absence of protein, avoiding certain problems associated with continuous values (S. McIlwain, personal communication 2005). From the discretized data, we can then learn a DBN model that best fits the data. Biologists can visually examine such a DBN structure (Fig. 1a) and interpret an arc from protein X_1 at time t to protein X_2 at

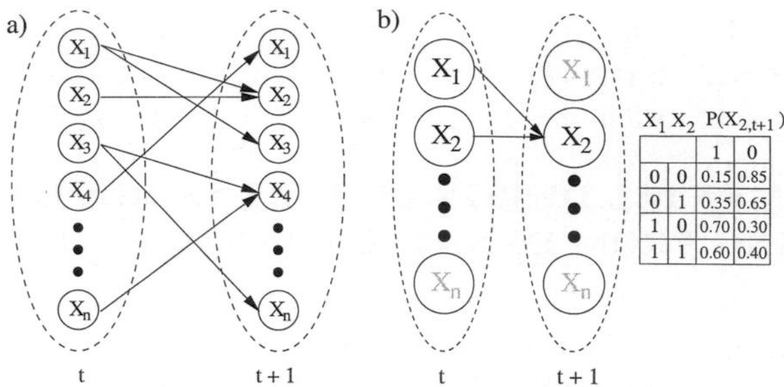

Figure 1. a) Simple DBN model. Labeled circles within a dotted oval represent our variables in one time slice. Formally, arcs connecting variables from one time slice to variables in the next have the same meaning as in a BN, but they intuitively carry a stronger implication of causality. b) Example of probabilistic CPTs.

time $t+1$ as evidence that X_1 influences X_2.

Following earlier experiments in learning DBNs from time series data[13], we consulted with biologists about the design of future time series experiments. While a number of design issues arise, the most common question is the following. "Given that we have resources to run r experiments, is it better to run many short time series or a few long time series?" Design issues also arise for our learning algorithms. For example, given a specific number of experiments r that will be run, and a given amount of time to learn a DBN model from the data, should we place a limit on the number of parents a node can have and, if so, what should this limit be? One way to answer these questions is to perform many runs with many time series data sets having different properties; unfortunately, few such data sets are available, and the cost of producing such a data set requires design insight now, before additional data sets are available. An alternative way to gain insight is to construct a formal model of the learning task, as realistic as possible though necessarily making some simplifying assumptions.

The present paper limits its attention to DBNs whose variables are Boolean, though the results extend naturally to non-Boolean discrete variables. Because of the use of Boolean variables, our DBNs also can be viewed as (deterministic or probabilistic) Boolean networks. Friedman and Yakhini[5] have examined the sample complexity of minimum description length based learning procedures for Bayesian network structures and Dasgupta[4] determined the sample complexity of learning fixed-structure Bayesian networks with and without hidden nodes using the PAC (probably approximately correct) framework. Furthermore, Akutsu *et al.*[1] formalized the task of constructing Boolean networks from data and others extended or improved their initial results[2,17,10]. Nevertheless, the results do not give guarantees about the accuracy of the learned network on new or unseen data, or the amount of data required to achieve a given level of accuracy. The present paper addresses the question of polynomial-time learnability using the PAC-learning framework[18] and its extension to probabilistic concepts[8]. Consequently, the novel aspect of this paper is its provision of (probabilistic) accuracy guarantees based on data set size. The formal results in this paper, imply the following practical advice for the design of time series MS experiments that will be analyzed using DBNs.

First, many short time series experiments, specifically two time steps, are preferable to a single long time series experiment.

Second, given the number of time series experiments that can feasibly be run with present-day costs, the number of parents per node in a DBN should be limited to at most three per node.

Third, even with only two parents per node, the worst-case number of examples required to guarantee a given level of accuracy with a given probability is cubic in the number of variables n, and this number typically is in the thousands. If we are concerned with gaining insight into—or accurate prediction of—only a small number m of the n variables, we can reduce this term to n^2m. This often is the case where we are interested in learning or refining a model of a particular pathway, and we know most of the key players (proteins). If in addition we have an over-estimate of the potential other players, and there are l of these, then we can reduce this term further to l^2m, reducing the number of required experimental data.

2. Definitions and Terminology

The PAC framework allows us to provide upper and lower bounds on the number of examples needed to learn a hypothesis class assuming that the learning algorithm is making use of all the data. We formulate our definitions for the PAC framework below in the style of Kearns and Vazirani[9].

Definition 2.1. A Boolean *dynamic Bayesian network* (DBN) is defined over the Boolean variables

$X_{1,1}, X_{2,1}, \ldots, X_{n,1},$
$X_{1,2}, X_{2,2}, \ldots, X_{n,2},$
$\ldots,$
$X_{1,T}, X_{2,T}, \ldots, X_{n,T}$

where $X_{i,t}$ denotes variable X_i at time t. For each $1 \leq i \leq n$ and $1 < t \leq T$ the value of variable $X_{i,t}$ is $f_i(X_{1,t-1}, \ldots, X_{n,t-1})$, where f_i is some (possibly stochastic) Boolean function.

Definition 2.2. We denote by $\mathcal{DBN}(\mathcal{C}_n)$ the class of Boolean DBNs for which each function f_i comes from Boolean concept class $\mathcal{C}_n$.

Any particular Boolean DBN in $\mathcal{DBN}(\mathcal{C}_n)$ is a set of functions $f_i(X_{1,t-1}, \ldots, X_{n,t-1})$, one for each variable X_i, $1 \leq i \leq n$ (f_i does not change with time). For example, the Boolean concept class $\mathcal{C}_n$ might be all stochastic functions of at most k variables. This class of functions corresponds to all possible conditional probability tables (CPTs) in a DBN for a node with at most k parents. An example of such a CPT is given in Fig. 1b. If the DBN is deterministic, $\mathcal{C}_n$ might be the set of all (non-stochastic) functions of at most k variables, that is, all truth tables over k variables. For such a CPT, each row in Fig. 1b would instead have one of the probabilities set to 1 and the other set to 0. A generalization of this class, allowing more than k parents in a still limited fashion would be to have as $\mathcal{C}_n$ the set of all functions that can be represented by a k disjunctive normal form (kDNF).

3. Results

3.1. *Boolean DBN from 2-slice data*

Before presenting the first of our related models, we establish some conventions. In practice a DBN model may contain some variables that cannot be observed or measured; such variables are known as hidden variables. The present paper does not consider hidden variables or missing data.

In generating a MS data set, the cost of r experiments (measuring the abundance of each protein in r samples) is the same regardless of whether the r samples are all part of a single, long time series or many different time series. Therefore, we treat our number of data points as the number of experiments rather than the number of time series.

In the ordinary PAC-learning model, one assumes each data point is drawn randomly, independently according to some probability distribution D. Our models cannot assume this, because in a time series each data point (after the first) depends on the previous data point. The most faithful we can remain to the original PAC-learning model is to specify that the first data point in each time series is drawn randomly according to some probability distribution D, and the first data points in different time series are drawn independently of one another.

For simplicity, we begin with a formal model of DBN learning that resembles the PAC-learning model as much as possible, by restricting consideration to deterministic concepts. Given a deterministic DBN and a specific (input) time slice, the next (output) time slice is fixed according to the DBN. We say that a DBN model and a target DBN disagree with one another on an input time slice iff, given the input time slice, the two DBNs produce different outputs. A DBN model is $(1-\epsilon)$-accurate with respect to a target model iff the sum of the probabilities, according to D, of input time slices on which the two DBNs disagree is at most ϵ. As is standard with the PAC model, we take $|T|$ to denote the size of the target concept (DBN model) $T \in \mathcal{DBN}(\mathcal{C}_n)$ in a "reasonable" encoding scheme. For concreteness, we specify $|T|$ as the number of bits required to encode, for each variable X_i, its parents and its function f_i. The following definition is an application of the PAC-learning model to DBNs.

Definition 3.1. An algorithm PAC-learns a deterministic $\mathcal{DBN}(\mathcal{C}_n)$ iff there exist polynomials $\text{poly}_1(\_,\_,\_,\_)$ and $\text{poly}_2(\_)$ such that for any target DBN T in $\mathcal{DBN}(\mathcal{C}_n)$, any $0 < \epsilon < 1$ and $0 < \delta < 1$, and any probability distribution D over initial data points for time series: given any $r \geq \text{poly}_1(n,|T|,\frac{1}{\epsilon},\frac{1}{\delta})$ data points, the algorithm runs in $\text{poly}_2(rn)$ and with probability at least $1-\delta$ outputs a model that is $(1-\epsilon)$-accurate wrt T.

Theorem 3.1. *For any fixed $k \in \mathbb{N}$ the class of $\mathcal{DBN}(kDNF)$ is PAC-learnable from 2-slice data.*

Proof. Algorithm A learns a kDNF formula to predict each of the n variables at time slice 2 from the values of the n variables at time slice 1. Each 2-slice time series (input and output) is used to generate one example for each $X_{i,2}$. For each $1 \leq i \leq n$ the output (class) is $X_{i,2}$ and input features are $X_{1,1}, \ldots, X_{n,1}$. Given a PAC learning algorithm L for kDNF expressions[7], we run L on n feature vectors to find a concept in $\mathcal{C}_n$.

Algorithm A iterates: for each variable $X_{i,2}$, $1 \leq i \leq n$, we call kDNF learning algorithm L with $\frac{\delta}{n}$ as the maximum probability of failure (i.e., with desired confidence of $1-\frac{\delta}{n}$) and with $\frac{\epsilon}{n}$ as the maximum error (i.e., with desired accuracy of $1-\frac{\epsilon}{n}$). Algorithm A's final model is the set of functions $f_i(X_{1,1}, \ldots, X_{n,1})$ returned by L, one per output variable $X_{i,2}$.

Algorithm A runs in polynomial time since $n * \text{poly}_1(n,|T|,\frac{n}{\epsilon},\frac{n}{\delta})$ yields a polynomial, and each call to L runs in time polynomial in the size of its input. It remains only to show that with probability $1-\delta$ the error is bounded by ϵ. The definition of union bound states that if A and B are any two events (i.e., subsets of a probability space), then $\Pr(A \cup B) \leq \Pr(A) + \Pr(B)$[9]. Since each call to L fails to achieve the desired accuracy with probability only $\frac{\delta}{n}$, by the union bound the probability that there exists *any* of the n calls to L that fails to achieve the desired accuracy is at most δ. If each call to L has a desired error bound of $\frac{\epsilon}{n}$, then the error of the model (probability according to D of drawing an input time slice on which the learned model and target will disagree for some variable $X_{i,2}$, $1 \leq i \leq n$) is the union of all n error expressions from L, i.e., $\Pr(Error) = \frac{\epsilon}{n} + \frac{\epsilon}{n} + \ldots + \frac{\epsilon}{n} \leq \epsilon$. □

kDNF is a richer representation than one usually uses in a DBN. Typically, each variable is a function (represented as a CPT) of up to k parents. We denote the class of such DBNs by $\mathcal{DBN}$(k-parents). While PAC-learnability of a more restricted class does not automatically follow from PAC-learnability of a more general class, in this case arguments very similar to those just given show that, for any fixed $k \in \mathbb{N}$, the class of deterministic $\mathcal{DBN}$(k-parents) is PAC-learnable from 2-slice data.

3.2. *Boolean DBN from r-slice data*

It is equally common in practice for time series measurements to yield one long time series instead of multiple time series of length 2, or somewhere between these two extremes. While the *total number of experiments* r is determined largely by budget, the choice of time series *lengths* for any *fixed total* number of MS experiments r usually is not driven by expense. Rather, researchers make the choice they believe will provide the most information, because r experiments will have the same cost regardless of whether they occur in one long time series or many shorter time series.

We now ask whether the class $\mathcal{DBN}$(k-parents) is PAC-learnable from a single time series, and if so, whether the total number of experiments required might be less. It is trivial to prove that no algorithm PAC-learns this class when all the data points are in a single time series; the algorithm simply cannot learn enough about the distribution D according to which the start of each time series is drawn. But such a trivial negative result in unsatisfying. In practice if we subject a cell to an experimental condition and run a long time series of measurements, it is because we wish to learn an accurate model of how the organism responds to *that particular condition.* Therefore, we next consider a natural variant of our first learning model, where this variant is tailored to data points in a single time series.

Definition 3.2. An algorithm learns a deterministic class $\mathcal{DBN}(\mathcal{C}_n)$ from a *single time series* iff there exists polynomials $\text{poly}_1(\_,\_,\_,\_)$ and $\text{poly}_2(\_)$ such that for any target DBN T in $\mathcal{DBN}(\mathcal{C}_n)$, any $0 < \epsilon < 1$ and $0 < \delta < 1$, and any starting point for the time series: given a time series of any length $r \geq \text{poly}_1(n,|T|,\frac{1}{\epsilon},\frac{1}{\delta})$, the algorithm runs in $\text{poly}_2(rn)$ and with probability at least $1-\delta$ outputs a model that with probability at least $(1-\epsilon)$ correctly predicts time slice $r+1$.

Notice that we do not require that the learning algorithm be capable of performing well for most starting points, but only for the one given. For deterministic DBN models, after some m time slices the time series must return to a previous state, from which point on the time series will cycle with some period length at most m. If for some class of DBN models m is only polynomial in the number of variables n then it will be possible to PAC-learn this class of models from a single time series, within the definition just given. Unfortunately, even for the simple class of deterministic $\mathcal{DBN}$(k-parents), the period is superpolynomial in n and the size of the target model, leading to the following negative result.

Theorem 3.2. *For any $k \geq 2$ the class of $\mathcal{DBN}$(k-parents) is not learnable from a single time series.*

Proof. Assume there exists a learning algorithm L for $\mathcal{DBN}$(k-parents). Then for any k-parent target DBN T, any $0 < \epsilon < 1$ and any $0 < \delta < 1$, given a time series of any length $r \geq \text{poly}_1(n,|T|,\frac{1}{\epsilon},\frac{1}{\delta})$, L will run in time polynomial in the size of its input and with probability at least $1 - \delta$ will output a model that will correctly predict time slice $r+1$ with probability at least $1-\epsilon$. Because any 2-parent DBN can be represented in a number of bits that is polynomial in n, we can simplify $\text{poly}_1(n,|T|,\frac{1}{\epsilon},\frac{1}{\delta})$ to $\text{poly}_1(n,\frac{1}{\epsilon},\frac{1}{\delta})$.

We consider a time series that starts from a point in which every variable is set to 0. For a suitable choice of n (any n such that $n - 1$ is divisible by 3) we can build two 2-parent deterministic DBNs T_1 and T_2 over variables $X_1, ..., X_n$ with the following properties when started from a time slice with variables set to 0: in both T_1 and T_2, X_n remains 0 for $r \geq 2^{\frac{n-1}{3}}$ steps and then X_n goes to 1 at step $r + 1$; in T_1 once X_n goes to 1 it remains 1; in T_2 when X_n goes to 1 it then reverts to 0 on the next step.

We choose $\epsilon = \delta = \frac{1}{4}$ and large enough n such that $2^{\frac{n-1}{3}} > \text{poly}_1(n,\frac{1}{\epsilon},\frac{1}{\delta})$. We present the algorithm L with a time series generated by T_1, of length r as specified in the previous paragraph, starting from the time slice in which all variables are set to 0. Then L must, with probability at least $1 - \frac{1}{4}$, return a model that will correctly predict time slice $r + 1$. Therefore, with probability at least $(\frac{3}{4})(\frac{3}{4}) > \frac{1}{2}$, L's output model predicts the value of X_n to be 1. Consider what happens when we give L exactly the same learning task, except that the target is T_2 instead of T_1. The time series of length r that L sees is identical to the previous one, so L will with probability greater than $\frac{1}{2}$ incorrectly predict the value of X_n at time slice $r+1$. Hence L will *not* produce, with probability at least $1-\delta$, a model that will predict time slice $r + 1$ with accuracy at least $1 - \epsilon$. □

3.3. *Stochastic Model of Boolean DBN from 2-slice data*

When we learn a Boolean DBN model, we are not only interested in learning the correct Boolean functions but also inferring a good model of probability wrt the target distribution. We therefore extend our theoretical framework to bring our model closer to practice. The foundation of this extension consists of the notions of *p-concept* (probabilistic concept) and (ϵ, γ)-*good models of probability*, defined as follows[8]. In these definitions X is the domain of possible examples and D is a probability distribution over X.

Definition 3.3. A *probabilistic concept (p-concept)* is a real-valued function $c : X \to [0, 1]$. When learning the p-concept c, the value $c(x)$ is interpreted as the probability that x exemplifies the concept being learned. A p-concept class $\mathcal{C}_p$ is a family of p-concepts. A learning algorithm for $\mathcal{C}_p$ attempts to learn a distinguished target p-concept $c \in \mathcal{C}_p$ wrt a fixed but unknown and arbitrary target distribution $\mathcal{D}$ over the instance space X[8].

Given this definition, it is easy to see that a function f_i in a (not necessarily deterministic) Boolean DBN, which gives the probability distribution over possible values for X_i at time $t + 1$ conditional on the values of the n variables at time t, is a p-concept. Therefore, a Boolean DBN as defined earlier is completely specified by a set of n p-concepts, one for each variable.

Definition 3.4. A p-concept h is an (ϵ, γ)-*good model of probability* of a target p-concept c with respect to D iff $\mathbf{Pr}_{x \in D}[|h(x) - c(x)| > \gamma] \leq \epsilon$.

We generalize this definition to apply to DBNs as follows. Given an input time slice, a DBN model defines a probability distribution over output time slices. We say that two DBNs M and T γ-disagree on an input time slice if they disagree by more than γ on the probability of an output time slice given that input. A learned DBN M is an (ϵ, γ)-*good model of probability* of a target DBN T with respect to a probability distribution D over input time slices if and only if the sum of the probabilities of input models on which M and T γ-disagree is at most ϵ.

The learning model we present next is a straightforward application of Kearns and Schapire's notion of *polynomially learnable with a model of probability* to DBNs, analogous to our earlier application of the PAC model to deterministic DBNs. In the following definition we take $\mathcal{C}_n$ to be any p-concept class. Thus for example $\mathcal{DBN}$(k-parents) is the set of DBNs in which each variable has at most k-parents; the p-concept class used here is the class of p-concepts of k-relevant variables, or the class of p-concepts representable by CPTs conditional on k parents.

Definition 3.5. Where $\mathcal{C}_n$ is a p-concept class, we say that an algorithm learns $\mathcal{DBN}(\mathcal{C}_n)$ with a model of probability iff there exist polynomials $\text{poly}_1(\_,\_,\_,\_,\_)$ and $\text{poly}_2(\_)$ such that for any target $T \in \mathcal{DBN}(\mathcal{C}_n)$, any $0 < \epsilon < 1$, $0 < \delta < 1$, and $0 < \gamma < 1$, and any probability distribution D over initial data points for time series: given $r \geq \text{poly}_2(n, |T|, \frac{1}{\epsilon}, \frac{1}{\delta}, \frac{1}{\gamma})$ data points, the algorithm runs in $\text{poly}_2(rn)$ and with probability at least $1 - \delta$ outputs an (ϵ, γ)-good model of probability of the target.

Theorem 3.3. *For any fixed $k \in \mathbb{N}$ the class $\mathcal{DBN}$(k-parents) is learnable with a model of probability from 2-slice time series data.*

Proof. We describe a learning algorithm B that is analogous to algorithm A of Theorem 3.1, and the correctness proof is analogous as well. In place of the kDNF learning algorithm used by the earlier algorithm A, algorithm B uses an algorithm P that with probability $1 - \delta$ learns an (ϵ, γ)-good model of probability for any p-concept with at most k relevant variables[8].

More specifically, where δ, ϵ and γ are the parameters provided to algorithm B, algorithm B calls algorithm P using instead $\frac{\delta}{n}$, $\frac{\epsilon}{n}$ and $\frac{\gamma}{2n}$. Algorithm B iterates: for each variable $X_{i,2}$, $1 \leq i \leq n$, algorithm B makes a call to algorithm P with the examples and parameters as specified. Algorithm B's final model of probability for each X_i, $1 \leq i \leq n$, is $\Pr(X_{i,t}|X_{1,t-1}, \ldots, X_{n,t-1}) = \Pr(X_{i,t}|\text{Pa}(X_i)_{t-1})$, where $\text{Pa}(X_i)_{t-1}$ denotes the (at most k) parents of X_i from the previous time step, as determined by algorithm P, and $\Pr(X_{i,t}|\text{Pa}(X_i)_{t-1})$ denotes the specific function (representable as a CPT) learned by algorithm P.

Algorithm B runs in polynomial time since $n{*}\text{poly}_1(n,|T|,\frac{n}{\epsilon},\frac{n}{\delta},\frac{2n}{\gamma})$ yields a polynomial, and each call to P runs in time polynomial in the size of its input. The remainder of the reasoning is analogous to that in the proof of Theorem 3.1, except that we must also note the following. If the learned DBN and target DBN agree within $\frac{\gamma}{2n}$ on the probability for a given setting for each variable X_i, $1 \leq i \leq n$, then they agree within γ on the probability of the entire setting. It follows that since *for any given variable* X_i the learned DBN with probability $1 - \frac{\delta}{n}$ has an $(\frac{\epsilon}{n}, \frac{\gamma}{2n})$-good model of probability compared with the target DBN, then with probability $1 - \delta$ the learned DBN is an (ϵ, γ)-*good model of probability* of the target DBN. □

We can also extend our model from a *single time series* to apply to probabilistic concepts. Since deterministic DBNs are a special case of probabilistic ones, it follows that the result for learning $\mathcal{DBN}$(2-parents) with a model of probability, from a single, long time series is a negative one.

4. Lessons and Limitations

The results show that, for natural definitions of learnability, DBNs are learnable from 2-slice time series data but not from a single, long time series. If we adopt a compromise, with k-slice time series for fixed k greater than two, we can again get positive results but the total number of time slices, e.g., experiments to be run, increases linearly with k. Hence the results imply that while time series are desirable, they should be kept as short as possible. The reasoning is that if we subject a cell to many different conditions and collect multiple 2-slice data, we can capture the change easily, whereas if we subject a cell to 1 condition and collect one long time series one would have to wait a very long time to see this condition.

Because PAC bounds are worst-case, the number of examples they imply are required, while polynomial in the relevant parameters, can be much greater than typically required in reality. Nevertheless, we can gain some insight into which factors most affect sample size required for a given degree of accuracy. The sample sizes required by the algorithms in this paper follow directly from those required by the learning algorithms they employ as subroutines. The sample sizes for those algorithms grow linearly with the number of variables n, the target concept size, and $\frac{1}{\epsilon}$ (and $\frac{1}{\gamma}$ where relevant), and logarithmically with $\frac{1}{\delta}$. But note that the sizes of our target concepts in $\mathcal{DBN}$(kDNF) and $\mathcal{DBN}$(k-parents) are at least $O(n^k n)$, because we must specify the choice of k out of n possible parents for each of n variables. Therefore, by far the most important factor in sample size is k, and the next most important is n. Because current costs limit MS data set sizes to around 1000 experiments (we know of no such data sets), a value of three for k seems the largest reasonable value, with $k = 2$ probably more sensible. The size of the target concept can be limited based on prior knowledge about particular pathways in which we are most interested.

The models defined in this paper are a natural application of existing PAC models to DBN learning. Nevertheless, several assumptions are inherited in this application—some from PAC modeling and some from DBNs—and several additional assumptions have been made. We now discuss these classes of assumptions in turn.

Inherent to the use of PAC modeling are the assumptions that (1) we must perform well on *all* target concepts, and (2) examples are drawn randomly, independently according to some probability distribution. Regarding assumption (1), numerous regulatory motifs have been identified to date, including logic gates and memory elements[11,15,19], giving credence to the simple DBN representation of difficult target concepts such as counters. For some real biological pathways that have very short periods, perhaps single, long time series will be more effective than our results imply. Regarding assumption (2), it seems plausible that an organism's environment imposes some probability distribution over states in which its regulatory machinery may find itself, and to which it will need to respond. Nevertheless, perhaps through careful selection of experimental conditions, active learning approaches may arise that will benefit more from a few long time series than from many short ones.

Inherent in the use of DBNs are several notable assumptions as well. First, the DBN framework assumes we are modeling a *stationary process*; while the state of an organism is not static (e.g., mRNA levels may change over time), the organism's regulatory network itself does not change over time. This assumption appears reasonable for the application to learning regulatory pathways. But more specific assumptions include the assumption of discrete time steps—that an organism, like a computer, operates on a fixed clock that governs when changes occur—and a first-order Markov assumption, that the organism's next state is a function of only its previous state inputs. These assumptions clearly are violated to some extent, and those violations present caveats to our lessons. For example, perhaps collecting longer time series, with a very fast sampling rate, could allow our algorithms to try different sampling rates (by skipping some of the time steps), to find optimal rates for providing insight into certain processes.

Finally, we have made additional simplifying assumptions beyond those of the PAC framework or DBNs. Specifically, we have assumed all Boolean variables and no missing values or hidden variables. While discretization is common, one may also use the continuous values in the data. We see no obvious reason why using such values should change the lessons in this paper, but such a change is possible. Missing values are rare in MS data, but one might wish to include hidden variables for unmeasured environmental factors. Extending the present work to handle hidden variables is an interesting direction for further work.

In addition to handling continuous-valued variables and hidden variables, another significant direction for further work is the use of background knowledge to limit the space of potential models and hence the sample complexity. Also, based on the notion of *membership queries*, perhaps the models in this paper can be extended to model active learning approaches. Finally, we intend to use the lessons from this paper to design a large number of short time series experiments aimed at gaining more detailed models of a few key pathways in human cells.

References

1. T. Akutsu, S. Kuhara, O. Maruyama, and S. Miyano. Identification of gene regulatory networks by strategic gene disruptions and gene overexpressions. In *Proc. of 9th ACM-SIAM Symp. on Discrete Algs.*, pages 695–702, 1998.
2. T. Akutsu, S. Miyano, and S. Kuhara. Identification of genetic networks from

a small number of gene expression patterns under the boolean network model. *Proc. of PSB*, 4:17–28, 1999.

3. A. Bernard and A.J. Hartemink. Informative structure priors: Joint learning of dynamic regulatory networks from multiple types of data. In *PSB*, pages 459–470, 2005.
4. S. Dasgupta. The sample complexity of learning fixed-structure Bayesian networks. *MLJ*, 29:165–180, 1997.
5. N. Friedman and Z. Yakhini. On the sample complexity of learning bayesian networks. In *Proc. of UAI*, pages 274–282, 1996.
6. D. Husmeier. Sensitivity and specificity of inferring genetic regulatory interactions from microarray experiments with dynamic bayesian networks. *Bioinformatics*, 19:2271–2282, 2003.
7. M. Kearns, M. Li, L. Pitt, and L.G. Valiant. On the learnability of boolean formulae. In *Proc. of ACM Theory of computing*, pages 285–295, 1987.
8. M. Kearns and R. Schapire. Efficient distribution-free learning of probabilistic concepts. In *Computational Learning Theory and Natural Learning Systems, Vol I: Constraints and Prospect.* MIT Press, 1994.
9. M.J. Kearns and U.V. Vazirani. *An Introduction to Computational Learning Theory.* MIT Press, Cambridge, MA, 1994.
10. H. Lähdesmäki, I. Shmulevich, and O. Yli-Harja. On learning gene regulatory networks under the boolean network model. *MLJ*, 52:147–167, 2003.
11. H.H. McAdams and A.P. Arkin. Simulation of prokaryotic genetic circuits. *Ann. Rev. of Biophy. and Biomol. Structure*, 27:199–224, 1998.
12. I. Nachman, A. Regev, and N. Friedman. Inferring quantitative models of regulatory networks from expression data. *Bioinformatics*, 20:i248–i256, 2004.
13. I.M. Ong, J.D. Glasner, and D. Page. Modelling regulatory pathways in E.coli from time series expression profiles. *Bioinformatics*, 18:241S–248S, 2002.
14. D. Pe'er. Bayesian network analysis of signaling networks: A primer. *Science's STKE*, 281:pl4, 2005.
15. C.V. Rao and A.P. Arkin. Control motifs for intracellular regulatory networks. *Ann. Rev. of Biomed. Eng.*, 3:391–419, 2001.
16. K. Sachs, O. Perez, D. Pe'er, D.A. Lauffenburger, and G.P. Nolan. Causal protein-signaling networks derived from multiparameter single-cell data. *Science*, 308:523–529, 2005.
17. I. Shmulevich, E.R. Dougherty, K. Seungchan, and W. Zhang. Probabilistic boolean networks: A rule-based uncertainty model for gene regulatory networks. *Bioinformatics*, 18:261–274, 2002.
18. L.G. Valiant. A theory of the learnable. *Comm. of ACM*, 27:1134–1142, 1984.
19. D.M. Wolf and A.P. Arkin. Motifs, modules and games in bacteria. *Curr. Op. in Microbio.*, 6:125–134, 2003.
20. Zhang Y., Wolf-Yadlin A., Ross P.L., Pappin D.J., Rush J., Lauffenburger D.A., and White F.M. Time-resolved mass spectrometry of tyrosine phosphorylatiopn sites in the EGF receptor signaling network reveals dynamic modules. *Mol. and Cell. Prot.*, 2005.

FINDING DIAGNOSTIC BIOMARKERS IN PROTEOMIC SPECTRA

PALLAVI N. PRATAPA[1], EDWARD F. PATZ, JR.[2], ALEXANDER J. HARTEMINK[1]
[1]*Duke University, Dept. of Computer Science, Box 90129, Durham, NC 27708*
`{pallavi,amink}@cs.duke.edu`

[2]*Duke University Medical Center, Depts. of Radiology & Pharmacology and Cancer Biology, Box 3808, Durham, NC 27710*
`patz0002@mc.duke.edu`

In seeking to find diagnostic biomarkers in proteomic spectra, two significant problems arise. First, not only is there noise in the measured intensity at each m/z value, but there is also noise in the measured m/z value itself. Second, the potential for overfitting is severe: it is easy to find features in the spectra that accurately discriminate disease states but have no biological meaning. We address these problems by developing and testing a series of steps for pre-processing proteomic spectra and extracting putatively meaningful features before presentation to feature selection and classification algorithms. These steps include an HMM-based latent spectrum extraction algorithm for fusing the information from multiple replicate spectra obtained from a single tissue sample, a simple algorithm for baseline correction based on a segmented convex hull, a peak identification and quantification algorithm, and a peak registration algorithm to align peaks from multiple tissue samples into common peak registers. We apply these steps to MALDI spectral data collected from normal and tumor lung tissue samples, and then compare the performance of feature selection with FDR followed by classification with an SVM, versus joint feature selection and classification with Bayesian sparse multinomial logistic regression (SMLR). The SMLR approach outperformed FDR+SVM, but both were effective in achieving good diagnostic accuracy with a small number of features. Some of the selected features have previously been investigated as clinical markers for lung cancer diagnosis; some of the remaining features are excellent candidates for further research.

1 Introduction and motivation

A diagnosis of cancer is often first suggested by radiological imaging. Unfortunately, imaging findings are not always specific so further evaluation with invasive procedures is typically required to establish a diagnosis. Many researchers have put great effort into developing alternative strategies for more effectively diagnosing cancer non-invasively, particularly through the identification of diagnostic biomarkers. While some groups have focused on genomics, others have pursued proteomics, hoping that protein expression profiles will lead to biomarkers that more accurately reflect disease phenotypes. Given the limitations of traditional methods involving 2D-GE, alternative proteomic platforms have been pursued. Over the last several years investigators have begun to explore the use of a variety of protein separation techniques followed by matrix-assisted laser desorption/ionization time-of-flight mass spectrometry (MALDI-TOF MS; henceforth just 'MALDI'). Although MALDI has traditionally been used for protein identification, several recent studies have suggested that direct analysis of MALDI data can provide diagnostic value.[1,2]

The data from MALDI is a list of mass-charge ratios ('m/z values') and corresponding measured intensities. If we plot the measured intensities as a function of m/z, we call the resultant curve a 'spectrum'. Peaks in the spectrum correspond to proteins in the tissue sample. Under ideal conditions, samples with similar protein composition would have peaks with identical intensities at identical m/z values. However, due both to variation in lysate preparation and limitations inherent in the measurement technology, not only is there noise in the measured intensity at each m/z value, but there is also noise in the measured m/z value itself. This makes it difficult to directly compare spectra between groups of patients and thus to identify specific protein expression patterns from complex biological samples. Because the spectral data we collect possess a hierarchical structure (multiple replicate spectra per sample, multiple samples per class), we develop a hierarchical strategy for solving this problem. We use a latent spectrum extraction algorithm to fuse information from multiple replicate spectra obtained from a single tissue sample, and then a peak registration algorithm to align peaks from multiple tissue samples into common 'peak registers'. These two algorithms are embedded in a longer data analysis pipeline (described below), designed to identify a small number of features as putatively meaningful diagnostic biomarkers. While previous methods have been developed for particular steps in this pipeline, and while very recent reviews have admirably and effectively summarized previously published methods,[3,4] our experience in implementing the entire pipeline has enabled us to test and compare both existing and novel methods at each step in the analysis, and in the context of the full hierarchical pipeline. Here, for reasons of limited space, we report the final choices that were made at each step.

2 Analytical methods

2.1 Overview

Our data analysis pipeline is hierarchically organized and consists of two levels of pre-processing followed by a third level of feature selection and classification. The first pre-processing level identifies and quantifies putatively meaningful peaks in each tissue sample from multiple replicate spectra. The second pre-processing level yields a matrix of comparable features across all the tissue samples. The steps in each of these levels of analysis are depicted in Fig. 1. In what follows, we discuss each of these steps in turn, devoting more attention to the more interesting steps.

2.2 Latent spectrum extraction

Ideally, the multiple replicate spectra collected from each tissue sample would be identical, but unfortunately, both X- and Y-axis measurements are noisy. As a model, we postulate that the measured spectra are noisy versions of some true 'latent spectrum'. We imagine that variability in the Y-axis arises from a combination of multiplicative and additive errors: a global scaling factor for each replicate, a local scaling

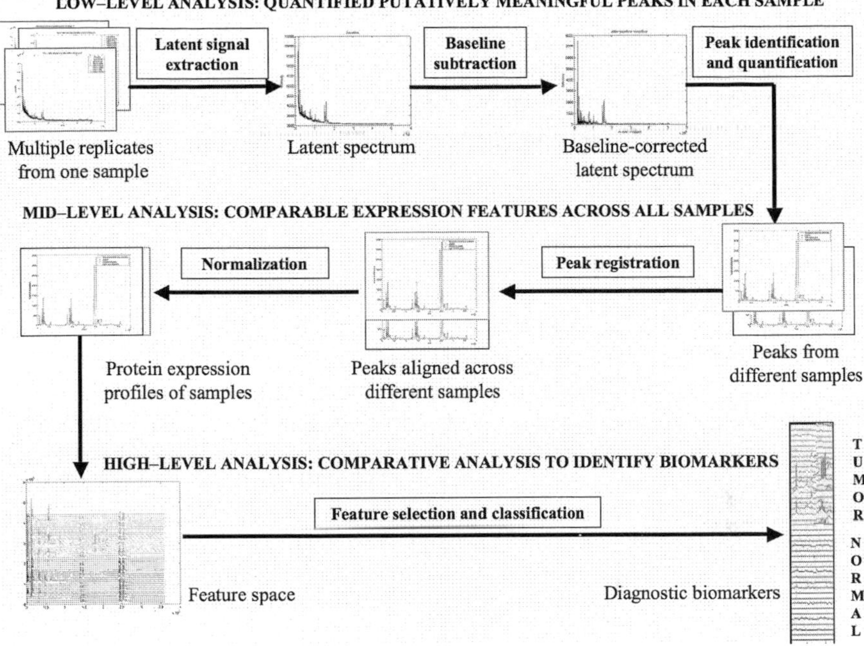

Figure 1. Overview of our hierarchical data analysis pipeline.

factor which varies smoothly within each replicate, and a local additive noise term. The global scaling factor can account for things like variation in the concentration of ions present at the laser location corresponding to each replicate spectrum. Likewise, the local scaling factor can account for signal suppression factors which influence the number of protein ions passing through the detector at each point in time. We further imagine that variability in the X-axis arises from non-uniform subsampling of the latent spectrum, which is different for each of the replicates. A latent spectrum extraction algorithm can then be used to find the latent spectrum that maximizes the likelihood of the measured replicate spectra. For this purpose, we use the recently-proposed continuous profile model (CPM),[5] implementing a learning algorithm with a few computational enhancements for our setting. Below, we formalize the model, provide brief details of the learning algorithm, and illustrate its operation with an example; readers interested in further details are encouraged to read the original paper.[5]

2.2.1 *Model description*

To ease comparison with the original paper, we adopt nearly identical notation, although we do correct a few errors. Assume we have K replicate spectra, indexed by $k \in \{1, 2, \ldots, K\}$. Let $X^k = [x_1^k, x_2^k, \ldots, x_{N^k}^k]$ denote the measured intensity values in the k-th replicate spectrum at the m/z values indexed by $i \in \{1, 2, \ldots, N^k\}$. Let $Z = [z_1, z_2, \ldots, z_M]$ denote the (unmeasured) intensity values in the latent spectrum at the m/z values indexed by $\tau \in \{1, 2, \ldots, M\}$. According to the model, each replicate spectrum is a non-uniformly subsampled version of the latent spectrum, to which global and local scaling factors have been applied and noise has been added. Hence, we have

$$x_i^k = z_{\tau_i^k} \phi_i^k u^k + \epsilon \tag{1}$$

where τ_i^k is the hidden time state (the value of τ in the latent spectrum that corresponds with i in replicate k), ϕ_i^k is the hidden local scale state, u^k is the global scaling factor for replicate k, and ϵ is drawn from a central normal distribution with variance σ^2, assumed to be the same for all replicates.

As previously described,[5] this is essentially a hidden Markov model (HMM), where the output is conditioned on the latent spectrum. Let π^k denote the hidden state sequence for the k-th replicate. Each state in this sequence consists of a time state and a local scale state: $\pi_i^k = \langle \tau_i^k, \phi_i^k \rangle$. The time states are selected from the m/z values of the latent spectrum ($\tau_i^k \in \{1, 2, \ldots, M\}$) and the local scale states are selected from an ordered set of P scale values ($\phi_i^k \in \{\phi^1, \phi^2, \ldots, \phi^P\}$). The HMM is specified by the following emission and transition probabilities:

$$\textbf{Emission:} \quad e_{\pi_i^k}(x_i^k|Z) = P(x_i^k|\pi_i^k, Z, u^k, \sigma^2) = \mathcal{N}(x_i^k; z_{\tau_i^k}\phi_i^k u^k, \sigma^2) \tag{2}$$

$$\textbf{Transition:} \quad T^k_{\pi_{i-1},\pi_i} = P^k(\pi_i|\pi_{i-1}) = P^k(\tau_i|\tau_{i-1})P(\phi_i|\phi_{i-1}) \tag{3}$$

Regarding the transition probabilities between time states, we impose the constraint that the time state must always advance at least one step and no more than J steps from the current state; transitioning $v \in \{1, \ldots, J\}$ steps is multinomial with probability d_v^k. Regarding the local scale states, we impose the constraint that the scale state must remain the same or change to the neighboring scale state above or below; transitioning $v \in \{-1, 0, 1\}$ steps is multinomial with probability $s_{|v|}$.

2.2.2 *Learning the latent spectrum using expectation-maximization*

Given the replicate spectra from a single tissue sample, we can train the model to learn the latent spectrum using Baum-Welch (EM). The M-step update rules are derived by solving for the values of the parameters Z, σ^2, and u^k that maximize the expected log likelihood (we ignore the smoothing prior), yielding the following:

$$z_j = \frac{\sum_{k=1}^K \sum_{\{s|\tau_s=j\}} \sum_{i=1}^N (\gamma_s^k(i) x_i^k \phi_s u^k)}{\sum_{k=1}^K \sum_{\{s|\tau_s=j\}} \sum_{i=1}^N (\gamma_s^k(i)(\phi_s u^k)^2)} \tag{4}$$

$$\sigma^2 = \frac{\sum_{k=1}^K \sum_{s=1}^S \sum_{i=1}^N \gamma_s^k(i)(x_i^k - z_{\tau_s}\phi_s u^k)^2}{KN} \tag{5}$$

$$u^k = \frac{\sum_{s=1}^S z_{\tau_s}\phi_s \sum_{i=1}^N \gamma_s^k(i) x_i^k}{\sum_{s=1}^S (z_{\tau_s}\phi_s)^2 \sum_{i=1}^N \gamma_s^k(i)} \tag{6}$$

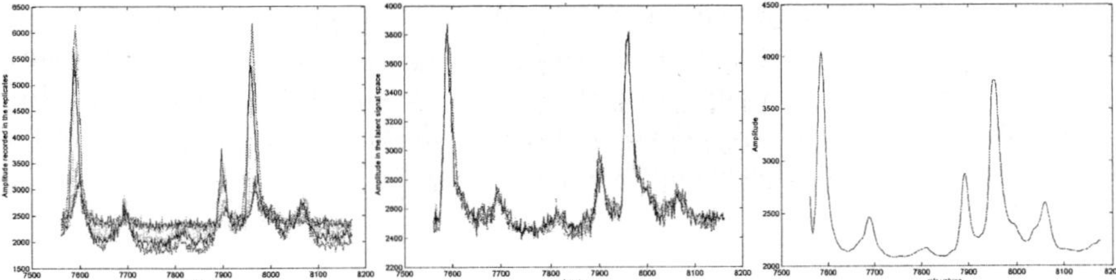

Figure 2. Latent spectrum extraction. (a) A subportion of length N=500 of seven unaligned replicate spectra from a single tissue sample; note variation in both the X- and Y-axes. (b) The replicate spectra after Viterbi alignment to the latent spectrum. (c) The latent spectrum extracted from the replicate spectra.

where $s \in \{1, 2, \ldots, S\}$ indexes the total number of possible states and $\gamma_s^k(i) = P(\pi_i = s|x_i^k)$ is computed during the forward-backward algorithm in the E-step. Similarly, the update rules for the various multinomial transition probabilities are shown below:

$$d_v^k = \frac{\sum_{s=1}^{S} \sum_{\{s'|\tau_s - \tau_{s'} = v\}} \sum_{i=2}^{N} \xi_{s,s'}^k(i)}{\sum_{s=1}^{S} \sum_{j=1}^{J} \sum_{\{s'|\tau_s - \tau_{s'} = j\}} \sum_{i=2}^{N} \xi_{s,s'}^k(i)} \tag{7}$$

$$s_v = \frac{\sum_{k=1}^{K} \sum_{s=1}^{S} \sum_{\{s' \in H(s,v)\}} \sum_{i=2}^{N} \xi_{s,s'}^k(i)}{\sum_{k=1}^{K} \sum_{s=1}^{S} \sum_{\{s' \in H(s,0), H(s,1)\}} \sum_{i=2}^{N} \xi_{s,s'}^k(i)} \tag{8}$$

where $H(s, v)$ indicates the set of states that are exactly v scale values away from s and $\xi_{s,s'}^k(i) = P(\pi_{i-1} = s, \pi_i = s' \mid x_i^k)$ is again computed during the E-step.

2.2.3 *Illustration of latent spectrum extraction*

We illustrate the results of applying the latent spectrum extraction algorithm to fuse information from multiple replicate spectra from a single sample. We consider a section of N=500 points of K=7 unaligned replicate spectra of a sample as shown in Fig. 2(a). For training the model, we set $M = 2.01 \times N$. The latent spectrum Z is initialized to be the median of the intensity values in the replicates, with Gaussian noise added; the standard deviation of the Gaussian noise is initialized to 30% of the difference between the minimum and maximum value in Z. The result is supersampled by a factor of two (repeating every value twice consecutively) with additional small values padding the beginning and end to achieve a total length of M. For each replicate, the global scale value u^k is initialized to unity. The largest time state transition, J, is set to 3. An ordered set of P=21 local scale values, $\{0.80, 0.82, 0.84, \ldots, 1.20\}$, is used. The multinomials defining the time state and scale state transition probabilities are initialized to uniform. The transition probabilities from the beginning state to π_1^k are set uniformly.

After training the model, we use the Viterbi algorithm to find the most likely path through the hidden states for each replicate. We then align each of the replicates to the latent profile, the output of which is shown in Fig. 2(b). As the figure illustrates, the replicates are now all well-aligned on both the X-axis and the Y-axis. Fig. 2(c) shows the latent spectrum, which offers a denoised summary representation of the information contained in the unaligned replicate spectra.

2.3 Baseline subtraction

To compensate for the gradually decreasing baseline of a complete latent spectrum, we find a monotone local minimum curve that lower-bounds it by computing the convex hull of the latent spectrum. We then subtract this from the latent spectrum to get the baseline-corrected latent spectrum, or simply latent spectrum henceforth. In addition to its extreme simplicity, an advantage of this method over previously proposed methods is that the resultant latent spectrum is everywhere non-negative.

2.4 Peak identification and quantification

Identifying peaks in the latent spectrum of a tissue sample is imperative because peaks represent the isolatable proteins or peptides that may be relevant in discriminating between the two classes of samples. Focusing our attention on the peaks rather than the entire spectrum eliminates from consideration potential features that are expected to have no biological meaning.

To identify the important peaks in a latent spectrum, a simple approach would be to identify all locally maximal points in the spectrum with height above a certain signal-to-noise threshold, T. However, a complication arises because some local maxima satisfying this criterion are simply noisy bumps on the side of a larger peak. For this reason, and because later we will quantify peaks by their area, we also need to determine an interval describing the support (on the m/z axis) of each putative peak. To accomplish this, we develop a peak picking algorithm that begins by sorting all local maxima in the latent spectrum into a priority queue by their signal-to-noise ratio, truncating the queue when the ratio drops below T. The extent of the support interval is determined for the peak at the head of the queue by moving left and right from the peak until a moving average of the gradient changes sign (the moving average prevents us from stopping at noisy bumps on the side of a larger peak). Elements of the queue falling within this interval are removed from the queue. Processing the entire queue gives us a list of putatively important peaks, along with a height and width for each.

The height of a peak is not good for quantifying its relative importance in the spectrum. For fixed laser energy, recorded intensity values are generally higher for lower m/z values; the peaks at lower m/z values are also narrower than those at higher m/z values. Hence, the area enclosed by a peak within its support interval has been suggested as perhaps a more useful measure for comparing peaks over a wide range of m/z values. This area is the measure we use for peak quantification.

2.5 Peak registration to align peaks across all samples

Just as m/z values for a peak can vary within replicates from the same tissue sample, they can also do so across samples. However, now it is not the case that we expect a single underlying latent spectrum to explain the spectra from different samples because the samples themselves may be biologically heterogeneous. To address this problem and enable peaks from different samples to be compared on an equal footing, we develop a simple peak registration algorithm, developed independently of but similar in flavor to one recently published.[6] Given a list of peaks from the latent spectra of all the samples, and an estimate of the mass resolution error of the MALDI instrument, we assign peaks in different latent spectra into the same 'peak register' if their m/z values are within the mass resolution of the instrument. Since the instrument's resolution is proportional to m/z, we first log-transform the m/z values and then perform hierarchical clustering in log-m/z space using complete linkage and a Euclidean distance metric. We can determine the number of registers by cutting the dendrogram at a depth given by the log-transformed mass resolution of the instrument. The m/z value that we associate with each peak register is the mean of the m/z values of the peaks that belong to the register.

2.6 Normalization

The areas of peaks belonging to one m/z register may have a high coefficient of variation across different samples of similar protein composition. The areas of the peaks for each sample need to be normalized to make a fair comparison of protein expression levels across different samples. We normalize by a global factor, computed so as to equalize either the mean peak area or the median peak area of all samples. In each case, the normalized peak areas are finally log-transformed so as to not overemphasize obvious peaks in relation to less obvious ones. This step could also be performed before the previous step because the previous step takes no account of the peak areas. More sophisticated strategies might merge these two steps.

2.7 Feature selection and classification

Once we have identified a matrix of comparable features across all the tissue samples, we can consider strategies for sparse feature selection and classification. If we rank each feature based on its Fisher discriminant ratio (FDR), one strategy is to start with the top ranking feature and sequentially add features until there is no further improvement in the leave-one-out cross-validation ('LOOCV') classification accuracy of a linear SVM. This strategy has the benefit of producing a classifier with a small number of features through the sequential combination of two common methods: FDR for feature selection and SVM for classification.

However, we may be able to do better by using a single method that jointly addresses the tasks of feature selection and classification, especially in a proteomic context where features so severely outnumber observations. Bayesian algorithms for learning sparse classifiers have recently been developed to learn simultaneously both a small subset of features relevant to classification and an optimal classifier.[7,8,9] Sparsity-promoting priors are used to regularize the feature weight vector, ensuring that weights are either significantly large or exactly zero, automatically removing irrelevant features from consideration. We use sparse multinomial logistic regression (SMLR).[9] The sparsity of the feature weight vector is controlled by the regularization parameter λ. Too high a value of λ will result in relevant features not being selected, thereby giving rise to more errors during training and cross-validation. On the other hand, too small a λ will cause more features to be selected than should be, resulting in over-fitting during training and more errors during cross-validation. Consequently, we can choose λ using LOOCV.

3 Results

3.1 Experimental procedure for collecting MALDI spectral data

Resections of lung tissue were obtained from 34 patients diagnosed with non-small cell lung cancer. In each case, normal and tumor lung tissue samples from the same patient were collected, yielding 68 total samples. Tissue samples were washed to remove blood contamination and then placed into a micro-centrifuge tube containing a protein extraction reagent and electrically homogenized. Cellular debris was removed by centrifugation and cell lysates were prepared. One microliter of lysate was deposited on the MALDI matrix and allowed to dry under ambient conditions. Spectra were acquired on a Voyager DE Biospectrometry Workstation using a nitrogen laser (337 nm). Multiple spectra were obtained from each tissue sample by focusing the laser on different sub-positions on the deposited lysate. By visually examining each spectrum for sufficient signal in terms of the number of peaks, ten replicate spectra were chosen for each tissue sample, for a total of 680 spectra.

3.2 Extraction of the latent spectrum from the replicates of a sample

Each replicate spectrum of each tissue sample contains N=27715 data points, with m/z values ranging from 1500 to 44000, approximately. We split the X-axis of the replicates into four sections to increase the speed of processing the data: we can run the latent spectrum extraction algorithm in parallel over the four sections of the replicate spectra. The spectra are split such that peaks exist in each section, but the tail end of each section contains only noise; this enables the sections to be recombined without loss of information about the peaks.

The various parameters specifying the latent spectrum extraction model and initializing the learning algorithm were chosen exactly as in the example of Sec. 2.2.3, except that for computational reasons, only three local scale values $\{0.8, 1, 1.2\}$ were used, representing local down-scaling, no scaling, and up-scaling of the latent spectrum, respectively. The model was trained using EM to learn the latent spectrum for

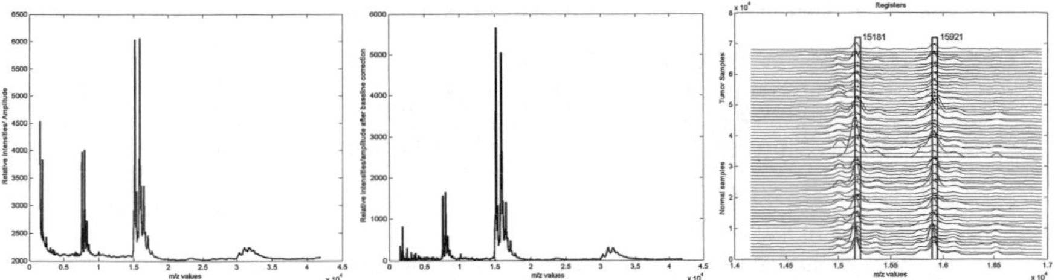

Figure 3. (a) The recombined latent spectrum of a sample obtained from ten replicate spectra. The four sections are shown in four different colors. (b) The baseline-corrected latent spectrum. (c) An example of peak registration, showing how peaks from multiple latent spectra are assigned to common peak registers.

each of the four sections. The four sections of the latent spectra were then combined by attaching them end to end and overlapping the last six points of one section with the first six points of the following section. As an example, the final recombined latent spectrum from the ten replicates of one tissue sample is shown in Fig. 3(a). The baseline is found using a segmented convex hull and subtracted from the latent spectrum to obtain the baseline-corrected latent spectrum, as shown in Fig. 3(b).

3.3 Peak registration

Cutting the complete-linkage hierarchical clustering dendrogram at a height equal to the log-transformed mass resolution error of the instrument produces 380 peak registers. In Fig. 3(c), the results of peak registration on two peak registers at m/z values 15181 and 15920 are shown, along with the boundary of each register; the peaks at these two m/z values are known to be present in all samples.

3.4 Normalization

We compared the two normalization strategies based on how much the standard deviations of the peak areas for the normal and tumor samples, taken as a class, were reduced by normalization in each case. Before normalization, the standard deviations of the peak areas for normal and tumor samples were 156 and 137, respectively. After normalization using mean peak area, these values became 196 and 110, respectively. After normalization using median peak area, these values became 128 and 72, respectively. Better performance using the median is reasonable since the mean is more susceptible to outliers; as a result of this consideration, and the corroborating numerical results, we proceed with normalization based on median peak areas. As mentioned above, the normalized peak areas were finally log-transformed so as to not overemphasize obvious peaks in relation to less obvious ones.

To summarize the two levels of pre-processing, whereas our initial data consisted of 680 replicate spectra of length N=27715, this has now been transformed

Table 1. LOOCV classification accuracy of SVM and SMLR based on an optimal set of selected features.

FDR+SVM		SMLR	
Accuracy	# features	Accuracy	# features
82% (9 errors)	7	92% (4 errors)	4

Table 2. m/z values of features selected by FDR and SMLR (ordered by rank).

Features selected by FDR	Features selected by SMLR
12386, 17932, 6206, 10884, 11215, 9781, 9109	12386, 5147, 16140, 15378

into a matrix with one row for each of the 68 tissue samples, one column for each of the 380 identified peak registers, and entries representing log-transformed normalized peak areas.

3.5 Feature selection and classification

We assess the performance of both feature selection and classification strategies discussed previously. For the first strategy, we sequentially add features based on their FDR ranking and continue in this manner until we no longer improve our generalization performance with a linear SVM classifier, in terms of LOOCV error. For the second strategy, we vary the value of λ over a range of values and select one that optimizes our generalization performance based on sparse multinomial logistic regression.

Table 1 summarizes the LOOCV classification accuracy of the two different classifiers with the subset of features selected to minimize LOOCV error. The performance of SMLR can be seen to be noticeably better than SVM on this particular data. Table 2 lists the m/z values of the features selected by both methods, ordered by rank. The top feature in each case is the same, indicating that it provides good discriminatory information between the two classes when used alone.

The plot on the left of Fig. 4 shows a grayscale representation of the features selected by FDR+SVM. The X-axis indexes the selected features and the Y-axis indexes the 68 samples, sorted by class. Each bar represents the log-transformed normalized peak area of the peak register in each sample. The features have clearly differential patterns of expression across the two different classes. The plot on the right shows the margin of each sample under SVM; blue open circles and red filled triangles represent normal and tumor classes respectively. Similar plots are shown for SMLR in Fig. 5. As can be seen from the plot on the left, SMLR selects features with non-redundant expression patterns, in contrast to FDR which picks features with redundant expression patterns; this is as expected. The plot on the right shows the probability of each sample being normal under SMLR.

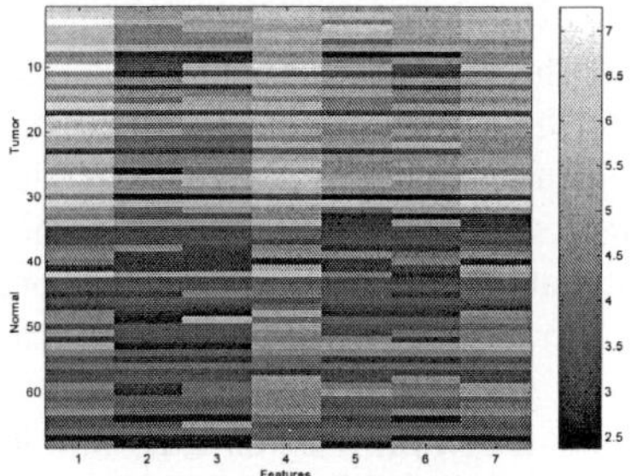

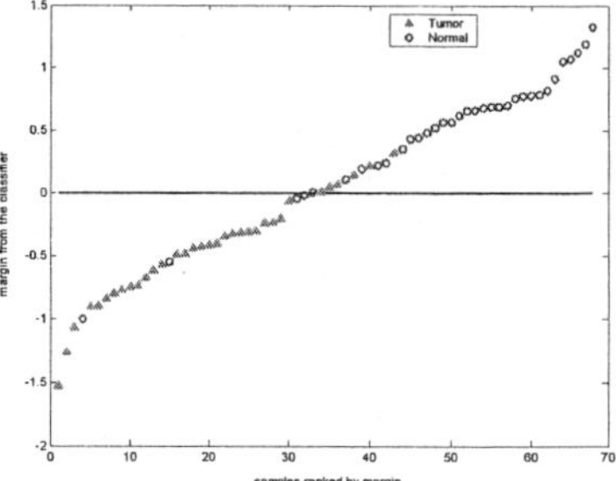

Figure 4. Features selected by FDR and classification by SVM using the selected features. The plot on the left depicts the differential patterns of the selected features across the samples belonging to the two different classes. The plot on the right shows the margin of each sample under SVM.

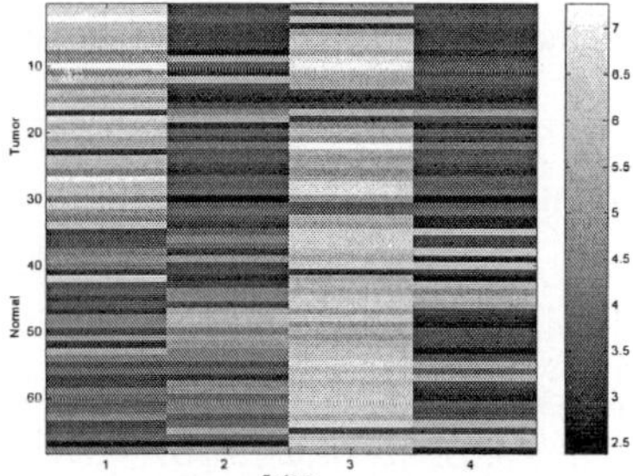

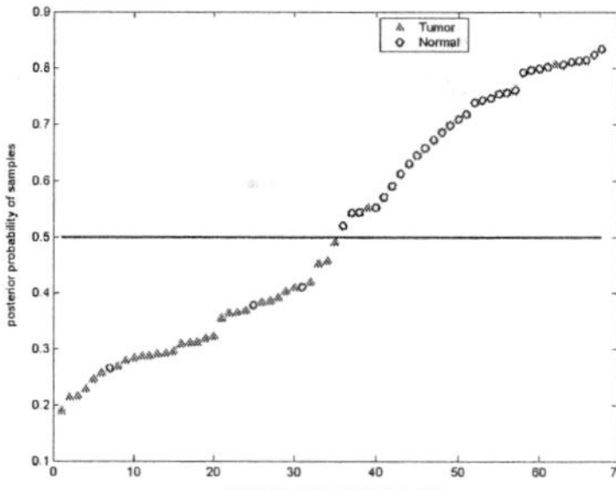

Figure 5. Feature selection and classification by SMLR. The plot on the left depicts the differential patterns of the selected features across the samples belonging to the two different classes. The plot on the right shows the probability of each sample being normal under SMLR.

4 Discussion

Even assuming that the data analysis pipeline presented here is successful in finding putatively meaningful discriminatory features, as it was in this case, isolation and identification of the proteins or peptides corresponding to these features must still be undertaken. In the case of this particular dataset, the two proteins found at m/z values 12386 and 17932 have already been identified by immunohistochemical analysis as macrophage migration inhibitory factor and cyclophilin A, respectively;[10] the prognostic value of these markers is currently under investigation. Proteins corresponding to certain other features are also being identified.

A number of improvements could be made upon the methods presented herein. Our methods ignore that ions can be multiply-charged, leading to the presence of 'harmonic' peaks in the spectra. For example, peaks at m/z values 12386 and 6206 were both found by FDR to be overexpressed in the tumor tissue samples, but these may be singly- and doubly-charged variants of the same protein. In addition, our methods also ignore isotope variants of a molecule[11] which, although chemically identical, can result in the presence of 'sister' peaks near the peak of the predominant isotope. Because SMLR selects non-redundant features, it would seem less prone to this sort of problem, but incorporation of this information into the peak identification and quantification step can only help.

Our peak registration algorithm aligns peaks based on only on their m/z locations, and ignores information about the heights, widths, or even shapes of peaks. To incorporate this information, alignment algorithms similar to multiple sequence alignment or dynamic time warping (DTW) for multiple alignment of speech signals could be used.

Finally, in the hierarchical strategy we proposed, the output of one step is piped as the input to the next step sequentially. Instead, a more unified model that combines different steps of analysis could provide a framework to share information across different steps, possibly leading to better results than is possible with our sequential approach.

[Larger versions of all figures are available from http://www.cs.duke.edu/~amink/]

References

1. G. L. J. Wright. *Expert Rev Mol Diagn*, 2:549–563, 2002.
2. S. A. Schwartz, M. L. Reyzer, and R. M. Caprioli. *J Mass Spectrom*, 38:699–708, 2003.
3. S. Hyunjin and M. K. Markey. *J Biomedical Informatics*, 38, 2005.
4. J. Listgarten and A. Emili. *Molecular and Cellular Proteomics*, 4:419–434, 2005.
5. J. Listgarten, R. M. Neal, S. T. Roweis, and A. Emili. In *Advances in Neural Information Processing Systems*, volume 17, Cambridge, MA, 2005. MIT Press.
6. R. Tibshirani, T. Hastiey, N. Balasubramanian, S. Soltys, G. Shi, A. Koong, and Q. T. Le. *Bioinformatics*, 2004.
7. M. Tipping. *J Machine Learning Research*, 1:211–244, 2001.
8. M. Figueiredo and A. Jain. In *Computer Vision and Pattern Recognition*, 2001.
9. B. Krishnapuram, M. Figueiredo, A. J. Hartemink, and L. Carin. *IEEE Transactions on Pattern Analysis and Machine Intelligence*, 27:957–968, 2005.
10. M. J. Campa, M. Z. Wang, B. Howard, M. C. Fitzgerald, and E. F. J. Patz. *Cancer Res*, 63:1652–1656, 2003.
11. K. R. Coombes, J. M. Koomen, J. S. Baggerly, K. A.and Morris, and R. Kobayashi. Technical report, Department of Biostatistics and Applied Mathematics, UT M.D. Anderson Cancer Center, 2004.

GAUSSIAN MIXTURE MODELING OF α-HELIX SUBCLASSES: STRUCTURE AND SEQUENCE VARIATIONS

ASHISH. V. TENDULKAR*

Kanwal Rekhi School of Information Technology,
Indian Institute of Technology Bombay,
Powai, Mumbai-400 076, India.
E-mail: ashish@it.iitb.ac.in

BABATUNDE OGUNNAIKE

Department of Chemical Engg.,
University of Delaware,
Newark, DE 19716.
Email: ogunnaik@che.udel.edu

PRAMOD P. WANGIKAR†

Department of Chemical Engg.,
Indian Institute of Technology Bombay,
Powai, Mumbai-400 076, India.
E-mail: pramodw@iitb.ac.in

Classification of helical structures and identification of class specific sequence features is of interest for protein structure modeling. We use geometric invariant based method to first select helix-like local conformations. These conformations are mapped in a principal component space and subjected to Gaussian mixture modeling. The largest Gaussian corresponds to the regular α-helix. Kinked helix and curved helix appear as a separate gaussians. Class conditional, position specific amino acid propensity analysis reveals striking difference among the three classes. In regular helix, proline propensity is significant only in the beginning and low in the rest of the region regardless of length of the helix. In kinked helix, the proline propensity has a sharp peak at the helix center, while in the curved helix, the proline propensity has a broad peak in the middle region.

*presenting author
†corresponding author

1. Introduction

α-Helices are the most important structural elements in globular proteins. Based on handedness of helix turn, they have been classified as right handed α-helices and left handed α-helices[1]. The right handed helices are most commonly occurring helices in globular proteins. Although, α-helices are certainly the most regular structural building blocks, they show significant imperfections[1,2,3]. The perturbation is caused by a variety of reasons such as occurrence of proline residue[4] in the middle producing a kink. Based on structural and geometric features, the helices are categorized as linear, curved and kinked α-helices[4,5].

It is well known that the structural and geometric differences give rise to different subclasses of α-helix. Moreover, the structural features of a helix is encoded in its sequence composition. Propensities of different amino acids for formation of a specific secondary structure is a basis of many secondary structure prediction methods[6]. Kumar and Bansal[5] have reported a strong correlation between propensities of individual amino acids and helix length, geometry and location on protein globe. Doig and co-workers[7] have reported amino acid propensities at N and C terminus of α-helices. Engel and DeGrado[8] have reported very strong position dependent propensities of different amino acids throughout the length of a helix. These methods first extract helices from protein dataset based on secondary structure assignment[9] or steriochemical punctuation marks[10]. These methods have limitations in terms of accuracy in determining beginning and endpoints of the helix.

In our earlier work, we have reported classification of overlapping octapeptide substructures in globular proteins[11]. To obtain more fine-grained classification of α-helices, we have combined all the octapeptides classified as helices and modeled the structure space as a mixture of Gaussian. The resulting clusters represents various helix subclasses. The overlapping octapeptides in a particular subclass are merged to form longer length peptides. Thus, we construct our dataset of helices of varying length. Thus, our method provides structure based unbiased way of extracting helices. The analysis of amino acid propensities for different subclasses of helices reveals that the amino acid propensities in helices are strongly dependent on the subclass and position in the helix structure.

2. Method

2.1. *Selection of Helices*

We had performed k-means clustering(k=150) on overlapping octapeptides local conformations drawn from ASTRAL_95 dataset, version(1.67)[12]. The octapeptides were described using a set of 29 non-redundant geometric invariants[13,2] such as edge, perimeter, volume, area of triangle etc.[11]. The geometric invariants were directly computed from x, y, z coordinates of C_α atom. Thus, we have approximated backbone geometry with C_α geometry[14]. Consensus secondary structure was calculated for the clusters resulting from K-means application[11]. To obtain a more finer level classification of α-helices, we have combined all the octapeptide local conformations, which have been classified in the clusters, which have consensus secondary structure as HHHHHHHH.

2.2. *Gaussian Mixture Modeling of α-helix Structure Space*

Each geometric invariant was normalized to mean-centric, unity standard deviation values[15]. Principal component analysis[15] was performed on the standardized geometric invariants of octapeptides helical structures. We have chosen first s principal components to represent an octapeptide helix structure. Thus, n octapeptide helical structures are represented as a vector in space spanned by the first s principal components.

Let $y = \{\vec{x}_1, \vec{x}_2, ..., \vec{x}_n\}$ be the set of n helix octapeptides. We model the data as a mixture of k Gaussian,

$$f(y) = \sum_{i=1}^{k} \phi_i f_i(y, \theta_i) \tag{1}$$

where, k is the number of components in the mixture, ϕ_i is the probability that a given helix octapeptides will come from ith component, also called as ith mixing proportion, θ_i is the vector of parameters describing the ith component. Since we're using Gaussian mixture model, θ_i consists of the mean vector, μ_i and covariance matrix, Σ_i. Thus ith component in the mixture is characterized by ϕ_i, μ_i and Σ_i. To estimate the parameters of the Gaussian mixture model, we used Expectation Maximization(EM) algorithm, fastmix[16], which automatically determines optimal number of components k required to maximize the expectation based on Bayesian information content[16].

Each helix octapeptide $\vec{x}$ is scored against each Gaussian i in the mixture using the following formula:

$$ln\ p(i|\vec{x}) = (-\frac{1}{2}\ ln\ |\Sigma_i| - \frac{1}{2}(\vec{x} - \vec{\mu_i})'\ \Sigma_i^{-1}\ (\vec{x} - \vec{\mu_i}) + ln\ \phi_i) \qquad (2)$$

The score signifies amount of influence each Gaussian exerts on the helix octapeptide. The octapeptides is assigned to the highest scoring Gaussian.

2.3. *Visualization of Clusters*

The clusters are visualized in form of bivariate distribution of the first two principal components conditioned on third and fourth principal component and marginal on the subsequent principal components. The detail procedure for obtaining such conditional bivariate distribution is given in [11]. The third and fourth principal components are divided into 4x4 equidensity grid. For each cell in the grid, we calculate bivariate density for the first two principal components. The bivariate distribution shows multiple peaks. Each peak has one to one corresponds to a mixture components and hence with clusters. Each peak is labeled with the cluster number and cartoon representing the structure of the helix octapeptide closest to its mean.

2.4. *Concatenation of octapeptides to form longer helices*

The octapeptide helices are assigned to different subclasses based on Gaussian mixture modeling. We carried out analysis of structural subclass assigned to the neighboring octapeptides i and $i+1$, which share an overlap of consecutive seven residues at the end of i and at the start of $i+1$. Suppose that the length of ith octapeptides is l. The neighboring octapeptides, assigned to the identical subclass, are merged to from helix of length $l+1$. Such kind of neighboring helices can be combined to form longer helices as they share similar geometric and structural properties.

2.5. *Amino Acid Propensity Analysis*

We analyzed position specific propensity of different amino acids for different subclasses. The position specific propensity is defined as [8],

$$P_{ij} = \frac{f_{ij}}{f_i} = \frac{n_{ij}/\sum_i n_{ij}}{N_i/\sum_i N_i} \qquad (3)$$

where, f_{ij} is the fraction of ith amino acid at jth helix position, n_{ij} is the number of ith amino acid at jth helix position, f_i is the fraction of ith amino acid over entire protein structure dataset and N_i is the number of ith amino acid over entire protein structure dataset.

3. Results

Approximately 0.4 million helices were selected from 1.7 million overlapping octapeptides drawn from ASTRAL_95 dataset(version 1.67)[12] based on the criteria defined in subsection 2.1. Principal component analysis of the dataset reveals that the first six principal components explain about 80% of variance in the data. The details about interpretation of principal components based on contributions from individual geometric invariants are documented in our earlier work[11]

3.1. *Visualization of Clusters*

Due to space constraints, we have shown the two most interesting cells from 4x4 equidensity grid in Figure 1. The fig. 1a shows two distinct and well separated peaks corresponding to gaussians 1 and 6. The cartoons of representative structures suggest that the Gaussian 1 is regular α-helix, whereas Gaussian 6 is kinked helix. The peak for regular α-helix is sharp and tall, whereas the peak corresponding to kinked helix is broad and short in height. The sharp nature of regular α-helix peak signifies highly regular nature of the class with lesser tolerance towards structural variations. The tall height of the peak signifies that the regular α-helix class is the most dominant class among helix subclasses. The broad nature of kinked helix peak suggests tolerance for structural variations. The short height of the peak denotes lesser probability of occurrence of kinked helices regular α-helices

The fig. 1b shows three distinct peaks corresponding to gaussians 1, 4 and 9. The cartoons denotes the structural differences between the classes corresponding to the peak. It further suggests that the peak for Gaussian 1 represents regular α-helix class, the peak for Gaussian 4 denotes a extended helix and the peak for Gaussian 9 denotes a helix class with distortion in the middle. The height of the corresponding peaks suggests that the regular α-helices have the highest probability of occurrence than the remaining two subclasses.

It is interesting to note that the helix structure space is sparse as shown by a lot of open area in both the bivariate plots in fig. 1a and fig. 1b.

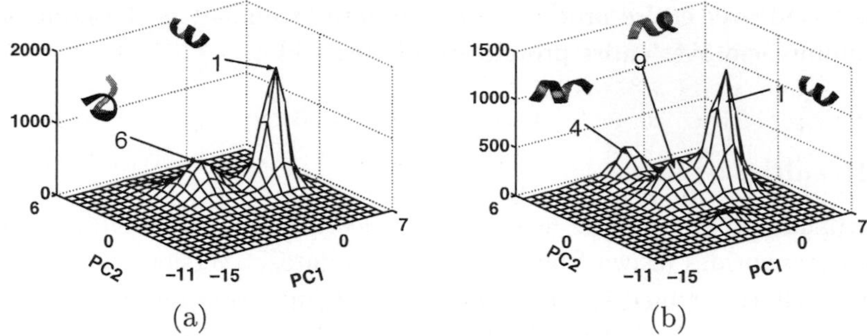

Figure 1. Representative conditional bivariate probability distribution of local helix conformations(A, B). Each panel shows a bivariate distribution on the first two principal component values, conditional on a specific range of the third and fourth principal components and marginal on the subsequent principal components values. (A) A bivariate distribution on the first two principal component values, conditional on the following values of the third and fourth principal components, $-\infty \leq PC_3 < -0.84$ and $-\infty \leq PC_4 < -0.70$. (B) A bivariate distribution on the first two principal component values, conditional on the following values of the third and fourth principal components, $+0.68 < PC_3 \leq +\infty$ and $+0.72 < PC_4 \leq +\infty$

3.2. *Properties of Gaussian Mixtures*

The summary of mixture parameters estimated by Expectation maximization algorithm[16] are provided in Table 1. Total of eleven gaussians have been detected with skewed mixing proportions. The gaussians are arranged in their descending values of mixing proportions. The most heavily represented Gaussian has mixing proportion of 76%. The lowest mixing proportion is 1%. The helix octapeptides are assigned to appropriate gaussians.

The first Gaussian represent regular right handed α helix subclass. Its centroid occupies positive value on the first principal component, a small positive value on fourth one, and small negative values on second, third, fifth and sixth principal components(Table 1). These values are in agreement with the interpretation of contributions of individual geometric invariants to the principal components. The covariance matrix of the first cluster is the tightest among all the clusters, suggesting highly regular nature of the helix subclass(Table 1).

The Gaussian number 6 and 10 have substantial negative values for their means along the first principal component. The Gaussian number 6 corresponds to kinked helix and number 10 corresponds to curved helix. The curved helix shows smaller length between the first and the last residue of helix octapeptide(Table 2). The decrease in length is accompanied by increase in the area of triangle formed by first, fifth and the last residue(Table 2). The kinked helices have mixing proportion of 2%, whereas the curved helix has a mixing proportion of 1%.

Comparison of covariance matrices of regular, kinked and curved helix reveals that regular helix has the tightest covariance matrix, kinked helix has moderate covariance matrix, whereas the curved helix had the largest covariance matrix.

Table 1. Gaussian mixture characteristics of the helix subclasses

Characteristics	Regular α-helix	Kinked Helix	Curved Helix
ϕ_i	0.76	0.02	0.01
μ_1	1.12	-5.94	-7.71
μ_2	-0.06	2.02	-5.33
μ_3	-0.14	-3.03	-4.27
μ_4	0.04	-4.82	2.51
μ_5	-0.03	-1.29	1.39
μ_6	-0.04	-2.24	-0.41
σ_{11}	1.20	4.28	7.23
σ_{22}	0.75	4.93	7.70
σ_{33}	0.93	2.94	6.33
σ_{44}	0.85	3.91	5.22
σ_{55}	0.73	2.82	7.04
σ_{66}	0.63	2.16	9.63

Note: a. $\sum_i^k \phi_i = 1$. ϕ_i represent probability of a helix octapeptides belonging to mixture i. The total number of helix subclasses, $k = 11$. The three most important helix subclasses are described here.
b. μ_{ij} represents mean of mixture i on jth principal component.
c. σ_{ii} represents variance of mixture i along ith principal component. The covariance matrix for each mixture contains non-zero values only along its diagonal.

3.3. *Finer structural differences between distinct subclasses*

The analysis reveals three distinct helix subclasses: regular, kinked and curved. We analyzed structural differences between these distinct subclasses based on structural descriptors used for representing the helix octapeptides. The structural differences between these subclasses have been summarized in Table 2.

The structural descriptors characterize differences between various subclasses. The distance between i and $i+3$ residues($d_{i,i+3}$) characterizes regularity of helix. The regular α-helix has the least $d_{i,i+3}$ with the least

Table 2. Mean and standard deviation of $d_{i,i+3}$, d_{18}, $Vol_{i,i+1,i+2,i+3}$, and $Area_{158}$ for vastly differing subclasses.

Structure Descriptors	Regular α-helix	Kinked Helix	Curved Helix
$d_{i,i+3}$	$5.15 +/- 0.20$	$5.51 +/- 0.45$	$5.64 +/- 0.57$
d_{18}	$10.64 +/- 0.33$	$11.38 +/- 0.74$	$8.64 +/- 1.59$
$vol_{i,i+1,i+2,i+3}$	$6.99 +/- 0.57$	$5.56 +/- 2.61$	$5.31 +/- 3.00$
$area_{158}$	$10.40 +/- 1.49$	$18.53 +/- 4.16$	$11.38 +/- 5.93$

Note: (i) $d_{i,i+3}$ denotes distance between i and $i + 3$ residues.
(ii) d_{18} denotes distance between first and the last residue.
(iii) $vol_{i,i+1,i+2,i+3}$ denotes volume of tetrahedron formed by $i, i+1, i+2$and $i+3$ residues.
(iv) $area_{158}$ denotes area of triangle formed by first, fifth and eighth residue.

standard deviation signifying regular nature of helices. The kinked helix show longer average $d_{i,i+3}$ with more deviation. The longest average $d_{i,i+3}$ is assumed by the curved helix. The larger values of mean and standard deviation of $d_{i,i+3}$ denotes significant departure from the regularity.

The distance between end to end residues d_{18} of helices characterizes extended structure. The kinked helix is the most extended structures among the distinct subclasses. The regular α-helix has moderate end to end length. The decrease in the end to end distance either denotes shrink or a curve in the helix octapeptide. The area of triangle($Area_{158}$) formed by first, fifth and the last residue differentiate between helices with a bend , kinked and regular ones when coupled with d_{18}. With reference to regular α-helix, the decrease in d_{18} and increase in $Area_{158}$ denotes a curve in the middle of the octapeptide. The increase in both d_{18} and $Area_{158}$ denotes a kinked helix.

The sign of $Vol_{i,i+1,i+2,i+3}$ differentiate between left handed and right handed systems[11]. The positive values of volumes in all three distinct subclasses means that all the helix subclasses make a right handed system.

3.4. *Concatenation of octapeptides to form longer helices*

For the regular α-helix subclass, the majority of neighboring octapeptides lies in the same subclass endorsing the strong regular nature of the subclass. This leads to formation of variable length helices. The distribution of length of regular helices shows a wide spectrum of values ranging from 8 to 82, with the maximum number(6909) of regular α-helix of length 8, while a singleton α-helix of length 82. The distribution of length of regular α-helix is shown in Fig. 2. The distribution shows roughly exponential trend. The maximum helix length in kinked and curved helix is 9.

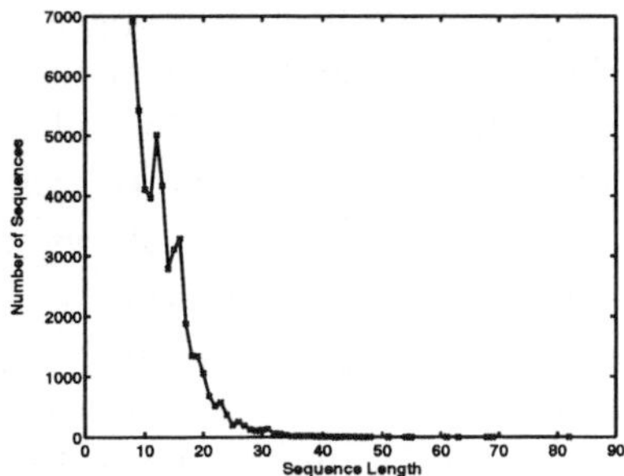

Figure 2. Distribution of length of regular α-helices in cluster 1 ($8 \leq length \leq 82$)

3.5. *Analysis of amino acid propensities in helix subclasses*

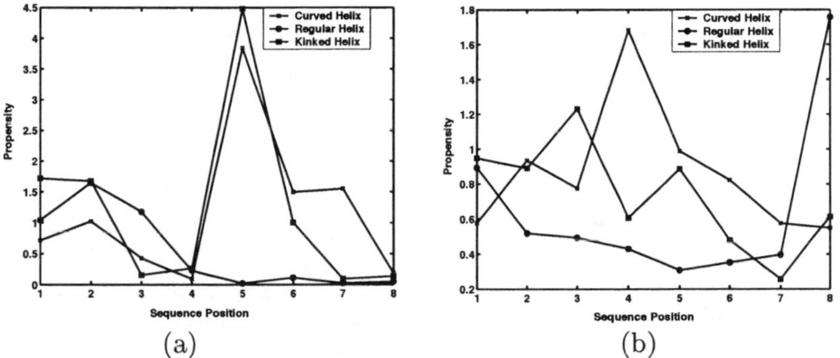

Figure 3. Propensity of (a) Proline and (b) Glycine in regular α-helix, kinked helix and curved helix.

The analysis of amino acid propensities at different positions in different helix subclasses was carried out on the helices having length 8. The propensities of an individual amino acids at various positions in different subclasses were plotted. We have shown propensity graphs for two representative amino acids, glycine and proline, in fig 3.

Proline is considered as a prominent helix breaker[4]. The propensity graph of proline (Fig. 3a) shows different propensity numbers at different positions of different helix classes. The proline has significant propensity up to third position in regular α-helix. The highest proline propensity was observed in kinked helix at fifth position, where the kink occur. The kinked helix also shows significant proline propensity at second position. The curved helix has highest proline propensity at fifth position. It also shows significant proline propensities at sixth and seventh position also. Thus, proline propensity is higher in the curving region of curved helix.

The propensity graph of glycine (Fig. 3b) shows that the propensity of glycine shows different trends based on position and helix subclass. The regular α-helices shows significant glycine propensity only at the end position of the helix. The other position shows moderate glycine propensities. Kinked helix shows significant glycine propensity at third position and moderate propensities at the remaining positions. The curved helix shows significant glycine propensities at second, fourth, fifth and sixth position position with highest propensity at fourth position.

3.6. *Analysis of lengthwise amino acid propensities for regular α helix*

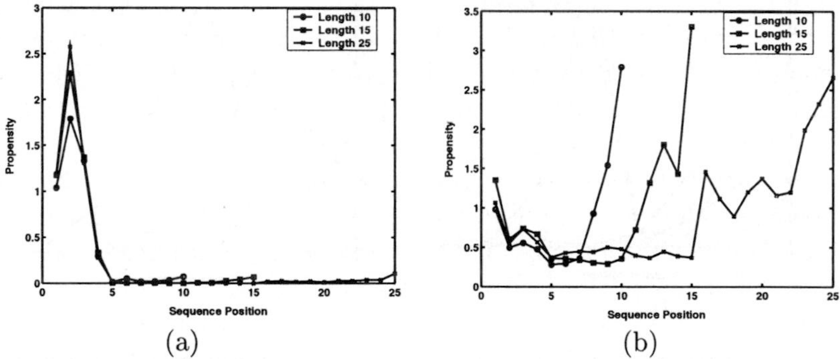

(a) (b)

Figure 4. Propensity of (a) Proline and (b) Glycine in regular α-helix of varying lengths

The lengthwise propensity analysis was done for the regular α-helix class. We have selected three groups of lengths: 10, 15 and 25 for analysis. All the sequences having length ≤ 10 were merged in first group, the sequences having length between 11 and 15 were merged in secondary group, while the sequences having length between 16 and 25 are merged in the third group. The lengthwise propensity analysis is carried out on all the three groups for all 20 amino acids. Here, We have included two represent propensity graphs of proline and glycine(Fig 4).

The propensity of proline is shown in fig 4a. and the propensity of glycine is shown in fig 4b. Regardless of the total length of helix, we have observed almost similar trends for both proline and glycine. Proline is having significant propensity up to first three positions in all three groups. For the rest of the positions the propensity is almost zero. Glycine is having significant propensity in the beginning of helix, then low propensity in the middle region and then high propensity at towards the end. The same trend is observed in all three groups. Since we have combined helices having length between 11 to 15 in group of 15, glycine propensity starts upward trend from 11th position for group of length 15, the same is true for the other two groups.

4. Discussion

We have reported a novel method for fine grain classification of helical substructures into its subclasses using geometric invariants and Gaussian mixture modeling. We also provide detailed explanation about roles of various structure descriptors in differentiating between various helix subclasses. It is interesting to note that the individual geometric invariants are capable of differentiating one or other feature of helix geometries(Table 2). The linear combination of these individual descriptors differentiates between various subclasses more efficiently.

Gaussian mixture modeling of helix local conformation space provides a formal framework for analyzing geometry of theoretical helix structures or newly formed helix structure. The modeling of this sort provides a formal method for detecting outliers in the data as well as subclass of a particular helix structure. The Bayesian information content based expectation maximization algorithm[16] ensures that the right number of mixture components are selected to model the Gaussian mixture accurately. The mixing proportion ϕ_i assigned to different subclasses matches well with literature reported mixing proportions(Table 1). The regular α-helix subclass has been assigned a mixing proportion of 76%, which is in accordance with 74% mixing proportion reported by Kumar and Bansal[5]. The visualization of helix local conformation space in form of conditional bivariate distribution plots(Fig 1) helps in getting quick idea about separation between various subclasses and differences in their geometry.

The merging of neighboring octapeptides having identical subclass provides more accurate and unbiased approach for extracting helices from protein structures. This is a fundamental shift from the literature reported methods, which depends on secondary structure assignment[5] or helix breaking signals in protein sequence[8]. The method provides structure based method for extracting helices more accurately.

Analysis of propensities of amino acids for different classes of helix reveals different position specific trends. It implies that the propensity of amino acid is also dependent on the class of helix. The analysis of propensity for different length of helices reveals similar trends regardless of length of helix.

Class conditional position specific analysis of amino acid propensities provides vital clues in better understanding sequence-structure relationship in various subclasses of α-helices, leading to better prediction of helix subclass in protein structure prediction. The results presented in the paper are also useful for designing artificial helices from a specific subclass.

References

1. J. Richardson, *Adv. Prot. Chem.* **34**, 167 (1981).
2. A. Tendulkar, A. Joshi, M. Sohono, P. Wagikar, *J. Mol. Biol.* **338**, 611 (2004).
3. E. Emberly, R. Mukhopadhyay, N. Wingrren, C. Tang, *J. Mol. Biol.* **327**, 229 (2003).
4. D. Barlow, J. Thornton, *J. Mol. Biol.* **201**, 601 (1988).
5. S. Kumar, J. Bansal, *Proteins: Struct. Funct. Genet.* **31**, 460 (1998).
6. S. Dasgupta, J. Bell, *Int. J. Pept. Protein Res.* **41**, 499 (1993).
7. A. Doig, R. Baldwin, *Protein Sci.* **4**, 1325 (1995).
8. D. Engel, W. DeGrado, *J. Mol. Biol.* **337**, 1195 (2004).
9. W. Kabsch, C. Sander, *Biopolymers.* **22**, 2577 (1983).
10. K. Gunasekaran, H. Nagrajaram, C. Ramkrishnan, P. Balaram, *J. Mol. Biol.* **275**, 917 (1998).
11. A. Tendulkar, M. Sohono, B. Ogunnaike, P. Wagikar, *Bioinformatics.* **21**, 18 (2005).
12. S. Brenner, P. Koehl, *Nucleic Acid Research.* **28**, 254 (2000).
13. A. Tendulkar, V. Samant, C. Mone, M. Sohono, P. Wagikar, *J. Mol. Biol.* **334**, 157 (2003).
14. T. Oldfield, R. Hubbard, *Proteins: Struct. Funct. Genet.* **18**, 324 (1994).
15. R. Johson, D. Wichern, *Prentice Hall of India.* (2003).
16. A. Moore, *Adv. Neural Information Processing Systems.* **11** (1999).

AN SVM SCORER FOR MORE SENSITIVE AND RELIABLE PEPTIDE IDENTIFICATION VIA TANDEM MASS SPECTROMETRY*

HAIPENG WANG[1,2†], YAN FU[1,2], RUIXIANG SUN[1], SIMIN HE[1], RONG ZENG[3], and WEN GAO[1,2]

[1]*Digital Technology Lab, Institute of Computing Technology, Chinese Academy of Sciences, Beijing 100080, China*

[2]*Graduate University of Chinese Academy of Sciences, Beijing 100039, China*

[3]*Research Center for Proteome Analysis, Key Lab of Proteomics, Institute of Biochemistry and Cell Biology, Shanghai Institutes for Biological Sciences, Chinese Academy of Sciences, Shanghai 200031, China*

Tandem mass spectrometry (MS/MS) has become increasingly important and indispensable in high-throughput proteomics for identifying complex protein mixtures. Database searching is the standard method to accomplish this purpose. A key sub-routine, peptide identification, is used to generate a list of candidate peptides from a protein database according to an experimental MS/MS spectrum, and then validate these candidate peptides for protein identification. Although currently there are many algorithms for peptide identification, most of them either lack an effective validation module or only validate the first-ranked peptide, thus leading to a low identification reliability or sensitivity. This paper proposes a new algorithm, named pepReap, to overcome the above drawbacks. It consists of a two-layered scoring scheme based on machine learning. The first layer is a rough scoring function which uses some simple and heuristic factors to measure the degree of the matches between an experimental MS/MS spectrum and the candidate peptides; thus a ranked list of candidate peptides is generated at a relatively low computational cost. The second layer is a fine scoring function which re-ranks the candidate peptides generated in the first layer and determines which one among them is the true positive. The fine scoring function was designed based on support vector machines (SVMs) using more comprehensive factors, such as the correlations between ions, the mass matching errors of fragment and peptide ions, *etc*. Consequently, the SVM classifier serves as not only a scorer but also a validation module. Experimental comparison with the popular SEQUEST algorithm coupled with threshold validation criteria on a reported dataset demonstrates that the pepReap algorithm achieves higher performance in terms of identification sensitivity with comparable precision.

1. Introduction

The essential mission of proteomics is to identify and quantify all the levels of proteins found in a cell or tissue under various physiological conditions [1]. Tandem mass spectrometry (MS/MS), which can measure the mass-to-charge ratios (*m/z*) of ionized molecules, has become increasingly important and indis-

* This work is supported by the National Key Basic R&D Program (Grant No. 2002CB713807) and the National Key Technologies R&D Program (Grant No. 2004BA711A21) of China.

† To whom correspondence should be addressed. E-mail: hpwang@jdl.ac.cn.

pensable for identifying complex protein mixtures in high-throughput proteomics experiments [2–4].

In a typical "bottom-up" experiment, protein mixtures are directly digested with a site-specific protease, usually the trypsin, into complex peptide mixtures which are subsequently separated by liquid chromatography (LC). The separated peptides eluted from LC are then ionized with one or more units of charges (to form precursor ions), selected according to their *m/z* values, and analyzed through fragmentation by MS/MS. In this process, hundreds of thousands of MS/MS spectra are produced which are then computationally interpreted to generate their candidate peptide sequences. Finally the candidate peptides are validated and the correct ones are grouped to identify the proteins from which the peptides derived. The peptide identification including peptide scoring and subsequent validation is a critical step in the process of protein identification [5,6].

Aiming at the drawbacks of existing algorithms for peptide identification, we developed a more robust algorithm, pepReap, which consists of a two-layered scoring scheme based on support vector machines (SVMs) using some elaborated features characterizing the matches between peptides and MS/MS spectra, to obtain a positive or negative score for each candidate peptide, explicitly distinguishing correct matches from incorrect matches dispensing with setting significant thresholds.

2. Background and Related Work

2.1. *Peptide MS/MS Spectrum*

The basic molecular building blocks of proteins are amino acids which are differentiated from each other by the side chain *R* (shown in Figure 1(a)). A protein or peptide is a chain that consists of amino acid residues linked together by peptide bonds formed by condensation reactions (shown in Figure 1(b)). For MS/MS, peptides are the products of enzymatic digestion of proteins.

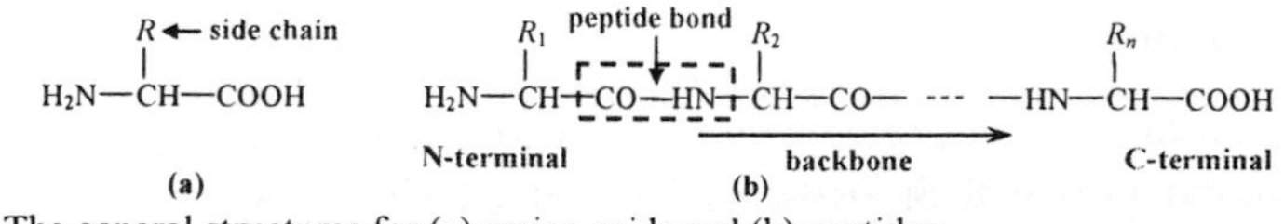

R ← side chain
$H_2N—CH—COOH$
(a)

R_1 peptide bond R_2 R_n
$H_2N—CH—CO—HN—CH—CO—$ ··· $—HN—CH—COOH$
N-terminal backbone C-terminal
(b)

Figure 1. The general structures for (a) amino acids and (b) peptides.

In a tandem mass spectrometer, precursor ions within a given range around a specific *m/z* value are selected and subjected to fragmentation by collision-induced dissociation (CID) resulting in various types of fragment ions. Fragment ions are measured to obtain a number of spectral peaks each comprising an *m/z* and an intensity value. Peaks plus the *m/z* value and charge state of the precursor ion constitutes a peptide MS/MS spectrum which normally corresponds to a unique peptide. Besides the peaks from the peptide to be identified, there are also many noisy peaks brought by chemical contaminants or electronic fluctuations.

Common fragmentation patterns of parent ions and nomenclature for fragment ions are shown in Figure 2. According to the fragmentation position relative to the peptide bond along the backbone and the terminal (N or C-terminal) where the charge(s) is (are) retained, fragment ions are classified as *a*-, *b*-, *c*-, *x*-, *y*-, or *z*-ions. The pairs of (*a*, *x*), (*b*, *y*) and (*c*, *z*) are complementary ion types. Fragment ions can be singly or multiply charged and possibly lose a neutral water (H_2O) or ammonia (NH_3).

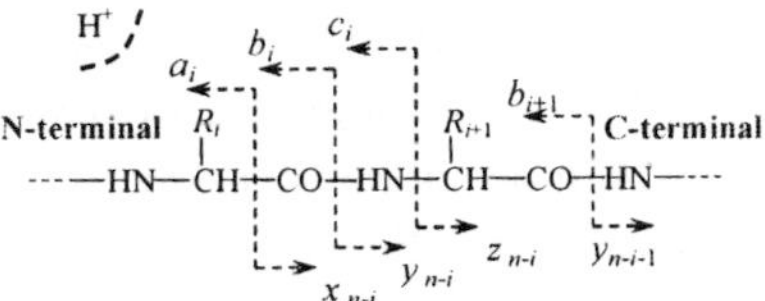

Figure 2. Fragmentation patterns of parent ions and nomenclature for various types of fragment ions.

The notations used for describing fragments are listed in Table 1. Based on the notations, $T_{\text{cnsc}} = \{t_i, t_{i+1}, \cdots, t_{i+k}\}$ denotes a set of consecutive ions, $T_{\text{cmpl}} = \{t_i, \overline{t_{n-i}}\}$ a pair of complementary ions, and $T_{\text{homo}} = \{t_i, t_i^{++}, t_i^{0}, t_i^{*}\}$ a set of homologous ions, with t denoting any ion type and $0 < k < (n\text{-}i)$. Ions of *a*, *b*, *y* and their homologues are dominant fragments observed in MS/MS spectra.

Table 1. Notations for fragments.

Notation	Meaning
i $(0 < i < n)$	Subscript denoting the cleavage site, i.e., the number of residues contained in the fragment; *n* is the number of residues in the precursor, i.e., the peptide length
-	Overline denoting the complement of an ion type, e.g., $\overline{b}$ is *y*
0	Superscript indicating a neutral loss of water
*	Superscript indicating a neutral loss of ammonia
++	Superscript denoting a double-charge state (single-charge state by default)

2.2. *Approaches to Peptide Identification from MS/MS Spectra*

There are mainly three approaches to peptide identification from MS/MS spectra: *de novo* sequencing [7–9], sequence tagging [10,11] and database searching [13–23].

De novo sequencing tries to infer the complete peptide sequence directly from the *m/z* differences between peaks in an MS/MS spectrum without any help of databases. It is capable of identifying the peptides which are not present in databases or suffer from unanticipated post-translational modifications (PTMs) or mutations; however, it is generally less tolerant of low-quality mass spectra than the database searching approach.

Sequence tagging yields one or more partial sequences (called sequence tag) by manual interpretation or *de novo* sequencing. Candidate peptides containing this sequence tag can be found by homologous sequence searching. The sequence tags can serve as an effective filter for candidate peptide generation in database searching especially when PTM identification is involved [12].

Database searching is the most widely used method in high-throughput proteomics experiments due to its sensitivity, rapidness, and tolerance for low-quality spectra. It compares an experimental MS/MS spectrum with theoretical ones predicted from the peptide sequences resulting from *in silico* digestion of proteins in databases, whereby the experimental spectrum is correlated by a scoring function with a ranked list of candidate peptides. The match of the highest score is normally regarded as the peptide corresponding to the spectrum. However, random matches often occur between theoretical fragments and noisy peaks, or between false isobaric theoretical fragments and signal peaks. In addition, unanticipated fragment ions or modifications can lead to missing matches. Therefore the best matching peptide may not be correct and inversely the correct peptide may not be the top-ranked one. Consequently the peptide identifications need to be carefully validated.

Scoring functions, validation method, and fragmentation model [24,25] are central and fundamental to all the above three approaches.

2.3. *Peptide Scoring and Validation Methods in Database Searching*

To measure the similarity between experimental and theoretical MS/MS spectra, two strategies are adopted based on the descriptive framework or probabilistic framework [13].

In the probabilistic strategy, a probability is calculated for the event that the match between a peptide and an experimental spectrum is completely random [18], or that a peptide actually generated the experimental spectrum [19–21]. The combination of the above two means leads to the likelihood-ratio score [22,23].

In the descriptive strategy, an experimental and a theoretical MS/MS spectrum are represented as vectors $\mathbf{S} = (s_1, s_2, \cdots, s_n)$ and $\mathbf{T} = (t_1, t_2, \cdots, t_n)$, respectively, where n denotes the number of predicted fragments, s_i and t_i are binary values or the observed and predicted intensity values of the ith fragment, respectively. The spectral dot product (SDP) between **S** and **T** serves as their similarity measure [15]. In SDP, correlative information among fragments is totally ignored. Many scoring algorithms are based on the SDP [14–16]. A representative of the descriptive strategy is the popular SEQUEST algorithm [14], in which consecutive ion pairs are considered in a preliminary scoring function and then the cross-correlation analysis is performed between the experimental and theoretical spectra. The KSDP algorithm adopted in the pFind software extends the SDP by using the kernel technique to comprehensively incorporate correlative information among ions [16,17].

In routine experiments, the threshold method is widely used to validate peptide identifications of SEQUEST. However, there are no uniform rules to set the cutoff values [3,4,26,27]. Some sophisticated validation algorithms based on probability (or pseudo-probability) [28–31] and machine learning [30,32,33] have been developed to improve the reliability of peptide identifications. Most algorithms take advantage of some outputs of SEQUEST, such as *XCorr*, ΔCn, *Sp*, *RSp*, and *Ions* [14]. Keller *et al.* discriminated between positive and negtive identifications according to a Gaussian and a gamma distribution [28]. MacCoss *et al.* proposed a scoring scheme which normalizes XCorr values to be independent of peptide length and then derived a confidence of an identification [29]. RScore combines the XCorr and matched intensity value to get a measurement of randomicity [31]. Anderson *et al.* and Baczek *et al.* performed classification tasks via machine learning approaches using outputs of SEQUEST and some additional factors (for example, the peak count, the ratios of matched peaks and matched intensities; isoelectric value, hydrophobicity, molecular weight, and charge state of peptides) [32,33]. These algorithms are remedies for the validation of SEQUEST results to some extent. However, they all focus on deciding whether the first-ranked peptide is correct while ignoring all other lower-ranked peptides that often include the correct one. Moreover, they do not fully exploit the features characterizing the quality of matches between MS/MS spectra and peptides.

In this paper, the scoring and validating processes are combined together by directly using an SVM classifier for scoring the peptide-spectrum matches based on a variety of matching features.

3. Methods

3.1. *The Framework of pepReap*

The pepReap algorithm comprises a rough scorer and a fine SVM scorer shown in Figure 3.

In the first step, a ranked list of candidate peptides is generated by the rough scoring function:

$$RS = (\sum_{i=1}^{N_{match}} I_i) \times N_{match} \Big/ L_{pep} \,, \tag{1}$$

where N_{match} is the number of the predicted fragment ions matching the peaks in the experimental spectrum, I_i is the intensity value of the ith matched peak and L_{pep} is the peptide length. Similar formulas have been adopted in SEQUEST [14] and pFind [16].

In the second step, an SVM-based scoring function gives a signed decision value according to the features constructed from the matching matrix (see section 3.2). The parameters of the SVM scorer are tuned on a training dataset by cross validation.

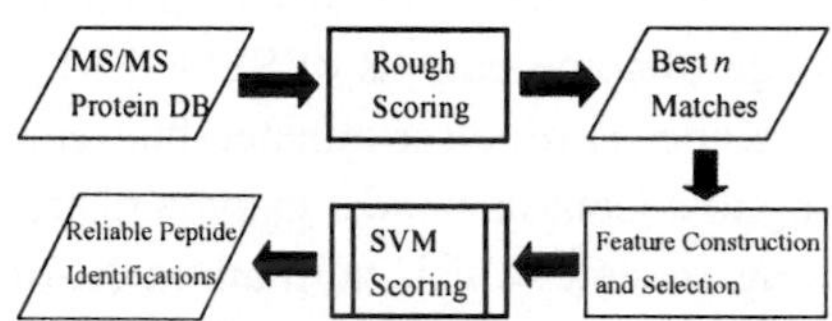

Figure 3. The framework of the pepReap algorithm.

3.2. *Matching Matrix*

To make data processing convenient, a matching matrix between a peptide and a spectrum is constructed, as shown in Figure 4. The matching matrix is an $m \times n$ array, where m denotes the number of different ion types under consideration, n is the length of a peptide, the column indexes $(1, 2, \cdots, n)$ represent the cleavage sites of a peptide, the row indexes $(a, b, \cdots, y)$ denote various ion types and the element p_{ti} ($t \in \{a, b, \cdots, y\}$) holds the information of the corresponding matched peak, or keeps null.

	1	2	$\cdots$	n
a	p_{a1}	p_{a2}	$\cdots$	p_{an}
b	p_{b1}	p_{b2}	$\cdots$	p_{bn}
$\vdots$	$\vdots$	$\vdots$	$\ddots$	$\vdots$
y	p_{y1}	p_{y2}	$\cdots$	p_{yn}

Figure 4. Matching matrix between a peptide and a spectrum.

3.3. *Support Vector Machines*

Support vector machines are developed by Vapnik and his coworkers based on the statistical learning theory [34]. The principle of structural risk minimization establishes the basis of the good generalization performance of SVMs. For a binary classification problem, the input to the SVM training algorithm is a set of n samples denoted as

$$D = \{(\mathbf{x}_1, y_1), (\mathbf{x}_2, y_2), \cdots, (\mathbf{x}_n, y_n)\}, \tag{2}$$

where $\mathbf{x}_i \in \mathbb{R}^d$ is the ith sample and $y_i \in \{-1, 1\}$ is its class label. The objective of SVMs is to find an optimal separating hyperplane that maximizes the margin between two classes in a high dimensional feature space into which the input vectors are mapped by a kernel function, as shown in Figure 5. The kernel function implicitly calculates a dot product in the feature space with all necessary computations performed in the input space. One advantage of it is that it can get linearly non-separable samples in the input space to be linearly separable in the feature space.

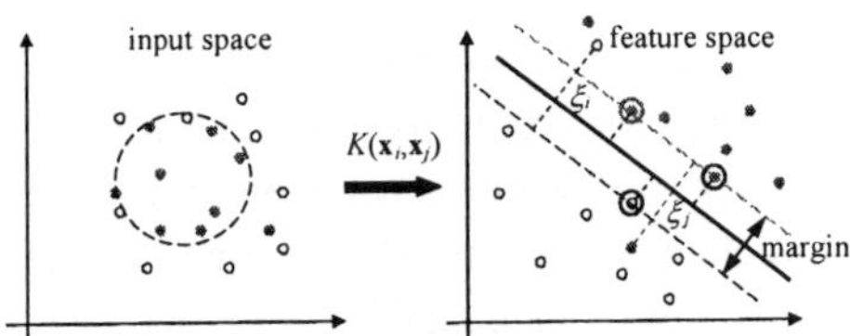

Figure 5. A linear separating hyperplane (the solid line in the right coordinates) in the feature space corresponding to a non-linear boundary (the dashed line in the left coordinates) in the input space. The data points in circles are support vectors (SVs).

The decision function of the SVM classifier is

$$f(\mathbf{x}) = \sum_{i \in \text{SVs}} y_i \alpha_i K(\mathbf{x}, \mathbf{x}_i) + b, \tag{3}$$

where the coefficients α_i are solved in the interval [0, C] by a convex quadratic programming. C is a tradeoff between maximizing the margin and minimizing the empirical risks and can be specified for positive and negative samples respectively in the case of unbalanced datasets. The radial basis function (RBF) kernel $K(\mathbf{x}_i, \mathbf{x}_j) = \exp(-\gamma \|\mathbf{x}_i - \mathbf{x}_j\|^2)$ is popular for practical use due to its approximate behaviors to other kernels under certain conditions and the less number of parameters to be tuned (only C and γ) [35].

3.4. *Performance Measurement*

For a binary classification problem, let tp, fp, tn and fn denote the number of true positives, false positives, true negatives and false negatives respectively. The Matthews correlation coefficient (MCC) [36] incorporates all four prediction indexes into a single statistic to measure the performance of classifiers:

$$MCC = \frac{(tp \times tn) - (fp \times fn)}{\sqrt{(tp + fp) \times (fp + tn) \times (tn + fn) \times (fn + tp)}}. \tag{4}$$

The MCC is a number in the interval [-1, 1], with 1 indicating completely correct classification, -1 indicating completely incorrect classification and 0 indicating no correlations between predictions and the true class labels. The MCC is superior to the accuracy which is defined as the proportion of correctly classified samples, especially when datasets are unbalanced, because the accuracy is dominated by the majority class and thus can be misleading. Therefore the MCC is employed in the cross-validation training process of SVMs.

Sensitivity ($SEN = tp / (tp + fn)$) and precision ($PRE = tp / (tp + fp)$) are used as the performance measures for comparing pepReap with SEQUEST.

3.5. *Features Characterizing the Quality of Matches*

In the step of SVM scoring, we constructed some features from the matching matrix to characterize the quality of matches between peptides and spectra. The features fall into seven categories: outputs of rough scoring, total matched intensities for various ion types, correlations bwtween matched ions, residue composition and properties of candidate peptides, statistics of cleavage sites, ratios and mass errors of matched peaks, and some other descriptive features such as missed proteolytic cleavage site and charge state.

The measures for consecutive, complementary, and homologous ions are

$$f_{\text{cnsc}} = \sum_{i \in \mathbf{T}_{\text{cnsc}}} \left(\prod_{t \in T_{\text{cnsc}}^{i}} I_t \right), \tag{5}$$

$$f_{\text{cmpl}} = \sum_{i \in \mathbf{T}_{\text{cmpl}}} \left(\sum_{t \in T_{\text{cmpl}}^{i}} I_t \times I_{\bar{t}} \right), \tag{6}$$

and

$$f_{\text{homo}} = \sum_{i \in \mathbf{T}_{\text{homo}}} \left(\prod_{t \in T_{\text{homo}}^{i}} I_t \right), \tag{7}$$

respectively, for all $I_t > 1$, where I_t is the intensity of a matched peak in the matching matrix, and $\mathbf{T}_j$ is a set comprising all T_j^i with $j \in \{\text{cnsc,cmpl,homo}\}$.

The average matching error is calculated by

$$f_{\text{ame}} = \sqrt{\frac{1}{M} \sum_{i=1}^{M} (emz_i - tmz_i)^2}, \tag{8}$$

where M is the number of matching fragments and emz_i and tmz_i are the observed and the theoretical m/z values of a predicted fragment respectively.

Fragmentation patterns of peptides are influenced not only by collision energy of the tandem mass spectrometer but by physical and chemical properties (gas-phase basicity, hydrophobicity, *etc.*) of amino acid composition, as is explained by the mobile proton model [37]. Therefore we believe that such features as the number of certain residues, the hydrophobicity of peptides, and the statistics of cleavage sites jointly provide better clues to describing the quality of matches. These statistics are normalized by the peptide length.

4. Experiments

4.1. *Datasets*

The ion trap MS/MS spectra reported in Ref. 38 were used for our experiments. These spectra were divided into two datasets, A and B, according to the different concentration of two mixtures of 18 purified proteins with known sequences which were digested by trypsin. All the spectra were searched using SEQUEST against a database combining the human proteins with the 18 proteins; then the peptide identification results were validated manually. Consequently, there are totally 2054 spectra identified correctly with their peptide terminus consistent with the substrate specificity of trypsin. In our experiments, 731 validated identifications in dataset B were used for training the SVM scorer in pepReap and 1323 validated identifications in dataset A were used for comparing pepReap with SEQUEST.

4.2. *Noise Reduction and Intensity Normalization*

To weaken the influence of noises on peptide identification and eliminate the diversities of total ion currents of different spectra, noise reduction and intensity normalization are performed. All the peaks lower than 2% intensity of the highest peak are removed and the intensities of the remaining peaks are normalized:

$$I_{Ni} = 100 \times \sqrt{I_{Oi} / \sum_j I_{Oj}} \tag{9}$$

where I_{Ni} is the normalized intensity and I_{Oi} is the original intensity.

4.3. *Protein Database and Search Parameters*

The protein database searched is a union of the SWISS-PROT protein database and the 18 known proteins. The search parameters used in pepReap and SEQUEST is: maximum number of missed cleavage sites: 2; tolerance of fragment ions: 1.0 Da; tolerance of precursor: 3.0 Da; ion types: b, b^{++}, b^{0}, y, y^{++}, y^{0}; and enzyme: trypsin.

4.4. *Feature Selection*

A total of 56 features are extracted from each match between a spectrum and its corresponding candidate peptides. These features are scaled into the interval [0,

1]. The best 20 features are selected by information gain ratio [39] on dataset B through cross validation, as listed in Table 2.

Table 2. The top 20 features selected by information gain ratio.

FEATURES	GAIN RATIO	FEATURES	GAIN RATIO
delta_rough_score	0.409 ± 0.096	*missed_cleavage_site*	0.045 ± 0.001
rough_score	0.409 ± 0.096	*total_intensity_b*	0.044 ± 0.003
match_ratio_intensity	0.274 ± 0.035	*intensity_b*	0.037 ± 0.001
rank_rough_score	0.275 ± 0.001	*average_match_error*	0.032 ± 0.001
intensity_y	0.156 ± 0.004	*homologous_y*	0.031 ± 0.004
match_ratio_peak	0.160 ± 0.026	*peptide_match_error*	0.027 ± 0.001
complementary_by	0.146 ± 0.013	*peptide_mass*	0.027 ± 0.001
total_intensity_y	0.092 ± 0.003	*number_HKR_pep*	0.016 ± 0.000
total_consecutive_y	0.088 ± 0.012	*hydrophobicity_pep*	0.012 ± 0.001
total_consecutive_b	0.056 ± 0.001	*clvgsite_median_b*	0.012 ± 0.001

4.5. *Results*

A ranked list of 500 candidate peptides was first generated by rough scoring, whereby we observed that all correct peptide identifications ranked in the top ten except for four spectra in dataset A and two spectra in dataset B. The SVM scorer was trained and tested on the top ten rough-scoring results of dataset B and dataset A, respectively. LIBSVM, an implementation of SVMs, with the RBF kernel, was employed [40]. A peptide is regarded as the correct answer if its SVM prediction value is the highest and above a given threshold; otherwise, peptides are considered incorrect answers. Performance comparison between the pepReap algorithm using the SVM scorer and the SEQUEST using threshold validation criteria are shown in Table 3.

Table 3. Performance comparison of pepReap and SEQUEST.

pepReap						SEQUEST(threshold[2])	
Training Set (Dataset B)				Test Set (Dataset A)		Test Set (Dataset A)	
weight[1]	best *C*	best γ	*MCC*	*SEN*	*PRE*	*SEN*	*PRE*
1:1	4.0000	0.03125	0.9212±0.0158	0.9106	0.9128	0.8715[a]	0.9107[a]
10:1	0.0625	0.12500	0.9269±0.0177	0.9116	0.9204	0.5397[b]	0.9420[b]
50:1	1.0000	0.12500	0.9300±0.0105	0.9175	0.9257	0.5548[c]	0.9410[c]
						0.6757[d]	0.9391[d]

Note: [1]The *weight* is used to set the parameter *C* of class 1 and -1 to *weight* × *C*. [2]The commonly used threshold criteria for evaluating SEQUEST identification results are, (a) $XCorr \geq 1.5, 2.0, 2.0$ [3], (b) $\Delta Cn \geq 0.1$ and $XCorr \geq 1.9, 2.2, 3.75$ [26], (c) $\Delta Cn \geq 0.1$ and $XCorr \geq 1.8, 2.2, 3.7$ [27], and (d) $\Delta Cn \geq 0.08$ and $XCorr \geq 2.0, 1.5, 3.3$ [4], for +1, +2, +3 charged fully tryptic peptides, respectively.

From Table 3, it can be seen that the high precision of SEQUEST is obtained at the cost of a very low sensitivity. In contrast, pepReap achieves much higher sensitivity than SEQUEST with some insignificant loss of precision. Both measures tend to increase when higher weight ratios for positives and negatives are applied to the SVM scorer.

5. Conclusions and Future Work

We have presented a novel and promising peptide identification algorithm, named pepReap, based on support vector machines. The characteristics distinguishing the pepReap from other algorithms lie in the flexible use of an SVM classifier both as the scoring function and the validation module and comprehensive features we used for measuring the match between a spectrum and a peptide. Preliminary experimental results on a dataset demonstrate that the pepReap algorithm can achieve much higher identification sensitivity without significant loss in identification precision compared with the popular SEQUEST algorithm that uses simple threshold validation criteria. A prerequisite of the pepReap algorithm is a set of mass spectra with known peptide sequences. Such training dataset for a given instrument can be obtained by first applying an independent identification algorithm to the spectra to be identified and then picking out high-confidence identifications. Our future work includes exploiting more informative features based on improved fragmentation models, testing the pepReap algorithm on more datasets and comparing it with sophisticated validation algorithms coupled to the SEQUEST (e.g. algorithms in Ref. [28], [32]).

Acknowledgments

We would like to thank Dr. Andrew Keller from the Institute for Systems Biology for providing the dataset of MS/MS spectra. We would also like to thank Dr. Chih-Jen Lin from the National Taiwan University for helpful discussions on support vector machines.

References

1. A. Pandey and M. Mann, *Nature* **405**, 837 (2000).
2. R. Aebersold and M. Mann, *Nature* **422**, 198 (2003).
3. A. J. Link *et al.*, *Nat. Biotechnol.* **17**, 676 (1999).
4. J. Peng *et al.*, *J. Proteome Res.* **2**, 43 (2003).
5. A. I. Nesvizhskii and R. Aebersold, *Drug Discov. Today* **9**, 173 (2004).
6. H. Steen and M. Mann, *Nat. Rev. Mol. Cell Biol.* **5**, 699 (2004).
7. B. Lu and T. Chen, *Drug Discov. Today: Biosilico* **2**, 85 (2004).
8. Z. Zhang, *Anal. Chem.* **76**, 6374 (2004).
9. A. Frank and P. Pevzner, *Anal. Chem.* **77**, 964 (2005).
10. M. Mann and M. Wilm, *Anal. Chem.* **66**, 4390 (1994).
11. D. L. Tabb, A. Saraf, and J. R. Yates III, *Anal. Chem.* **75**, 6415 (2003).
12. A. Frank, *et al.*, *J. Proteome Res.* **4**, 1287 (2005).
13. R. G. Sadygov, D. Cociorva and J. R. Yates III, *Nat. Methods* **1**, 195 (2004).

14. J. K. Eng, A. L. McCormack and J. R. Yate III, *J. Am. Soc. Mass Spectrom.* **5**, 976 (1994).
15. H. I. Field, D. Fenyö and R. C. Beavis, *Proteomics* **2**, 36 (2002).
16. Y. Fu *et al.*, *Bioinformatics* **20**, 1948 (2004).
17. D. Li *et al.*, *Bioinformatics* **21**, 3049 (2005).
18. D. N. Perkins *et al.*, *Electrophoresis* **20**, 3551 (1999).
19. V. Bafna and N. Edwards, *Bioinformatics* **17**, S13 (2001).
20. N. Zhang, R. Aebersold and B. Schwikowski, *Proteomics* **2**, 1406 (2002).
21. M. Havilio, Y. Haddad and Z. Smilansky, *Anal. Chem.* **75**, 435 (2003).
22. J. Colinge *et al.*, *Proteomics* **3**, 1454 (2003).
23. J. E. Elias *et al.*, *Nat. Biotechnol.* **22**, 214 (2004).
24. F. Schütz *et al.*, *Biochem. Soc. Trans.* **31**, 1479 (2003).
25. Z. Zhang, *Anal. Chem.* **76**, 3908 (2004).
26. M. P. Washburn, D. Wolters and J. R. Yates III, *Nat. Biotechnol.* **19**, 242 (2001).
27. X.-S. Jiang *et al.*, *Mol. Cell. Proteomics* **3**, 441 (2004).
28. A. Keller *et al.*, *Anal. Chem.* **74**, 5383 (2002).
29. M. J. MacCoss, C. C. Wu and J. R. Yates III, *Anal. Chem.* **74**, 5593 (2002).
30. J. Razumovskaya *et al.*, *Proteomics* **4**, 961 (2004).
31. F. Li *et al.*, *Rapid Commun. Mass Spectrom.* **18**, 1655 (2004).
32. D. C. Anderson *et al.*, *J. Proteome Res.* **2**, 137 (2003).
33. T. Baczek *et al.*, *Anal. Chem.* **76**, 1726 (2004).
34. V. N. Vapnik, *Statistical Learning Theory*, New York: John Wiley and Sons (1998).
35. S. S. Keerthi and C.-J. Lin, *Neural Comput.* **15**, 1667 (2003).
36. B. W. Matthews *et al.*, *Biochim. Biophys. Acta* **405**, 442 (1975).
37. A. R. Dongré *et al.*, *J. Am. Chem. Soc.* **118**, 8365 (1996).
38. A. Keller *et al.*, *OMICS* **6**, 207 (2002).
39. R. Quinlan, *C4.5: Programs for Machine Learning*, San Mateo, CA: Morgan Kaufmann Publishers Inc. (1993).
40. C.-C. Chang and C.-J. Lin, *LIBSVM: a library for support vector machines*, Software available at http://www.csie.ntu.edu.tw/~cjlin/libsvm (2001).

NORMALIZATION REGARDING NON-RANDOM MISSING VALUES IN HIGH-THROUGHPUT MASS SPECTROMETRY DATA

PEI WANG[§], HUA TANG, HEIDI ZHANG, JEFFREY WHITEAKER, AMANDA G PAULOVICH, MARTIN MCINTOSH

Fred Hutchinson Cancer Research Center,
Seattle, WA, 98109
[§]*E-mail: pwang@fhcrc.org*

We propose a two-step normalization procedure for high-throughput mass spectrometry (MS) data, which is a necessary step in biomarker clustering or classification. First, a global normalization step is used to remove sources of systematic variation between MS profiles due to, for instance, varying amounts of sample degradation over time. A probability model is then used to investigate the intensity-dependent missing events and provides possible substitutions for the missing values. We illustrate the performance of the method with a LC-MS data set of synthetic protein mixtures.

1. Introduction

High-throughput mass spectrometry (MS) technology offers a powerful means of analyzing biological samples. The ability of MS to identify and precisely quantify thousands of proteins from complex samples is expected to broadly affect biology and medicine[3]. However, MS systems are subject to considerable noise and variability that is not fully characterized or accounted for. Thus, it is important and necessary to properly conduct data-preprocessing steps such as signal filtering, peak detection, alignment in time (and mass charge ratio), and amplitude normalization before reliable conclusions can be made from the data[1].

In this paper, we focus on the normalization step, and propose a probability model for intensity-dependent missing events in MS-based data sets. In MS experiments, the instrument may have trouble detecting the weak signals of low-abundance peptides. Even if the instrument detects the signal, the peak intensities may be too low to be distinguished from background noise during data processing. Therefore, the lower the ion abundance, the more likely the peptide will be "missing" in the MS output data. Ignoring such non-random missing pattern may introduce significant bias into subsequent analyses. In this paper, we propose a novel probability model to describe the missing behavior, which accounts for this type of intensity-dependent missing events.

The rest of the paper is organized as follows: Section 2 provides a brief description of a data example illustrating the problem. Section 3 introduces a global normalization step, which adjusts systematic trends. The missing model, which represents our major contribution, is described in Section 4. Section 5 applies the proposed methods to an example data set and Section 6 is the conclusion.

2. Experiment and Data

In this section, we describe an experiment, in which replicates of two protein mixtures were analyzed on three consecutive days. We find that the samples processed in later days experienced higher levels of protein degradation due to, for instance, longer storage time as well as more freeze-thaw cycles[2]. Such variations are often unavoidable in real disease studies involving human samples.

2.1. *Sample preparation*

Two mixtures of proteins were assembled as part of an exploratory study to understand the performance of our MS instrument. One mixture (denoted as A) consisted of four proteins: bovine albumin, bovine transferrin, bovine alpha lactalbumin and bovine catalase. The other mixture (denoted as B) consisted of the same four proteins plus bovine beta lactoglobulin (proteins were selected based on their length and abilities to produce tryptic peptides). All five recombinant proteins were purified with reversed-phase high performance liquid chromatography (VisionWorkstation Applied Biosystems, Framingham, MA, USA). The collected protein fractions were dried in SpeedVac (Thermo Savant, San Jose, CA, USA). The purified proteins were denatured individually with 60% MeOH, reduced with 10 mM DTT at 60°C for 1 hr, and alkylated with 50 mM iodoacetamide in the dark at room temperature for 30 min. The polypeptides were trypsinized for 6 hr at 37°C with a protein/enzyme of 50/1.

2.2. *LC-MS system*

The LC-MS system comprised an 1100 Series Nanoflow LC system (Agilent Technologies, Palo Alto, CA, USA), a binary capillary pump, a C18 Symmetry NanoEase trapping column (Waters Corporation, Milford, MA, USA), a C18 PepMap nano LC column(LC Packings, Sunnyvale, CA, USA), and an LCT Premier time-of-flight mass spectrometer (Waters Corporation). The flow rates are 20 uL/min in the trapping column, and 400 nL/min in the LC column. The solvents were A(0.1% formic acid in water) and B (0.1% formic acid in acetonitrile). Linear gradient elution was applied from 0 to 40% B in 30 min. Mass spectra were acquired every 1.0 s with a 0.1 s interscan delay time. The instrument was mass-calibrated with a sodium formate solution prior to analysis.

2.3. *Data and problem*

The raw data is first processed using a program developed in our group,*msInspect.*[a], which includes modules for detecting and aligning peptide features. The output peptide array reports the intensities of all peptide features in each sample (an LC-MS experiment). Denote the intensity of the ith feature in the kth sample as y_i^k. If the ith feature is detected in the kth sample, then y_i^k is set to 0.

The total number of non-zero intensity peptides in each sample is summarized in Table 1. Clearly, more features were detected in experiments performed on day 1 than those performed on day 3.

Table 1. Number of peptide features in each sample. Mixture A consists of four proteins, while mixture B consists of five proteins.

Day 1	Sample Index	A1	B2	B3	A4		
	Feature Number	660	648	789	495		
Day 2	Sample Index	B5	B6	A7	B8	A9	A10
	Feature Number	609	339	386	492	384	413
Day 3	Sample Index	A11	B12	B13	A14		
	Feature Number	237	302	406	178		

Note: In the sample indexes, A=4 protein mixtures, B=five protein mixtures, and the number indicators the experimental order.

If we further compare Sample A1 (with 660 features) and Sample A14

[a]Available at http://proteomics.fhcrc.org/CPL/home.html

(with 178 features), as illustrated in Figure 1, we see there is an overall decrease in intensity in sample A14 compared to sample A1.

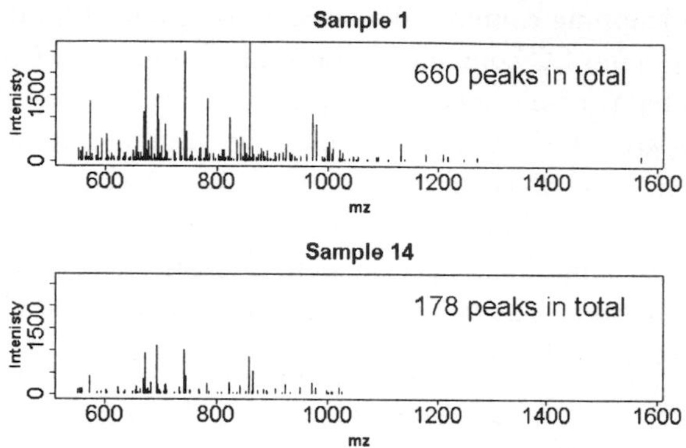

Figure 1. **Compare Sample A1 and Sample A14.** The two plots compare the intensities of all features in the two samples. The x coordinate is the mass charge ratio (mz), and the y coordinate is the intensity value. Each feature is represented by a vertical line at its mz position, with the length of the line equal to its intensity.

Given this kind of variation, it is crucial to normalize intensities before different samples can be properly compared.

3. Global Normalization

By globally normalizing signal intensities across multiple samples, we aim to identify and remove systematic variation arising because of differential amounts of sample loaded into the LC-MS system, protein degradation over time, or variation in the sensitivity of the instrument detector.

It is natural to assume that the sample intensities are all related by a constant factor[5]. A common choice for this re-scaling coefficient is the sample mean or median. This choice is based on the assumption that the number of features whose measurements change is few compared to the total number of features. So the distribution of the measurements of all the features should be roughly the same across different experimental runs[4].

However, in MS experiments, because of the limitation of detector sensitivity and the unavoidable instrument noise, ions below a certain intensity level may hardly be detected, which leads to non-random missing of peptide features in the result. Thus, it is not appropriate to use overall mean

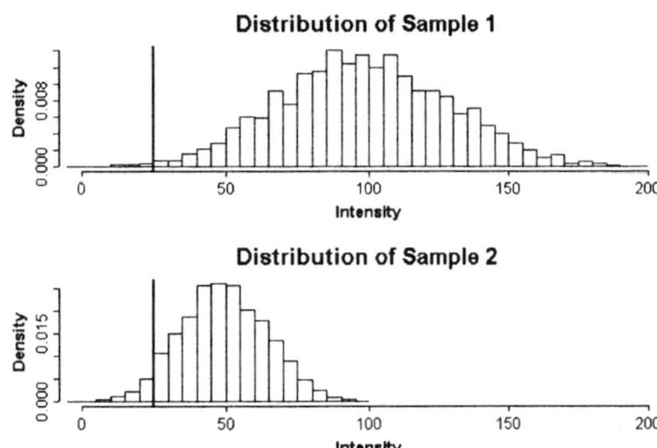

Figure 2. **The effect of non-random missing.** Suppose the overall peptide abundance of sample 1 are twice as great as the overall peptide abundance of sample 2. The histograms shows the true intensity distribution of all peptides in sample 1 and sample 2 respectively. The minimal detection level of the instrument is represented by the vertical line (features on the left side of the line can not be observed in the experiment). Using the mean or median intensity of the observed features in each sample leads to biased estimate of the scaling coefficients.

or median for re-scaling. This is illustrated in Figure 2. In order to avoid the possible bias due to non-random missing events, we propose to use the top L ordered statistics of feature intensities in each sample, where L is a parameter chosen by users.

For the simple case of two samples, denote the intensity measurements of one sample as $X = (x_1, x_2, ..., x_n)$ and of the other sample as $Y = (y_1, y_2, ..., y_m)$, whose order statistic can be represented as $x_{(1)} > x_{(2)} > ... > x_{(n)}$ and $y_{(1)} > y_{(2)} > ... > y_{(m)}$ respectively. Then, for a chosen number $L(L < \min(n, m))$, the scaling coefficient of X versus Y can be estimated as $\lambda = \sum_{i=1}^{L} x_{(i)} / \sum_{j=1}^{L} y_{(j)}$ or more robustly,

$$\lambda = \text{median}(x_{(1)}, ..., x_{(L)}) / \text{median}(y_{(1)}, ..., y_{(L)}). \tag{1}$$

For the case of $K(K > 2)$ samples, denote the intensity measurements of the kth sample as $X^k = (x_1^k, x_2^k, ..., x_{n_k}^k)$. For a given number $L(L < \min(\{n_k\}_{k=1}^{K}))$, define the population median as

$$\mu_0 = \frac{1}{K} \sum_k \text{median}(x_{(1)}^k, x_{(2)}^k, ..., x_{(L)}^k).$$

Then the scaling coefficient for the kth sample is

$$\lambda^k = \frac{1}{\mu_0} \text{median}(x_{(1)}^k, x_{(2)}^k, ..., x_{(L)}^k) \tag{2}$$

4. Model of Missing Events

We can make inferences on the missing events of one sample based on the information from other samples. The idea is illustrated in Figure 3. Suppose Sample 1 and Sample 2 are identical mixtures, but due to experimental factors, the overall peptide abundance of Sample 1 is smaller than the overall peptide abundance of Sample 2. Peptide 1 cannot be observed in Sample 1 because its intensity falls below the minimum detectable level. However, based on the intensities of those peptides observed in both samples (*e.g.* Peptide 2), the scale difference of the overall abundances between Sample 1 and Sample 2 can be estimated. Therefore, the "missing" intensity of Peptide 1 in Sample 1 can be reasonably approximated with the intensity measured in Sample 2 divided by a scale coefficient.

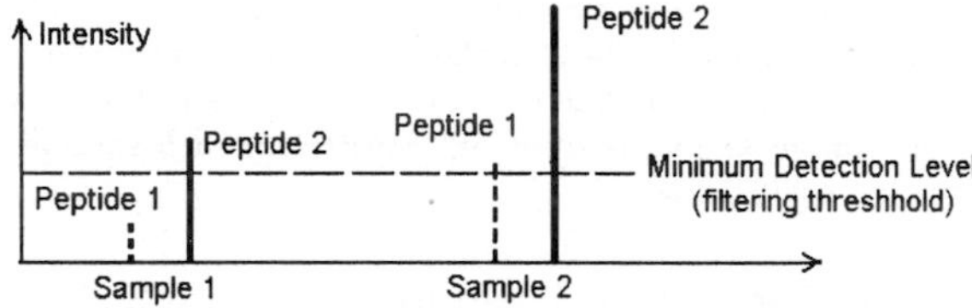

Figure 3. **Missing model.** The heights of the vertical bars indicate the true intensities of different peptides in the two samples. The dashed vertical lines represent Peptide 1, while the solid vertical lines represent Peptide 2. The horizontal dashed line indicates the minimal detection level of the instrument.

More general, we use a probability model to describe such missing events, which is described in below.

4.1. *Probability Model*

In one sample, we introduce a latent variable z_i for the ith peptide, which indicates whether this peptide exists in the sample or not:

$$z_i = \begin{cases} 1, & \text{if } i\text{th peptide exists in the sample;} \\ 0, & \text{if } i\text{th peptide does not exist in the sample.} \end{cases} \tag{3}$$

Given $z_i = 1$ (the ith peptide exists in the sample), the abundance of this peptide x_i can be deemed as a random variable:

$$x_i \begin{cases} = 0, & \text{if } z_i = 0, \\ \sim f_i, & \text{if } z_i = 1, \end{cases} \tag{4}$$

where f_i is the density function of some probability distribution. It is reasonable to assume that z_i and $x_i|z_i = 1$ are independent with each other.

Suppose the minimum detectable level of the instrument is d. Then, given the value (x_i, z_i, d), the observed abundance y_i of this peptide satisfies

$$y_i|(x_i, z_i, d) = \begin{cases} 0, & \text{if } z_i = 0; \\ 0, & \text{if } z_i = 1 \text{ and } x_i < d; \\ x_i, & \text{if } z_i = 1 \text{ and } x_i \geq d. \end{cases} \tag{5}$$

We say that a missing event happens to the ith peptide if the ith peptide exists in the sample but no signal has been detected (denoted as $M_i = \{z_i = 1, y_i = 0\}$). We are interested in the probability of missing event when no signal is observed, *i.e.* $P(M_i|y_i = 0)$, which can be calculated as follows:

$$\begin{aligned} P_d((z_i = 1, y_i = 0)|y_i = 0) &= \frac{P_d(z_i=1, y_i=0)}{P_d(y_i=0)} \\ &= \frac{P_d(y_i=0|z_i=1)P(z_i=1)}{P_d(y_i=0|z_i=1)P(z_i=1)+P_d(y_i=0|z_i=0)P(z_i=0)} \\ &= \frac{P_d(x_i<d|z_i=1)P(z_i=1)}{P_d(x_i<d|z_i=1)P(z_i=1)+P(z_i=0)}, \end{aligned} \tag{6}$$

where

$$P_d(y_i = 0|z_i = 1) = P_d(x_i < d, z_i = 1|z_i = 1) = P_d(x_i < d|z_i = 1)$$

and

$$P_d(y_i = 0|z_i = 0) = P_d(z_i = 0|z_i = 0) = 1$$

comes from Equation (5); $P(z_i)$ does not depend on d.

In addition, if $P(x_i > d|z_i = 1) > 0$, we have

$$P(z_i = 1) = \frac{P_d(z_i = 1, x_i > d)}{P_d(x_i > d|z_i = 1)} = \frac{P_d(y_i > 0)}{P_d(x_i > d|z_i = 1)}. \tag{7}$$

Therefore, given the detectable level parameter d, the distribution function f_i, and the observed abundance y_i, we can estimate the probability $P(M_i|y_i = 0)$ with Equation (6) and (7).

Moreover, a natural choice for imputing the intensity of a missing peak is $E(x_i|y_i = 0)$, which can be calculate as

$$\begin{aligned} E(x_i|y_i = 0) &= E(x_i|y_i = 0, z_i = 1)P(z_i = 1|y_i = 0) \\ &\quad +E(x_i|y_i = 0, z_i = 0)P(z_i = 0|y_i = 0) \\ &= E(x_i|x_i < d, z_i = 1)P(z_i = 1|y_i = 0) + 0 \\ &= E(x_i|x_i < d, z_i = 1)P(M_i|y_i = 0). \end{aligned} \tag{8}$$

Note $E(x_i|x_i < d, z_i = 1)$ only depends on the detector level parameter d and the distribution function f_i.

4.2. *Model Fitting*

4.2.1. *Detectable level d*

A reasonable estimate of the parameter, d, is the background noise level in each MS profiles, since those peaks with height below this value can not be confidently distinguished from noise signals. For a set of profiles from the same instrument, we assume that the same detectable level. Hence, we estimate d using all raw profiles. After the global normalization described in section (3), the detectable level of the kth profile becomes $\widetilde{d^k} = \frac{\hat{d}}{\lambda^k}$, where λ^k is the normalization scale coefficient in Equation (2).

4.2.2. *Abundance distribution f_x*

A. When Biological Replicates Available

For K replicates of the same biology samples, we assume that $\{\frac{x_i^k}{\lambda^k}|z_i^k = 1\}_{k=1}^K$ are independently identically distributed as $N(\mu_i, \sigma_i^2)$ for some parameter μ_i and σ_i, where x_i^k is the true abundance of the ith peptide in the kth profile.

Since $x_i^k|z_i^k = 1$ and z_i^k are independent from each other, it is easy to see that $y_i^k|(y_i^k > 0)$ and $x_i^k|(x_i^k > d, z_i^k = 1)$ are equal in distribution. Thus,

$$\begin{aligned} \frac{y_i^k}{\lambda^k}|(y_i^k > 0) \sim \widetilde{f_i^k}(t) &= \frac{P(x_i^k/\lambda^k \in dt, x_i^k > d, z_i = 1)}{P(x_i^k > d, z_i = 1)} \\ &= \frac{\varphi_{\mu_i,\sigma_i}(t)}{P(x_i > d|z_i = 1)}, \text{ for } t > \widetilde{d^k}. \end{aligned} \tag{9}$$

where φ_{μ_i,σ_i} is the density function of $N(\mu_i, \sigma_i^2)$.

For the simple case where $\sigma_i \ll \left|\widetilde{d^k} - \mu_i\right|$, we can approximate $P(x_i^k > d|z_i^k = 1)$ with $I\left(\widetilde{d^k} < \mu_i\right)$. It follows

$$\widetilde{f_i^k} \approx \varphi_{\mu_i,\sigma_i}, \text{ when } \widetilde{d^k} < \mu_i. \tag{10}$$

Thus, the mean intensity of the ith peptide can be estimated as the average of the observed signals:

$$\widehat{\mu}_i = \frac{\sum_k y_i^k/\lambda^k}{\sum_k I(y_i^k > 0)}. \tag{11}$$

Together with Eq.(7), we have

$$\widehat{P}(z_i = 1) = \frac{\sum_k I(y_i^k > 0)}{\sum_k I(\hat{\mu}_i > \widetilde{d^k})}. \tag{12}$$

Therefore

$$\widehat{P}(M_i^k|y_i^k = 0) = \begin{cases} \hat{P}(z_i = 1), & \text{if } \hat{\mu}_i < \widetilde{d^k}, \\ 0, & \text{if } \hat{\mu}_i > \widetilde{d^k}. \end{cases} \tag{13}$$

And then,

$$\widehat{E}(x_i^k|y_i^k = 0) = \begin{cases} \hat{\mu}_i\hat{P}(z_i = 1), & \text{if } \hat{\mu}_i < \widetilde{d^k}, \\ 0, & \text{if } \hat{\mu}_i > \widetilde{d^k}. \end{cases} \tag{14}$$

with Eq.(8), Eq.(11) and Eq.(12).

B. When Biological Replicates Not Available

Because the biological samples are limited, a large number of MS replicates are not always available for each sample. In such cases, a natural solution is to use the nearest K "neighbor samples" as pseudo replicates to fit the missing model. Here nearest K "neighbors" refers to the K closest profiles to the target profile under certain distance metrics (*i.e.* L_2 norm). However, if the missing rate is relatively high, the distance measured with the raw data could be misleading. Thus, we propose the following iteration procedure to try to recover the true "neighborhood" structure:

(1) Begin with K=N, where N is the total number of samples. Denote the original peptide array data matrix as Pep^0.
(2) (a) Based on Pep^{N-K}, calculate the distance between each two samples.
(b) For each sample, estimate the missing features by using its nearest K neighbors. Denote the new peptide arrays as Pep^{N-K+1}.
(c) K=K-1.
(3) Repeat step 2 until $K = K_0$, where K_0 is a pre-selected number.

If we aim to separate the samples into two clusters, a possible choice for K_0 is $N/2$.

5. Result

5.1. *Global normalization*

The scale coefficients of global normalization are estimated with the top 80 order statistics of each sample according to Eq.(2). Fig.4 shows the relationships between the top 80 order statistics of Samples 11 – 14 (the four samples on the third day) and the top 80 order statistics of Sample 1.

Table3 shows the scale coefficients for the four pairs of samples in Fig.4. Compared to the estimators derived with the order statistics, the estimators

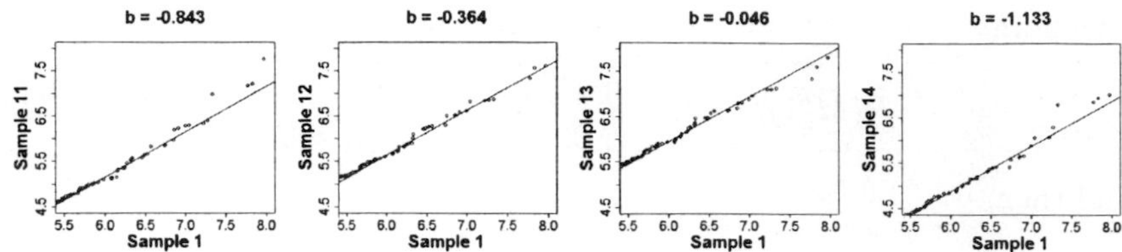

Figure 4. **The top 80 order statistics of Sample** 11 − 14 *v.s.* **Sample 1.** The y and x coordinates represent the log intensities of the 80 most abundant features in the corresponding samples. The good linear relationship with slope= 1 justifies the assumption that the sample intensities are related by a constant factor (model $x = \lambda y$ is equivalent to model $log(x) = log(y) + b$, where b and λ are parameters).

derived with overall medians dramatically overestimate the scale change between these sample pairs. This demonstrates the necessariness of using the top order statistics to conduct the global normalization when non-random missing is a concern in the study.

Table 2. Scale Coefficients *v.s.* Sample 1.

Sample Index	11	12	13	14
λ (based on the order statistics)	0.43	0.69	0.95	0.32
λ (based on overall median)	1.19	1.10	0.99	1.09

5.2. *Study of Missing Events*

We consider 12 of the 14 samples whose non-zero features are at least 10% of the total.

5.2.1. *Supervised analysis*

Treating all 4-protein samples as replicas and all 5-protein samples as replicas, using Equation (13) we can estimate the total number of possibly missing features $\sum_i I(\widehat{P}(M_i^k|y_i^k = 0) > 0)$ for each sample. The result is shown in Table 3. Again, we can see that the missing trend is more severe in some samples than in others. Ignoring such trend may bring unexpected bias into downstream analysis.

Table 3. Number of peptide features in each sample.

Sample	A1	B2	B3	A4	B5	B6	A7	B8	A9	A10	B12	B13
Missing	8	0	0	8	0	187	118	63	116	69	116	17

5.2.2. *Unsupervised analysis*

The goal here is to use the MS profiles to recover the 4-protein and 5-protein group labels for each sample. First, based on the data after global normalization, we perform hierarchical clustering analysis using the R[b] function *hclust* with complete linkage. The dendrogram is illustrated in the top plot of Figure 5. The two main sub-clusters are separated according to when the MS experiments were conducted (the first four samples were processed on day 1 while the others on the day 2 and 3).

Next, we use the iterative procedure described in section 4.2.2 to substitute the possible missing measurements with their expected values, and perform the hierarchical clustering on the resulting data. The new dendrogram is illustrated in the bottom plot of Figure 5, in which the 4-protein samples and the 5-protein samples are correctly clustered into two groups. This suggests that properly modelling the missing events would prevent the analysis from being driven by experimental variation rather than biological variation.

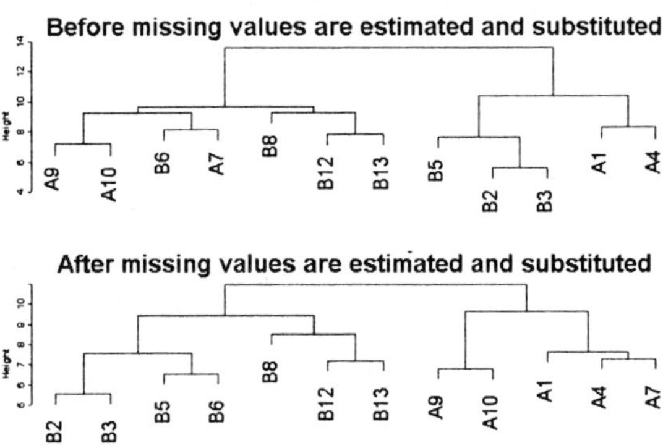

Figure 5. **Tree Structures of Unsupervised Hierarchical Clustering.** Each leaf in the tree represents one sample. A = 4 protein mixture; B = 5 protein mixture.

[b] R is a free statistics software, which can be downloaded at: http://www.r-project.org/

6. Conclusion

In this paper, we have shown that ignoring the intensity-dependent missing events in MS experiments may result in severe biases in the data analysis. To address this problem, we developed a probability model for the missing events and implemented a few normalization schemes to remove the negative effects. The missing rate estimates can also be used as a quality control of the data.

In the probability model, given that one peptide exists in the sample, a normal density is used to approximate the distribution of the intensity of this peptide. This approximation is supported by the synthetic data example: the Kolmogorov-Smirnov distance between $N(0, 1)$ and the observed distribution of intensities (centered to $mean = 0$ and scaled to $sd = 1$) is 0.0392, which corresponds to a p-value of 0.1742.

When we estimate the missing values with nearest-neighbor scheme, the iteration number need to be carefully controlled to avoid problem of over-fitting.

Acknowledgments

We would like to thank three referees and Andrea E Detter for the comments that improved this manuscript. This work was funded by National Cancer Institute contract $\sharp 23XS144A$. HT was partially supported by NIH-CA86368. HZ, JW and AP were partially supported by philanthropy from Listwin Foundation/Canary Fund, Paul G. Allen Family Foundation and Keck Foundation.

References

1. J. Listgarten and A. Emili, *Molecular and Cellular Proteomics* **4.4**, 2005.
2. B.L. Mitchell, Y. Yasui, C.I.Li, A.L.Fitzpatrick and P.D.lampe, *Cancer Informatics* **1(1) 25-31**, 2005.
3. M. Man and R.Aebersold, *Nature* **422**, 2003.
4. J. Quackenbush, *Nat. Genet.* **32**, 2002.
5. A. Sauve and T. Speed, *Proceedings Gensips*, 2005.
6. M. Wagner, D. Naik and A. Pothem, *Proteomics* **3**, 1692-1698, 2003.
7. K.A. Baggerly, J.S. Morris, J. Wang, D. Gold, L.C.Xiao and K.R. Coombes, *Proteomics* **3**, 1667-1672, 2003.
8. M. Anderle, S. Roy, H. Lin, C. Becker and K. John, *Bioinformatics* **20**, 3575-3582, 2004.
9. R. Tibshirani, T. Hastie, and et.al. *Bioinformatics* **20**, 3034-3044, 2004.
10. W. Wang, H. Zhou,and et.al. *Anal. Chem.* **75**, 4818-4826, 2003.

A POINT-PROCESS MODEL FOR RAPID IDENTIFICATION OF POST-TRANSLATIONAL MODIFICATIONS

BO YAN[1], TONG ZHOU[2], PENG WANG[1], ZHIJIE LIU[1], VINCENT A. EMANUELE II[2], VICTOR OLMAN[1], YING XU[1]

[1]*Department of Biochemical and Molecular Biology, University of Georgia, GA, USA*

[2]*School of Electrical and Computer Engineering, Georgia Tech, Atlanta, GA, USA*

Post-translational modifications (PTMs) are very important to biological function, and yet are notoriously difficult to detect and identify, especially in a high-throughput manner. Most of the existing approaches rely on exhaustive searches which are highly time consuming and thus are currently limited to handling of a few types of PTMs. In this paper, we present a point-process model that aims to find the optimal mass shifts to maximize the spectra alignment between an experimental MS/MS spectrum and a candidate theoretical spectrum, through cross-correlation calculation, yields a rapid search for all types of PTMs in a *blind* mode, i.e., without giving the types of the searching PTMs in advance. The test results show that our new approach's performance is comparable to or better than the other *blind* search methods, but is more efficient computationally and simpler in its concept.

1. Introduction

Post-translational modifications (PTMs) are chemical alterations in protein structures that change the properties of protein by proteolytic cleavage or by modification of amino acids [1]. They play key roles in many important cellular functions and regulatory processes. However, accurate identification of PTMs through analysis of high-throughput MS/MS data represents a highly challenging problem [2, 3]. The main difficulty lies in that the occurrences of PTMs change the molecular weights and the fragmentation patterns of peptides, which make them difficult to detect using the classical MS/MS data interpretation methods. Currently, great amount of tandem mass spectra, possibly ranging from tens of thousands to millions of mass spectra, are being collected daily across many proteomic centers and labs for functional studies of proteins. However often only a small fraction of these data could be successfully interpreted using popular analysis tools such as Sequest [4], Mascot [5], PepFrag [6] and ProteinProspector [7], etc. This can be attributed to several factors, including technical reasons such as poor peptide fragmentation, contaminants and others. Among the biological reasons, PTMs are generally believed to be a major contributor [8].

Theoretically all possible PTMs can be identified by exhaustively searching through all types of (known) PTMs and their combinations. However, such a strategy is very time consuming; only a few types and a very small number of PTMs can be taken into account in real applications [5, 9, 10]. While recently Tanner *et al.* [11] reported a fast PTMs search method which uses peptide sequence tags [12, 13] as efficient filters to reduce the size of the database by a few orders of magnitude, this algorithm also has to "guess" the types of PTMs in advance and the time complexity depends exponentially on the number of allowed PTM types. On the other hand, most *de novo* sequencing algorithms can be modified to identify PTMs by regarding the PTMs as pseudo amino acids in additional to the 20 basic ones [14-22]. However, the requirement of relative high quality spectra such as perfect fragmentations, has seriously limited their applications.

Recently, an approach called *blind* PTM identification has been proposed by Pevzner's group [23], which allows to search for all possible types of PTMs without looking up a set of pre-specified PTMs. They reported an interesting dynamic programming approach to performing the optimal spectral alignment. The idea is that each peak segment in a theoretical spectrum is allowed to shift by one or more "appropriate" values, such that the resulting spectrum optimally matches the experimental spectrum, and from which PTMs and their locations could be derived [23]. This method could also reveal some still unknown modifications. While encouraging results have been documented, the demand for computing power to obtain the optimal spectral alignment may be too high to be practically applicable.

In this paper, we describe a point-process model [24], a time-delay estimation framework, to *blindly* search for all possible PTMs in an efficient manner. Through analyzing the cross correlation function between a query spectrum and a candidate theoretical spectrum from peptide database, as modeled by two point processes, we are able to detect all good local and global alignments between the two processes at once straightforwardly, and thus to infer the types and locations of PTMs efficiently. Spectral similarity is measured by the optimal common mass peaks shared between two spectra, short peptide segments due to missing peaks or resulting from PTMs thus can be tolerated.

We have implemented the algorithms and tested their performance on both simulated and real experimental spectra. Our approach is able to conduct *blind* PTMs search in a few seconds. For simulated spectra with 0, 1, 2 and 3 PTMs, it achieves 100%, 97%, 86% and 75% success rates respectively. The performance on experimental spectra is comparable with or better than Tsur *et al.*'s result [23].

2. Algorithm

2.1. *A point process model*

We model a tandem mass spectrum by a point process [24]:

$$x(t)=\sum_{i=1}^{N}\delta(t-t_i)$$

where $\{t_i\}$ is a set of mass peak locations with N peaks and $\delta(t)$ is the Kronecker delta function [25]. It follows easily that

$$\delta(t-t_i)\delta(t+\tau-t_j)=\begin{cases}\delta(t-t_i) & \text{if } \tau=t_j-t_i\\ 0 & \text{otherwise}\end{cases}$$

To deal with PTMs, we assume, without loss of generality, that $\{t_i\}$ can be clustered into K+1 groups (to model K mass shifts) such that

$$x(t)=\sum_{k=0}^{K}x_k(t)$$

where each $x_k(t)$ is described by a point process model $x_k(t)=\sum_{i=1}^{N_k}\delta(t-t_i^{(k)})$, $N_k\geq 1$ and $\bigcup_{k=0}^{K}\{t_i^{(k)}\}=\{t_i\}$. N_k is the number of peaks in group k.

We further introduce a measure $C[\cdot]$ to be the total number of non-zero values in a point process, i.e., $C[x_k(t)]=N_k$ and $C[x(t)]=\sum_{k=0}^{K}N_k=N$.

When a PTM happens, a particular shift occurs to $x_k(t)$ to produce $y_k(t)$:

$$y_k(t)=x_k(t-\Delta_k)=\sum_{i=1}^{N_k}\delta(t-\Delta_k-t_i^{(k)})$$

and the resulting PTM spectrum is $y(t)=\sum_{k=0}^{K}y_k(t)$. Obviously, $C[y(t)]=N$.

Note that here we have separated the spectral peaks into K+1 different groups $\{x_k(t),y_k(t)\}$ according to their shift patterns in light of PTMs.

To simplify the analysis, we assume that all pair-wise differences among $\{t_i^{(k)}\}$ are different for all i, k and Δ_k. Then, we have

$$C[x(t-\Delta_k)y(t)]=C[y_k(t)y_k(t)]=N_k$$

Define $c_{xy}(\tau)\equiv C[x(t-\tau)y(t)]/N$, we infer that

$$c_{xy}(\tau)\,|_{\tau=\Delta_k}=\frac{N_k}{N}$$

$$c_{xy}(\tau)\,|_{\tau=(t_j^{(k)}+\Delta_k)-t_i^{(l)}}=\frac{1}{N}$$

$$c_{xy}(\tau)=0,\ \text{if } \tau\neq\Delta_k,\tau\neq(t_j^{(k)}+\Delta_k)-t_i^{(l)}$$

In practice, N_k is the number of the mass peaks shifted by one PTM, which is often much larger than 1. Therefore, the largest $K+1$ values of $c_{xy}(\tau)$ will be at $\tau = \{\Delta_k\}_{k=0}^{K}$, i.e., correspond to K mass shifts introduced by multiple PTMs (and their combinations), plus the local alignment between the unaffected portions of the two spectra. In other words, if a candidate peptide is a good match and no PTM exists in the query spectrum, the maximum of $c_{xy}(\tau)$ shall occur at $\tau = \Delta_0 \equiv 0$ (the rule used in Sequest [4]). On the other hand, if a candidate is the correct hit and the query peptide contains PTMs, the highest peaks of $c_{xy}(\tau)$ have a very good chance to be the right PTMs (and/or their combinations)[1]. This forms the basis of our approach to inferring PTMs.

2.2. *Implementation*

We have implemented two modes to perform *blind* PTMs search, (a) a homology search mode and (b) a strict match mode. We first calculate the cross-correlations between the query experimental spectrum and the theoretical one of each candidate from a peptide database, from which we feasibly obtain all non-zero $c_{xy}(\tau)$. Obviously those $c_{xy}(\tau)$ values contain all possible mass shifts to optimize the spectral alignment. The homology search mode reports the best hit which has the best spectral alignment with a set of optimal mass shifts. The strict match mode further requires that the hit peptide's molecular weight, or its molecular weight after modifications (if there are any PTMs), must be equal to that of the query peptide (at some tolerance). The search procedure is only applied to the candidates whose parent mass is at most Δ Da away from the query peptide (in this paper, Δ is set to 160 Da which is large enough to cover three typical PTMs).

Homology search mode: For each spectral alignment between the query spectrum and a peptide candidate, we record $c_{xy}(0)$ and the three additional largest values of $c_{xy}(\tau)$ at $\tau \neq 0$, say $c_{xy}(\Delta_1)$, $c_{xy}(\Delta_2)$ and $c_{xy}(\Delta_3)$. To find the best hits from our target database, we screen two best candidates, one with the highest value of $c_{xy}(0)$, and the other with the highest value of $c_{xy}(0)+c_{xy}(\Delta_1)$. We first check the best hit with the highest value of $c_{xy}(0)$. If this value is higher than a threshold, say 0.5, we consider there is a good match between the query and the candidate, and no PTM exists. Otherwise, either the correct peptide is not in the database or the query peptide has been modified. Then we check the best hit with the highest value of $c_{xy}(0)+c_{xy}(\Delta_1)$. If this value is

[1] We have observed in the dataset MOD1 (2620 experimental spectra with one PTM, *see* **Results**) that for 96.30%, 2.48% and 0.38% of cases, the PTMs correspond to the highest, second highest and third highest peaks of $c_{xy}(\tau)$, respectively.

higher than a threshold, say 0.7, we consider that there is a good match between the query spectrum and the candidate peptide if the candidate is modified by Δ_1, and we regard Δ_1 as the right PTM.

One might consider using $c_{xy}(0)+c_{xy}(\Delta_1)+c_{xy}(\Delta_2)$ (and so on) as the criterion. Our test results show that using $c_{xy}(0)+c_{xy}(\Delta_1)$ has a better performance, even for the cases with more than one PTM. One possible reason is that no matter how many PTMs exist, the first highest peaks of $c_{xy}(\tau)$ ($\tau \neq 0$) has the largest probability to be one of the right PTMs or of their combinations. However it may not hold for the rest highest peaks of $c_{xy}(\tau)$.

Strict match mode: This mode uses a parameter K to guess the number of PTMs.

For K=0, we implement it as a simple version of Sequest [4].

For K=1 (i.e., with one PTM), we report the top candidate with a Δ such that $| PW_{\text{exp}} - PW - \Delta | \leq \varepsilon$ and $c_{xy}(0)+c_{xy}(\Delta)$ is maximized, where Δ represents the mass of a possible PTM which could be any value ranging from 0 to 160 Da in this paper. PW_{exp} and PW are the molecular weights of the query and the candidate peptides, respectively, and ε is their maximal difference allowed after modification (4 Da is used).

For K=2, we report the best candidate with a pair $\{\Delta_i, \Delta_j\}$ such that $| PW_{\text{exp}} - PW - (\Delta_i + \Delta_j) | \leq \varepsilon$ and $c_{xy}(0)+c_{xy}(\Delta_i)+c_{xy}(\Delta_j)+c_{xy}(\Delta_i+\Delta_j)$ is maximized. Where Δ_i and Δ_j ($\Delta_i \neq \Delta_j \neq 0$) represent the masses of two possible PTMs, respectively.

Since the relationship $\sum_{k=0}^{K} c_{xy}(\Delta_k)=1$ always holds, the individual signal (peak) at $c_{xy}(\Delta_k)$ will diminish as K increases and will ultimately disappear into the background noise as K increases beyond certain value. We have found that strict match model is unsuitable to deal with the case with more than two PTMs.

2.3. *Determination of PTM positions*

The above calculation procedure itself does not provide information about the location of PTMs, if there are any. A tracing back procedure has been developed to locate the actual location of each predicted PTM based on the starts of peak shifts. Further details of this algorithm are omitted in this extended abstract.

2.4. *Statistical significance measurement*

We use the following z-score to measure the significance of the best hit,

$$\text{z}-\text{score} = \frac{\text{raw\_score} - <\text{raw\_score}>}{\sigma_{\text{raw\_score}}}$$

where raw_score represents a c_{xy} score , <raw_score> and $\sigma_{\text{raw\_score}}$ are the mean or standard deviation of the raw_scores derived from all peptide candidates.

We consider that a hit is significant if the hit has a high z-score. We found that in general, a correct peptide and its homologs have both high raw_score and z-score. To further classify them, we introduce a δ-score which is the difference in raw_score between the best two matches (equivalent to Δ_{cn} used in Sequest [4]). For a best hit with both high raw_score and z-score, if it has a large δ-score as well, we consider that the hit could be the correct peptide; otherwise, the hit is predicted to be a homolog of the correct peptide.

3. Results

Annotated tandem mass spectra with known PTMs are currently very limited in proteomics community. In this paper, we have tested our approach on three datasets: a large set of simulated spectra with PTMs, a set of annotated experimental spectra with added PTMs, and a small set of annotated experimental spectra with real PTMs.

3.1. *Datasets*

SIM_SET: Consisting of 4 subsets of simulated spectra, with 0, 1, 2, or 3 PTMs respectively. Each subset contains 10,000 spectra. The peptides are randomly chosen from yeast peptide database which is tryptically digested (allowed up to 2 missing cleavages). The lengths of chosen peptides are required to be at least 6, 8, 10, and 12 aa's for cases of 0, 1, 2 and 3 PTMs, respectively.

The set of simulated PTMs is acetylation of Lysine (+42), hydroxylation of Proline (+16), methylation of Aspartic acid or Glutamic acid (+14), oxidation of Methionine (+16), and phosphorylation of Serine (+80).

MOD_SET: Annotated high-quality yeast mass spectra from the Open Proteomics Database [26] (charge 2, Sequest Xcorr score ≥ 2.5). We constructed three subsets of modified spectra by adding 0, 1 or 2 PTMs selected from above PTM pool. We shift the peaks (here b, y ions only) of a spectrum to the selected modifications. If certain peaks are absent in the spectrum, we just skip these missing peaks. For the three subsets, we got 2657, 2620 and 2422 spectra with 0, 1, 2 PTMs respectively.

EXP_SET: 47 annotated high-quality spectra with real PTMs from Strader *et al.* [27]. 42 out of 47 are associated with one PTM, and most of them are oxidation of Methionine while a few are methylation of Lysine or Arginine. The 47 spectra are from 26 peptides of *R.* Palustris. We perform *blind* search against yeast peptide database mixed up with the 26 peptide sequences of *R.* Palustris.

All the experimental mass spectra were LCQ data which had a relative low mass resolution. We run a data preprocessing procedure as described in PepNovo [26] to filter tiny noise peaks and isotopic peaks. For cross-correlation calculation, we regard peak shifts in the range of $(\Delta - 0.5, \Delta + 0.5)$ as having the same nominal shift Δ, where Δ is an integer. However, our approach doesn't require that Δ is an integer.

3.2. *Search results on simulated spectra*

Table 1 shows the *blind* search results on the *SIM_SET*. Both search modes obtained a similar performance: for 0, 1, 2 and 3 PTMs, we got 100%, 97%, 85% and 72% of the spectra correctly identified, respectively. We consider a (best) hit as correct only if it matches the original peptide sequence exactly.

Table 1: Search results against the simulated spectra by homology search mode (a) and by strict match mode (b). SIM*k* refers to the sub datasets with *k* PTMs. Values at rank *i* are the percentages of the correct candidates reported at Top *i*. CPU is the average amount of time the program needs to analyze a spectrum (on a PC with a 2.8GHz CPU).

(a)

	rank 1	rank 2	rank 3	rank 4	rank 5	CPU(s)
SIM0	100%	0	0	0	0	1.516
SIM1	97.45%	1.53%	0.57%	0.21%	0.10%	1.563
SIM2	85.52%	3.83%	1.85%	1.15%	0.91%	1.467
SIM3	72.04%	7.12%	3.16%	1.85%	1.33%	1.556

(b)

	rank 1	rank 2	rank 3	rank 4	rank 5	CPU(s)
SIM0	100%	0	0	0	0	0.764
SIM1	97.90%	1.39%	0.39%	0.17%	0.03%	1.321
SIM2	85.60%	5.55%	2.10%	1.25%	0.66%	1.532

3.3. *Search results on experimental spectra*

Table 2 lists the search results against the *MOD_SET*. For spectra without PTMs, both *blind* search modes achieved 99% success rate, very close to the performance of Sequest [4] (note that the experimental spectra were annotated by Sequest). For spectra with one or two PTMs, the identification rates were relative lower. However, we found that there were certain correct candidates not reported exactly at the number one hit but within top five hits. If we regard them correct as well, then homology search mode achieved 65% and 20% success rates for spectra with one or two PTMs respectively, while strict match mode achieved 81% and 17% accuracies respectively. These results are comparable to or even better than Tsur *et al.*'s dynamic programming approach, which obtained 57.3% and 15.6% accuracies for spectra with one or two PTMs respectively [23].

Table 2: Search results against the experimental spectra by homology search mode (a) and by strict match mode (b). MODk refers to the datasets with k PTMs. Total column is the percentage of the correct candidates reported within rank 5 in the hit list. *Search results against the dataset MOD2 by assuming peptides containing one PTM.

(a)

	rank 1	rank 2	rank 3	rank 4	rank 5	total	CPU(s)
MOD0	99.28%	0.41%	0.08%	0.04%	0	99.81%	0.823
MOD1	44.39%	9.77%	5.04%	3.05%	2.40%	64.65%	1.563
MOD2	11.44%	3.51%	2.31%	1.16%	1.32%	19.74%	1.604

(b)

	rank 1	rank 2	rank 3	rank 4	rank 5	total	CPU(s)
MOD0	99.21%	0.56%	0.04%	0.08%	0	99.89%	0.765
MOD1	60.38%	11.95%	4.31%	2.52%	2.01%	81.17%	1.467
MOD2	5.86%	3.43%	2.56%	2.52%	2.27%	16.64%	1.645
MOD2*	16.23%	5.08%	2.48%	1.65%	1.61%	27.05%	1.523

The technical reason for that strict match mode havs a much lower success rate for two PTMs may lie in that, finding a pair $\{\Delta_i, \Delta_j\}$ with maximization of $c_{xy}(0)+c_{xy}(\Delta_i)+c_{xy}(\Delta_j)+c_{xy}(\Delta_i+\Delta_j)$ might not be a good criterion to screen the correct candidate. Since not all the four signals can be observed simultaneously in some cases (depending on the positions of the two PTMs). Thus we searched MOD2 again by assuming k=1 (equivalent to maximizing $c_{xy}(0)+c_{xy}(\Delta_i+\Delta_j)$), a significant improvement was then achieved — 16% of spectra were identified correctly as top 1 and 27% within top 5.

Test results on the 47 experimental spectra with real PTMs were similar to those obtained for the experimental spectra with added PTMs. Since most of spectra have one PTM, only k=1 was searched by strict match mode. We got 27 spectra (i.e., 57.45% of spectra) identified correctly (ranked at top 1) by strict match mode and 24 spectra (i.e., 51.06%) correctly by homology search mode.

3.4. *Hits of homologs*

We have observed that some of the top hits have both relatively high z-score and raw_score. However they don't match the original peptide sequence exactly. We found that many of them are the homologs of the original peptides which have very similar sequences. For example, for the query peptide DGKYDLDFKNpESDK (where the lower case letter p indicates hydroxylation of Proline), our homology search mode reported a very similar peptide DGKYDLDFKNPNSDK with one mutation pE12PN. Table 3 lists some examples of the top one hits being the homologs of the query peptides. We estimated that ~20% of the poorly performed results by homology search mode are due to the reason of homologous proteins, which should be considered as partially correct. In Table 3, we also listed several important features of the

Table 3: Partial lists of homologs reported on the MOD1. [a] The lower case letter indicates the type and location of one PTM; [b] Change in mass; [c]Cross-correlation value at τ =0; [d] Accumulative intensities of common peaks shared between two spectra at 0 mass shift. Note that here the logarithm value of the relative intensity was used; [e] The first optimal mass shift (Δ_1) that maximizes the spectral alignment.

Query peptide		Hits of homologs				Correct candidates				
sequence[a]	PTM[b]	sequence	z-score	cxy(0)[c]	ΣI(0)[d]	cxy(0)	ΣI(0)	Δ1[e]	cxy(Δ1)	ΣI(Δ1)
AIPGeYVTYALSGYVR	14	AIPGEYITYALSGYVR	8.82	0.57	3.51	0.33	1.54	14	0.37	-0.4
DGKYDLDFKNpESDK	16	DGKYDLDFKNPNSDK	8.95	0.39	2.04	0.36	1.05	16	0.43	8.38
DYIMSPVGNPEGPEKpNKK	16	DYIMSPVGNPEGPEKPNK	8.56	0.41	-7.45	0.28	-4.83	16	0.44	8.24
IINEPTAAAIAYGLdKK	14	IINEPTAAAIAYGLDK	8.32	0.40	-2.49	0.25	-2.45	14	0.47	7.99
IINEPTAAAIAYGLGAGk	42	IINEPTAAAIAYGLDK	8.51	0.40	1.70	0.41	1.23	42	0.38	13.57
KEDeEDKFDAMGNK	14	EDEEDKFDAMGNK	7.72	0.54	3.92	0.35	4.82	14	0.46	4.89
KGEQeLEGLTDTTVPK	14	GEQELEGLTDTTVPK	9.38	0.46	0.85	0.27	2.62	14	0.43	7.14
LIDLTQFPAFVTPmGK	16	LIDLTQFPAFVTPLGK	9.35	0.53	-7.62	0.33	0.86	16	0.47	5.53
LIDLTQFPAFVTpMGK	16	LIDLTQFPAFVTPLGK	8.95	0.53	-7.45	0.30	1.21	16	0.47	4.04
LIEAFNEIAEDSEQFdK	14	LIEAFNEIAEDSEQFEK	12.28	0.72	10.61	0.28	-2.81	14	0.50	9.94
LNKETTYDeIKK	14	LNKETTYDEIK	7.55	0.55	-4.28	0.41	-3.35	14	0.55	9.93
NFNDPEVQGdMK	14	NFNDPEVQADMK	9.31	0.73	12.86	0.36	2.03	14	0.45	7.88
NIVEFHsDHMK	80	NIVEFHSDHIK	8.10	0.55	0.65	0.30	3.69	80	0.50	8.99
NQAAMNpANTVFDAK	16	NQAAMNPSNTVFDAK	11.94	0.79	9.70	0.39	4.58	16	0.50	2.64
NQAAmNPANTVFDAK	16	NQAAMNPSNTVFDAK	9.10	0.61	2.42	0.39	7.88	16	0.50	-2.65
NQAAMNPsNTVFDAK	80	NQAAMNPANTVFDAK	9.51	0.61	1.46	0.43	4.37	80	0.61	2.76
RPkYFITANDVK	42	RPEYFITANDVK	13.79	0.95	11.58	0.41	1.58	1	0.59	9.99
SeVFSTYADNQPGVLIQVFEGER	14	SETFSTYADNQPGVLIQVFEGER	8.74	0.39	6.89	0.34	8.99	14	0.25	-1.83
SQIdEVVLVGGSTR	14	SQIDEIVLVGGSTR	9.79	0.69	8.09	0.35	4.14	14	0.46	-0.35
sQVDEIVLVGGSTR	80	SQIDEIVLVGGSTR	6.94	0.50	1.92	0.42	4.09	80	0.42	1.38
TAGIQIVADDLTVTNpAR	16	TAGIQIVADDLTVTNPK	10.11	0.41	1.97	0.38	-1.11	16	0.47	14.5
VATTGEWdKLTQDK	14	VATTGEWEKLTQDK	11.59	0.77	9.13	0.31	0.76	14	0.50	9.15
VdIIANDQGNR	14	VEIIANDQGNR	11.57	0.90	0.98	0.40	6.65	14	0.55	-7.65
YLdQVLDHQR	14	YLEQVLDHQR	11.65	0.94	10.74	0.39	10.83	14	0.56	-0.09

correct candidates. As we have expected that the vast majority of the inferred optimal modifications for the correct candidates correspond to the correct PTM.

4. Discussion and Conclusion

We have presented a point-process model for rapid *blind* PTMs search without the need of a list of pre-specified PTMs. Our test results show that its performance is comparable to or better than Tsur *et al.*'s dynamic programming approach [23]. Since both approaches aim to find a set of optimal mass shifts to maximize the spectral alignment, it is not surprising that they have a similar *blind* search performance. However, our algorithm is able to feasibly obtain all possible mass shifts (naturally includes the optimal mass shifts) using one round of cross-correlation calculation, thus it is conceptually simple and more computationally efficient than otheirs. Moreover, the computing time of our algorithm is independent of the number of PTMs, one major merit compared to most of the existing approaches, for which the computing time grows exponentially at the size of the set of pre-specified PTMs. We also implemented a homology search mode which is able to find the homologs of a query peptide. This feature is also found to be useful in mass spectra interpretation. Furthermore, since the similarity between two spectra is measured by the shared common mass peaks, our algorithm can tolerate short peptide segments resulted from multiple PTMs or missing peaks.

Cross correlation function has long been used to measure the similarity between two time-dependent signals. Both our approach and Sequest [4, 9] use it to measure the spectral similarity between two spectra. However our approach is significantly different from that employed in Sequest. First, Sequest performs an exhaustive search to identify PTMs for which a set of pre-specified PTMs must be given in advance. Second, Sequest considers the cross correlation value at $\tau = 0$ only. If the experimental spectrum contains PTMs, Sequest enumerates all possible PTM modifications for each candidate peptide, shifts mass peaks in a theoretical spectrum for each PTM and then compares the modified theoretical spectrum with the query spectrum. Third, Sequest doesn't remove isotopic peaks before spectral comparison. Instead, it adds artificial satellite peaks in theoretical spectrum to mimic the experimental spectrum which unnecessarily increases the computing time, and even worse it might increase the false positive identification rate. Our approach has extended the overall functionality of Sequest while maintaining its sensitivity.

In this paper, our algorithm has only considered b and y ions with +1 charge state for spectral alignment, thus it is suitable to handle MS/MS with +1 and +2 charge states, but unsuitable for spectra with +3 charge state for which the daughter ions with +1 and +2 charge states are tangled together. A simple solution could be to consider b and y ions with both +1 and +2 charge states together. However, an elaborative model should be developed to deal with this case in the future. In addition, we didn't explicitly utilize peak intensities and the number of consecutive peaks, etc. in our work. A more sophisticated scoring system which incorporates these features may further increase the success rate. Note that considering neutral mass losses doesn't improve the performance of pattern match, since contaminants generally have the same neutral mass loss patterns as well [28].

Acknowledgement

This research was supported in part by National Science Foundation (#NSF/DBI-0354771 and #NSF/ITR-IIS-0407204), by Georgia Cancer Coalition under Distinguished Cancer Clinicians & Scientists Program, and by the US Department of Energy's Genomes to Life program (http://doegenomestolife.org/) under project, "Carbon Sequestration in Synechococcus sp.: From Molecular Machines to Hierarchical Modeling".

References

1. A.A. Gooley and N.H. Packer, in *Proteome Research: New Frontiers in Functional Genomics*, M.R. Wilkins, et al., Editors. Springer-Verlag. p. 65-91 (1997).
2. M. Mann and O.N. Jensen, *Nat Biotechnol.* **21**(3):255-261 (2003).
3. O.N. Jensen, *Curr Opin Chem Biol.* **8**(1):33-41 (2004).
4. J.K. Eng, A.L. McCormack, and J.R. Yates, 3rd, *J Am Soc Mass Spectrom.* **5**(11):976-989 (1994).
5. D.N. Perkins, D.J. Pappin, D.M. Creasy, and J.S. Cottrell, *Electrophoresis.* **20**(18):3551-3567 (1999).
6. D. Fenyo, J. Qin, and B.T. Chait, *Electrophoresis.* **19**(6):998-1005 (1998).
7. K.R. Clauser, P. Baker, and A.L. Burlingame, *Anal Chem.* **71**(14):2871-2882 (1999).
8. A.I. Nesvizhskii and R. Aebersold, *Drug Discov Today.* **9**(4):173-181 (2004).
9. J.R. Yates, 3rd, J.K. Eng, and A.L. McCormack, *Anal Chem.* **67**(18):3202-3210 (1995).
10. M.R. Wilkins, E. Gasteiger, A.A. Gooley, B.R. Herbert, M.P. Molloy, P.A. Binz, K. Ou, J.C. Sanchez, A. Bairoch, K.L. Williams, and D.F. Hochstrasser, *J Mol Biol.* **289**(3):645-657 (1999).

11. S. Tanner, H. Shu, A. Frank, L.C. Wang, E. Zandi, M. Mumby, P.A. Pevzner, and V. Bafna, *Anal Chem.* **77**(14):4626-4639 (2005).
12. M. Mann and M. Wilm, *Anal Chem.* **66**(24):4390-4399 (1994).
13. D.L. Tabb, A. Saraf, and J.R. Yates, 3rd, *Anal Chem.* **75**(23):6415-6421 (2003).
14. J.A. Taylor and R.S. Johnson, *Rapid Commun Mass Spectrom.* **11**(9):1067-1075 (1997).
15. P.A. Pevzner, V. Dancik, and C.L. Tang, *J Comput Biol.* **7**(6):777-787 (2000).
16. J.A. Taylor and R.S. Johnson, *Anal Chem.* **73**(11):2594-2604 (2001).
17. P.A. Pevzner, Z. Mulyukov, V. Dancik, and C.L. Tang, *Genome Res.* **11**(2):290-299 (2001).
18. B. Ma, K. Zhang, C. Hendrie, C. Liang, M. Li, A. Doherty-Kirby, and G. Lajoie, *Rapid Commun Mass Spectrom.* **17**(20):2337-2342 (2003).
19. Y. Han, B. Ma, and K. Zhang. in *Proceedings of 2004 IEEE Computational Systems Bioinformatics (CSB)*. 206-215 (2004)
20. B.C. Searle, S. Dasari, M. Turner, A.P. Reddy, D. Choi, P.A. Wilmarth, A.L. McCormack, L.L. David, and S.R. Nagalla, *Anal Chem.* **76**(8):2220-2230 (2004).
21. B. Ma, K. Zhang, and C. Liang, *Journal of Computer and System Sciences*. **70**:418-430 (2005).
22. B. Yan, Y. Qu, F. Mao, V.N. Olman, and Y. Xu, *J Comput Sci Technol.* **20**:483-490 (2005).
23. D. Tsur, S. Tanner, E. Zandi, V. Bafna, and P.A. Pevzner. in *Proceedings of 2005 IEEE Computational Systems Bioinformatics (CSB)*. 157-166 (2005)
24. D.L. Snyder and M.I. Miller, *Random point processes in time and space, 2nd edition*. 1991: Springer-Verlag.
25. A.V. Oppenheim and R.W. Schafer, *Discrete-time signal processing*. 1989: Prentice Hall.
26. J.T. Prince, M.W. Carlson, R. Wang, P. Lu, and E.M. Marcotte, *Nat Biotechnol.* **22**(4):471-472 (2004).
27. M.B. Strader, N.C. Verberkmoes, D.L. Tabb, H.M. Connelly, J.W. Barton, B.D. Bruce, D.A. Pelletier, B.H. Davison, R.L. Hettich, F.W. Larimer, and G.B. Hurst, *J Proteome Res.* **3**(5):965-978 (2004).
28. Y. Fu, Q. Yang, R. Sun, D. Li, R. Zeng, C.X. Ling, and W. Gao, *Bioinformatics*. **20**(12):1948-1954 (2004).

A NEW APPROACH FOR ALIGNMENT OF MULTIPLE PROTEINS

XU ZHANG TAMER KAHVECI

Department of Computer and Information Sciences and Engineering, University of Florida, Gainesville, FL, USA, 32611

E-mail: {xuzhang, tamer}@cise.ufl.edu

We introduce a new graph-based multiple sequence alignment method for protein sequences. We name our method HSA (Horizontal Sequence Alignment) for it horizontally slides a window on the protein sequences simultaneously. Current progressive alignment tools build up final alignment by adding sequences one by one to existing alignment. Thus, they have the shortcoming of order-dependent alignment. In contrast, HSA considers all the proteins at once. It obtains final alignment by concatenating cliques of graph. In order to find a biologically relevant alignment, HSA takes secondary structure information as well as amino acid sequences into account. The experimental results show that HSA achieves higher accuracy compared to existing tools on BAliBASE benchmarks. The improvement is more significant for proteins with low similarity.

1. Motivation

Multiple sequence alignment (MSA) of protein sequences is one of the most fundamental problems in computational biology. It is an alignment of three or more protein sequences. MSA is widely used in many applications such as phylogenetic analysis [17] and identification of conserved motifs [22].

The alignment of two sequences with maximum score can be found in $O(L^2)$ time using dynamic programing [14], where L is the length of the sequences. This algorithm can be extended to align N sequences, but requires $O(L^N)$ time [12,18]. A variety of heuristic MSA algorithms have been developed. Most of them are based on progressive application of pairwise alignment. They build up alignments of larger numbers of sequences by adding sequences one by one to existing alignment [5]. We call this a *vertical alignment* since it progressively adds a new sequence (i.e., row) to a consensus alignment. These methods have the shortcoming that the order of sequences to be added to existing alignment significantly affects the quality of the resulting alignment. This problem is more apparent when the percentage of identities among amino acids falls below 25%, called the *twilight zone* [3]. The accuracies of most progressive sequence alignment methods drop considerably for such proteins.

In this paper, we consider the problem of alignment of multiple proteins. We develop a graph-based solution to this problem. We name this algorithm HSA (Horizontal Sequence Alignment) as it *horizontally* aligns sequences. Here, horizontal alignment means that all proteins are aligned simultaneously, one column at a time. HSA first constructs a directed-graph. In this graph, each amino acid of the input sequences maps to a vertex. An edge is drawn between pairs of vertices that may be aligned together. The graph is then adjusted by inserting gap vertices. Later, this graph is traversed to find high scoring cliques. Final alignment is obtained by concatenating these cliques. The experimental results show that HSA finds better alignments on average than existing popular tools. The quality improvement is much greater for low similarity sequences.

The rest of the paper is organized as follows. Section 2 discusses related work. Section 3 introduces the algorithm in detail. Section 4 presents experimental results. Section 5 concludes with a brief discussion.

2. Related Work

Finding the multiple sequence alignment that maximizes the SP (Sum-of-Pairs) score is an NP-complete problem [26]. A variety of heuristic algorithms have been developed to overcome this difficulty [21]. These heuristic methods can be classified into four groups: progressive, iterative, anchor-based and probabilistic.

Progressive methods find multiple alignment by iteratively picking two sequences from this set and replacing them with their alignment (i.e., consensus sequence) until all sequences are aligned into a single consensus sequence. Thus, progressive methods guarantee that never more than two sequences are simultaneously aligned. This approach is sufficiently fast to allow alignments of virtually any size. ClustalW [21,23], T-coffee [16], Treealign [7] and POA [11] can be grouped into this class [20].

Iterative methods start with an initial alignment. They then repeatedly refine this alignment through a series of iterations until no more improvements can be made. Depending on the strategy used to improve the alignment, iterative methods can be deterministic or stochastic. Muscle [4] and DIALIGN [13] can be grouped into this class.

Anchor-based methods use local motifs (short common subsequences) as anchors. Later, the unaligned regions between consecutive anchors are aligned using other techniques. MAFFT [10], Align-m [25], L-align [8], Mavid, PRRP [6], DIALIGN [13] belongs to this class.

Probabilistic methods pre-compute the substitution probabilities by analyzing known multiple alignments. They use these probabilities to maximize the substitution probabilities for a given set of sequences. Probcons [3], Hmmt [19], SAGA [15], and Muscle [4] can be grouped into this class.

3. Proposed method

The underlying assumption of HSA is that the residues that have same SSE types have more chance to be aligned compared to the residues that have different SSE types. This assumption is verified by a number of real experiments and observations [1,2,9,24].

HSA works in five steps: (1) An initial directed graph is constructed by considering residue information such as amino acid and secondary structure type. (2) The vertices are grouped based on the types of residues. The residue vertices in each group are more likely to be aligned together in the following step. (3) Gap vertices are inserted to the graph in order to bring vertices in the same group close to each other in terms topological position in the graph. (4) A window is slid from beginning to end. The clique with highest score is found in each window and an initial alignment is constructed by concatenating these cliques. (5) The final alignment is constructed by adjusting gap vertices of the initial alignment. Next, we describe these five steps in detail.

3.1. *Constructing initial graph*

This step constructs the initial graph which will guide the alignment later. Let s_1, s_2, $\cdots$, s_k be the protein sequences to be aligned. Let $s_i(j)$ denote the jth amino acid of protein s_i. A vertex is built for each amino acid. The vertices corresponding to different proteins are marked with different colors. Thus, the vertices of the graph span k different colors. If available, Secondary Structure Element (SSE) type (α-helix , β-sheet) of each residue is also stored along with the vertex. For simplicity, SSE types include α-helix , β-sheet, and no SSE information, as shown in Figure 1. Two types of edges are defined. First, a directed edge is included from the vertex corresponding to $s_i(j)$ to $s_i(j+1)$ for all consecutive amino acids. Second, an undirected edge is drawn between pairs of vertices of different colors if their substitution score is higher than a threshold. HSA gets the substitution score from BLOSUM62 matrix. A weight is assigned to each undirected edge as the sum of the substitution score and *typeScore* for the amino acid pair that make up that edge. The *typeScore* is computed from the SSE types. If two residues belong to the same SSE type, then their *typeScore* is high. Otherwise, it is low. We discuss this in more detail in Section 3.2. This policy of weight assignment lets residues with same SSE type or similar amino acids have higher change to be aligned in following steps. We will discuss this in Section 3.4. Figure 1 demonstrates this step on three proteins. The amino acid sequences and the SSEs are shown at the top of this figure. The dotted arrows represent the undirected edges between two vertices of different color, the solid arrows only appear between the vertices corresponding to consecutive amino acids of the same protein and they only have one direction, from left to right.

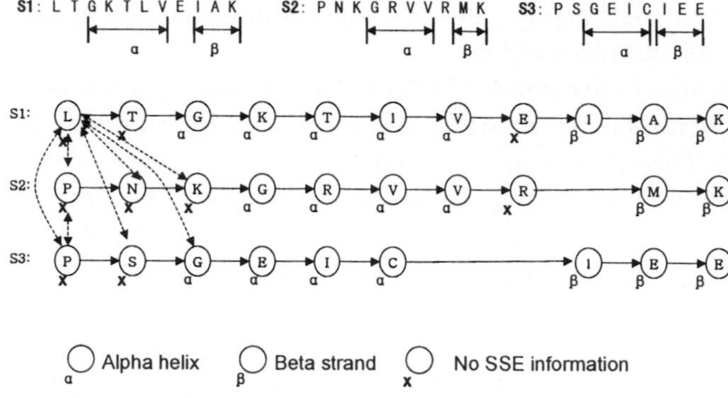

Figure 1. The initial graph constructed for sequence S_1, S_2 and S_3. Each residue maps to a vertex in this graph. The figure shows some edges between the first vertices of the sequences, indicated by dashed arrows. The vertices for different sequences are marked with different colors (colors not shown in figure).

3.2. *Grouping Fragments*

The graph constructed at the first step shows the similarity of pairs of residues. However, multiple alignment involves alignment of groups of amino acids rather than pairs. In this step, we group the *fragments* that are more likely to be aligned together. Here, a fragment is defined by the following four properties: 1) It is composed of consecutive vertices. 2) All the vertices have the same color. 3) All vertices have the same SSE type. 4) There is no other fragment that contains it. For example, in Figure 2, S_1 consists of four fragments: f_1 = LT, f_2 = GKTIV, f_3 = E, and f_4 = IAK. Thus, S_1 can be written as $S_1 = f_1 f_2 f_3 f_4$.

With the knowledge that the fragments with the same SSE type are more likely to be aligned, all sequences are scanned to find fragments with known SSE types. The fragments are then clustered into groups, where each group consists of one fragment from each sequence. To group fragments, we align the fragments first. We use a simplified dynamic programming algorithm by considering each fragment as a residue in the basic algorithm [14]. The score of two fragment pairs is computed from the following formula:

$$totalScore = typeScore - positionPenalty - lengthPenalty$$

The *typeScore* is computed from the SSE types. Fragments with the same SSE type contribute a high score whereas fragments of different SSE types incur penalty. This is because of our assumption that residues with the same SSE type have higher chance to be aligned. Thus *typeScore* is calculated as follows: we check the types of two fragment first and return a number according to the following 5 different conditions. 1) They are the same type of α-helix, we return 4; 2) They are the same type of β-sheet, we return 2; 3) They are the same type of no SSE type, we return 1; 4) They

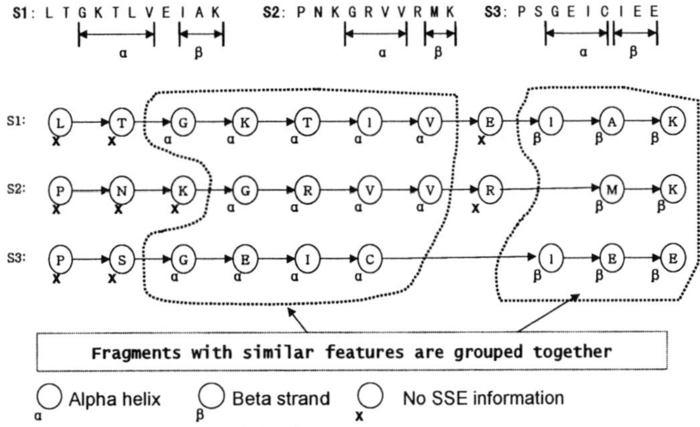

Figure 2. The fragments with similar features, such as SSE types, lengths and positions in original sequences are grouped together.

are α-helix and β-sheet, we return -4; 5) Otherwise, we return 0. The *positionPenalty* is computed as the difference between the positions of two fragments. Here the position of a fragment is the topological position in the original sequence. If two fragments are far away in their sequences, then the pair of them gets a higher penalty. This is because the alignment of such fragments introduce many gaps. The *lengthPenalty* is computed as the difference between the lengths of the two fragments. The length of a fragment is the number of residues it contains. Fragment pairs with similar length will be given smaller penalty. This is because as the lengths the fragment pairs differ more, the number of gap vertices that need to be inserted in the later alignment increases.

Figure 2 demonstrates how HSA groups fragments. Using the example of Figure 1, fragments with same SSE type, similar positions and lengths are clustered into the same group. Two such groups with α-helix and β-sheet are circled in Figure 2.

3.3. *Fragment Position Adjustment*

Once the groups of fragments are determined, we update the graph to bring the fragments in same group close to each other in terms of *vertical* position. Here, *vertical* position corresponds to a position in the topological order of the vertices of the same color. For example, in Figure 3, vertex L in S_1, vertex P in S_2, and vertex P in S_3 are at the same vertical position 1, similarly, vertex T in S_1, vertex N in S_2, and vertex S in S_3 are at the same vertical position 2, etc. As we will discuss later, this process increases the possibility that the vertices in these fragments are aligned.

We update the graph by inserting gap vertices, as shown in Figure 3. First, we compute the number of gap vertices to be inserted based on two

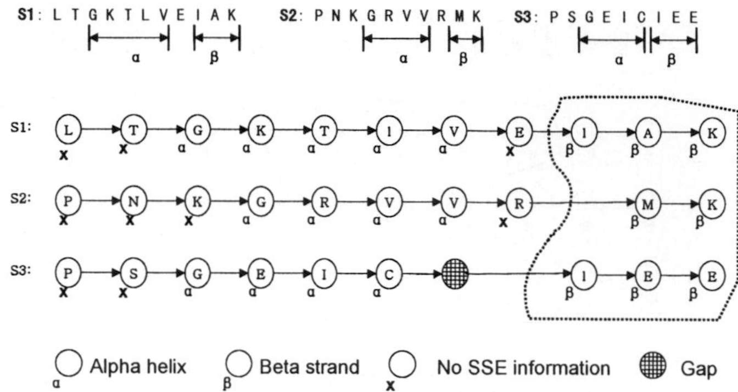

Figure 3. A gap vertex is inserted to let the fragments in same group close to other each other vertically.

factors: 1) The number of residues in fragments. 2) The relative positions of fragments in the same group. Here a good relative position of fragments means that the positions of fragments lead to a high scoring alignment of the vertices in these fragments. We align the vertices in fragments of the same group to compute those positions. Then, we randomly select a position between two consecutive fragment groups. Finally, for each sequence we insert gap vertices at these positions to bring the fragments within the same group together. In Figure 3, a gap vertex is inserted before residue I in S_3 to bring fragments in the group with β-sheet type close to each other.

3.4. *Alignment*

So far, we have prepared the graph for actual alignment by two means. (1) We determined vertex pairs that can be a part of the alignment; (2) We brought sequences to roughly the same size by inserting gap vertices, while keeping similar vertices vertically close. In this step, the sequences are actually aligned by scanning the updated graph in topological order.

As demonstrated in Figure 4, we start by placing a window of width w at the beginning of each sequence. This window defines a subgraph of the graph. Typically, we use $w = 4$ or 6. The example in Figure 4 uses $w = 3$. Next, we greedily choose a clique with the best *expectation score* from this subgraph. We will define the expectation score of a clique later. A clique here is defined as a complete subgraph that consists of one vertex from each color. In other words, if K sequences are to be aligned, a clique corresponds to the alignment of one letter from each of the K sequences. Thus, each clique produces one column of the multiple alignment. For each clique, we align the letters of that clique, and iteratively find the next best clique that 1) does not conflict with this clique, and 2) has at least one letter next to a letter in this clique. This iteration is repeated t times to find t columns.

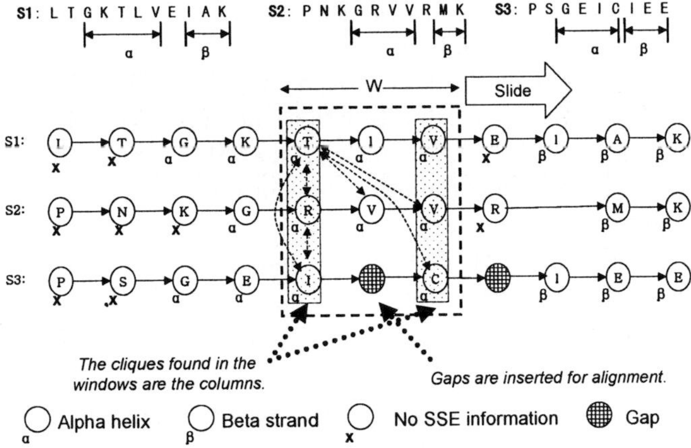

Figure 4. Cliques found in the sliding window (window size = 3) are the columns of the resulting alignment. Gaps are inserted to concatenate these columns.

Typically, $t = 4$. These t cliques define a local alignment of the input sequences. The expectation score of the original clique is defined as the SP score of this local alignment. After finding the highest expectation score clique, we add this clique as a column to existing alignment. We then slide the window to the location which is immediately after the clique found and repeat the same process until it reaches the end of sequences. Each clique defines a column in the multiple alignment. The columns are concatenated and gaps are inserted to align them. Figure 4 illustrates this step, in the window (circled by the dotted rectangle), the highest expectation score clique (the left shadow background marked column) consists of residues T, R, and I in S_1, S_2 and S_3 respectively. Then, the window slides to next location toward the right of the graph (this window is not shown in the Figure 4), and the highest expectation score clique (the right background marked column) in the window consists of residue V, V, and C in S_1, S_2 and S_3 respectively. The two cliques found (marked by shadow background) are two columns in resulting alignment. The resulting alignment is obtained by inserting a gap vertex to S_3.

As mentioned in section 3.1, due to the policy of edge weight assignment, cliques that contain vertices of the same SSE type or similar amino acids have higher score than other possible cliques. Since a clique contains one vertex of each color, finding the best clique does not assure any order for traversal of vertices of different colors. Thus, unlike existing tools, our method is order independent.

3.5. *Gap Adjustment*

After concatenating the cliques in previous step, short gaps may be scattered in the sequence. In this step, the alignment obtained in the previous step is adjusted by moving the gaps as follows. The sequences are scanned from left to right to find isolated gaps. If a gap is inside a fragment of type α-helix or β-sheet, it is moved outside of that fragment, either before or after. We choose the direction that produces higher alignment score. If a gap is inside a fragment with no SSE type, it is moved next to the neighboring gap only if the movement produces a higher score than the current alignment.

The final alignment is obtained by mapping each vertex in the final graph back to its original residue.

4. Experimental Result

In order to demonstrate the feasibility of our method, we ran it on BAliBASE benchmarks [22] (`http://www-igbmc.u-strasbg.fr/BioInfo/BAliBASE/`). We chose the benchmarks that contain SSE information since our algorithm needs SSE information of sequences. We downloaded ClustalW [21,23], Probcons [3], Muscle [4] and T-Coffee [16] for comparison since they are the most commonly used and the most recent tools. We ran all experiments on a computer with 3 GHz speed, Intel pentium 4 processor, and 1 GB main memory. The operating system is Windows XP.

4.1. *Evaluation of alignment quality*

Alignment of dissimilar proteins is usually harder than the alignment of highly similar proteins. Figures 5, 6 and 7 show the BAliBASE scores of HSA, ClustalW, Probcons, Muscle and T-Coffee on benchmarks with low, medium, and high similarity respectively. From Figure 5, we conclude that for low similarity benchmarks, our method outperforms all other tools. On the average HSA achieves a score of 0.619, which is better than any other tool. HSA finds the best result for 14 out of 21 reference benchmarks. HSA is the second best in 5 of the remaining 7 benchmarks. Figure 6 shows that for sequences with 20-40% identity, HSA is comparable to other tools on average. The average score is not the best one. However, it is only slightly worse than the winner of this group (0.909 versus 0.901). HSA performs best for 2 cases out of 7, including a case for which HSA gets full score. In Figure 7, HSA is higher than other tools on average. HSA performs best on 2 cases out of 7, including a case for which HSA gets full score. High scores of existing methods for sequences with high percentage of identity (Figures 6 and 7) show that there is little room for improvement for such sequences. Proteins at the twilight zone (Figure 5) pose a greater challenge. These results show that our algorithm performs best for such sequences. For medium and high similarity benchmarks, our results are comparable to existing tools.

Figure 8 shows the SP scores of HSA, ClustalW, Probcons, Muscle, T-Coffee and original BAliBASE alignment. On the average, ClustalW,

		ClustalW	Probcons	Muscle	T-Coffee	HSA
Short	1aboA	0.693	0.624	0.616	0.320	0.833
	1idy	0.546	0.679	0.354	0.183	0.700
	1r69	0.655	0.655	0.345	0.234	0.772
	1tvxA	0.223	0.439	0.239	0.235	0.462
	1ubi	0.607	0.464	0.478	0.445	0.648
	1wit	0.630	0.690	0.660	0.707	0.675
	2trx	0.660	0.705	0.712	0.667	0.756
avg		0.573	0.608	0.486	0.398	0.692
Medium	1bbt3	0.512	0.373	0.488	0.440	0.539
	1sbp	0.467	0.585	0.587	0.548	0.590
	1havA	0.222	0.397	0.293	0.256	0.352
	1uky	0.531	0.498	0.535	0.441	0.596
	2hsdA	0.482	0.606	0.748	0.573	0.614
	2pia	0.624	0.700	0.691	0.579	0.608
	3grs	0.377	0.355	0.309	0.383	0.487
avg		0.459	0.502	0.521	0.460	0.541
long	1ajsA	0.388	0.411	0.370	0.379	0.472
	1cpt	0.697	0.719	0.765	0.726	0.810
	1lvl	0.368	0.590	0.451	0.528	0.532
	1pamA	0.405	0.534	0.439	0.461	0.524
	1ped	0.678	0.717	0.746	0.638	0.746
	2myr	0.394	0.568	0.386	0.454	0.630
	4enl	0.664	0.573	0.526	0.582	0.652
avg		0.513	0.587	0.526	0.538	0.624
Avg all		0.515	0.565	0.511	0.465	0.619

Figure 5. The BAliBASE score of HSA and other tools. less than 25 % identity

Muscle, and T-Coffee find the highest SP score for low, medium, and high similarity sequences respectively. However, according to Figures 5 to 7, those methods have relatively low BAliBASE scores. This means that, the alignment with the highest SP score is not necessarily the most meaningful alignment. The SP score of HSA is comparable to other tools on the average. For low similarity sequence benchmarks, the average SP score of HSA is higher than the average SP score of the reference alignment.

4.2. *Performance Evaluation*

The time complexity of our algorithm is $O(W^K N + K^2 M^2)$, where K is the number of sequences, W is the sliding window size, N is the sequence length and M is the number of fragments in a protein sequence. The complexity is computed as follows. The clique, in a window, with the highest expectation score is found in W^K time, and there are N positions for the sliding window. $K^2 M^2$ time is required for aligning fragments. Usually, $M \ll N$. Thus, the total time complexity, in practice, is $O(W^K N)$. Typically W is a small number such as 4. For reasonably small K, $W^K N = O(N)$. Therefore, for small K, the complexity is $O(N)$. As K increases, the complexity increases quickly. However, this complexity is observed

	ClustalW	Probcons	Muscle	T-Coffee	HSA
1fjlA	0.994	0.989	0.971	0.991	1.000
1csy	0.861	0.897	0.799	0.887	0.871
1tgxA	0.833	0.760	0.679	0.817	0.782
1ldg	0.920	0.939	0.954	0.956	0.941
1mrj	0.853	0.925	0.894	0.894	0.925
1pgtA	0.941	0.926	0.912	0.955	0.924
1ton	0.718	0.898	0.865	0.867	0.867
avg	0.874	0.904	0.867	0.909	0.901

Figure 6. The BAliBASE score of HSA and other tools. 20%-40% identity

	ClustalW	Probcons	Muscle	T-Coffee	HSA
1amk	0.978	0.984	0.986	0.988	0.986
1ar5A	0.953	0.956	0.969	0.947	1.000
1led	0.900	0.931	0.950	0.956	0.929
1ppn	0.987	0.983	0.983	0.984	0.981
1thm	0.898	0.900	0.899	0.893	0.910
1zin	0.955	0.975	0.985	0.958	0.978
5ptp	0.948	0.963	0.950	0.961	0.957
avg	0.945	0.956	0.960	0.955	0.963

Figure 7. The BAliBASE score of HSA and other tools. more than 35% identity

only if the subgraphs inside a window is highly connected. It is possible to get rid of the W^K term in the complexity by using longest path methods rather than clique finding methods. The experimental results in Figure 9 coincides with the above conclusion. In general, ClustalW performs best. However, ClustalW achieves this at expense of low accuracy (see Figures 5 to 7). HSA is slower than ClustalW and Muscle. It is, however, faster than Probcons and T-Coffee.

5. Conclusion and Future Work

We developed a new algorithm called HSA for alignment of multiple proteins. HSA is graph-based and differs from existing progressive multiple sequence alignment methods since it builds up the final alignment by considering all sequences at once. HSA first constructs a graph based on the amino acid and SSE types of the residues of the input proteins. It then groups the vertices of this graph with the guide of SSE type information. Next, HSA slides a window from the beginning to the end of the graph and finds cliques in the window. The concatenation of these cliques defines an alignment. HSA obtains the final alignment by adjusting the positions of the gap vertices in the graph.

Experimental results show that HSA achieves high accuracy and still maintains competitive running time. The quality improvement over existing tools is more significant for low similarity sequences. For high or medium similarity sequences, HSA produces comparable accuracy. The running time of HSA is comparable to existing tools.

	REF	ClustalW	Probcons	Muscle	T-Coffee	HSA
Short, <25%	-602	-453	-594	-496	-912	-599
Medium, <25%	-2036	-1466	-2516	-1543	-2461	-1617
Long, <25%	-2989	-1964	-3266	-2291	-2991	-2436
Short, 20%-40%	456	499	508	480	491	493
Medium, 20%-40%	1238	1119	1138	1231	1191	1138
Medium, >35%	3474	3477	3479	3526	3528	3468
AVG overall	-76	202	-208	151	-192	74

Figure 8. The SP score of HSA and other tools.

	ClustalW	Probcons	Muscle	T-Coffee	HSA
Short, <25%	69	238	98	915	194
Medium, <25%	133	638	297	1890	535
Long, <25%	308	1564	584	3240	1191
Short, 20%-40%	62	265	83	1187	421
Medium, 20%-40%	171	695	175	2316	613
Medium, >35%	154	629	136	2502	660
AVG overall	149	672	229	2008	602

Figure 9. The running time of HSA and other tools (measured by milliseconds).

The running time of HSA can be further improved by employing longest path methods rather than cliques. Another future direction is to iteratively refine the alignment by the updating the graph after alignment [6].

References

1. Phil Bradley, Peter S. Kim, and Bonnie Berger. Trilogy: Discovery of sequence-structure patterns across diverse proteins. In *Annual Conference on Research in Computational Molecular Biology*, pages 77 – 88, 2002.
2. Luonan Chen. Multiple Protein Structure Alignment by Deterministic Annealing. In *IEEE Computer Society Bioinformatics Conference (CSB'03)*, volume 00, page 609, 2003.
3. C. Do, M. Brudno, and S. Batzoglou. PROBCONS: Probabilistic Consistency-based Multiple Alignment of Amino Acid Sequences . In *Intelligent Systems for Molecular Biology (ISMB)*, 2004.
4. R.C. Edgar. MUSCLE: multiple sequence alignment with high accuracy and high throughput. *Nucleic Acids Research*, 32(5):1792–1797, 2004.
5. D.F. Feng and R.F. Doolittle. Progressive Sequence Alignment As A Prerequisite To Correct Phylogenetic Trees. *Journal Of Molecular Evolution*, 25(4):351–360, 1987.
6. O. Gotoh. Significant Improvement in Accuracy of Multiple Protein Sequence Alignments by Iterative Refinement as Assessed by Reference to Structural Alignments. *Journal of Molecular Biology*, 264(4):823–838, 1996.
7. J Hein. A new method that simultaneously aligns and reconstructs ancestral sequences forany number of homologous sequences, when the phylogeny is given. *Molecular Biology and Evolution*, 6(6):649–668, 1989.
8. X. Huang and W. Miller. A time-efficient, linear-space local similarity algorithm. *Advances in Applied Mathematics*, 12:337–357, 1991.
9. Gibrat JF, Madej T, and Bryant SH. Surprising similarities in structure comparison. *Current Opinion in Structural Biology*, 6(3):377–385, 1996.

10. K. Katoh, K. Misawa, K. Kuma, and T. Miyata. MAFFT: a novel method for rapid multiple sequence alignment based on fast Fourier transform. *Nucleic Acids Research*, 30(14):3059–3066, 2002.
11. C. Lee, C. Grasso, and M.F. Sharlow. Multiple sequence alignment using partial order graphs. *Bioinformatics*, 18(3):452–464, 2002.
12. D.J. Lipman, S.F. Altschul, and J.D. Kececioglu. A Tool for Multiple Sequence Alignment. *Proceedings of the National Academy of Sciences of the United States of America (PNAS)*, 86(12):4412–4415, 1989.
13. B. Morgenstern, K. Frech, A. Dress, and T. Werner. DIALIGN: Finding Local Similarities by Multiple Sequence Alignment. *Bioinformatics*, 14(3):290–294, 1998.
14. S. B. Needleman and C. D. Wunsch. A General Method Applicable to the Search for Similarities in the Amino Acid Sequence of Two Proteins. *Journal of Molecular Biology*, 48:443–53, 1970.
15. C Notredame and DG Higgins. SAGA: sequence alignment by genetic algorithm. *Nucleic Acids Research*, 24(8):1515–1524, 1996.
16. C. Notredame, D.G. Higgins, and J. Heringa. T-coffee: a novel method for fast and accurate multiple sequence alignment. *Journal of Molecular Biology*, 302(1):205–217, 2000.
17. A. Phillips, D. Janies, and W. Wheeler. Multiple Sequence Alignment in Phylogenetic Analysis. *Molecular Phylogenetics and Evolution*, 16(3):317–330, 2000.
18. Gupta SK, Kececioglu JD, and Schaffer AA. Improving the Practical Space and Time Efficiency of the Shortest-paths Approach to Sum-of-pairs Multiple Sequence Alignment. *Journal of Computational Biology*, 2(3):459, 1995.
19. Eddy SR. Multiple Alignment Using Hidden Markov Models. In *Intelligent Systems for Molecular Biology (ISMB)*, volume 3, pages 114–120, 1995.
20. S.-H. Sze, Y. Lu, and Q. Yang. A polynomial time solvable formulation of multiple sequence alignment. In *International Conference on Research in Computational Molecular Biology (RECOMB)*, pages 204–216, 2005.
21. J.D. Thompson, D.G. Higgins, and T.J. Gibson. CLUSTAL W: Improving the Sensitivity of Progressive Multiple Sequence Alignment through Sequence Weighting, Position-specific Gap Penalties and Weight Matrix Choice. *Nucleic Acids Research*, 22(22):4673–4680, 1994.
22. J.D. Thompson, H. Plewniak, and O. Poch. A comprehensive comparison of multiple sequence alignment programs. *Nucleic Acids Research*, 27(13):2682–2690, 1999.
23. Rene Thomsen, Gary B. Fogel, and Thiemo Krink. Improvement of Clustal-Derived Sequence Alignments with Evolutionary Algorithms. In *Congress on Evolutionary Computation*, volume 1, pages 312–319, 2003.
24. Simossis V.A. and Heringa J. A new method for iterative multiple sequence alignment using secondary structure prediction. In *Intelligent Systems for Molecular Biology (ISMB)*, 2002.
25. I.V. Walle, I. Lasters, and L. Wyns. Align-m–a new algorithm for multiple alignment of highly divergent sequences. *Bioinformatics*, 20(9):1428–1435, 2004.
26. L Wang and T. Jiang. On the complexity of multiple sequence alignment. *Journal of Computqtional Biology*, 1(4):337–348, 1994.

PROTEIN INTERACTIONS AND DISEASE

MARICEL KANN
National Center for Biotechnology Information, NIH
Rockville, MD 20814, U.S.A.

YANAY OFRAN
Department of Biochemistry & Molecular Biophysics, Columbia University
New York, NY 10032, U.S.A.

MARCO PUNTA
Department of Biochemistry & Molecular Biophysics, Columbia University
New York, NY 10032, U.S.A.

PREDRAG RADIVOJAC
School of Informatics, Indiana University
Bloomington, IN 47408, U.S.A.

One of the ultimate goals of biological sciences, and certainly one with a high impact on society, is to improve our understanding of the processes and events related to diseases. Molecular biologists, who traditionally study the structure and function of individual proteins and genes, have gained insight and introduced several discoveries that have ultimately reached the bedside. However, biological processes are not realized by a single molecule, but rather by the complex interaction of proteins with their environment, including nucleic acids, ions, lipids, membranes and, of course, other proteins. Thus, while the analysis of the structure and function of individual proteins is crucial for the understanding of their role in biological processes, it has a limited capability to explain the processes themselves.

Some diseases are caused by simple genomic events that mutate or eliminate specific genes (e.g. frame shift mutations or insertion of viral genes). On the other end of the spectrum are "complex diseases", such as heart or psychiatric diseases that are multifactorial and arguably could not be fully accounted for by simple molecular processes. Between these two ends lie many, if not most, of the pathologies and illnesses, including AIDS, cancer, and Alzheimer's disease.

The goal of this session is to discuss the latest progress in the study of protein interactions and disease. The eight papers accepted cover a wide range of approaches and demonstrate readiness of the community to address this problem.

Troncale et al. used hybrid Petri nets to model the regulation of human hematopoieses. Their work demonstrates a noteworthy role of *in silico* approaches for developing an understanding of the subtle events that determine the fate of stem cells (differentiation vs. self-renewal) and that underlie many hematological diseases and their treatment. Bandyopadhyay et al. used a network analysis of gene expression and protein interaction data to identify active pathways related to HIV pathogenesis. A functional analysis of the detected subnetworks provides useful insights into various stages of the HIV replication cycle.

The papers by Ye et al. and Singh et al. explore the relationship between a disease and protein interactions using structural analysis of the interacting proteins. The work by Ye et al. analyzed distributions of non-synonymous SNPs and found that disease-related mutations tend to cluster at the protein surface. They hypothesized that these sites may be involved in protein binding and suggested that one could use comparative modeling to elucidate the mechanism of protein malfunctions. Singh et al. combined structural data with other sources of information to improve the quality of protein-protein interaction predictions. Their results suggest that the predictive power of structural information is high and that structural information alone can be used for successful inference of protein interactions.

Chen et al. developed a framework to mine disease-related proteins from OMIM and protein interaction data. They demonstrate the power of their method by applying it to Alzheimer's disease. The key to their method is a scoring function that ranks proteins according to their relevance to a particular disease pathway. Terribilini et al. used a machine learning approach to classify amino acids on binding sites. In particular, they applied their methodology to Rev proteins of HIV and found good agreement with experimental data.

The remaining two papers focused on computational aspects of predicting protein-protein and protein-nucleic acid interactions. Gunewardena et al. addressed the problem of predicting binding affinity of protein-DNA interactions, while Robinson et al. explored computational aspects related to the correct prediction of protein interactions.

In recent years, the emergence of new experimental protocols and techniques such as RNA, DNA and protein microarrays, two-hybrid systems, and mass spectrometry, as well as the explosion on the number and size of sequence and structure databases, have changed biomedical science. By taking advantage of the enormous amount of data generated by all these techniques computational biology can now attempt to capture more of the complexity of a biological process. The increasing number of computational studies of protein networks, pathways, protein-protein, protein-metabolite and protein-DNA/RNA interactions indicates that it is now possible to address the connections between protein interactions and diseases. We are confident that the papers presented in this session will contribute to further advance this field, which may soon become a major area of biomedical research.

Acknowledgements
The session co-chairs would like to thank the numerous reviewers for their help in selecting the best papers among many excellent submissions.

DISCOVERING REGULATED NETWORKS DURING HIV-1 LATENCY AND REACTIVATION

SOURAV BANDYOPADHYAY, RYAN KELLEY, TREY IDEKER[1,2]

[1]*Program in Bioinformatics, University of California at San Diego, 9500 Gilman Drive, La Jolla, CA 92093, USA.* [2]*Department of Bioengineering, University of California at San Diego, 9500 Gilman Drive, La Jolla, CA 92093, USA*

{sourav,rkelley,trey}@bioeng.ucsd.edu

Human immunodeficiency virus (HIV) affects millions of people across the globe. Despite the introduction of powerful anti-viral therapies, one factor confounding viral elimination is the ability of HIV to remain latent within the host genome. Here, we perform a network analysis of the viral reactivation process using human gene expression profiles and curated databases of both human-human and human-HIV protein interactions. Based on this analysis, we report the identification of active pathways in both the latent and early phases of reactivation. These active pathways suggest host functions that are altered and important for HIV pathogenesis.

1. Introduction

Human Immunodeficiency Virus (HIV-1) infects T lymphocytes and macrophages, resulting in the depletion of CD4+ T cells , which is the defining feature of acquired immune deficiency syndrome (AIDS). Recently, highly active anti-retroviral therapy (HAART) has led to a dramatic decrease in morbidity and mortality due to HIV and, when successful, results in undetectable levels of HIV-1 RNA in blood plasma. This stage of latency represents a period of proviral integration with little to no viral replication [1]. HIV-1 persists in a small reservoir of latently infected resting memory CD4+ T cells, which shows minimal decay even in patients on HAART and can persist for the lifetime of the patient [2, 3]. Latent HIV reservoirs are the principal barriers preventing the eradication of HIV infection and viral reactivation is necessary for targeting by antiviral drugs [2, 4]. Although HIV has been the subject of much study, the precise mechanisms by which the virus reactivates within the host cell remain unclear. Evidence points to the alteration of cellular transcription machinery in order to maximize the viral replication process [5].

Recently, microarray analysis has been employed to survey the changes of host cell transcription [6-8]. In particular, Krishnan and Zeichner have studied the changes in cellular gene expression associated with reactivation and completion of the lytic viral cycle in cell lines chronically infected with HIV-1 [6]. The viral lytic cycle follows distinct mechanistic changes corresponding to stages of HIV synthesis. In the early stage, fully spliced mRNAs for Rev, Tat and Nef are exported from the nucleus for translation. In the late stage, the accumulation of Rev protein in the cytoplasm triggers the export of unspliced viral RNAs from the nucleus to form new viral particles. [9]. Results indicated that uninduced, latent cells had an altered gene expression program and that the host cell underwent specific and ordered changes in gene expression upon reactivation that corresponds to different stages of the viral life cycle.

The authors arrived at these conclusions through application of hierarchical clustering and functional categorization of differentially expressed genes. Clustering gene expression data allows similarly expressed genes to be grouped together, but does not provide any functional explanation for the mechanisms of regulation. Functional categorization identifies known pathways and functional categories that are enriched for differentially expressed genes. This type of analysis provides limited functional insights by constraining analysis to known pathways and reactions [10]. Ideally, we would like to integrate the clustering of similarly expressed genes with an incorporation of a wide variety pathway and biological protein interaction information in a coherent fashion [11, 12]. Network based analysis can improve upon this approach by identifying interesting groups of genes which have not been specifically delineated as a pathway in an ontological framework. One such approach, "Active modules", is a method for searching networks to find subnetworks of interactions with unexpectedly high levels of differential expression [13]. In this approach, gene expression data are mapped onto biological networks, a statistical measure is used to score sub-networks based on gene expression data, and a search algorithm is used to find sub-networks with high score.

Commonly, gene expression analysis using biological networks has been limited to lower organisms for which large scale experimentally derived protein-protein interactions networks are available. The generation of literature-based protein interaction networks from text mining has shown utility in the interpretation of gene expression [14]. However, manually curated protein interaction data is preferable to automated prediction of interactions through natural language processing. The Human Protein Reference Database (HPRD) includes information on protein–protein interactions, post-translational modifications, enzyme–substrate relationships and disease association of human genes which was derived manually by a critical reading of the published literature by expert biologists [15]. Another manually curated database, the HIV-1, human interaction database, provides protein-protein interaction data among 7 of the 9 (excluding *env* and *nef*) HIV-1 genes and human host cell genes[16]. These interactions are traceable to primary literature and are annotated by an ontology of terms describing the nature of the interaction. Together, these networks strive to encompass all of the information that has been published concerning protein inter-relationships between both HIV-1 and human proteins.

Here, we report the identification of protein interaction modules that are significantly activated or repressed across different stages of the latent HIV-1 replication cycle. Our results indicate multiple significant clusters of gene expression in which genes are linked together through established interactions in the literature. This computational analysis allows for the evaluation of a mechanism of the observed changes in gene expression. Analysis of the observed differences in active networks between HIV-1 life cycle stages suggest that these differences are associated with the movement from latent to actively replicating HIV *in vivo*.

2. Methods

2.1. *Microarray Data*

Microarray data was taken from Krishnan et. al. [6] Each array was normalized and the fold changes reported. 131 genes showed altered expression before induction and 1,740 spots showed significant altered gene expression at some point though the lytic replication cycle. Data were averaged over time points corresponding to specific stages in the HIV-1 life cycle which were determined by RT-PCR analysis of specific HIV-1 mRNA fragments. The early stage gene expression of the lytic cycle was taken as the mean over the 0.5,3,6 and 8 hour time points. The intermediate stage gene expression was taken as the mean over 12, 18 and 24 hours post induction. The late stage was the mean over 48, 72 and 96 hours post induction. Each stage included at least 18 arrays including replicates. In total, 1,334 genes were differentially expressed during the early time points, 756 during the intermediate stage, and 566 during the late stage ($P<0.001$). P-values were assigned to each probe for each of the four stages using the t-test using log ratios for all arrays for a specific stage, including replicates against a mean of 0 (no differential expression).

2.2. *Network Generation*

Human – human protein interaction data was culled from HPRD [15] (June 2005 download) and protein fragments were BLAST matched to Entrez Gene and unigene identifiers. HIV-human interactions were taken from the HIV-1, human interaction database[16]. Each protein –protein link can be traced to a specific literature citation which assists in any hypothesis generation.

2.3. *Algorithm Implementation*

Briefly, the ActiveModules algorithm attempts to identify connected regions of a network, which have an unexpectedly high occurrence of genes with significant changes in expression. These network regions represent putative "active modules" in response to a particular test condition. The score of a subgraph is defined as the sum of expression Z-values divided by the square root of the number of nodes in the subgraph.

$$\text{Score}(V) = \frac{1}{\sqrt{|V|}} \sum_{v \in V} z_v$$

Here, V is a set of nodes which define a subgraph, while Z_v refers to the Z score for node v. Individual Z scores are determined by application of an approximation of the inverse normal CDF to the individual expression p-values. This scoring system ensures that if the original Z scores are distributed according to the normal distribution, the expected mean and variance of the subgraph scores are independent of subgraph size. In order to find high scoring regions according to this criterion, a greedy search is initiated from each protein in the network. At each step of the search, all adjacent proteins are considered for inclusion in the result network. The search is executed with a search depth of one node and a maximum diameter of three nodes (corresponding to a local search of "depth"=2 and "max depth"=2 with the jActiveModules plugin available for the Cytoscape Network Modeling package at http://www.cytoscape.org). In order to reduce the influence of network topology on the significance of our final result, we employed a "neighborhood scoring" method [17]. In this method, the search procedure is required to add either all or no node neighbors at each step in the search process. This prevents the selection of a few highly scoring adjacent nodes in a large neighborhood. In order to assess the significance of our result, the search was repeated one-hundred times with the assignment of expression significance values to proteins randomly permuted in each trial. The top scoring result from each of these trials was retained. Those networks which scored higher than 95% of these retained networks were considered significant. To produce smaller subnetworks for visualization in Figures 1 and 2 we repeated the Active Modules search within each original subnetwork with a local search of "depth"=1 and "max depth"=1 to identify singleton nodes which had a significant number of neighbors with differential expression.

2.4. *GO ontology analysis*

We utilized the BiNGO plugin for Cytoscape to determine which Gene Ontology (GO) Molecular Function and Biological Process categories are statistically over-represented in a set of genes [18]. We applied a hypergeometric test to determine which categories were significantly represented (p-value cutoff of 0.01). This significance value was adjusted for multiple hypothesis testing using the Bonferroni Family-wise error rate correction. Only those over-represented terms present at the 8th level of the GO hierarchy were reported.

3. Results

3.1. *An integrated network of protein – protein interactions.*

We sought to elucidate mechanisms of HIV-1 latency and reactivation through integration of biological networks based on literature with measurements of cellular gene expression. Recently, the curation efforts of the Human Protein Reference Database (HPRD) have produced high confidence protein-protein interaction networks derived from literature on a scale such that systems biology based modeling is possible [15]. Additionally, the NCBI has produced a HIV-1-human protein interaction database, providing a summary of known interactions of HIV-1 proteins with those of the host cell [18]. The human protein-protein interaction data consisted of 17,558 interactions among 6,050 genes. HIV-human interactions consisted of 2,420 total interactions over 796 human genes. Representing HIV-1 and human proteins as nodes and interactions between those nodes as edges, we constructed an integrated network which summarizes the corpus of literature-based knowledge about human and HIV interactions.

3.2. *Discovering regulated subnetworks*

The study of Krishnan and Zeichner [6] assayed cellular gene expression of human cell lines chronically infected with HIV, before and during activation of the lytic viral replication cycle. In this study, latently infected ACH-2 cells (derived from a human T-cell line) were treated with phorbol myristyl acetate (PMA), which induces the lytic replication cycle. The changes in cellular gene expression were assayed and compared to uninfected cells exposed to equal amounts of PMA. The authors defined several time points which correspond to different stages of the lytic replication cycle (early, intermediate, late) by comparative analysis of spliced to unspliced mRNAs and cell viability. Their analysis of these stages indicates significant changes of host gene expression in latently infected cells as compared to uninfected cells, as well as systematic, synchronous changes of gene expression in reactivated cells compared to uninfected controls.

To further characterize expression changes at various stages of the viral reactivation cycle, we used an integrated approach of expression clustering and network analysis to find "activated modules" of connected proteins with significant levels of differential activity. We identified highly significant subnetworks for both the latent (uninduced) and early (up to 8 hour post-induction) stages. Both the intermediate and late stages did not produce significant networks ($P<0.05$), which may be due to any combination of factors, notably loss of synchronization, the broad effects of cytopathicity, lack of adequate interaction data or this particular grouping of time points into phases.

In the uninduced stage of the HIV latent infection we found a single active network of 116 Tat-interacting proteins with a score of 10.9 ($P<0.01$). The overview in Figure 1 shows the active subnetwork with differentially expressed neighbors ($P<0.05$) and all HIV interactions removed for clarity. The network was significantly enriched for proteins associated with various aspects of HIV replication, including genes involved in apoptosis and cell death regulation [19] (see Table 1). To condense the network and facilitate interpretation, we ran the algorithm again on the significant subnetwork to find local regions of significant differential expression. The top five modules from this analysis are shown in Figure 1 (a-d). In this stage there is significant down regulation of collagen and fibronectin associated genes (Figure 1c). These are mostly upregulated by Tat [20] during the intermediate stage of the lytic cycle (data not shown) corresponding to their roles in cell-cell adhesion. In the tubulin associated network (Figure 1a), TubA3 interacts with multiple differentially expressed genes. TubA3 expression levels were not available, but due to its in interactions with both Rev and Tat (Rev acts to depolymerize microtubules that are formed by tubulin [21] and Tat binds tubulin[22]) and other differentially expressed neighbors, one can infer its role in the maintenance of the HIV latent phase.

In the early stages of HIV reactivation we found a single active network with a score of 10.7 ($P<0.01$) composed of 79 proteins which all interact with Tat. The subnetwork was enriched for proteins involved in transcription. The apparent importance of this process is consistent with the considerable transcription of integrated viral genes which occurs at this state of viral reactivation (Table 1). The overview Figure 2 shows the active subnetwork with differentially expressed neighbors ($P<0.01$) and all HIV links removed for clarity. We ran the active modules algorithm again and returned the top five proteins within the module which had significant numbers of differentially

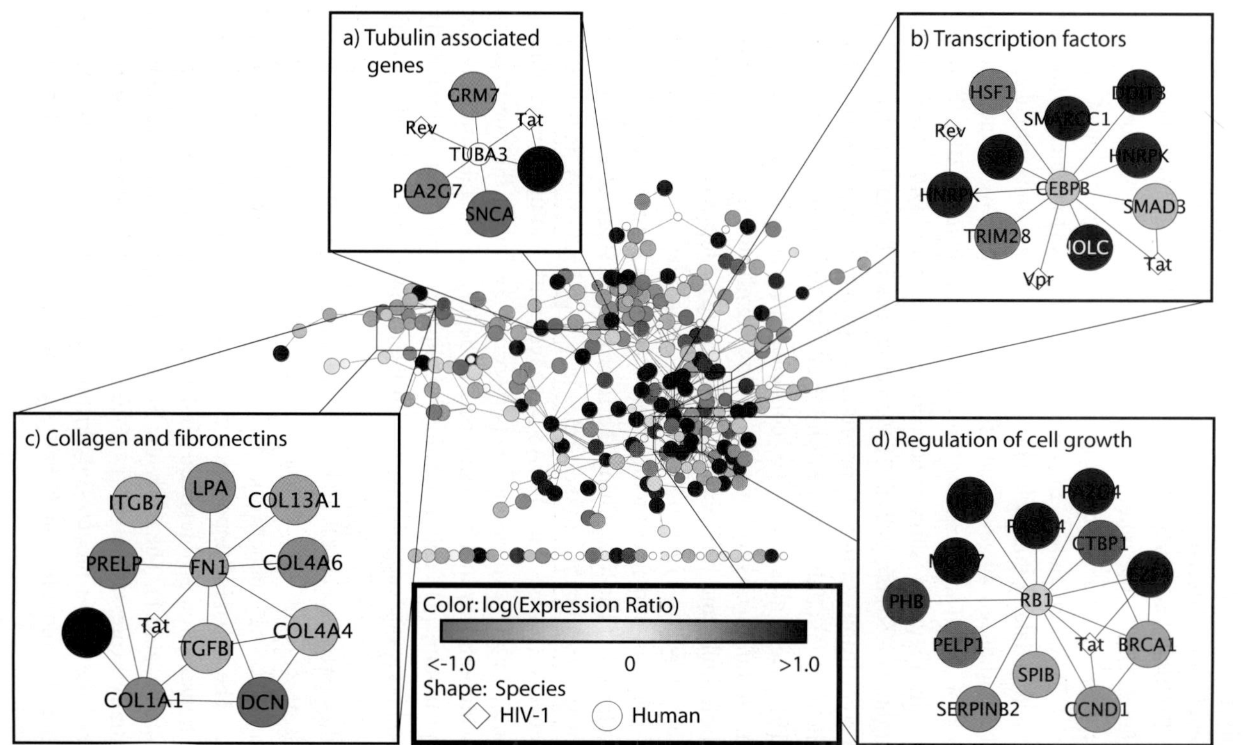

Figure 1: Active subnetwork of gene expression during HIV latency as monitored by gene expression. The significant differentially expressed subnetwork is further broken down into the top five genes with differentially expressed neighbors. These networks correspond to A) alpha-Tubulin (TubA3) and tubulin associate genes; B) CEBPB and other transcription factors; C) FN1, COL1A1 and other collagens and fibronectins; and D) Retinoblastoma (Rb1) and other regulators of cell growth. The color of each node corresponds to the mean change in gene expression, the size of a node is inversely related to its significance of expression during the early phase, and the shape corresponds to the species (HIV - diamond and human- circle).

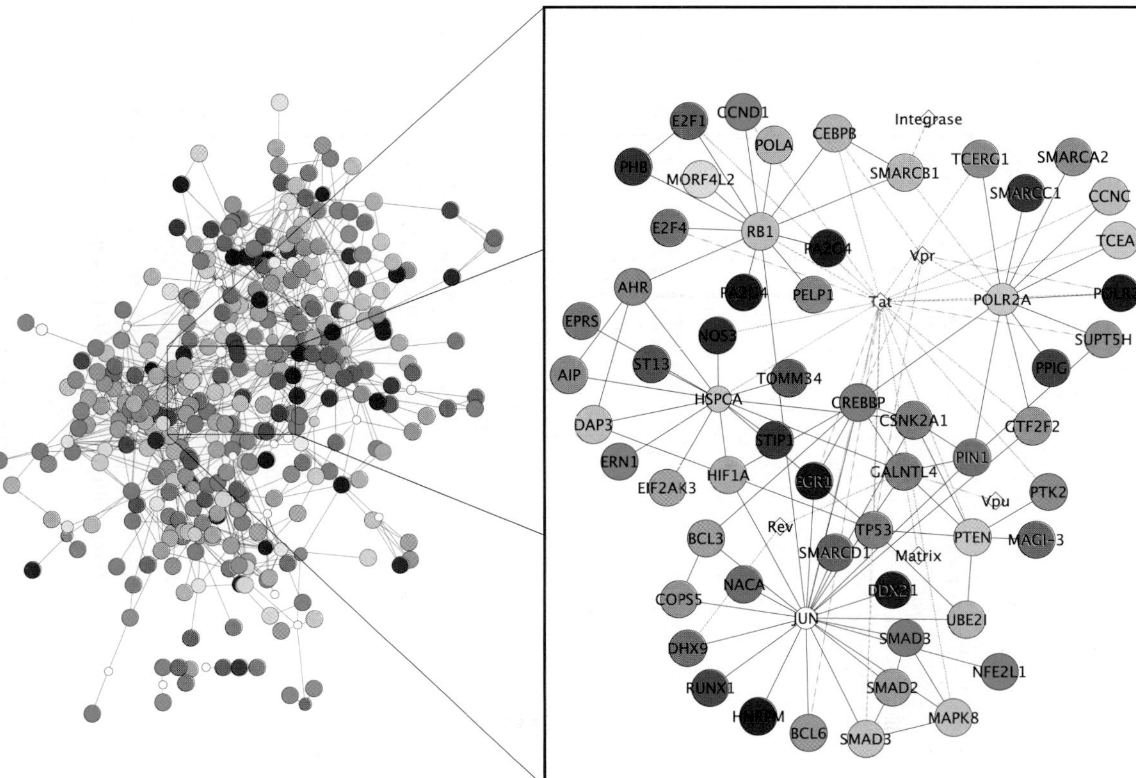

Figure 2: Active subnetwork of gene expression during the early stage of HIV-1 reactivation. The top five proteins associated with proteins with differential expression for one connected component (inset). The proteins retinoblastoma (Rb1), RNA Polymerase II (Polr2a), Jun, HSP 90 (HSPca), and PTEN show interactions with a significant number of differentially expressed genes.

expressed neighbors (Rb1, Polr2a, Jun, HSPca and Pten). The resulting network formed a single connected component, indicating the specific nature of expression changes in this subnetwork. Expectedly, a large portion of this subnetwork was geared toward the expression of HIV-1 proteins including p53 (TP53) which cooperates with Tat in the activation of HIV-1 gene expression [23]. Additionally, casein kinase 2 (Csnk2a1) is activated by Rev in the activation of the HIV-1 replication process [24, 25]. Together these genes interact with 'hubs' of gene expression PTEN and Hsp90 (HSPca) both of which have been shown to associate with Tat [26, 27].

Table 1: GO Categories significantly enriched in the active networks for latent and early reactivation stages of the HIV lytic cycle. The latent stage shows enrichment for apoptotic regulators and the early stage shows enrichment for genes involved in control of transcription.

Stage	GO Term	Aspect	Adjusted Significance
Latent	Induction of programmed cell death	Biological Process	5.44E-08
	Positive regulation of apoptosis	Biological Process	9.03E-08
	Protein kinase C activity	Molecular Function	6.90E-07
	Protein amino acid phosphorylation	Biological Process	3.66E-06
Early Induction	Transcription from RNA polymerase II promoter	Biological Process	8.51E-18
	Regulation of transcription, DNA-dependent	Biological Process	1.09E-14
	Establishment and/or maintenance of chromatin architecture	Biological Process	1.28E-07
	Negative regulation of apoptosis	Biological Process	1.26E-03

Another goal of our analysis was to examine the differences between active subnetworks identified in different stages, to better understand the temporal progression from latent integrated HIV to active replication. We searched the active modules for members interacting with a large number of unique differentially expressed neighbors in each stage. We report a sampling of such human proteins from each stage in Table 2. Most these proteins are known to be involved in HIV related processes. Proteins such as Grb2, Cav1, and Cbl have been loosely implicated with HIV replication previously, but no definite role or interaction for them has been established in the literature. Specifically, it has only recently been identified that Cbl phosphorylation was upregulated by Nef [28]. The dramatically changing active partners with Cbl suggest a role for Cbl in the early stages of viral reactivation. Using the network and a similar scheme, a potential mechanism for the involvement of other proteins in the transition between stages of the HIV-1 cycle can be postulated.

Table 2: A selection of proteins within both control and early induction modules that show both a large number of differentially expressed neighbors and unique differentially expressed neighbors between stages. Shown is the number of differentially expressed neighbors found in either only the latent or early stages as well as the total number of differentially expressed neighbors found in each stage. Proteins annotated with an asterisk represent genes that had no HIV-human interaction in the network.

		Number differentially expressed neighbors					
Stage	Protein	Only in latent stage	Only in early reactivation stage	Total during latent stage	Total during early reactivation stage	Function	HIV link
Latent	Src	7	15	17	25	tyrosine-protein kinase	Phosphorylated by Tat[29]
	Smad4	4	7	8	11	transcription factor	Inhibits Smad-3 and Tat transcription [30]
	Smad3	6	6	10	10	transcription factor	Stimulates Tat transcription[30]
Early Induction	Grb2	8	25	19	36	Cell signaling	Isoform Grb3-3 is upregulated in CD4+ cells[31]*
	Nr3c1	5	14	16	25	nuclear transport, transcription factor	Member of complex targeted by Vpr[32]
	Cav1	7	9	8	10	cell cycle progression	Co-expression blocks virion production[33]*
	Cbl	7	6	8	7	ubiquitination protein ligase	Nef-induced phosphorylation[28]*

4. Discussion

In this report, we have identified a number of genes and pathways involved in HIV latency and reactivation. This identification utilizes an integrated, literature curated network of protein-protein interactions combined with time series expression data for viral reactivation. There are several advantages to an integrated network-based approach. First, the analysis suggests the involvement of genes which are not differentially expressed or may even not have been profiled under a given condition. It may be the case that these genes are post-transcriptionally regulated which cannot be detected by the current approach. In this case, the implication is made based on the combined evidence of many interaction associations with other differentially expressed genes. For example, TUBA3 was involved in one of our most significant networks, but was not itself differentially expressed. Second, this analysis allows more freedom in the identification of active pathways. Given that some pathways may exist in response only to a small number of conditions, it is advantageous to be able to identify such pathways as they are revealed in the interaction data.

One major difference between this approach and previous work is the application of an integrated literature curated interaction network. By using a HIV-human and human-human interaction map, the observed changes in gene expression can be interpreted by identifying altered human pathways stemmed by the invasive HIV-human interactions. Traditional high-throughput networks are known to have a high rate of false positives [34]. Here, since the gene associations are more certain, we can view these subnetworks as providing additional context to the types of conditions under which this association becomes active, such as is the case of genes with a changing neighborhood of differentially expressed genes in Table 2. An additional benefit of such a curated network is that we can interpret those interactions identified in active subnetworks in the context of primary literature citations.

One interesting facet of this research was the examination of how these active networks change from one stage to the next. By comparing and contrasting the members of the significant subnetworks for both latent uninduced and early induction stages of HIV replication we can also find genes who may or may not be differentially expressed themselves but whose neighborhood of differentially expressed genes changes between stages, implicating them in the transitional process. Genes such as Grb2, Cav1, Cbl and Smad2 have no annotated HIV-1 interaction but can be postulated to play a role in the transition from latent to actively replicating HIV. Although we have presented examples that show a dramatic difference in the regulation of their interacting partners, our current approach does not address this question in a systematic fashion. Further development is needed to achieve this goal. With the emergence of an increasing number of large scale protein-protein interaction networks either experimentally or computationally, this type of analysis, while still in its infancy, has the potential to have an immense impact on the investigation of regulatory circuits involved in drugs and disease.

References

1. Pomerantz, R.J., Reservoirs of human immunodeficiency virus type 1: the main obstacles to viral eradication. Clin Infect Dis, 2002. **34**(1): p. 91-7.

2. Blankson, J.N., D. Persaud, and R.F. Siliciano, The challenge of viral reservoirs in HIV-1 infection. Annu Rev Med, 2002. **53**: p. 557-93.
3. Finzi, D., et al., Latent infection of CD4+ T cells provides a mechanism for lifelong persistence of HIV-1, even in patients on effective combination therapy. Nat Med, 1999. 5(5): p. 512-7.
4. Brooks, D.G., et al., Molecular characterization, reactivation, and depletion of latent HIV. Immunity, 2003. **19**(3): p. 413-23.
5. Arendt, C.W. and D.R. Littman, HIV: master of the host cell. Genome Biol, 2001. **2**(11): p. REVIEWS1030.
6. Krishnan, V. and S.L. Zeichner, Host cell gene expression during human immunodeficiency virus type 1 latency and reactivation and effects of targeting genes that are differentially expressed in viral latency. J Virol, 2004. **78**(17): p. 9458-73.
7. Vahey, M.T., et al., Impact of viral infection on the gene expression profiles of proliferating normal human peripheral blood mononuclear cells infected with HIV type 1 RF. AIDS Res Hum Retroviruses, 2002. **18**(3): p. 179-92.
8. van 't Wout, A.B., et al., Cellular gene expression upon human immunodeficiency virus type 1 infection of CD4(+)-T-cell lines. J Virol, 2003. **77**(2): p. 1392-402.
9. Alberts, B., *Molecular biology of the cell.* 4th ed. 2002, New York: Garland Science. 1 v. (various pagings).
10. Kurhekar, M.P., et al., *Genome-wide pathway analysis and visualization using gene expression data.* Pac Symp Biocomput, 2002: p. 462-73.
11. Hanisch, D., et al., *Co-clustering of biological networks and gene expression data.* Bioinformatics, 2002. **18 Suppl 1**: p. S145-54.
12. Tornow, S. and H.W. Mewes, *Functional modules by relating protein interaction networks and gene expression.* Nucleic Acids Res, 2003. **31**(21): p. 6283-9.
13. Ideker, T., et al., *Discovering regulatory and signalling circuits in molecular interaction networks.* Bioinformatics, 2002. **18 Suppl 1**: p. S233-40.
14. Jenssen, T.K., et al., *A literature network of human genes for high-throughput analysis of gene expression.* Nat Genet, 2001. **28**(1): p. 21-8.
15. Peri, S., et al., *Human protein reference database as a discovery resource for proteomics.* Nucleic Acids Res, 2004. **32**(Database issue): p. D497-501.
16. NCBI, *HIV-1, Human Interaction Database.* 2005.
17. Haugen, A.C., et al., *Integrating phenotypic and expression profiles to map arsenic-response networks.* Genome Biol, 2004. **5**(12): p. R95.
18. Maere, S., K. Heymans, and M. Kuiper, *BiNGO: a Cytoscape plugin to assess overrepresentation of Gene Ontology categories in biological networks.* Bioinformatics, 2005.
19. Ashburner, M., et al., *Gene ontology: tool for the unification of biology. The Gene Ontology Consortium.* Nat Genet, 2000. **25**(1): p. 25-9.

20. Taylor, J.P., et al., Activation of expression of genes coding for extracellular matrix proteins in Tat-producing glioblastoma cells. Proc Natl Acad Sci U S A, 1992. **89**(20): p. 9617-21.

21. Watts, N.R., et al., HIV-1 rev depolymerizes microtubules to form stable bilayered rings. J Cell Biol, 2000. **150**(2): p. 349-60.

22. Chen, D., et al., HIV-1 Tat targets microtubules to induce apoptosis, a process promoted by the pro-apoptotic Bcl-2 relative Bim. Embo J, 2002. **21**(24): p. 6801-10.

23. Ariumi, Y., et al., Functional cross-talk of HIV-1 Tat with p53 through its C-terminal domain. Biochem Biophys Res Commun, 2001. **287**(2): p. 556-61.

24. Meggio, F., et al., HIV-1 Rev transactivator: a beta-subunit directed substrate and effector of protein kinase CK2. Mol Cell Biochem, 2001. **227**(1-2): p. 145-51.

25. Ohtsuki, K., et al., Biochemical characterization of HIV-1 Rev as a potent activator of casein kinase II in vitro. FEBS Lett, 1998. **428**(3): p. 235-40.

26. Cook, J.A., A. August, and A.J. Henderson, Recruitment of phosphatidylinositol 3-kinase to CD28 inhibits HIV transcription by a Tat-dependent mechanism. J Immunol, 2002. **169**(1): p. 254-60.

27. Nielsen, B.B., S. McMillan, and E. Diaz, Instruments that measure beliefs about cancer from a cultural perspective. Cancer Nurs, 1992. **15**(2): p. 109-15.

28. Yang, P. and A.J. Henderson, Nef enhances c-Cbl phosphorylation in HIV-infected CD4+ T lymphocytes. Virology, 2005. **336**(2): p. 219-28.

29. Ganju, R.K., et al., Human immunodeficiency virus tat modulates the Flk-1/KDR receptor, mitogen-activated protein kinases, and components of focal adhesion in Kaposi's sarcoma cells. J Virol, 1998. **72**(7): p. 6131-7.

30. Coyle-Rink, J., et al., Interaction between TGFbeta signaling proteins and C/EBP controls basal and Tat-mediated transcription of HIV-1 LTR in astrocytes. Virology, 2002. **299**(2): p. 240-7.

31. Li, X., et al., Grb3-3 is up-regulated in HIV-1-infected T-cells and can potentiate cell activation through NFATc. J Biol Chem, 2000. **275**(40): p. 30925-33.

32. Refaeli, Y., D.N. Levy, and D.B. Weiner, The glucocorticoid receptor type II complex is a target of the HIV-1 vpr gene product. Proc Natl Acad Sci U S A, 1995. **92**(8): p. 3621-5.

33. Llano, M., et al., Blockade of human immunodeficiency virus type 1 expression by caveolin-1. J Virol, 2002. **76**(18): p. 9152-64.

34. von Mering, C., et al., Comparative assessment of large-scale data sets of protein-protein interactions. Nature, 2002. **417**(6887): p. 399-403.

MINING ALZHEIMER DISEASE RELEVANT PROTEINS FROM INTEGRATED PROTEIN INTERACTOME DATA

JAKE YUE CHEN[†]

Indiana University School of Informatics
Purdue University School of Science, Dept. of Computer and Information Science
Indianapolis, IN 46202, USA

CHANGYU SHEN

Division of Biostatistics, Indiana University School of Medicine
Indianapolis, IN 46202, USA

ANDREY Y. SIVACHENKO

Ariadne Genomics, Inc., 9700 Great Seneca Hwy
Rockville, MD 20850, USA

Huge unrealized post-genome opportunities remain in the understanding of detailed molecular mechanisms for Alzheimer Disease (AD). In this work, we developed a computational method to rank-order AD-related proteins, based on an initial list of AD-related genes and public human protein interaction data. In this method, we first collected an initial seed list of 65 AD-related genes from the OMIM database and mapped them to 70 AD seed proteins. We then expanded the seed proteins to an enriched AD set of 765 proteins using protein interactions from the Online Predicated Human Interaction Database (OPHID). We showed that the expanded AD-related proteins form a highly connected and statistically significant protein interaction sub-network. We further analyzed the sub-network to develop an algorithm, which can be used to automatically score and rank-order each protein for its biological relevance to AD pathways(s). Our results show that functionally relevant AD proteins were consistently ranked at the top: among the top 20 of 765 expanded AD proteins, 19 proteins are confirmed to belong to the original 70 AD seed protein set. Our method represents a novel use of protein interaction network data for Alzheimer disease studies and may be generalized for other disease areas in the future.

1. Introduction

Alzheimer Disease (AD) is a progressive neurodegenerative disease with 4.5 million patients in the United States today. This number of AD patients is

[†] To whom correspondence should be sent. Email: jakechen@iupui.edu.

expected to increase to 11 to 16 million by 2050 when the baby boomers age. The mental status of an AD patient deteriorates irreversibly over time and complete care is required for basic daily activities at late stages of the disease. In 2000, health care costs for AD patients in the United States totaled approximately $31.9 billion, which is expected to reach $49.3 billion by 2010 (above statistics can be found at http://www.alz.org/). Therefore, AD has become a major public health concern.

The exact molecular mechanisms leading to massive damage associated with AD remain unclear. Typically, AD patients exhibit selective brain neuronal loss, extracellular amyloid (senile) plaques, and intracellular neurofibrillary tangles (NFT) of hyperphosphorylated *tau* protein [*1, 2*]. According to the well-accepted "amyloid hypothesis" [*3, 4*], the unusual accumulation of beta-amyloid peptide (Aβ), the cleavage products of amyloid precursor protein (APP), is the major cause of AD early development. In *Familial Alzheimer Disease* (FAD), genetic defects of either APP or presenilin (PSEN1, PSEN2)—proteins that are normally responsible for breaking down Aβ proteins—often lead to abnormal formation of Aβ as "protofibrils" [*5*]. Aβ protofibrils can incite inflammatory response through cytotoxic cytokines and disrupt intracellular Ca^{2+} homeostasis through over-activation of glutamate receptors, therefore leading cells to oxidative stress and mitochondrial injury. Aβ protofibril deposit in the extracellular space can also cause neuronal cell damage by blocking axonal transport. Aberrant Aβ accumulation further causes aberrant accumulation of *tau* protein, which is essential to the initiation and stabilization of neuronal microtubules. As time going by, gradual breakdown of neuronal cytoskeleton eventually leads to neuron apoptosis in AD patients (For a comprehensive review, see [*1, 2*] and references therein). Complex and broad range of these cellular and biochemical responses make researchers believe that there must be a sophisticated network of AD signal transduction, gene regulation, and protein-protein interaction events. Therefore, deciphering AD related molecular network "circuitry" can help researchers understand AD disease model details and propose treatment ideas.

In this work, we will conduct initial AD protein interaction network analysis and demonstrate how to gain protein functional knowledge not directly implied from sequence information. We will organize the main body of the work by presenting our computational data analysis methods and results. We will discuss how to interpret our results and why this study proves to be significant at the end.

2. Computational Method

We introduce the computational techniques and procedures developed for AD protein interaction sub-network analysis, which can be summarized as the following. First, we searched the Online Mendelian Inheritance in Man (OMIM) database [*6*] to obtain an initial collection of AD related genes. Second, we used the HUGO Gene Nomenclature Committee (HGNC) [*7*] database to map the initial AD related genes to AD related proteins identified by their SwissProt IDs. Third, we used a nearest-neighbor expansion method to build an expanded AD protein interaction sub-network. Fourth, we developed and applied a bioinformatics software tool, ProteoLens [*8*], to visualize and annotate the AD interaction sub-network. Fifth, we performed statistical analysis to assess the significance of the subnetwork extracted. Sixth and lastly, we developed a heuristic algorithm and scoring method, which we used to obtain a rank-ordered list of proteins significantly related to the AD disease. A detailed description of our method is provided below.

2.1. *Initial Collection of AD Related Genes*

We used the OMIM database [*6*] as the starting point to retrieve an initial collection of AD related genes. In OMIM, human genes associated with genetic disorders are recorded in a mini-review format, along with additional information such as their functions, participating molecular pathways, and other disease-related information. To obtain a list of AD related genes, we performed a search of the OMIM database (integrated into our biological data warehouse in early 2004), retrieving each OMIM gene record in which the "description" field contains the term "Alzheimer". 65 OMIM gene records are retrieved. Note that since the retrieval method is coarse (i.e., based on simple term matches), the 65 collected AD related gene records may suffer from both *false positives* (containing retrieved genes that are not actually functionally relevant to AD) and *false negatives* (missing genes that are indeed functionally related to AD but not retrieved). Soon in subsequent protein interaction network analysis, we will use protein interaction network neighborhood information and show that these concerns can be ameliorated.

2.2. *Mapping of Initial AD Related Genes to Proteins*

We used the HUGO Gene Nomenclature Committee (HGNC) [*7*] database of gene symbols and proteins to map gene symbols to their correct protein identifiers. HGNC is an international standard repository of officially approved gene symbols. For each gene, the HGNC database provides its standard gene symbol and gene mappings to various IDs used in common public databases, e.g., Swiss-Prot, NCBI RefSeq, NCBI Locuslink, and KEGG enzyme. For our work, we started with 65 sets of OMIM gene records, some of which were associated with more than one gene symbol. After mapping all the gene symbols to protein SwissProt IDs using the HGNC gene mapping table, we obtained 70 AD-related proteins. The slight increase in protein count is due to one-to-many mapping between a gene and its multiple splice variant forms at the protein level.

2.3. *Collection and Expansion of AD Related Protein Interactions*

We used the Online Predicted Human Interaction Database (OPHID) [9] to collect AD related protein interaction data. OPHID is a web database of more than 40,000 human protein interactions involving ~9,000 human proteins. It is a comprehensive repository of known human protein interactions, both from curated literature publications and from high-throughput experiments. It also contains predicted interactions inferred from eukaryotic model organisms, e.g., yeast, worm, fly, and mouse. The prediction was performed by mapping interacting protein pairs from available model organisms onto their orthologous protein pairs in human, or by making inference from interacting domain co-occurrence, co-expression, and GO semantic distance evidence. More than half of OPHID's records are predicted human protein interactions; however, not all OPHID human protein interactions carry the same level of significance. In general, those derived from real human protein interaction experiments should be much more trustworthy than those derived from predictive methods applied on yeast data sets. Therefore, to assign an estimated interaction confidence score, we developed the following *heuristic* scoring rules:

1. Protein interactions from human experimental measurement or from literature curation are assigned a **high** confidence score of 0.9;
2. Human protein interactions inferred from high-quality interactions in mammalian organisms are assigned a **medium** confidence score of 0.5;
3. Human protein interactions inferred from low quality interactions or non-mammalian organisms are assigned a **low** confidence score of 0.3.

With the initial AD-related protein list and a comprehensive OPHID protein interaction data set, we can now derive the AD-related protein interaction sub-network using a **nearest-neighbor expansion method**. Here, we denote the initial 70 AD-related proteins as the **seed-AD-set**. To build AD sub-networks, we pull out protein interacting pairs in OPHID such that at least one member of the pair belongs to the seed-AD-set. The set of interacting pairs pulled out will be called the **AD-interaction-set**. We denote the new set of proteins expanded from initial seed-AD-set by new proteins involved in the AD-interaction-set as the **enriched-AD-set** (a superset of seed-AD-set). In our study, the AD-interaction-set contains 775 human protein interactions; the enriched-AD-set contains 657 human proteins identified by Swissprot IDs.

2.4. *Visualization of AD Protein Interaction Sub-Network*

We developed ProteoLens [*10*], a visual biological network data mining and annotation tool that can be freely downloaded at http://bio.informatics.iupui.edu/proteolens/, to help us analyze AD-related protein interaction sub-network. ProteoLens has native built-in support for relational database access and manipulations. It allows expert users to browse database schemas and tables, filter and join relational data using SQL queries, and customize any combination of data fields. The reconfigured view of data can be immediately visualized in the ProteoLens network viewer without needing to be exported as flat files first. Note that network nodes and edges can be used to represent proteins and protein interactions, whereas node/edge size, width, shape, and color can all be used to dynamically bind to customized data fields (such as gene symbol, functional category, and confidence score) to be visualized. Once a visual network layout is generated, the layout, visual annotation, and network member proteins/protein interactions can be tweaked without file editing.

2.5. *Statistical Evaluation of Sub-network*

We performed statistical data analysis tests to examine the significance of the connected sub-network formed by AD-interaction-set. Our hypothesis for this statistical evaluation is that if the enriched-AD-set indeed consists of functionally related proteins involved in the same process—even if the process were complex and broad—then we should expect that the **connectivity** among the enriched-AD-set proteins to be higher than that among a set of randomly selected proteins.

To formulate our hypothesis precisely, we introduce three concepts. First, we define a **path** between two proteins A and B as a set of proteins P1, P2,…, Pn such that A interacts with P1, P1 interacts with P2, …, and Pn interacts with B. Note that if A directly interacts with B, then the path is the empty set. Second, we define the **largest connected sub-network** of a network as the largest subset of proteins and interactions, among which there is at least one path between any two proteins in the subset. Third, we define the **index of aggregation** of a network as the ratio of the size of the largest sub-network that exists in this network to the size of this network. Note that size is calculated as the total number of proteins within a given network/sub-network.

To test the hypothesis that the enriched-AD-set proteins are "more connected" than a randomly selected set of protein, we develop the null hypothesis test using the following resampling procedure [*11*]:

1) Randomly select from the OPHID database the same number of human proteins as in the seed-AD-set.
2) Build the superset of the selected set by using the same nearest-neighbor expansion method described earlier.
3) Find the largest sub-network of the superset.
4) Compute the index of aggregation of the superset.
5) Repeat steps 1 through 4 1,000 times to generate a distribution of the index of aggregation under random selection.
6) Compare the index of aggregation of the enriched-AD-set with the distribution obtained in 5 and calculate the *p*-value.

2.6. *Scoring of Significant Proteins in the Sub-network*

In the final step, we present a scoring method to rank proteins in the sub-network, based on their overall roles and contribution to the AD related protein interaction sub-network. The **role** of a protein in the sub-network can be qualitatively defined as its ability to connect to many protein partners in the network with high specificity (the less promiscuously connected, the better) and high fidelity (the higher the interaction confidence, the better). To define this role quantitatively, we introduce a heuristic **relevance score** function s_i for each protein i from the sub-network:

$$s_i = k * \ln\left(\sum_{j\in N(i)\cap A} p(i,j)\right) - \ln\left(\sum_{j\in N(i)\cap A} N(i,j)\right), \tag{1}$$

In Eq. 1, i and j are indices for proteins in the sub-network, k is an empirical constant (k>1; we set k=2 here), $N(i)$ is the set of interaction partners of protein i in the network, A is the set of proteins in enriched-AD-set, $p(i,j)$ is the initial confidence score that we assigned to each interaction between proteins i and j (described in section 2.3), and $N(i,j)$ holds the value of 1 if protein j belongs to the intersection of $N(i) \cap A$ or 0 otherwise. Empirically assessing the relevance score function, we can tell that the score s_i ranks favorably in situations where interacting proteins with many high confidence interactions among its neighbors will fare out better than those with many low-quality interactions and those with only a few interactions. To avoid showing a negative score, in this work, we further converted s_i to the exponential scale using the transformation $t_i = exp(s_i)$, and report t_i as the final protein ranking score.

3. Results

By following the data analysis steps outlined in the Method section, we obtained the following results.

3.1. *AD Related Proteins and Protein Interactions*

In the AD seed set, we have an initial list of 65 AD related OMIM gene records. These records are later mapped to 70 seed-AD-set proteins using gene-to-protein mapping tables from HGNC. As explained earlier, this discrepancy was due to the one-to-many mapping relationships between genes and their protein products. Using OPHID and the nearest-neighbor expansion method, we obtained 775 AD-related human protein interactions (as the **AD-interaction-set**). This expanded AD-interaction-set contains an expanded 657 human proteins (as the **enriched-AD-set**).

The proteins in the enriched-AD-set form 16 sub-networks, with a size ranging from 2 to 586 (or, a relative size from 0.3% to 89.2%). Therefore, the largest connected sub-network of the enriched-AD-set contains 586 proteins and the **index of aggregation** is 82.9%. This suggests that the majority of AD related proteins are closely related by physical interaction—a phenomena that we would like to test for statistical significance (see Section 3.3 soon)

3.2. *Visualization of AD Expanded Protein Interaction Network*

Figure 1 shows the enriched-AD-set proteins and the AD-interaction-set of human protein interactions in visualized network. All the seed-AD-set proteins (shown as nodes) are colored dark gray, while the non seed-AD-set proteins (also shown as nodes) are colored light gray. All the protein interactions (shown as edges) are also color-labeled, with high-quality interaction in black, medium-quality interactions in dark gray, and low-quality interactions in gray. We observe that interactions tend to "fan out" from a few protein hubs in the network, and that there are relatively few interactions among the proteins extending from the seed-AD-set. One expects that true AD-related proteins would interact with many seed-AD-related proteins with high degree of confidence and specificity.

3.3. *Statistical Significance*

The empirical distribution of the index of aggregation obtained after 1000 random re-samplings is shown in Figure 2. Only 8 runs out of 1000 resulted in an index of aggregation value greater than 89.2%. Therefore, the p-value of the observed index of aggregation of the enriched-AD-set is 0.008. It is not surprising to observe such a significant result since the AD-set is selected in a way that proteins inside the set share certain level of connection since all of them are associated with AD.

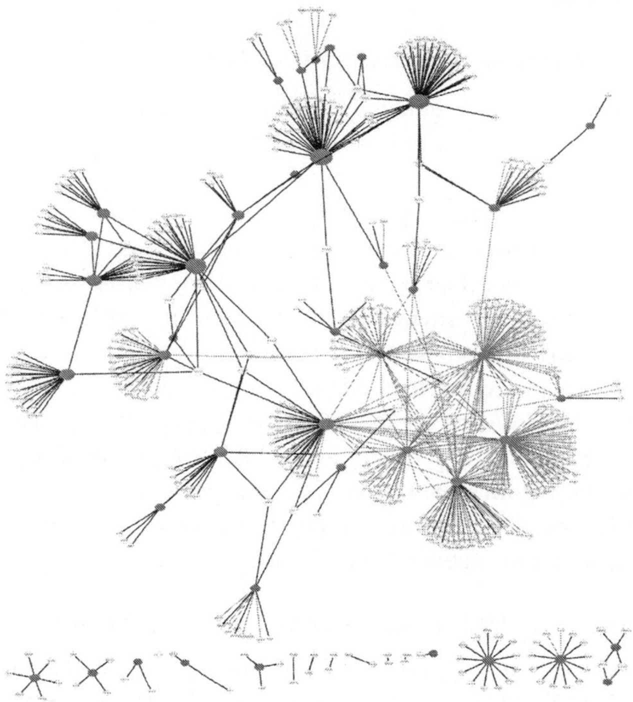

Figure 1. A network of OPHID human interactions expanded from initial 70 seed-AD-set (colored in dark gray). Protein interactions are also colored differently according to confidence level assigned (see text for details).

Next, the relevance score was calculated for each protein in the enriched-AD-set. The results (Table 1) show that our scoring function exhibits very high specificity: out of 20 top-scoring proteins, all but one (β-catenin, CTNNB1) are known AD-related proteins according to OMIMM annotation. Further literature study (see discussion) suggests that even CTNNB1 could be involved in the AD disease development process [*12*]. This result opens up exciting new possibilities to identify novel or previously ignored members of the AD pathway(s) for subsequent protein drug target investigations or disease biomarker studies.

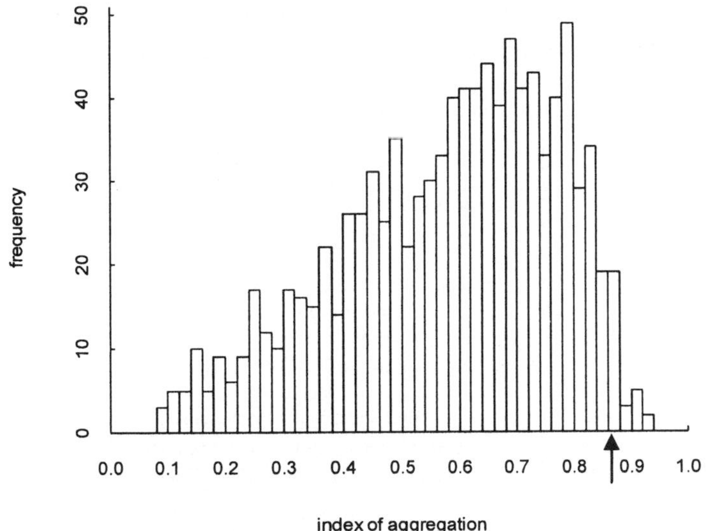

Figure 2. Histogram of the index of aggregation distribution for the enrichments of sets of proteins (size=48) randomly selected from OPHID. The arrow indicates the index of aggregation value for the enriched-AD-set.

4. Discussion

The integrated approach to the analysis of human interaction data and Alzheimer disease proteins allowed us to validate existing disease protein targets and predict novel ones not present in the initial list of disease protein targets that we started with. The result can be interesting to Alzheimer disease biologists, and our method can be generalized to other disease biology areas. By further examining our discoveries in the top-ranked proteins (top 20 are tabulated in Table 1, other will be made available on our web site http://bio.informatics.iupui.edu/ soon), we can make the following comments:

First, one of the important Alzheimer disease proteins, tau protein (MAPT, ranked 31), a well-known participants of AD-linked degeneration pathway(s), was not initially retrieved by our automated procedure from OMIM data but was later recovered from the interaction data analysis. Therefore, at least in a few isolated cases, our method can allow the recovery of false negatives.

Second, the amyloid beta A4 precursor-protein binding protein, APPB1 (ranked 33), represents another interesting case. It is a well-known interaction partner of APP, but genetic link to AD was reported in OMIM only for the other member of the family, APPB2 (ranked 32). Our method still predicts that APPB1 also plays some role in AD. A recent literature report [*13*], shows that

APPB1 indeed directly associates with tau and may provide the crucial missing link between tau and APP proteins in Alzheimer disease.

Table 1. Top 20 rank-ordered AD relevant proteins.

Score	Gene	Description	AD Relevance
43.01	APP	amyloid beta (A4) precursor protein (protease nexin-II, Alzheimer disease)	Known
36.98	PSEN1	presenilin 1 (Alzheimer disease 3)	Known
35.64	LRP1	low density lipoprotein-related protein 1 (alpha-2-macroglobulin receptor)	Known
21.87	PSEN2	presenilin 2 (Alzheimer disease 4)	Known
20.89	PIN1	protein (peptidyl-prolyl cis/trans isomerase) NIMA-interacting 1	Known
19.37	FHL2	four and a half LIM domains 2	Known
15.39	S100B	S100 calcium binding protein, beta (neural)	Known
12.96	FLNB	filamin B, beta (actin binding protein 278)	Known
12.37	CTNND2	catenin (cadherin-associated protein), delta 2 (neural plakophilin-related arm-repeat protein)	Known
12.15	CLU	clusterin (complement lysis inhibitor, SP-40,40, sulfated glycoprotein 2, testosterone-repressed prostate message 2, apolipoprotein J)	Known
11.34	APBA1	amyloid beta (A4) precursor protein-binding, family A, member 1 (X11)	Known
10.00	NAP1L1	nucleosome assembly protein 1-like 1	Known
9.54	GTPBP4	GTP binding protein 4	Known
9.48	NCOA6	nuclear receptor coactivator 6	Known
9.15	CDK5	cyclin-dependent kinase 5	Known
7.44	CTSB	cathepsin B	Known
7.29	ASL	argininosuccinate lyase	Known
4.86	**CTNNB1**	**catenin (cadherin-associated protein), beta 1, 88kDa**	**Novel**
4.86	NCKAP1	NCK-associated protein 1	Known
4.86	AGER	advanced glycosylation end product-specific receptor	Known

Third, β-Catenin (CTNNB1, ranked 18 as shown in Table 1), was not previously associated with AD in OMIM or in the general biomedical community, therefore representing a "clear" case of computational prediction results. Interestingly, while the exact role of β-catenin in AD is not well understood, it is known that Wnt signaling pathway (which β-catenin is a part of) is a target of Aβ toxicity [*14*]. Moreover, the Wnt-3a ligands and other agents that are reported to overcome beta amyloid toxicity stabilize CTNNB1 levels in cytoplasm [*15, 16*]. It should be stressed that while the OPHID interactions between β-catenin and AD-set proteins are of high-quality, *i.e.* derived from the literature, one could only speculate about the potential role of β-catenin, since, for instance, both CTNNB1 and its interaction partners's expressions are far from being limited to neurons. It is the *pattern* of β-catenin interactions revealed through the analysis of combined evidence that resulted in high AD-relevance score for β-catenin.

In all, our method incorporated protein interaction data and helped us to successfully carry out Alzheimer disease related biological studies. The computational results, which began with inputs that are not necessarily highly reliable, showed high biological relevance. Going down the ranked protein targets, one may generate many new biological hypotheses about the new functions of proteins in the protein interaction network context beyond the scope of this work. We are currently developing collaborations with industrial partners who has accumulated experimental human protein interactome to further conduct these computational investigations on high-confidence data sets. Meanwhile, we are also in the process of developing better scoring functions and applying these methods to the study of other disease areas.

Acknowledgments

This work was supported in part by systems obtained by Indiana University through its relationship with Sun Microsystems Inc. as a Sun Center of Excellence. We would like to thank Stephanie Burks who helps us maintaining a robust computer systems and Oracle 10g database server at Indiana University.

References

1. E. Bossy-Wetzel, R. Schwarzenbacher and S. A. Lipton, *Nat Med* **10 Suppl**, S2 (2004).
2. A. Kowalska, *Pol J Pharmacol* **56**, 171 (2004).
3. J. Hardy, *Trends Neurosci* **20**, 154 (1997).
4. D. J. Selkoe, *Nature* **399**, A23 (1999).
5. M. Hutton, J. Perez-Tur and J. Hardy, *Essays Biochem* **33**, 117 (1998).
6. . McKusick-Nathans Institute for Genetic Medicine, Johns Hopkins University (Baltimore, MD) and National Center for Biotechnology Information, National Library of Medicine (Bethesda, MD); Web URL: http://www.ncbi.nlm.nih.gov/omim/ 2000.
7. S. Povey, et al., *Hum Genet* **109**, 678 (2001).
8. A. Sivachenko and J. Y. Chen, *Bioinformatics* (2005), submitted.
9. K. R. Brown and I. Jurisica, *Bioinformatics* **21**, 2076 (2005).
10. J. Y. Chen and A. Sivachenko, *IEEE Magazine in Biology and Medicine* **24**, 95 (2005).
11. W. J. Ewens and G. Grant, "Statistical Methods in Bioinformatics: An Introduction" Springer, New York, 2004.

12. M. Nishimura, et al., *Nat Med* **5**, 164 (1999).
13. C. Barbato, et al., *Neurobiol Dis* **18**, 399 (2005).
14. R. A. Fuentealba, et al., *Brain Res Brain Res Rev* **47**, 275 (2004).
15. A. R. Alvarez, et al., *Exp Cell Res* **297**, 186 (2004).
16. R. A. Quintanilla, et al., *J Biol Chem* **280**, 11615 (2005).

ACCOUNTING FOR STRUCTURAL PROPERTIES AND NUCLEOTIDE CO-VARIATIONS IN THE QUANTITATIVE PREDICTION OF BINDING AFFINITIES OF PROTEIN-DNA INTERACTIONS

SUMEDHA GUNEWARDENA AND ZHAOLEI ZHANG

Charles H. Best Institute, University of Toronto,
112 College Street, Toronto, Ontario,
M5G 1L6, Canada
E-mail: Sumedha@cantab.net

We describe a quantitative model for predicting the binding affinity of protein-DNA interactions. The described model is based on *templates* capable of providing a global representation of the modelled transcription factor (TF) binding sites. Templates can capture non independent nucleotide variations and structural properties present in these sites. Tests carried out on the *p50p50* and *p50p65* variants of the transcription factor *NF-κB* demonstrate a high correlation between the observed binding affinities and the binding affinities predicted by the templates. Only a small subset of training data spanning the space of the binding sites is required to train the templates.

1. Introduction

In human and other higher eukaryotes, gene expression is regulated by the binding of various modulatory transcription factors (TF) onto *cis*-regulatory elements near genes. Binding of different combinations of transcription factors may result in a gene being expressed in different tissue types or at different developmental stages. To fully understand a gene's function, therefore, it is essential to identify the transcription factors that regulate the gene and the corresponding TF binding sites.

TF binding sites are relatively short (10-20bp) and highly degenerate sequences, which makes their effective identification a computationally challenging task. Early methods for identifying TF binding sites were mainly non-quantitative binary classifiers. They ranged from consensus sequences[24] and position specific weight matrices[20,27] to approaches such as rule-based systems[28], Gibbs sampling[12], expectation maximisation[8], neural networks[14,10] and comparative genomics[35,33].

Transcription factors, unlike restriction enzymes for example, display a wide variation of sequence specific binding affinities characterizing the strength of their interaction with different *cis*-regulatory elements that control the transcriptional mechanism[32,27]. A need for quantitative models for predicting the strength of protein-DNA interactions arise from this variation of binding affinities displayed by different sites to a given transcription factor. There is evidence to suggest that the binding energy of a protein-DNA interaction is to some extent intrinsic in the base composition of the operator DNA. Berg and von Hippel[2] for example showed, using statistical-mechanical theory, that given a set of regulatory sites, the logarithm of the base frequencies of those sites were proportional to their binding affinity. A more refined version of their calculation, taking into account the base composition of the genome in question, was introduced by Stormo & Fields[27] and Sarai & Takeda[22] showed that the binding energy of a site was additive in the free energy changes of individual bases. The binding affinity of a protein-DNA interaction can be measured experimentally as an equilibrium constant of its binding reaction[25,21].

Studies carried out on various domains of TF binding sites have shown correlated nucleotide variations to exist between different nucleotide positions of those sites[15,30,34]. This has led many researchers to question the base independence assumption on which methods such as consensus sequences and weight matrices for identifying TF binding sites are based[4,15]. This issue has been addressed by different authors with different techniques ranging from non-quantitative models such as improved weight matrices with prior information on correlated nucleotide positions[36,37], biophysical approaches[5], non-parametric models[11] and neural networks[14,10] to quantitative models such as principal coordinates analysis[31]. The method introduced in this paper, among other things, accounts for nucleotide co-variations to improve the quantitative prediction of binding affinities of protein-DNA interactions.

Another feature that plays a role in protein-DNA interactions is nucleotide structure[1,16,7]. It is reasonable to expect, given the multitude of binding sites recognized by the transcription machinery, that there exists other factors beside sequence similarity that influence the binding process. It has been shown that the binding of transcription factors cause a significant distortion to the regular twist and bending of the DNA double helix[9,23,26]. This often results in the bound DNA strand changing conformation from its B-form to A- or Z- forms[13]. It is conceivable that the binding affinity of a site to a given transcription factor will depend, at least to some extent, on its ability to tolerate such structural distortion from the classical B-form.

Nussinov[16], for example, demonstrated the presence of structural homology in regions with weak sequence homology at sites −10, −35 and −16, of the Escherichia coli promoter. Structural properties of a DNA helix can be expressed in terms of its conformational parameters in di-nucleotide and tri-nucleotide models. There are many different such parameters reported in the literature. The Property database[18], for example, lists 38 such parameters. Many non-quantitative algorithms have been developed to analyze binding sites based on their structural homology[19,17,29]. The method described in this paper uses a combination of different structural parameter representations of a site to account for sequence structure in the prediction of its binding affinity to a given transcription factor.

2. Method

The key to our method is the use of numerical *templates* to capture certain key features of TF binding sites. One of the principle drawbacks of base-independent models of TF binding sites is their inability to account for non-independent nucleotide variations. One problem of modelling nucleotide substitutions in a general model of TF binding sites is that the nucleotide positions that exhibit such correlations vary from factor to factor. As the exact positions on the TF binding sites which are correlated are unknown in the general case, one would need a model that accounts for all pairs of positions on the sites to fully represent them, which will need a very large number of parameters (e.g. a fully connected HMM). Templates present a compromise between the base independent model and the fully connected model. They model the correlation of an individual position relative to the rest of the positions on the site. By restricting the expression of correlation of a given position on the sites to all the other positions, instead of individual pairs of positions, templates are able to reduce the number of parameters required to the length of the sites, while still capturing a global representation of the positional correlations present in them.

In the template model, each template (defined by its template parameters $\mathbf{t}$) is modelled on a given numerical encoding of the nucleotides forming the training set of binding sites. The numerical encoding can be some value assigned to individual nucleotides or a value assigned to a combination of them. Values can be assigned to single nucleotides to capture sequence properties (e.g. sequence homology) of the sites. Values can be assigned to di- and tri- nucleotides to capture geometric and structural properties (e.g. propeller twist, stacking energy, protein induced deformability, DNAse I sensitivity, etc.) of the sites. For a given nucleotide sequence, s, and a given nucleotide parameter p, the resulting numerical vector will be denoted $\mathbf{r}^p(s)$ (see Figure 1.). Each nucleotide sequence is first converted into a table

$p =$ Slide

aa	at	ag	ac	ta	tt	tg	tc	ga	gt	gg	gc	ca	ct	cg	cc
0.1	-0.7	-0.3	-0.6	0.1	0.1	0.4	0.1	0.1	-0.6	-0.1	-0.3	0.4	-0.3	0.7	-0.1

$s =$ g g c g t g g c $\quad$ ($\mathbf{r}^p(s) = -0.1, -0.3, +0.7, -0.6, +0.4, -0.1, -0.3$)

Figure 1. The figure shows the encoding $\mathbf{r}^p(s)$ of sequence s by the dinucleotide step parameter values $p =$'Slide'.

of numerical representations. For each nucleotide sequence s, the representations of s that we work with will be denoted $\mathbf{r}^1(s), \mathbf{r}^2(s), \ldots, \mathbf{r}^m(s)$ where m is the number of parameters selected.

The global representation of positional correlations of a TF binding site s, encoded with parameter p (where p can be for example a mono-, di- or tri-nucleotide parameter), having encoded length L, is captured by a template t and is given by the following equations. As we do not know the contribution of individual bases towards the binding energy of a site, we make a simplifying assumption that the bases make a uniform contribution towards it, hence in these equations, the binding energy of a site is modelled as an external potential f equally distributed across its bases.

$$(\mathbf{Q} \ \mathrm{diag}(\mathbf{r}^p(s))) \ \mathbf{t} = \mathbf{r}^p(s) - \mathbf{f} - \mathbf{e} \tag{1}$$

Where $\mathbf{t} = (t[1], t[2], \ldots, t[L])^T$, is the vector of template parameters, $\mathbf{r}^p(s)$, is the vector representing the encoding of nucleotide sequence s with parameter p, $\mathbf{f} = (f, f, \ldots, f)^T_{(1\times L)}$, is an equally distributed vector of the binding affinity of site s, $\mathbf{e} = (e[1], e[2], \ldots, e[L])^T$, the residual error and $\mathbf{Q}_{(L\times L)}$ a square matrix with zeros on the diagonal and ones every where else.

For any numerical vector $\mathbf{r}^p = (r^p[1], r^p[2], \ldots, r^p[L])$, the *template error* of $\mathbf{r}^p$ with respect to a template $\mathbf{t}^p$, denoted as $E(\mathbf{r}^p, \mathbf{t}^p)$, is defined as the sum of squared residual errors.

$$E(\mathbf{r}^p, \mathbf{t}^p) = e[1]^2 + e[2]^2 + \ldots + e[L]^2 \tag{2}$$

Given a numerical vector $\mathbf{r}^p$, we can find a set of template parameters

$\mathbf{t}^p$ that minimises the template error $E(\mathbf{r}^p, \mathbf{t}^p)$ for that vector. This minimisation process is referred to as *'training the template'*. The template $\mathbf{t}^p$ that minimises $E(\mathbf{r}^p, \mathbf{t}^p)$ for the vector $\mathbf{r}^p$ is obtained as follows:

$$\begin{aligned}
E(\mathbf{r}^p, \mathbf{t}^p) &= \arg\min_{\mathbf{t}^p} \left(e[1]^2 + e[2]^2 + \ldots + e[L]^2\right) \\
&= \arg\min_{\mathbf{t}^p} \left(\mathbf{e}^T \mathbf{e}\right) \\
&\textit{making the substitution } \mathbf{e} = (\mathbf{r}^p \ - \ \mathbf{f}) \ - \ (\ \mathbf{Q}\ \operatorname{diag}(\mathbf{r}^p)\ )\ \mathbf{t}^p \\
&= \arg\min_{\mathbf{t}} \left(\ (\ \mathbf{r}^p - \mathbf{f} - \mathbf{Q}_\mathbf{r}\ \mathbf{t}^p\ )^T \ \ (\ \mathbf{r}^p - \mathbf{f} - \mathbf{Q}_\mathbf{r}\ \mathbf{t}^p\ )\ \right)
\end{aligned}$$

Where $\mathbf{Q}_\mathbf{r} = (\ \mathbf{Q}\ \operatorname{diag}(\mathbf{r}^p)\ )$.

For any **set** of numerical vectors, $\{\mathbf{r}_1^p, \mathbf{r}_2^p, \ldots, \mathbf{r}_n^p\}$, the mean value of the template error with respect to a **fixed** template $\mathbf{t}^p$ is given by

$$\frac{1}{n}\sum_{k=1}^{n} E(\mathbf{r}_k^p, \mathbf{t}^p) \tag{3}$$

The template that minimises this mean error value for this set of vectors can be obtained by calculating the partial derivatives of Equation 3 with respect to $t^p[1], t^p[2], \ldots, t^p[L]$ and setting each of these equal to zero. This gives the following set of L linear equations:

$$\mathbf{t} \ = \ \left[\ \sum_{k=1}^{n} \mathbf{Q}_{\mathbf{r}k}^T\ \mathbf{Q}_{\mathbf{r}k}\ \right]^{-1} \left[\ \sum_{k=1}^{n} \mathbf{Q}_{\mathbf{r}k}^T\ (\mathbf{r}_k^p \ - \ \mathbf{f}_k)\ \right] \tag{4}$$

Where $\mathbf{Q}_{\mathbf{r}k} = \mathbf{Q}\ \operatorname{diag}(\mathbf{r}_k^p)$.

These equations are symmetric and can be solved efficiently to find the set of template parameters $t^p[1], t^p[2], \ldots, t^p[L]$ that minimises the mean template error for the set of vectors. These parameters represent the template that best describe the relationship between the encoded sites and their binding affinity. Templates are created for all the different parametric encodings p, of the sites in the training data. Given a template $\mathbf{t}^p$ and any site of the appropriate length $\mathbf{x}$, we can compute the projected binding affinity $\tilde{f}^p$ of that site from

$$\tilde{\mathbf{f}}^p = \mathbf{r}^\mathbf{p}(\mathbf{x}) \ - \ (\mathbf{Q}\ \operatorname{diag}(\mathbf{r}^\mathbf{p}(\mathbf{x})))\ \mathbf{t}^p \tag{5}$$

Where the projected binding affinity $\tilde{f}^p$ is taken as the mean value of the vector $\tilde{\mathbf{f}}^p$.

The predictive power of a set of templates depends on the specific nucleotide parameters the templates are modelled on. As mentioned before, for a given set of training data, we first create templates from all available nucleotide parameters. We then use a greedy approach for selecting the best subset of templates from this set for the final predictor. The predictor output will be the average over all selected templates. The templates are selected based on the degree of correlation displayed between the predicted and observed values of binding affinity of the training data. The correlation coefficient between the predicted and observed values of binding affinity is computed for each template representing a specific nucleotide parameter. Then, starting with the template with the highest correlation coefficient of prediction vs. observed, we add templates, one at a time in descending order of their correlation coefficient of prediction with the observed values to an expanding set of templates. This process is continued until the correlation coefficient of prediction of the combine set of templates drop as a result of the addition of a new template to the set. In that case we select all the templates excluding the last template added, as our set of templates.

3. Results

The di-nucleotide parameters, representing structural properties of the sequences, were obtained from the Property database[18]. In its current release this database lists *38* different parameter values. We used all *38* of these parameters. The tri-nucleotide parameters were obtained from Brukner et al.[3]

There isn't much published quantitative experimental data available on the binding affinities of different protein-DNA interactions. One study is reported in Udalova *et. al.*[31] which reports on the binding affinities of *NF-κB* binding sites. We tested the predictive capability of the templates for predicting the binding affinities of these sites. The authors list the binding affinities of *52* of the possible *256* variants of the *'GGRRNNYYCC' NF-κB* motif to the recombinant *p50p50* homodimer and *p50p65* heterodimer complexes. There were two estimations of the experimental binding affinities listed for each oligo varying in rang from *0* to *2431* normalised to the control sequence *'GGGGTTCCCC'* which was given the value *227*. We used the average of these two measurements as the observed binding affinities of the sites. Templates were modelled on the log binding affinities of the sites. Figure 2 (a) shows the predicted log binding affinities of the *p50p50* variant of the *NF-κB* sites plotted against the observed log binding affinities of those sites. The templates were trained on the first *12* sites listed in Udalova *et. al.*[31] (the data had no particular arrangement so can be considered random). The correlation coefficient of the test data (i.e. observed

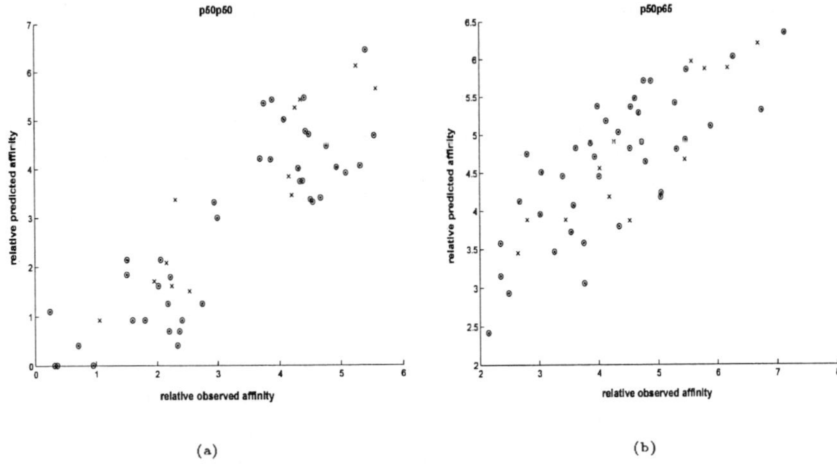

(a)

(b)

Figure 2. The predicted binding affinity (y axis) plotted against the observed values (x axis) of the (a) *p50p50* binding sites (b) *p50p65* binding sites. These predictions are based on templates modelled using *12* sites (the sites shown with crosses).

vs predicted binding affinities of sites *13* to *52*) was *0.8104* ($p = 10^{-9}$). The templates selected by the predictor in obtaining the above results were the two twist parameters (P0000018, P0000026), the two roll parameters (P0000014, P0000028), the propeller twist parameter (P0000030) and the tilt parameter (P0000016). The values listed in brackets refer to the Property database[18] ID of these parameters.

Figure 2 (b) shows the predicted log binding affinities of the *p50p65* variant of the *NF-κB* sites plotted against the observed log binding affinities of those sites. As with the *p50p50* sites, the templates were trained on the first *12* sites listed in Udalova *et. al.*[31]. For this variant, the correlation coefficient of the test data (i.e. sites *13* to *52*) was *0.7547* ($p = 10^{-6}$). The templates selected by the predictor in obtaining the above results were the two twist parameters (P0000018, P0000026), the two roll parameters (P0000014, P0000028), and the propeller twist parameters (P0000030). It is interesting that the predictor selected templates modelled on similar parameters in both cases except for the tilt parameter in the prior case (i.e. for the *p50p50* sites). It could be that the di-nucleotide step values of roll and twist express the flexibility of the DNA strand which facilitates its orientation when binding to proteins. We believe, as observed from these results, that this flexibility of the DNA strand, to orient itself around the binding protein, in tern may have a direct correlation to the binding affinity of the protein-DNA interaction. It is also interesting to note that templates modelled on sequence were not selected by the predictor.

To further ascertain the robustness of the templates to predict the binding affinity of a given site, we performed a recursive randomised prediction experiment for both the *p50p50* and *p50p65* variants of the *52* *NF-κB* binding sites. We forecasted the binding affinities of the *52* sites for both these variants *100* times with each time seeing its prediction based on a new template created from a set of *12* randomly selected sites and tested on the remaining sites. The results of these experiments are listed in Table 1. Also shown in Table 1 are the results of the above experiment carried out with an increase from *12* to *21* of the number of sites used to train the templates. The results produced by the *template* method described here are compared with three other methods. The first is the matrix similarity score computed as described in Quandt *et. al.*[20]. The second is the logarithm of the base frequencies as described by Berg and von Hippel[2] and the third is the non-parametric model described by King *et. al.*[11](with default parameters). These tests were done exactly as the tests carried out for the templates where predictions were performed *100* times with a new frequency matrix created from the training sites (both for *12* and *21* sites) selected randomly from the *52* sites and tested on the remaining sites, for each test. These results are also listed in Table 1. The standard deviation of each experiment is given in brackets.

Table 1. Performance statistics of templates for the *NF-κB* sites with training sets of *12* and *21* examples. The mean values are taken over *100* randomised trials. The standard deviations are given in brackets.

Mean correlation coefficient (Predicted vs. Observed)

	*p*50*p*50	
Number of training examples	*12*	*21*
Templates	0.7809 (0.0469)	0.8124 (0.0370)
Matrix similarity score[20]	0.3535 (0.3167)	0.4559 (0.2851)
Logarithm of base frequencies[2]	0.2033 (0.3087)	0.2880 (0.2781)
Non-parametric model[11]	0.2686 (0.1691)	0.2758 (0.1727)
	*p*50*p*65	
Number of training examples	*12*	*21*
Templates	0.6444 (0.0869)	0.6941 (0.0883)
Matrix similarity score[20]	0.1896 (0.2732)	0.3338 (0.2121)
Logarithm of base frequencies[2]	0.1302 (0.2583)	0.1719 (0.2397)
Non-parametric model[11]	0.2413 (0.1589)	0.2414 (0.1601)

4. Discussion

We have described a novel approach for predicting the binding affinity of protein-DNA interactions. The approach described is based on templates that are sensitive to positional co-variations. These can be co-variations expressing sequence or structural polymorphisms as described by the different parametric encodings of the nucleotide sequence. Templates work in sets, usually containing more than one element, with each template characterising a different sequence or structural property of the sites. The amalgamation of different templates optimally selected to work in unison endows a synergic effect on the predictive capabilities of the system.

The training phase of the system requires a subset of binding sites along with their experimentally verified binding affinities. One advantage of the method described above, unlike other approaches such as, for example, those based on base frequencies which are susceptible to small-sample uncertainties[2], is its ability to learn quite well from a minimal number of training data. This is a feature that has many practical advantages when we a dealing with a dearth of properly annotated examples.

Binding assays of transcription factors such as *NF-κB*, *Zif268* zinc fingers and *Mnt* repressor-operator proteins suggest strong evidence to the existence of non-independent effects on positional interactions when at least some proteins bind to DNA[30,15,4]. The exact positions that exhibit such interdependent effects vary from one factor to another, and there is no evidence that all transcription factors exhibit a similar pattern of behaviour. This makes it difficult to capture such properties in a general model. The requirement is for models that can learn such variations from a set of training data.

The sensitivity of templates described above to positional co-variations is not based on any prior knowledge of which positions exhibit correlated behaviour. This is an important characterisation, especially in the absence of such prior knowledge individualising a family of binding sites, which is usually the case. It is not always practical to build exhaustive models detailing the different co-variations present between individual positions. Models such as neural networks and HMMs that are able to account for such information suffer from the practical drawback of balancing between the complexity of the systems and the number of examples required to train them well. In these systems, the complexity of the model architecture imposes lower bounds on the number of examples required to form a good training set. These bounds usually increase exponentially with the increase in complexity of the system.

There is evidence[1,16,6] that suggests the presence of structural homologies in DNA sequences that interact with some transcription factors. What these structural homologies are and exactly what geometric features play a part in them is not always very clear or easy to ascertain. Programs that incorporate such features do so with an implicit assumption of the presence of these properties in the sequences that they analyse. This is a weak assumption that may be tentative in the absence of specific knowledge of their presence and would not hold for the general case. It is possible for different binding sites to exhibit different structural properties intrinsic to the particular factor that they bind to. It is also possible for some binding sites not to display any significant structural homology for any of the known structural parameters. In such cases, one has only got sequence homology to rely on.

Templates used here for predicting binding affinity can model both sequence and structural homology. The important fact when modelling templates for a particular family of TF binding sites is that we do not make any prior decision on which structural parameters to use. The selection of the best set of parameters is done automatically during the training phase of the system, though in a greedy fashion.

References

1. T. Aoyama and M. Takanami. Essential structure of E. coli promoter II. Effect of the sequences around the RNA start point on promoter function. *Nucleic Acids Res.*, 13 (11):4085–4096, 1985.
2. O. G Berg and P. H von Hippel. Selection of DNA binding sites by regulatory proteins. Statistical-mechanical theory and application to operators and promoters. *J Mol Biol*, 193(4):723–750, Feb 20 1987.
3. I. Brukner, R. Sanchez, D. Suck, and S. Pongor. Sequence-dependent bending propensity of DNA as revealed by DNase I: parameters for trinucleotides. *EMBO Journal*, 14:1812–1818, 1995.
4. M. L. Bulyk, P. L. F. Johnson, and G. M. Church. Nucleotides of transcription factor binding sites exert interdependent effects on the binding affinities of transcription factors. *Nucleic Acids Research*, 30(5):1255–1261, 2002.
5. M Djordjevic, A. M Sengupta, and B. I Shraiman. A biophysical approach to transcription factor binding site discovery. *Genome Res.*, 13(11):2381–90, Nov 2003.
6. M. A. El Hassan and C. R. Calladine. Propeller-twisting of base-pairs and the conformational mobility of dinucleotide steps in DNA. *Journal Molecular Biology*, 259(1):95–103, 1996.
7. M. A. El Hassan and C. R. Calladine. Two distinct modes of protein-induced bending in DNA. *Journal Molecular Biology*, 282(2):331–343, 1998.

8. W. N. Grundy, T. L. Bailey, and C. P. Elkan. ParaMEME: A parallel implementation and a web interface for a DNA and protein motif discovery tool. *Computer Applications in the Biological Sciences (CABIOS)*, 12(4):303–310, 1996.
9. T Gustafson, A Taylor, and L Kedes. DNA bending is induced by a transcription factor that interacts with the human c-FOS and alpha-actin promoters. *Proc Natl Acad Sci U S A*, 86(7):2162–6, 1989.
10. P. B. Horton and M. Kanehisa. An assessment of neural network and statistical approaches for prediction of E. coli promoter sites. *Nucleic Acids Research*, 20:4331–4338, 1992.
11. O. D King and F. P Roth. A non-parametric model for transcription factor binding sites. *Nucleic Acids Res.*, 31(19):e116, Oct 2003.
12. C. E. Lawrence, S. F. Altschul, M. S. Boguski, J. S. Liu, A. F. Neuwald, and J. C. Wootton. Detecting subtle sequence signals: a Gibbs sampling strategy for multiple alignment. *Science*, 262(5131):208–14, 1993.
13. S. Lisser and H. Margalit. Determination of common structural features in Escherichia coli promoters by computer analysis. *Eur J Biochem.*, 223(3):823–830, 1994.
14. I. Mahadevan and I. Ghosh. Analysis of E.coli promoter structures using neural networks. *Nucleic Acids Res.*, 22 (11):2158–2165, 1994.
15. T. K. Man and G. D. Stormo. Non-independence of Mnt repressor-operator interaction determined by a new quantitative multiple fluorescence relative affinity (QuMFRA) assay. *Nucleic Acids Research*, 29(12):2471–8, 2001.
16. R. Nussinov. Promoter helical structure variation at the Escherichia coli polymerase interaction sites. *Journal of Biological Chemistry.*, 259:6798–6805, 1984.
17. U. Ohler, H. Niemann, G. C. Liao, and G. M. Rubin. Joint modeling of DNA sequence and physical properties to improve eukaryotic promoter recognition. *Bioinformatics*, 17:199–206., 2001.
18. J. V. Ponomarenko, M. P. Ponomarenko, A. S. Frolov, D. G. Vorobyev, G. C. Overton, and N. A. Kolchanov. Conformational and physicochemical DNA features specific for transcription factor binding sites. *Bioinformatics*, 15(7/8):654–668, 1999.
19. M. P. Ponomarenko, J. V. Ponomarenko, A. E. Kel, and N. A. Kolchanov. Search for DNA conformational features for functional sites. Investigation of the TATA box. . *In: Biocomputing: proceedings of the 1997 Pacific Symposium. (Altman, R., et al., eds.), Word Sci. Publ., Singapore*, pages 340–351., 1997.
20. K. Quandt, K. Frech, H. Karas, E. Wingender, and T. Werner. MatInd and Matinspector: new fast and versatile tools for detection of consensus matches in nucleotide sequence data. *Nucleic Acids Res.*, 23:4878–4884, 1995.
21. E. Ragnhildstveit, A. Fjose, P. B. Becker, and J. P. Quivy. Solid phase technology improves coupled gel shift/footprinting analysis. *Nucleic Acids Research*, 25(2):453–454, 1997.
22. A Sarai and Y Takeda. Lambda repressor recognizes the approximately 2-fold symmetric half-operator sequences asymmetrically. *Proc Natl Acad Sci U S*

A, 86(17):6513–6517, Sep 1989.

23. R Schreck, H Zorbas, E. L Winnacker, and P. A Baeuerle. The NF-kappa B transcription factor induces DNA bending which is modulated by its 65-kD subunit. *Nucleic Acids Res.*, 18(22):6497–502, 1990.
24. J. Schug and G. C. Overton. TESS: Transcription Element Search Software on the WWW. *Technical Report CBIL-TR-1997-1001-v0.0, of the Computational Biology and Informatics Laboratory, School of Medicine, University of Pennsylvania*, 1997.
25. S. E. Shadle, D. F. Allen, H. Guo, W. K. Pogozelski, J. S. Bashkin, and T. D. Tullius. Quantitative analysis of electrophoresis data: novel curve fitting methodology and its application to the determination of a protein-DNA binding constant. *Nucleic Acids Research*, 25(4):850–860, 1997.
26. V. Y Stefanovsky, D. P Bazett-Jones, G Pelletier, and T Moss. The DNA supercoiling architecture induced by the transcription factor xUBF requires three of its five HMG-boxes. *Nucleic Acids Res.*, 24(16):3208–15, 1996.
27. G. D. Stormo and D. S. Fields. Specificity, energy and information in DNA-protein interactions. *Trends Biochemical Sciences*, 23:109–113, 1998.
28. G. D. Stormo, T. D. Schneider, L. Gold, and A. Ehrenfeucht. Use of the 'Perceptron' algorithm to distinguish translational initiation sites in E. coli. *Nucleic Acids Research*, 10:2997–3011, 1982.
29. K. M. Thayer and D. L. Beveridge. Hidden Markov models from molecular dynamics simulations on DNA. *Proceedings of the National Academy of Sciences*, 99(13):8642–8647, 2002.
30. I. A. Udalova, R. Mott, D. Field, and D. Kwiatkowski. Quantitative prediction of NF-kB DNA-protein interactions. *Proceedings of the National Academy of Sciences USA*, 99:8167–8172, 2002.
31. I. A. Udalova, R. Mott, D. Field, and D. Kwiatkowski. Quantitative prediction of NF-kB DNA-protein interactions. *Proceedings of the National Academy of Sciences USA*, 99:8167–8172, 2002.
32. I. A Udalova, A Richardson, A Denys, C Smith, H Ackerman, B Foxwell, and D Kwiatkowski. Functional consequences of a polymorphism affecting NF-kappa B p50-p50 binding to the TNF promoter region. *Molecular and Cellular Biology*, 20(24):9113–9119, Dec 2000.
33. W. Wasserman and A Sandelin. Applied bioinformatics for the identification of regulatory elements. *Nat Rev Genet.*, 5(4):276–87, Apr 2004.
34. S. A. Wolfe, H. A. Greisman, E. I. Ramm, and C. O. Pabo. Analysis of Zinc Fingers Optimized Via Phage Display: Evaluating the Utility of a Recognition Code. *Journal of. Molecular Biology*, 285:1917–1934, 1999.
35. X Xie, J Lu, E. J Kulbokas, T. R Golub, V Mootha, K Lindblad-Toh, E. S Lander, and M Kellis. Systematic discovery of regulatory motifs in human promoters and 3' UTRs by comparison of several mammals. *Nature*, 434(7031):338–45, Mar 2005.
36. Q. M. Zhang and T. G. Marr. A weight array method for splicing signal analysis. *Computer Applications in Biosciences*, 9(5):499–509, 1993.
37. Q Zhou and J. S Liu. Modeling within-motif dependence for transcription factor binding site predictions. *Bioinformatics*, 20(6):909–16, Apr 2004.

IMPROVING COMPUTATIONAL PREDICTIONS OF *CIS*-REGULATORY BINDING SITES

MARK ROBINSON*, YI SUN, RENE TE BOEKHORST, PAUL KAYE, ROD ADAMS, NEIL DAVEY

Science and Technology Research Institute, University of Hertfordshire, College Lane Hatfield, Hertfordshire AL10 9AB, UK
{m.robinson, y.2.sun, R.TeBoekhorst, p.h.kaye, r.g.adams, n.davey}@herts.ac.uk

ALISTAIR G. RUST

Institute of Systems Biology, 1441 North 34th Street
Seattle, WA 98103, USA
arust@systemsbiology.org

The location of *cis*-regulatory binding sites determine the connectivity of genetic regulatory networks and therefore constitute a natural focal point for research into the many biological systems controlled by such regulatory networks. Accurate computational prediction of these binding sites would facilitate research into a multitude of key areas, including embryonic development, evolution, pharmacogenemics, cancer and many other transcriptional diseases, and is likely to be an important precursor for the reverse engineering of genome wide, genetic regulatory networks. Many algorithmic strategies have been developed for the computational prediction of *cis*-regulatory binding sites but currently all approaches are prone to high rates of false positive predictions, and many are highly dependent on additional information, limiting their usefulness as research tools. In this paper we present an approach for improving the accuracy of a selection of established prediction algorithms. Firstly, it is shown that species specific optimization of algorithmic parameters can, in some cases, significantly improve the accuracy of algorithmic predictions. Secondly, it is demonstrated that the use of non-linear classification algorithms to integrate predictions from multiple sources can result in more accurate predictions. Finally, it is shown that further improvements in prediction accuracy can be gained with the use of biologically inspired post-processing of predictions.

1 Introduction

Gene regulatory networks control, to a large extent, many important biological systems, including: the accurate, and stable, expression of a subset of the proteins encoded by a genome that determine the character and properties of a cell type; the intricate program of sequential organization and subsequent cellular specialization during embryonic development; and the inherently complex dynamics of metabolic responses to pharmaceuticals. Additionally, it has become clear in recent years that much of the genetic change underlying

---

* Corresponding author (Mark Robinson)

morphological evolution must have occurred in gene regulatory regions [1]. To gain a functional understanding of genetic regulatory networks, along with an ability to accurately predict their topological structure and dynamics, is a research goal promising far reaching ramifications into many important biological fields.

The primary determinant of connectivity in genetic regulatory networks is the presence, or absence, of *cis*-regulatory binding sites in the regions proximal to each gene's promoter. The accurate computational prediction of the location of *cis*-regulatory binding sites is therefore a highly desirable research goal, and a key step towards the ability to reverse engineer genetic regulatory networks at a genomic scale. Such predictions could significantly streamline the, costly and time consuming, process of annotating regulatory regions by focusing attention on sequences associated with a high probability of functionality. However, prediction of *cis*-regulatory binding sites is a non-trivial problem. The rules determining which DNA sequences functionally bind transcription factors specify position dependant preferences for interactions between amino acids and nucleotide bases, rather than a simple deterministic sequence identity. In many cases, contextual information, in the form of proximally located binding sites, may play a key role in determining whether a potential binding site is in fact functional *in vivo*, further complicating computational predictions of such sites.

Many algorithms have been developed to exploit the various sources of experimental information available and the various statistical properties that appear to distinguish regulatory regions from the genome in general. These algorithms can typically be classified into four main groups based on the approach to the problem. *Scanning algorithms* attempt to generate a model, such as a position weight matrix, for each binding site from available experimental data. These models can then be used to scan potential regulatory sequences for good matches to the model. *Statistical algorithms* typically attempt to detect motifs that are considered statistically unlikely in the context of a model of the background base-pair distribution. *Co-regulatory algorithms* rely on the hypothesis that genes clustered on the basis of their expression profiles are likely to be regulated by the same transcription factors. Iterative techniques, such as Expectation Maximization, are used to generate and refine predictive models for the most over-represented motifs in the set of upstream sequences for such gene clusters. *Phylogenetic algorithms* exploit the conservation of functional DNA sequences against the background of random mutational noise during evolution. Homologous regulatory sequences from appropriately related species are compared and significant sequence alignments are predicted to act as functional *cis*-regulatory binding sites.

In spite of the wealth of research performed in the area of binding site prediction, and the many insights gained, the current state of the art in this area is still far from perfect. In fact, results presented in this study agree strongly with other assessments of the performance of prediction algorithms [2], in showing that typically 70-80% of predictions are false positives. Interpretation of such results to guide the experimental analysis of gene regulatory regions, or the modeling of gene regulatory networks, is a difficult problem, further exasperated by technicalities of choosing an appropriate algorithm given the available data, and subsequent selection of appropriate algorithmic parameters. The utility of algorithms that scan for putative binding sites using experimentally determined weight matrices, or that require knowledge about the identity of co-regulated sets of genes, are obviously of limited use for exploring systems where that data is not available. The study of such systems is often limited to the statistical class of algorithms; statistical algorithms are, unsurprisingly, typically unable to achieve the levels of prediction accuracy observed with other classes of algorithms.

In this paper it is demonstrated, firstly, that algorithm performances can be improved by species specific optimization of algorithmic parameters. Secondly, that integration of multiple sources of algorithmic predictions, using non-linear classification techniques, can significantly improve prediction accuracy while at the same time circumventing the experimental data dependences. Finally, in order to ensure that the integration process produces biologically feasible predictions, it is necessary to perform some post-processing, and we show that this step can further improve prediction accuracy.

2 Methods

2.1 Description of the Data

Generation of appropriate data sets for use in evaluating the performance of binding site prediction algorithms is a challenging problem with no clear solution [2]. The use of promoter sequences that have been experimentally annotated is commonly used, although with no assurance of the completeness of sequence annotations, penalization of some correctly predicting algorithms is inevitable. An alternative strategy is for the stochastic generation of random sequences embedded with examples of binding sites, but, our current lack of knowledge of the stochastic processes underlying the sequences found in nature renders this strategy open to unknown biases.

For the purposes of this study we chose the annotated sequence strategy, attempting to minimize the error by using promoter sequences from one of the most well studied model organisms, the *S.cerevisiae* promoter database[†]. 120 annotated promoter sequences were selected for training and testing the algorithms, a total of 68910 bp of sequence data. In addition, homologous promoter sequences for 59 of the sequences were collected from *S.paradoxus*, *S.mikatae* and *S.bayanus*, and 69 of the sequences were determined, by the use of micro-array studies, to be likely candidates for co-regulation.

For integration of multiple algorithmic predictions a matrix was generated, consisting of 68910 12-ary real valued vectors, each associated with a binary label indicating the presence or absence of a binding site annotation at this sequence position. Each 12-ary vector represents the predictions, at this position in the sequence dataset, for each of the twelve algorithms. All predictions are normalized as real values in the range [-1, 1] with 0 allocated to sequence positions where algorithm predictions were not possible.

In this work we divided our dataset into a training set and a test set: the first 2/3 for training and the final 1/3 for testing. Additionally, we contextualize the training and test datasets to ensure that the classification algorithms have data on contiguous binding site predictions. This is achieved by windowing the vectors. We use a window size of 7, providing contextual information for 3 bps either side of the position of interest. This procedure carries the considerable benefit of eliminating a large number of repeated or inconsistent vectors which are found to be present in the data and would otherwise pose a significant obstacle to the training of the classifiers.

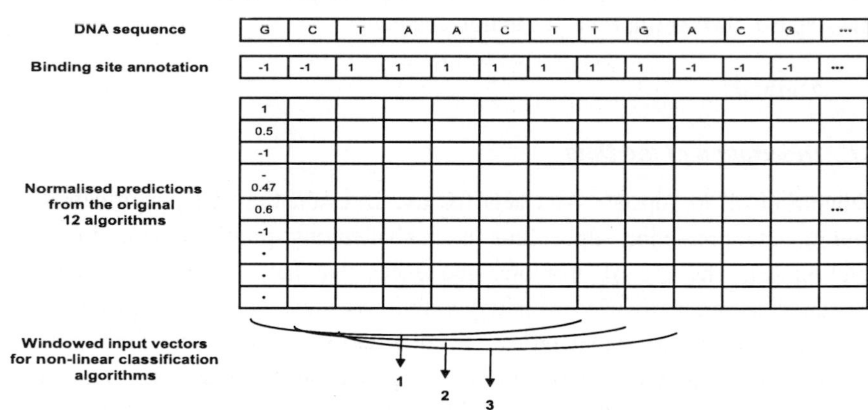

Figure 1: Organization and structure of the dataset used to train classification algorithms. Windowing of the input data is show, for example, the first input vector consists of the first seven columns of predictions concatenated together. The target output is the binding site annotation corresponding to the middle column of the window (i.e. the fourth column for input vector 1)

[†] SCPD: http://cgsigma.cshl.org/jian/

2.2 Performance Metrics

Approximately 7% of our dataset, consisting of 68910 data points, are labelled as annotated binding sites, making this an *imbalanced* dataset [3]. Supervised classification algorithms would be expected to over predict the majority class in an imbalanced dataset, i.e. in this instance a success rate of 93% could be achieved by only predicting the majority class, namely the non-binding site class. In this work we deal with this issue in two ways: firstly, with the use of appropriate metrics for evaluation of algorithmic performance and secondly, with the use of data-based methods during classifier training (see 2.5).

Several common performance metrics such as *Recall*, *Precision* and *F-score* [4] are defined to allow evaluation of performance on the minority class. The *Correlation Coefficient* [2] is also defined, providing a measure of the correlation of predicted binding sites to the annotated data. Each of these metrics is defined using a confusion matrix (see Table 1):

Table 1: A confusion matrix – TN is the true negative count, FP is the false positive count, FN is the false negative count and TP is the true positive count.

TN	FP
FN	TP

$$\text{Recall} = \frac{TP}{(TP + FN)}, \quad \text{Precision} = \frac{TP}{(TP + FP)}$$

$$\text{F-Score} = \frac{2 \cdot \text{Recall} \cdot \text{Precision}}{\text{Recall} + \text{Precision}}, \quad \text{FP-rate} = \frac{FP}{(FP + TN)}$$

$$\text{Correlation Coefficient (CC)} = \frac{TP \cdot TN - FN \cdot FP}{\sqrt{(TP + FN)(TN + FP)(TP + FP)(TN + FN)}}$$

2.3 Description of Prediction Algorithms Used

The binding site prediction tools evaluated in this study were selected, either from the research literature or as tools developed in-house (PARS & DREAM) or by collaborators (Sampler), to include representatives for all the major prediction strategies, see Table 2. The aim in selecting these disparate algorithms was to maximize the relevant information with the full set of binding site predictions. Where possible, algorithmic parameters were set to those reported in the literature; for the remainder default parameter settings were used.

Table 2: Categorization of algorithms used in study

Strategy	Algorithm
Scanning	Fuzznuc[‡] Motif Scanner [5] Ahab [6]
Statistical	PARS[§] Dream (over and under represented motifs) Verbumculus [7]
Co-Regulatory	MEME [8] AlignACE [9] Sampler[**] (Institute for Systems Biology)
Evolutionary	SeqComp [10] Footprinter [11]

2.4 Species Specific Optimization of Prediction Algorithm Parameters

The many algorithms available for *cis*-regulatory binding site prediction have typically been developed, and suitable operating parameters selected, for a specific model organism. It is an open question as to whether such operational parameter settings would be expected to be optimal across a wide range of organisms, although in practice this is often the assumption. It was decided to search the parameter space of each algorithm to find optimal settings for binding site detection in the yeast dataset.

The parameter space consisted of an assemblage of various data types: Boolean, discrete and real valued types of varying ranges. An implementation of an efficient simulated annealing schedule [12] was used to search the parameter space. All optimization runs were performed with a single algorithm and were initialized with the default parameters. Evolution of novel solutions in the parameter space was achieved by the random selection of one parameter per iteration with the subsequent selection of a new random point with the range of the selected parameter. The user-specified variable, λ [12], that determines the rate at which stocasticity deceases over time, was set to a value of 0.001. The training set was divided into two equal parts for training and validation of performance during the optimization process. The fitness function was implemented using the F-Score performance metric.

---

‡ http://www.hgmp.mrc.ac.uk/Software/EMBOSS/

§ http://sourceforge.net/projects/pars/

** http://sourceforge.net/projects/netmotsa

2.5 *Sampling Techniques for Learning Imbalanced Datasets*

To ensure efficient training of classifiers on this imbalanced dataset, data based sampling techniques [13, 14] were employed, namely under-sampling of the majority class (negative examples) and over-sampling of the minority class (positive examples). For under-sampling, we randomly selected a subset of data points from the majority class. The more complex issues that arise with over-sampling [3] are addressed by the use of *synthetic minority over-sampling* as proposed in [13]. In the absence of these sampling techniques, the supervised classifiers achieved negligible rates of true positive predictions. The number of items in the minority class is doubled and degree of under-sampling is chosen so as to ensure the final ratio of minority and majority members is one half. Preliminary cross-validation experiments were used to set these parameters, this parameter space will be explored more thoroughly in future work.

2.6 *Supervised Classifiers*

A variety of supervised classification algorithms were used to explore their relative merits for improvement of prediction accuracy by the integration of predictions from multiple algorithmic sources [15]. A single layer neural network (SLN) was used in this study to provide a standard for baseline performance. A Support Vector Machine (SVM), an effective, contemporary kernel based classification algorithm was utilized. The final algorithm used was the Adaboost algorithm [16], a powerful, recently proposed method for producing a strong classifier from a sequence of weak classifiers. The algorithm begins by training a weak classifier, here an SLN, on the original dataset. A new dataset is then produced by increasing the frequency of data points poorly classified. This process then iterates until a strong classifier has been produced.

2.7 *Biologically Constrained Post-Processing*

Observation of the predictions from the supervised classifiers used here, suggest that many of their false positive predictions could be ruled out based on known, or suspected, biological constraints of functional binding sites. One possible constraint is that a binding motif must be of sufficient length to make randomly occurring copies unlikely. Predictions that fall below some threshold length are therefore prime candidates for post-processing, either to filter them out, or to extend their size. This is a particularly pertinent step as the meta-predictions generated from the original, noisy, algorithmic predictions can produce fragmented predictions as an artefact of the integration process. In this study a post-processing step is incorporated filtering out predictions that do not reach a minimum threshold for contiguous length. Classification performance is evaluated for threshold values of 5 bp and 6 bp, as shown in Section 3.3.

3 Results

3.1 Comparison of Performance Using Default and Species Specific Optimized Parameters

Each of algorithms used in this study were initially evaluated on the annotated *S.cerevisiae* sequence test set of 22967 bp, producing a set of scores using their respective default parameters. These scores were used as a baseline for the evaluation of performance using parameter sets identified as conferring a performance improvement during training on the *S.cerevisiae* training set of 42919 bp. It is important to note that performance was evaluated over the entire test dataset; an algorithm is effectively penalized when it is unable to make predictions for specific sequences due to lack of supplementary data. When evaluated on the subset of sequences where predictions were made, MEME, for example was able to achieve an optimized F-Score of over 45%, although, when evaluated on the entire dataset its performance dropped to an F-Score value of 18.21%, as shown in Table 3. However, as we are interested in evaluating the functional usefulness of the algorithms, with an aim to overcome these limitations by integrating multiple sources of information, the full test dataset is most appropriate.

Table 3: A comparison of algorithmic performance using default vs. optimized parameters. Dashes indicate that no improved parameters settings could be found or that optimization was not possible.

Algorithm	Default			Optimized		
	F-Score	CC	FP-rate	F-Score	CC	FP-rate
Fuzznuc	24.59	19.02	10.61	-	-	-
PARS	10.41	2.42	12.45	-	-	-
Verbumculus	19.24	12.79	12.23	-	-	-
Ahab	14.31	7.45	48.86	25.60	22.59	3.36
Dream (over)	9.21	-1.02	24.13	-	-	-
Dream (under)	8.25	-2.58	25.19	13.40	5.54	15.53
Motif Scanner	16.19	9.41	9.95	-	-	-
Sampler	4.72	1.19	2.56	6.86	11.72	3.14
MEME	18.21	15.61	2.43	18.83	45.23	0.94
AlignACE	13.09	11.08	2.03	-	-	-
SeqComp	9.56	2.11	9.81	-	-	-
Footprinter	10.39	3.19	9.20	-	-	-

F-Score performance, with default parameters, was typically below 20% with even lower scores for the correlation coefficient. The simple scanning algorithm, Fuzznuc, easily outperforms the others with an F-score of nearly 25%. The last 5 algorithms in Table 2, the co-regulatory and phylogenetic algorithms, produced notably low FP-rates. This is likely due to their predictions, where possible, being of high acuity but of a conservative nature.

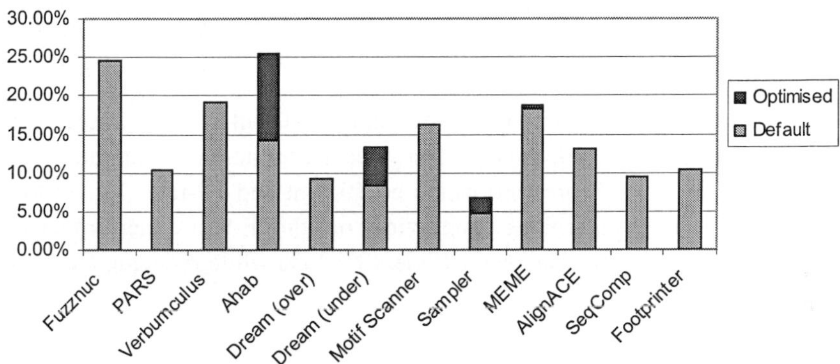

Figure 2: Comparison of F-score performance of algorithms using default and optimized parameters

Optimized parameters were found for Ahab, Dream (over), Sampler and MEME that improve performance on the test set, as can be seen in Figure 2. The performance improvement seen with Ahab was particularly impressive, with an 80% increase in F-score while the false positive rate was reduced by 93%. It is intriguing to note that both Ahab, and Dream, were developed using *D.melanogaster* as a model. Conversely, for Verbumculus and AlignACE, both known to have been developed with *S.cerevisiae* as a model, no parameter improvement could be found. The possibility is certainly raised that species specific parameter optimization may be necessary for optimal algorithmic predictions; it remains to be seen whether this situation will in fact prove to be the case for other organisms and if so what the underlying causes in terms of *cis*-regulatory organization and structure might be.

3.2 Integration of Multiple Algorithmic Predictions Using Supervised Classifiers

Another important question is whether performance can be further refined by the integration of multiple algorithm predictions. To this end, three non-linear supervised classifiers were trained using the predictions of the original algorithms on the training set. Cross-validation was used to select appropriate parameter settings for the classifiers, with each parameter setting being trained on 4/5 of training set and validated on the final 1/5. The algorithm, Fuzznuc, was chosen to provide a baseline performance based on its high performance across the entire test dataset. The optimized version of Ahab, which achieved even higher levels of performance, was not available in time to be included in this study.

Table 4: Comparison of Fuzznuc performance vs. integration strategies performance

Algorithm	F-Score	CC	FP-rate
Fuzznuc	24.59	19.0	10.1
SLN	25.0	19.5	7.3
SVM	27.2	21.8	8.1
Adaboost	27.0	21.7	6.3

The results in Table 4 show a clear and consistent picture. Integration of multiple algorithm predictions consistently results in more accurate predictions, as measured by the F-Score, correlation coefficient and FP-rate. The SVM out performs all other algorithms, improving on the F-Score performance of Fuzznuc by 10%, the correlation coefficient by 14% while reducing the FP-rate by 38%.

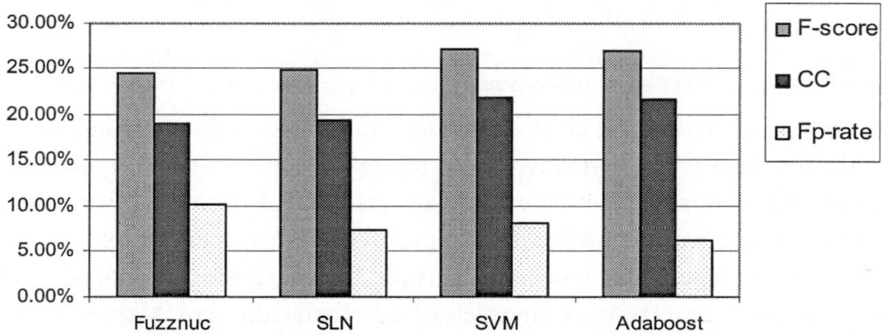

Figure 3: Performance statistics comparing the accuracy of different classification strategies

3.3 Refinement of Results Using Biologically Constrained Post-Processing

The final step in our refinement of binding site predictions is the conceptually simple one of ensuring that all predictions are biologically viable. Table 5 details the results of an experiment designed to explore whether small, fragmented predictions were artefacts of the meta-analysis shown in Section 3.2.

Table 5: Performance improvements with a range of minimum word size filter thresholds

Algorithm	No Filtering			Filter < 5			Filter < 6		
	F-Score	CC	FP-rate	F-Score	CC	FP-rate	F-Score	CC	FP-rate
SLN	25.0	19.5	7.3	25.6	20.3	5.9	26.0	20.8	5.5
SVM	27.2	21.8	8.1	28.2	23.0	6.7	28.4	23.2	6.2
Adaboost	27.0	21.7	6.3	28.0	23.2	4.7	27.3	22.8	4.3

It can be seen that in all cases filtering out predictions less than 5 bp in extent, improves prediction accuracy, by all performance measures, for all classifiers. Filtering predictions that are less than 6 bp, further improves performance for both the SLN and SVM but causes a considerable drop in accuracy for the Adaboost algorithm. In the best case, the combination of integration using the SVM combined with the post-processing filtering with threshold < 6 improves the F-Score performance relative to Fuzznuc by 15%.

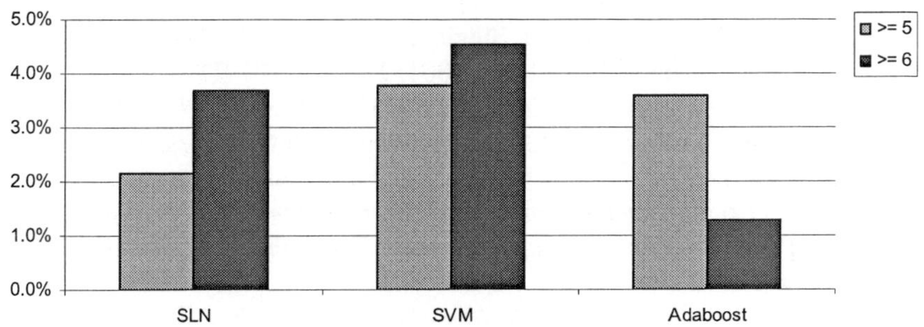

Figure 4: Percentage change in F-Score, relative to unfiltered results, after post-processing of fragmented results using thresholds 5 and 6 respectively

4 Conclusions

The important, and significant, result presented here is that an incremental approach to algorithmic refinement can produce considerable improvement in prediction accuracy. In the best case, the combination of integrating predictions using the SVM followed by post-processing filtering using a threshold < 6, improves the F-Score to 28.4%, an improvement of 15% relative to the performance achieved by the best algorithm, Fuzznuc.

The performance improvements achieved by parameter optimization, most notably those of Ahab, are highly suggestive; optimal computation prediction of cis-regulatory binding sites may require species specific optimization of parameter sets.

The use of supervised classification techniques for integrating predictions from multiple sources is shown to be a particularly promising approach. The success of integrating these multiple prediction sources indicates that there is additional information to be exploited, collectively, in these prediction sets.

Initial attempts at post-processing meta-predictions were worthwhile and present many opportunities for future work in this area. Other important biological constraints that might be explored in future work include, clustering of predicted sites, and bias in the base pair distributions within predicted sites.

References

1.Davidson, E.H., *Genomic Regulatory Systems: Development and Evolution.* 2001, San Diego: Academic Press.
2.Tompa, M., et al., *Assessing computational tools for the discovery of transcription factor binding sites.* Nat Biotechnol, 2005. **23**(1): p. 137-44.
3.Japkowicz, N. *Class imbalances: Are we focusing on the right isse?* in *Workshop on learning from imbalanced datasets, II, ICML.* 2003. Washington DC.
4.Bussemaker, H.J., H. Li, and E.D. Siggia, *Regulatory element detection using correlation with expression.* Nat Genet, 2001. **27**(2): p. 167-71.
5.Thijs, G., et al., *A higher-order background model improves the detection of promoter regulatory elements by Gibbs sampling.* Bioinformatics, 2001. **17**(12): p. 1113-22.
6.Rajewsky, N., et al., *Computational detection of genomic cis-regulatory modules applied to body patterning in the early Drosophila embryo.* BMC Bioinformatics, 2002. **3**(1): p. 30.
7.Apostolico, A., et al., *Efficient detection of unusual words.* J Comput Biol, 2000. **7**(1-2): p. 71-94.
8.Bailey, T.L. and C. Elkan, *Fitting a mixture model by expectation maximization to discover motifs in biopolymers.* Proc Int Conf Intell Syst Mol Biol, 1994. **2**: p. 28-36.
9.Hughes, J.D., et al., *Computational identification of cis-regulatory elements associated with groups of functionally related genes in Saccharomyces cerevisiae.* J Mol Biol, 2000. **296**(5): p. 1205-14.
10.Brown, C.T., et al., *New computational approaches for analysis of cis-regulatory networks.* Dev Biol, 2002. **246**(1): p. 86-102.
11.Blanchette, M. and M. Tompa, *FootPrinter: A program designed for phylogenetic footprinting.* Nucleic Acids Res, 2003. **31**(13): p. 3840-2.
12.Lam, J. and J. Delosme, *Performance of a New Annealing Schedule.* Proceedings 25th ACM/IEEE Design Automation Conference, 1988: p. 306-311.
13.Chawla, N.V., et al., *SMOTE: Synthetic minority over-sampling Technique.* Journal of Artificial Intelligence Research, 2002. **16**: p. 321-357.
14.Radivojac, P., et al., *Classification and knowledge discovery in protein databases.* J Biomed Inform, 2004. **37**(4): p. 224-39.
15.Sun, Y., et al. *Integrating binding site predictions using non-linear classification methods.* in *Machine Learning Workshop.* 2005. Sheffield: LNAI.
16.Freund, Y. and R.E. Schapire, *A decision-theoretic generalization of on-line learning and an application to boosting.* Journal of Computer and Systems Sciences, 1997. **55**(1): p. 119-139.

STRUCT2NET: INTEGRATING STRUCTURE INTO PROTEIN-PROTEIN INTERACTION PREDICTION

ROHIT SINGH* JINBO XU*‡ BONNIE BERGER†‡

Computer Science and Artificial Intelligence Laboratory
Massachusetts Institute of Technology
Cambridge MA 02139
E-mail: {rsingh, j3xu, bab}@theory.csail.mit.edu

This paper presents a framework for predicting protein-protein interactions (PPI) that integrates structure-based information with other functional annotations, e.g. GO, co-expression and co-localization, etc. Given two protein sequences, the structure-based interaction prediction technique threads these two sequences to all the protein complexes in the PDB and then chooses the best potential match. Based on this match, structural information is incorporated into logistic regression to evaluate the probability of these two proteins interacting. This paper also describes a random forest classifier which can effectively combine the structure-based prediction results and other functional annotations together to predict protein interactions. Experimental results indicate that the predictive power of the structure-based method is better than many other information sources. Also, combining the structure-based method with other information sources allows us to achieve a better performance than when structure information is not used. We also tested our method on a set of approximately 1000 yeast genes and, interestingly, the predicted interaction network is a scale-free network. Our method predicted some potential interactions involving yeast homologs of human disease-related proteins.
Supplementary Information: `http://theory.csail.mit.edu/struct2net`

1. Introduction

Proteins are the workhorses of the cell, performing a wide variety of functions. Most often, they perform these functions by interacting with other proteins. Indeed, many diseases can be traced to undesirable or malfunctioning protein-protein interactions (e.g.: viral-host interactions[14], prion formation[11]). Clearly, the study of such interactions is very important.

Protein-protein interactions (PPIs) can be studied from two different perspectives. In the traditional view of PPIs, the aim has been to understand the physical mechanism of interaction between two proteins by using experimental and/or computational methods to study each interaction individually.

*These authors contributed equally to the work
†Corresponding author
‡Also in the MIT Dept. of Mathematics

In contrast, the more-recent "high-throughput" view of PPIs treats proteins simply as logical entities and visualizes their interactions as a network, aiming to understand the system of interactions as a whole. This paper describes a computational technique that applies insights gleaned from the older perspective to independently supplement experimental methods designed for the newer, systems-level perspective of PPIs.

We consider the problem of predicting if two proteins interact, given their sequence information and, optionally, other genomic and proteomic information. Such computational prediction of PPIs can supplement experimental methods for elucidating PPIs. When mapping very large interactomes (e.g., human), such PPI predictions– even if only partially accurate– would be valuable in prioritizing the set of interactions to experimentally test. Moreover, experimental techniques are quite error-prone; as prediction methods gain accuracy, they can be used to double-check the results of the experiments.

Contributions: This paper proposes to use structure-based methods, in conjunction with high-throughput information, to predict interactions. We describe a fully-automated structure-based method for computing the likelihood of an interaction, solely from sequence data. A key idea here is that if a potential interaction is sufficiently favorable energetically, it is likely to be true. As part of our method, we introduce a novel algorithm for computing the most-likely structure of the complex formed by two given proteins and describe the use of logistic regression[2] for evaluating if the putative complex corresponds to a true interaction. Furthermore, to the best of our knowledge, this paper is the first to describe a framework for predicting PPIs that integrates structure-based insights with other functional annotation (e.g., co-expression, GO description). Finally, our methods predict new potential interactions involving yeast homologs of human disease-related proteins.

Algorithm Overview: We employ a structure-based method to answer the following question: "*assuming* two given proteins interact, what is the interaction energy of the formed complex[a]?" The method exploits homology between the given protein-pair and complexes with known structure. Then we use logistic regression to identify those pairs for which the interaction energy is low enough and, hence, an interaction is likely. To combine PPI predictions made by our structure-based method with other kinds of functional information we have used a random-forest classifier[7] (see Fig 1).

Related Work: Existing work on predicting PPIs has mostly followed a "guilt-by-association" approach, the idea being that if two proteins share

[a]In this paper, when referring to protein *complexes*, we consider only those with exactly two components.

functional characteristics (co-expression, similar GO annotations etc.) they are likely to interact[15]. These methods employ a variety of functional information, using them to classify an interaction as 'true' or 'false'. Many different machine learning techniques have been used for classification: Bayesian networks[5], random forests[12], probabilistic decision trees[18], and kernel canonical analysis[17]. Qi *et al.*[12], in particular, incorporated a large variety of functional information. More recently, Lin *et al.*[8] have ranked various information sources to identify the strongest predictors of an interaction. However, some of these approaches[5,8] also use high-throughput experimental PPI data itself as a predictor[b]. In contrast, our goal is to predict PPIs *completely independently* of experimental PPI data. In other work, Deng *et al.*[3] have used sequence-based domain signatures, derived from low-throughput data, to identify interacting domains between proteins. None of these methods incorporate structure-based approaches.

Our work is different from existing work in the introduction of structure-based methods as additional predictors of PPIs. The use of such methods provides several advantages. First, these methods can provide *insight* into how, if at all, an interaction happens, unlike guilt-by-association methods which do not. Second, for many protein pairs very little functional annotation is available and structure-based methods might often be the only available predictors. Third, as we show, these methods can be used in addition to existing methods, allowing us to improve upon current performance. We note that Lu, Lu & Skolnick[13] have explored the use of purely structure-based methods to predict PPIs. In comparison, our structure-based method has several advantages (described later) and we describe how it can be integrated with other information sources.

A possible concern might be that current structure prediction methods are not sufficiently accurate and may not work well for every protein-pair. In response, we note that our framework is modular so that better methods can be substituted in, as they become available. Second, our method is homology-based and will improve in performance and coverage as the recent NIH-funded push to elucidate more structures gains momentum.

Another concern might be that just because two protein structures interact *in-silico*, they might not interact *in-vivo*. This risk can be mitigated by combining inferences based on structural-techniques with other kinds of data. Also, note that this concern is equally applicable to existing approaches. Similarly, like many previous approaches, we restrict ourselves

[b]The usual reasoning in such cases is that high-throughput PPI determination methods are noisy enough that they only *indicate* an interaction, not *confirm* it.

to pairwise protein interactions, even though more than two proteins may simultaneously interact *in vivo*.

2. Problem Formulation

We now provide a precise formulation of the two problems we address here:
Problem [STRUCTONLY] Given two proteins p and q, and their associated sequences S_p and S_q, compute the probability that p and q interact.
Problem [STRUCT&OTHERINFO] Given two proteins, p and q, their associated sequences S_p and S_q, and optional annotation information $\{X_p^1, X_p^2, \ldots\}$ and $\{X_q^1, X_q^2, \ldots\}$, compute the probability that p and q interact.

In STRUCTONLY, note that we only require the protein sequences, and not structures. If necessary, the protein sequences can themselves be inferred from the corresponding gene sequences. In STRUCT&OTHERINFO, different kinds of annotation information can be incorporated, as available. Our method for solving this problem can be used with as many information sources as desired, but here we have restricted ourselves to a few:

#	Name	Description
1	Coexpression	Similarity between expression levels of the corresponding genes
2	Colocalization	Co-localization information for the two proteins
3	GO	Similarity between Gene Ontology(GO) terms for the two genes
4	MIPS	Similarity between MIPS terms for the corresponding genes
5	Domain	Seq. motifs indicating the presence of interacting domains
6	Coessentiality	Whether one, both, or none of the corresponding genes are *essential*

Table 1: The various kinds of functional annotation used in STRUCT&OTHERINFO. These benchmark annotations have previously been found to be particularly relevant in PPI predictions (see Supp. Info. for details).

3. Algorithms

3.1. *Problem #1:* STRUCTONLY

Here, we follow a two-staged process (see Fig 1(a)). The advantage of this two-staged process is that as structure-based methods improve in accuracy, better ones can be plugged into the first stage.

3.1.1. Stage 1: Computing Interaction Energies

Here we introduce DBLRAP ("DouBLe RAPTOR"), a novel algorithm that exploits the idea that if the homologs of a pair of proteins interact in a specific way, the latter will also interact in a similar way. The algorithm consists of two major components: (1) construction of the complex template database, and (2) threading the two sequences to each potential complex template.

The complex template database is derived from the latest SCOP[9] database (i.e., SCOP v1.67) as follows: we first check if two protein domains can form a complex as per the following rule. For any pair of SCOP

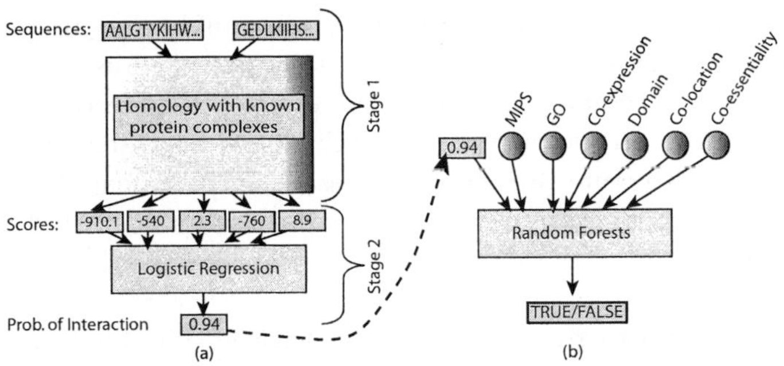

Figure 1: Schematic of our method for (a) StructOnly (b) Struct&OtherInfo

domains with the same PDB ID, we calculate their interfacial contacts, using the same method described in Lu, Lu & Skolnick[13]. If there are more than 10 interfacial contacts between two domains, then we assume that they form a complex. Next, we remove redundant complexes to improve computational efficiency. We use the following clustering method. Suppose we have two protein complexes C_1 and C_2, which are composed of domains A_1 and B_1 and domains A_2 and B_2, respectively. We classify C_1 and C_2 into the same cluster if one of the following two conditions are satisfied: i) A_1 and A_2 are in the same SCOP family and so are B_1 and B_2; ii) A_1 and B_2 are in the same SCOP family and so are A_2 and B_1. We randomly choose one representative from each complex cluster. All the representatives together form a complex template database. In total, the complex template database contains 2443 complexes, which are composed of 4142 unique SCOP domains.

After constructing the complex template database, we then thread each sequence pair to all the complex templates to find the best potential match. We align each sequence pair to the best-matching complex. Using this alignment and the interaction pattern between the complex's constituent subunits, we can also calculate the interfacial energy between our input proteins. The interfacial potential parameters are taken from Lu, Lu, & Skolnick's[13] paper. For computational efficiency, in the actual implementation we did some preprocessing first, the details of which are in the Supp. Info.

In summary, for any given sequence pair (p and q), the threading-based interaction prediction method will generate two alignment scores (E_p, E_q), their associated z-scores (z_p, z_q) , and an interfacial energy (E_{pq}). These are fed into the logistic regression model to predict interaction.

DBLRAP circumvents the docking problem: searching for the optimal orientation of two proteins in a complex. But it has a limitation that the number of complexes with known structures is not yet sufficiently large, though it is increasing. An alternative approach would be to use homology only to predict structures for individual proteins and then use methods for protein docking to compute and score the optimal relative orientation of the two structures. In theory, this approach should have greater coverage: homology-based structure prediction is possible, and reasonably accurate, for many proteins now. However, our limited exploration indicated that this method does not work very well, possibly because docking programs are not yet sufficiently good.

3.1.2. Stage 2: From Energy Values to Interaction Probabilities

We use binary logistic regression[2] to classify whether a set of scores corresponds to an interaction or not. In binary logistic regression, the goal is to predict a binary output variable Y, given a set of r predictor variables $\mathbf{X} = \{X_1, X_2, \ldots, X_r\}$. For an instance i, suppose y_i and $\mathbf{x_i} = \{x_{i1}, x_{i2}, \ldots, x_{ir}\}$ are the random variables corresponding to Y and $\mathbf{X}$, respectively. Let $\theta_i = P(y_i = 1|\mathbf{x_i})$. In this model, the dependence of θ_i on $\mathbf{x_i}$ is expressed by the logit function:

$$logit(\theta_i) = log(\frac{\theta_i}{1-\theta_i}) = \alpha + \beta^t \mathbf{x_i} = \alpha + \beta_1 x_{i1} + \beta_2 x_{i2} + \ldots + \beta_r x_{ir} \quad (1)$$

This can be rewritten as:

$$\frac{P(y_i = 1|\mathbf{x_i})}{P(y_i = 0|\mathbf{x_i})} = e^{\alpha+\beta^t \mathbf{x_i}} \quad \text{or} \quad P(y_i = 1|\mathbf{x_i}) = \frac{e^{\alpha+\beta^t \mathbf{x_i}}}{1+e^{\alpha+\beta^t \mathbf{x_i}}} \quad (2)$$

Logistic regression was performed by the standard Iterative Re-weighted Least Squares algorithm (the R package: `http://www.r-project.org`).

We now describe how we have set up the logistic regression problem for our case. The output variable Y is the probability of interaction of two proteins p and q. The predictor variables come from the first stage. For proteins p and q, DBLRAP provides their interfacial energy E_{pq}, their respective alignment scores E_p and E_q, as well as associated z-scores z_p and z_q. In addition to these, we also put in $\sigma_{pq} = N_p + N_q$ and $\pi_{pq} = \sqrt{N_p N_q}$, where N_p and N_q are the sequence lengths of p and q. Finally we introduce, as separate predictor variables, various functions and combinations of the existing terms: e.g. $\frac{E_p}{N_p}, \frac{E_q}{N_q}, \frac{E_{pq}}{\sigma_{pq}}, \frac{E_{pq}}{\pi_{pq}}, \sqrt{E_{pq}}$, etc.

We intentionally built an initial model with an excessively large set of predictor variables: one of our goals was to identify the most informative subset of predictors, using Akaike Information Criterion (AIC) to determine the subset with the optimal trade-off between prediction accuracy and subset-size. The AIC score for a logistic regression model is defined by:

$$AIC = -2\text{log-likelihood} + 2k/N$$

where k = number of predictor variables, N = number of instances in the dataset, and the log-likelihood of data under the model is computed using Eq. 2. The subset of predictor variables with the lowest AIC score was chosen.

The use of logistic regression for prediction confers certain advantages. It allows us to combine multiple scores (interaction energies, z-scores etc.), possibly from different methods. Functions of these scores can also be considered. We can then use logistic regression to identify the most relevant subset of predictors. Compared to Lu, Lu, & Skolnick[13], who only compared the interfacial energy against some threshold, the use of logistic regression allows us to make more sophisticated decisions.

3.2. *Problem# 2:* STRUCT&OTHERINFO

For classification purposes one can associate, with each pair of proteins p and q, a data-vector $D_{pq} = (d_1, \ldots, d_6)$ that contains information from the six non-structure-based information sources described in Table 1. To add structure-based information to this, we simply add one more feature d_7 to D_{pq}. Here, d_7 is the probability of interaction between proteins p and q as computed using logistic regression. Given some training data consisting of known true and likely false interactions, we then train a random forest to classify a possible interaction based on its data-vector (see Fig 1b).

Random Forests: Random forests[7] (RF) generalize the intuition behind decision trees. Given a dataset $\mathbf{D}$ of N data-vectors $D_1, \ldots, D_N$, κ decision trees are constructed. For each tree, only a subset of the feature-space is used to train the tree using the data $\mathbf{D}$. For example, for tree T_{12}, only the features d_1, d_3, and d_7 might be used to create it. Given a test data-vector D_t, the predicted class is determined by running down D_t on each tree and then taking the majority vote over the predicted classes. Random forests can handle missing data. The procedure for handling missing data is somewhat involved; please see the original reference[7] for details.

Our use of random forests is rather straightforward. Our feature space consists of the 7 features described earlier.We then trained a random forest with 500 trees over this space.

Though random forests have only recently been introduced, they have quickly become very popular. They have many desirable characteristics: they rarely overfit the data; they allow classification when features are not independent; they allow for missing values. Lastly, their output is easy to analyze in terms of identifying the strongest predictors and the relationships between the different features. We also note that their usefulness in prediction and analysis of PPIs has previously been demonstrated[12].

Dataset	Interactions				Motivation behind creating the dataset	Post Filtering Interctns.
	Pos.	Notes	Neg.			
LT	100	From high-quality low-throughput experiments	400		Low-throughput interactions provide "gold-standard" pos.s	69
HTFEWANNOT	508	Between 1000 proteins with little functional annotation	2000		Existing guilt-by-assoc. methods do not work well with these	332
HTMANYANNOT	489	Between proteins with a lot of functional annotation	300		Test how to combine structure-based methods with other info.	160

Table 2: The construction of three datasets for yeast PPI data. The positive interactions (#'s shown in table) were retrieved from GRID while (putative) negative interactions were generated by randomly pairing two yeast proteins. The difference between the datasets is primarily in how different positive sets were picked. The datasets were filtered to keep only those interactions for which homologous models could be found.

4. Results

Datasets: In this work, we have focused on predicting PPIs in yeast (*S. cerevisiae*). The list of experimentally discovered PPIs for yeast was retrieved from GRID[1]. From this database, three datasets were created: LT, HTFEWANNOT, and HTMANYANNOT (see Table 2). The datasets differed in how their positive examples (true interactions) were selected (see Notes in Table 2). Note that because of the significant error-rate[15] in high-throughput experiments, some of the positive examples in HTFEWANNOT and HTMANYANNOT are likely to be incorrect.

Collecting negative examples (false interactions) is difficult: experimentally confirmed false interactions are rare. As such, we had to design our own— a problem faced by other researchers as well[15,12]. We followed Qi *et al.*'s strategy of considering a random pair of proteins as non-interacting. Since, on average[12], only 1 in 600 possible interactions is true, the chances of a random pair being truly non-interacting are $> 99\%$.

However, not all interactions in the datasets corresponded to protein-pairs for which homologous complexes could be found. Therefore, we had to filter out a subset of the dataset. As discussed before, as more structures become available, the coverage of the homology-based methods will increase and fewer pairs will be filtered out.

Using Only Structure-based Method (STRUCTONLY):

Using the AIC criterion as described before, we discovered that the subset of predictors of interaction with the optimal balance between model complexity and goodness-of-fit were: $\{\frac{E_{pq}}{\pi_{pq}}, z_p, z_q, \pi_{pq}\}$, where π_{pq} is the square root of the product of sequence lengths of p and q. Of these, $\frac{E_{pq}}{\pi_{pq}}$ ($p < 0.001$) and

z_p, z_q ($p < 0.05$) were the more significant predictors.

In hindsight, it does seem reasonable that E_{pq}/π_{pq} is a stronger predictor of interaction than E_{pq} itself: for large proteins, even relatively weak interactions will have a large (negative) interfacial energy, simply because of there being more interacting entities. Thus, it makes sense that the energy score should be normalized by the sequence length of the two proteins.

We tested our method by 4-fold cross-validation on the LT dataset. In addition, the method was trained on the entire LT dataset and tested on the combined HTFEWANNOT + HTMANYANNOT dataset. By comparing against some threshold value (say $p_{thresh} = 0.5$), the probabilities of interaction predicted by logistic regression can be interpreted as true/false interactions. By varying p_{thresh}, we can plot the sensitivity-vs.-specificity (ROC) curve of the method (see Fig 2a). As can be seen, the structure-based method provides significant signal for prediction purposes. The performance of the method is better on the low-throughput (LT) dataset than on the high-throughput datasets. A possible cause might be that the high-throughput datasets have more errors, i.e., negative examples mis-labeled as positive. Of course, the LT dataset is smaller, and the better performance on it needs more validation. It is also possible that the Skolnick potentials work better for LT dataset. In future work, we plan to explore these issues further.

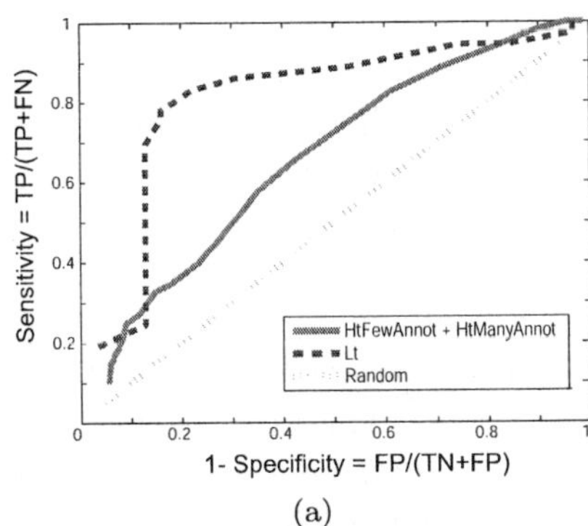

(a)

Features	Error
All	8.4%
All - Coexpression	28.4%
All - Structure	11.6%
All - Domain	10.9%
All - Coessentiality	8.4%
All - GO	8.4%
All - MIPS	7.7%
All - Colocation	5.8%

(b)

Figure 2: (a) STRUCTONLY: Specificity-vs.-Sensitivity curve when using only the structure-based approach. TP=True Pos., FP=False Pos., TN=True Neg., FN=False Neg. The dotted diagonal line indicates the baseline, a method with zero predictive power. The performance of our method is better for LT than for HTFEWANNOT +HTMANYANNOT. A possible reason might be that the latter datasets themselves might have mislabeled instances. (b)STRUCT&OTHERINFO: Classification error, and its dependence on the various features. "All - X" indicates that all features, except X, have been used for classification. As can be seen, the classification error increases if the structure-based method is not used.

Combining Various Information Sources (STRUCT&OTHERINFO):

We tested our entire framework on the HtManyAnnot dataset, a dataset specifically chosen for proteins with lots of functional annotation available. We used 5-fold cross-validation to evaluate our method, using the cross-validation error (CVE) as the quality metric.[c]

With average sensitivity = 94.1% and specificity = 92.1%, the overall performance of our method is better than that of existing work, e.g., Zhang *et al.*'s[18] (sensitivity = 81% at specificity = 80%, approximately)[d]. Even when experimental PPI data itself has been used as one of the predictors by others (e.g., Lin *et al.*[8]: sensitivity = 98%, specificity = 92%, approximately), our method— which is *completely* independent of experimental PPI information— performs comparably.

One interesting question is: "do structure-based methods contribute to the predictive power, compared to other features?" To quantify a feature's importance, we removed it from the mix and recomputed the CVE. The difference between this CVE and the baseline CVE (with all the features present) indicates the increase in accuracy offered by including that feature. As the table in Fig 2 shows, coexpression is the most important feature, followed by the information provided by our method. Some of the other features, e.g. colocation, do not seem to be particularly important.

4.1. *Novel Predictions*

Predictions on Less-Characterized Proteins: The proteins in the HtFewAnnot have very little functional annotation and very few known PPIs (see Supp. Info. for more details). For these, there isn't enough functional annotation for "guilt-by-association" methods to work; in contrast, our structure-based method will still work.

We tested all possible pairs in this set for interaction, using our structure-based method, without any additional functional annotation. The probabilities of interaction, as computed by logistic regression, were used to rank the pairs and the top 2000 pairs were chosen. The network formed by these predicted set of interactions (see Supp. Info. for the predicted set) shows some intriguing properties. It has a scale-free character[15], just like the experimentally-determined yeast PPI network, i.e., the node degree distribution follows the power law. Moreover, the two power-law coefficients

[c]Computing 5-fold cross-validation error (CVE): data was randomly partitioned into five equal parts. Four of the parts constituted the training set while the fifth one made up the test set. The error was computed as the classification error on this test set. By repeating this error computation for each of the classes, five error values were computed and averaged to compute the CVE.

[d]We compared against Zhang *et al.*'s performance in the case when they did not use experimental PPI data as a predictor

are comparable (1.9 for predicted network; 2.3 for the yeast interactome).

In the predicted network, the protein CHS2 is a hub (86 interactions), and the set of its partners is enriched for genes involved in amino acid and amine transporter activity. So, we hypothesized that this protein would have similar functions. This turns out to be true— CHS2 is involved in transferring N-acetylglucosamine to chitin. It is also relevant in disease-treatment; some recent work on developing antiprotozoal drugs has focused on targeting the chitin-synthesis pathway[6]. Similarly, for DSF2—a hitherto uncharacterized gene—the set of its predicted interaction partners is enriched for genes related to DNA transposition and retrotransposons ($p < 0.001$), indicating DSF2's possible function.

Disease-Related Proteins: In the predictions, we also specifically looked for homologs of human disease-related genes. We describe a few findings here; the rest are in Supp. Info.

The human homolog of RAD28 has been implicated in Cockayne Syndrome (related to malfunctions in DNA-repair machinery). Currently, there are only two known PPIs involving RAD28. Our method predicts 19 additional PPIs, and 6 of the predicted partners are involved in DNA repair.

Similarly, the human homolog of PAT1 is Adrenoleukodystrophy– a neurodegenerative disease caused by a malfunctioning fatty-acid transporter protein. There are only three known PPIs involving PAT1; our method predicts 26 more. Moreover, the set of its interaction partners is enriched for proteins involved in lipid and fatty acid transport ($p < 0.01$).

Genome-scale Predictions: We used the structure-based method, without any additional functional annotation, to perform an all-vs-all prediction of the interactions in the yeast genome (see Supp. Info. for predictions). The predicted network has a scale-free character similar to the known yeast interactome and has about 9% overlap with it. This is significantly better than overlap achieved by Lu, Lu & Skolnick's[13] method and is comparable to the overlap between large-scale experimental PPI datasets.

5. Discussion

We have described how structure-based methods can be integrated with other genomic and proteomic information for predicting PPIs. Structure-based methods can be used by themselves when other functional annotation is not available. When used in conjunction with functional annotation, their addition improves prediction accuracy over existing methods.

Our future efforts will focus on (1) applying this method to mammalian genomes, (2) incorporating other kinds of functional annotation (e.g., corre-

lated mutations[10]), and (3) using docking programs as an additional way of computing interfacial energies. As mentioned before, our brief exploration indicated that current docking programs did not perform satisfactorily. However, more work might suggest ways to improve them for our purposes.
Acknowledgments: The authors thank Dr. Ying Xu and Dr. Fengluo Mao at the Univ. of Georgia for allowing us the use of their Linux cluster.

References

1. B.J. Breitkreutz, C. Stark, M. Tyers. The GRID: the General Repository for Interaction Datasets *Genome Biol*, 4(3):R23, 2003
2. T. Hastie and R. Tibshirani. *Generalized Additive Models*. Chapman and Hall
3. M. Deng, S. Mehta, F. Sun, and T. Chen. Inferring domain-domain interactions from protein-protein interactions. *Genome Research*, 12(10):1540–8, 2002.
4. A. Jaimovich, G. Elidan, H. Margalit, and N. Friedman. Towards an Integrated Protein-Protein Interaction Network. *Proceedings of RECOMB*, 2005.
5. R. Jansen *et al.* A Bayesian networks approach for predicting protein-protein interactions from genomic data. *Science*, 302(5644):449–53, 2003.
6. E.L. Jarroll and K. Sener. Potential drug targets in cyst-wall biosynthesis by intestinal protozoa. *Drug Resist Update*, 6(5), 2003.
7. L. Breiman. Random Forests. *Machine Learning Journal*, 45(1), 2001.
8. N. Lin, B. Wu, R. Jansen, M. Gerstein, H. Zhao. Information assessment on predicting protein-protein interactions. *BMC Bioinformatics* 18;5:154, 2004.
9. A. Murzin, S. Brenner, T. Hubbard, C. Chothia. SCOP:A structural classification of proteins database *J Mol Biol* 247, 536-540, 1995
10. F. Pazos *et al.* Correlated mutations contain information about protein-protein interaction *J Mol Biol*, 271(4):511-23, 1997
11. D.L. Price, D.R. Borchelt, and S.S. Sisodia. Alzheimer Disease and the Prion Disorders *Proc Natl Acad Sci USA*, 90(14):6381-4, 1993
12. Y. Qi, J. Klein-Seetharaman, and Z. Bar-Joseph. Random Forest Similarity for Protein-Protein Interaction Prediction. *PSB*, 2005.
13. H. Lu, L. Lu, and J. Skolnick. Development of Unified Statistical Potentials Describing Protein- Protein Interactions *Biophysical J*, 84:1895-1901, 2003.
14. P. Uetz, S.V. Rajagopala, Y.A. Dong, and J. Haas. From ORFeomes to protein interaction maps in viruses. *Genome Research*, 14(10B):2029–33, 2004.
15. C. von Mering *et al.* Comparative assessment of large-scale data sets of protein-protein interactions. *Nature*, 417(6887):399–403, 2002.
16. J. Xu, M. Li, D. Kim, and Y. Xu. RAPTOR: optimal protein threading by linear programming. *J. of Bioinformatics and Comp. Biol.*, 1(1):95–117, 2003.
17. Y. Yamanishi *et al.* Extraction of correlated gene clusters from multiple genomic data. *Bioinformatics*, 19:323–30, 2003.
18. L.V. Zhang, S.L Wong, O.D. King, F.P. Roth. Predicting co-complexed protein pairs using genomic and proteomic data integration. *BMC Bioinformatics*, 5(38), 2004.

IDENTIFYING INTERACTION SITES IN "RECALCITRANT" PROTEINS: PREDICTED PROTEIN AND RNA BINDING SITES IN REV PROTEINS OF HIV-1 AND EIAV AGREE WITH EXPERIMENTAL DATA

MICHAEL TERRIBILINI[1,3†], JAE-HYUNG LEE[1,3], CHANGHUI YAN[1,2,4], ROBERT L. JERNIGAN[1,3,5], SUSAN CARPENTER[6], VASANT HONAVAR[1,2,5], DRENA DOBBS[1,4,5]

[1]Bioinformatics and Computational Biology Graduate Program and L.H. Baker Center for Bioinformatics and Biological Statistics, [2]Department of Computer Science, [3]Department of Biochemistry, Biophysics and Molecular Biology, [4]Department of Genetics, Development and Cell Biology, [5]Artificial Intelligence Research Laboratory, and Center for Computational Intelligence, Learning and Discovery, Iowa State University, Ames, IA, 50010, USA

[6]Department of Veterinary Microbiology and Pathology, Washington State University, Pullman, WA, 99164, USA

Protein-protein and protein nucleic acid interactions are vitally important for a wide range of biological processes, including regulation of gene expression, protein synthesis, and replication and assembly of many viruses. We have developed machine learning approaches for predicting which amino acids of a protein participate in its interactions with other proteins and/or nucleic acids, using only the protein sequence as input. In this paper, we describe an application of classifiers trained on datasets of well-characterized protein-protein and protein-RNA complexes for which experimental structures are available. We apply these classifiers to the problem of predicting protein and RNA binding sites in the sequence of a clinically important protein for which the structure is not known: the regulatory protein Rev, essential for the replication of HIV-1 and other lentiviruses. We compare our predictions with published biochemical, genetic and partial structural information for HIV-1 and EIAV Rev and with our own published experimental mapping of RNA binding sites in EIAV Rev. The predicted and experimentally determined binding sites are in very good agreement. The ability to predict reliably the residues of a protein that directly contribute to specific binding events - without the requirement for structural information regarding either the protein or complexes in which it participates - can potentially generate new disease intervention strategies.

† Corresponding author

1. Introduction

The human AIDS virus, Human immunodeficiency virus Type 1 (HIV-1), is closely related to a number of lentiviruses that cause persistent, insidious infections in other primates and domestic animals. Recent advances in molecular virology have resulted in novel antiviral therapies that inhibit specific proteins required for the replication of lentiviruses and other important retroviruses. Rev is a multifunctional regulatory protein that plays an essential role in the production of infectious virus (1, 2) and, as such, is an attractive target for new antiviral therapies. To date, however, no Rev-targeted drugs for AIDS therapy are available.

Rev is known to participate in protein-protein interactions with several cellular proteins as well as in RNA-protein interactions with lentiviral RNAs (3, 4). It is required for the transition to the late stage of viral replication and facilitates export of incompletely spliced viral RNAs from the nucleus to the cytoplasm. After its import into the nucleus, HIV-1 Rev binds a structure in the viral pre-mRNA called the Rev-responsive element (RRE) (5, 6), multimerizes (6, 7), then utilizes the CRM1 nuclear export pathway to redirect movement of incompletely spliced viral RNA out of the nucleus (8). As shown in Figure 1, functional domains within HIV-1 Rev are known to mediate interactions with viral RNA and with host cell proteins that are required for nuclear localization, RNA binding, multimerization, and nuclear export (3).

Efforts to develop inhibitors of Rev activity have been hampered by a lack of information regarding Rev protein structure. A major stumbling block for structural analysis is the tendency of Rev to aggregate at concentrations needed for crystallization or solution NMR studies (9). The only high resolution information available is for short peptide fragments of HIV-1 Rev. In an NMR solution structure of a 23 amino acid fragment of Rev bound to a 34 nucleotide RRE RNA fragment, the Rev peptide adopts an α-helical conformation and is bound in the major groove of the RNA (10). Structures of other critical functional domains of Rev (e.g., nuclear localization, multimerization, export) have not been reported. Furthermore, it has not been possible to apply homology modeling approaches to gain insight into Rev structure because Rev has no detectable sequence similarity to any protein of known structure. Indeed, despite their apparently conserved functions, Rev protein sequences are highly variable between species, with < 10% sequence identity between HIV-1 and one of the most divergent Rev proteins, equine infectious anemia virus, (EIAV) Rev (11).

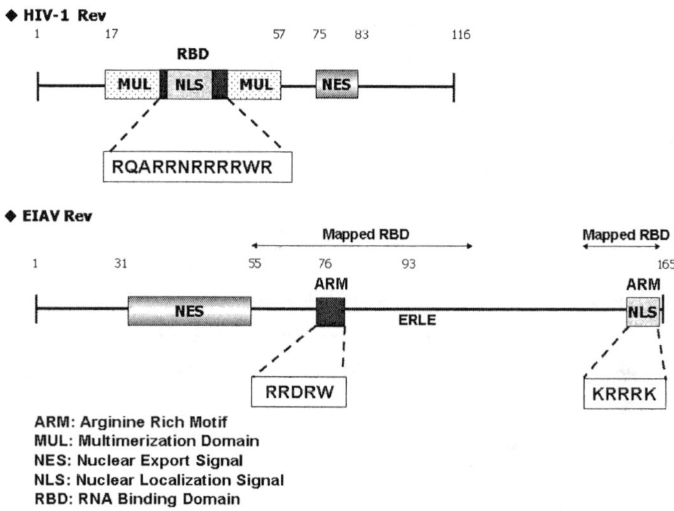

Figure 1. Functional domains of HIV-1 and EIAV Rev proteins. The linear organization of functional domains within the two Rev proteins differs significantly, but both have been shown to contain specific sequences involved in Rev interactions with proteins (MUL, NLS, NES) or RNA (RBD, ARMs).

When protein structures cannot be solved using experimental approaches, computational analyses can provide valuable insight into protein structure-function relationships and aid in identification of key functional residues that may offer tractable targets for therapeutic intervention in disease (12). Here we describe the identification of critical residues that mediate protein-protein and protein-RNA interactions in Rev, using machine learning approaches that rely on the primary amino acid sequence of Rev, but do not require any information regarding its structure or the sequence or structure of its interaction partners. Our predictions are in good agreement with previously published biochemical, biophysical and genetic data for HIV-1 and EIAV Rev as well as with our recent experimental mapping of RNA binding sites in EIAV Rev (13). Taken together, these results demonstrate the utility of sequence-based approaches for identifying putative binding sites of proteins with potential therapeutic value that are, at present, recalcitrant to experimental structure determination.

2. Datasets, Materials and Methods

2.1. *Datasets*

Protein-protein binding site dataset (PBS). We extracted individual proteins from a set of 70 protein–protein heterocomplexes used in the study of Chakrabarti and Janin (14). After removal of redundant proteins and molecules with fewer than 10 residues, we obtained a dataset of 77 individual proteins with sequence identity <30%. The dataset contains a total of 12,719 amino acids, of which 2340 (18.4%) are interface residues (positive examples).

RNA-protein binding dataset (RBS). A dataset of protein-RNA interactions was extracted from structures of known protein-RNA complexes in the Protein Data Bank (PDB) (15). Proteins with >30% sequence identity or structures with resolution worse than 3.5Å were removed using PISCES (16). This resulted in a set of 109 non-redundant protein chains containing a total of 25,118 amino acids. Amino acids in the protein-RNA interface were identified using ENTANGLE (17). Using default parameters, 3518 (14%) of the amino acids in the dataset are defined as interface residues (positive examples).

2.2. *Protein-protein interface residue prediction*

We have previously developed a two-stage classifier for predicting interface residues in protein-protein complexes (18). In the first stage, a Support Vector Machine (SVM), trained on the PBS dataset, is used to classify each residue as interface or non-interface. Input to the SVM is a window of nine amino acid identities. Because interface residues tend to be clustered in primary sequence, a second stage was introduced to take advantage of this to improve predictions. In the second stage, a Bayesian Network classifier is trained based on the predictions of the target residue and its neighbors from the first stage SVM. Let $C \in \{0,1\}$ denote the actual class label of a residue; $X \in \{0,1\}$ be the prediction of the SVM classifier; Y denote the number of predicted interface residues within 4 amino acids of the target residue. For each residue, the likelihood that it is an interface residue given the SVM predictions for itself and its neighbors is calculated and compared to a chosen threshold θ as formula 1.

$$\frac{P(C=1\mid X,Y)}{P(C=0\mid X,Y)} > \theta \qquad (1)$$

The residue is predicted to be an interface residue if the likelihood is larger than θ and non-interface otherwise. The conditional probability table P(C|X,Y) is derived from training datasets. To determine θ, the classifier was applied to the training set and different values of θ ranging from 0.01 to 1 were tested, in increments of 0.01. The value of θ for which the classifier yields the highest correlation coefficient was used to make predictions on the Rev proteins.

2.3. *Protein-RNA interface residue prediction*

We have previously developed a Naïve Bayes (NB) classifier for predicting which amino acids in a given protein are likely to be found in protein-RNA interfaces (19), using the NB classifier from the Weka package (20). The input is a window of 25 contiguous amino acid identities. The output is an instance where + indicates that the target residue is an interface residue and – indicates a non-interface residue. A training example is an ordered pair (x, c) where $x = (x_{-n}, x_{-n+1}, ..x_{T-1}, .x_T, x_{T+1}, ...x_{n-1}, x_n)$ and c is the corresponding class label (interface or non-interface). A training data set D is a collection of labeled training examples.

Let $X = (X_{-n}, ...X_T, ...X_n)$ denote the random variable corresponding to the input to the classifier and C denote the binary random variable corresponding to the output of the classifier. The Naïve Bayes classifier assigns input x the class label + (interface) if:

$$\frac{P(C=+)\prod_{i=-n}^{i=n} P(X_i = x_i \mid C=+)}{P(C=-)\prod_{i=-n}^{i=n} P(X_i = x_i \mid C=-)} \geq \theta$$

and the class label – (non interface) otherwise. θ was set to the value that optimized the correlation coefficient (21) on the *training set* in each leave-one-out cross validation experiment.

2.4. *Experimental mapping of RNA binding sites*

Details of our experimental mapping of RNA binding sites are provided in Lee et al., (13). Briefly, Maltose Binding Protein-EIAV Rev (MBP-ERev) constructs containing deletions or point mutations in the EAIV Rev coding region were cloned in pHMTc, based on the pMal-c2x expression vector, which enhances solubility of Rev fusion proteins. MBP-ERev fusion proteins were expressed in *E. coli*, purified prior to use in RNA binding experiments. UV cross linking experiments were used to quantitate the effects of mutations on Rev RNA binding activity (13).

3. Results

3.1. *Binding site predictions on datasets of known protein-protein and protein-RNA complexes*

In previous work, we have developed classifiers for predicting interface residues in protein-protein, protein-DNA and protein RNA complexes (18, 19, 22), typically using a combination of sequence and structure-derived information as input. In choosing classifiers for the task of predicting protein-protein and protein-RNA interface residues in Rev proteins, we compared several types of classifiers for predicting each type of interface residue (data not shown). Table 1 shows an example of the classification performance values obtained for protein binding site prediction using the PBS dataset, which contains 77 proteins used in our previous study (18) and for RNA binding site prediction using the RBS dataset, which contains 109 RNA-binding proteins (19).

Table 1. Classification performance in predicting protein-protein and RNA-protein binding site residues, using leave-one-out experiments

Classification Performance Measure	Protein Interface Residues (2-stage classifier)	RNA Interface Residues (NB classifier)
Accuracy	72%	85%
Specificity	58%	51%
Sensitivity	39%	38%
Correlation coefficient	0.30	0.35

These results were obtained using a modified 2-stage classifier developed in this work to predict protein interface residues (see Methods) and a Naive Bayes classifier published previously (19) to predict RNA interface residues. The results of the latter study are reproduced here for comparison.

3.2. *Predicted binding sites in wildtype HIV-1 and EIAV Rev proteins*

Using classifiers trained on the datasets described above, we predicted protein-protein and protein-RNA interface residues in Rev proteins from HIV-1 and EIAV. As shown in Figure 2A, the 2-stage protein classifier predicted a total of 56 protein-protein interface residues (indicated by "**p**") within the 116 amino acid HIV-1 Rev sequence. These are primarily located in 5 clusters consisting of 6-15 amino acids. The Naive Bayes classifier predicted a total of 26 RNA-protein interface residues (indicated by "**r**"), located in a single large cluster near the N-terminus of the protein. The predicted RNA binding site sequence is PPNPEGTRQARRNRRRRWRERQRQIHSIG, corresponding to amino acids 28-56. Ile26 and Ile29 are the only two residues within this sequence that are predicted to be non-interface residues.

The prediction results for EIAV Rev, using the same classifiers, are shown in Figure 2B. A total of 79 protein-protein interface residues were predicted in the 165 amino acid protein. In EIAV Rev, most of these predicted protein-binding residues are also located in 5 clusters that are somewhat larger (8-24 amino acids) than those predicted in HIV-1. There are two predicted clusters of RNA-protein interface residues, one consisting of 15 contiguous amino acids, located in the central region and a second consisting of 19 contiguous residues at the C-terminus of the protein. The predicted RNA binding site sequences are RHLGPGPTQHTPSRR, (aa 63-77) and QSSPRVLRPGDSKRRRKHL (aa 147-165. The only other predicted interface residues are 5 scattered amino acids in the region of aa 113-133.

3.3. *Comparison of predicted Rev binding sites with experimental data*

Functional domains in HIV-1 Rev have been extensively interrogated through the analysis of sequence variants and mutants generated both *in vivo* and *in vitro* (4). These experimental results are summarized in Figure 1 and mapped onto amino acid sequence of HIV-1 Rev for comparison with our predicted RNA and protein interface residues in Figure 2A. Notably, the single cluster of

RNA interface residues predicted by the Naive Bayes classifier closely matches the experimentally mapped RNA binding domain (RBD), which in HIV-1 also includes an Arginine Rich Motif (ARM) that also functions as a nuclear localization signal (NLS). Three predicted clusters of protein interface residues also characterized protein binding sites: one cluster (aa 22-32) maps to Rev multimerization domain, and two clusters are located within a large C-terminal domain (aa 87-116) that has been shown to play multiple roles in nuclear export, dimerization and transactivation activities of HIV-1 Rev (23). One of these clusters (aa75-93) also overlaps with the modular nuclear export signal (NES), which is interchangeable between various lentiviruses, including HIV-1 and EIAV (24).

Although the functional domains in EIAV Rev have been studied in less detail than those in HIV-1 Rev, previous biochemical and genetic studies had localized the NLS and NES domains and implicated two motifs in the central region in RNA binding, RRDRW and ERLE (Figure 1) (13, 25-28). In predictions generated *before* we initiated our experimental mapping of EIAV RNA-binding domains, the Naïve Bayes classifier identified one potential RNA-binding region overlapping the RRDRW motif and another overlapping a KRRRK motif within the mapped C-terminal NLS domain, but did not predict any interface residues near the ERLE motif. Our recent direct mapping of the RNA binding domain of EIAV Rev by UV cross linking showed that two separate regions of Rev are necessary for RNA binding: a central region encompassing aa 75-127 and a region comprising the 20 C-terminal residues of EIAV Rev (13). These experiments also demonstrated critical roles for both the central RRDRW motif and the KRRRK motif within the NLS in RNA-binding (13). Interestingly, however, the ERLE motif was not required for RNA-binding, in agreement with our predictions. Thus, our biochemical RNA-binding site mapping studies for EIAV Rev have provided direct experimental validation of the RNA interface residue predictions of the Naive Bayes classifier.

Of the five clusters of predicted protein binding residues in EIAV Rev, two overlap with known or putative protein interaction domains (the NES and the NLS, respectively), one is located in the non-essential "hypervariable" region (13), one is located near the N-terminus of the protein, and one overlaps within the central RNA binding domain (Figure 3B). There is no available biochemical data regarding the possibility that the central region of EIAV Rev binds both RNA and protein, but it is interesting that the classifier predicted binding of the

NLS region to both protein and RNA. The same residues could directly interact with both the nuclear import machinery and RNA because these interactions occur at different times and in different cellular compartments. Also, by analogy with HIV-1 Rev, it is likely that some of the protein interactions that occur when EIAV Rev multimerizes after binding RNA involve additional residues located near the RNA binding region that initiates the specific interaction between Rev and the RRE in unspliced EIAV RNA.

A.

```
        1.........11........21........31........41........51........
SEQ     MAGRSGDSDEELIRTVRLIKLLYQSNPPPNPEGTRQARRNRRRRWRERQRQIHSISERIL
PRO     ....pppppppp.........ppppppppppp...........................p
RNA     ...r............................r.rrrrrrrrrrrrrrrrr.rr.r.r..

        61........71........81........91........101.......111...
SEQ     GTYLGRSAEPVPLQLPPLERLTLDCNEDCGTSGTQGVGSPQILVESPTVLESGTKE
PRO     .pppppppppp.....p..pppppppppppppp.pp..ppp.........pppppp
RNA     ........................................................
```

B.

```
        1.........11........21........31........41........51........
SEQ     MAESKEARDQEMNLKEESKEEKRRNDWWKIDPQGPLESDQWCRVLRQSLPEEKISSQTCI
PRO     .....pppppppp.........pppppppppppppppppp.....pp.............
RNA     ............................................................

        61........71........81........91........101.......111.......
SEQ     ARRHLGPGPTQHTPSRRDRWIREQILQAEVLQERLEWRIRGVQQVAKELGEVNRGIWREL
PRO     ppppppppppppppp....................................ppppppppp
RNA     ..rrrrrrrrrrrrrr.....................................r......

        121.......131.......141.......151.......161..
SEQ     HFREDQRGDFSAWGDYQQAQERRWGEQSSPRVLRPGDSKRRRKHL
PRO     ppppppppppppppp.............pp...ppp...pppppp
RNA     ............r.............rrrrrrrrrrrrrrrrrrr
```

Figure 2. Predicted interface residues in Rev proteins. The protein sequences (SEQ) for **A)** HIV-1 Rev & **B)** EIAV Rev are shown on top line, with binding site residues for protein (PRO) and RNA shown by "p" or "r" on the lines below. Important functional domains boxed in the sequence are: NES, NLS/ARM, RBD, MULTIMERIZATION, MULTIFUNCTIONAL, ARM, UNKNOWN.

3.4. *Comparison of predicted and biochemically mapped RNA binding sites in EIAV mutant Rev proteins*

Site-specific mutagenesis, coupled with functional assays, has identified functional domains of EIAV Rev (13, 25, 26). As mentioned above, an NLS/ARM at the C-terminus was identified at the EIAV Rev C-terminus and our cross-linking analyses of the RRDRW and KRRRK motifs indicated that both are likely to contact RNA. To investigate whether our classifiers are capable of detecting mutations that give rise to differences in RNA binding, we performed predictions on several mutant EIAV Rev sequences. As shown in Figure 3, changes in RNA interface predictions are seen in sequences in which Ala residues are substituted for positively charged residues in the RRDRW and KRRRK motifs (to AADAA and KAAAK). These mutations result in >80% reduction in RNA binding activity (13). The predicted RNA binding sites no longer overlap these motifs. In contrast, predicted protein interface residues remain unchanged, consistent with the experimental results.

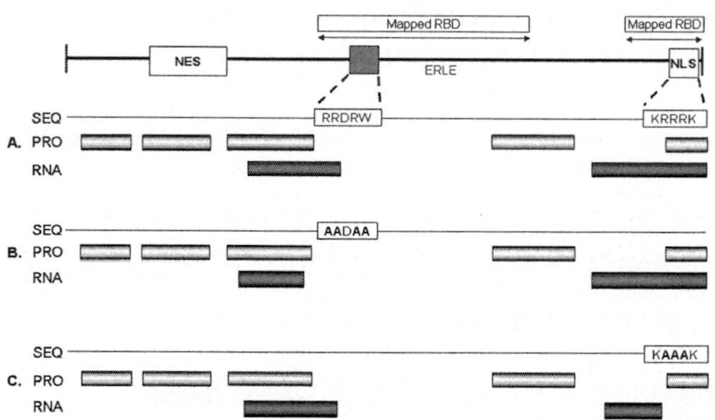

Figure 3. RNA binding site predictions differ for "wildtype" and mutant EIAV Rev sequences. Predicted protein (PRO) and RNA binding sites are indicated along the sequence (SEQ). **A**. Wildtype, **B**. & **C**. Mutant EIAV Rev sequences. RNA binding activity is reduced by >80% in both mutants (see text for details).

4. Summary and Discussion

Many effective antiviral drugs are directed at blocking the interaction between regulatory proteins and their binding partners or small effector ligands. HIV-1 Rev is one of several clinically important proteins that are "experimentally recalcitrant," i.e., for which it has not been possible to obtain high resolution structural information. Identifying critical functional residues in Rev is further complicated by the fact that Rev proteins have no significant sequence similarity to any protein with known structure, and that Rev sequences from different species have very little similarity to one another.

Our comparison of predictions with experimental data on the Rev proteins from HIV-1 and EIAV demonstrates that sequence-based computational methods can identify residues in "recalcitrant" proteins that interact with other proteins or nucleic acids. When structural information *is* available for a protein of interest, enhanced prediction accuracy can be achieved (18, 29). Developing improved methods for predicting binding sites will contribute to our understanding of how proteins recognize their targets in cells and may significantly decrease the time needed to precisely map binding sites in the laboratory. The level of accuracy obtained using the sequence-based methods presented here suggests that they could expedite the design of experiments to explore the function of key regulatory proteins, even when no structural information is available, with obvious implications for developing new therapies for both genetic and infectious diseases.

Acknowledgments

This Research was supported in part by grants NIH, GM 066387 (VH, DD, & RLJ) and CA97936 (SC), by an ISU Center for Integrated Animal Genomics grant (DD, VH & RLJ), and by USDA Formula Funds (SC & DD). We thank Sijun Liu for technical assistance and Jeffrey Sander for useful comments.

References

1. De Clercq, E. (2002) *Med Res Rev* **22,** 531-65.
2. Moore, J. P. & Stevenson, M. (2000) *Nat Rev Mol Cell Biol* **1,** 40-9.
3. Hope, T. J. (1999) *Arch Biochem Biophys* **365,** 186-91.
4. Pollard, V. W. & Malim, M. H. (1998) *Annu Rev Microbiol* **52,** 491-532.
5. Cook, K. S., Fisk, G. J., Hauber, J., Usman, N., Daly, T. J. & Rusche, J. R. (1991) *Nucleic Acids Res* **19,** 1577-83.
6. Zapp, M. L., Hope, T. J., Parslow, T. G. & Green, M. R. (1991) *Proc Natl Acad Sci U S A* **88,** 7734-8.
7. Olsen, H. S., Cochrane, A. W., Dillon, P. J., Nalin, C. M. & Rosen, C. A. (1990) *Genes Dev* **4,** 1357-64.

8. Cullen, B. R. (1992) *Microbiol Rev* **56,** 375-94.
9. Turner, B. G. & Summers, M. F. (1999) *J Mol Biol* **285,** 1-32.
10. Battiste, J. L., Mao, H., Rao, N. S., Tan, R., Muhandiram, D. R., Kay, L. E., Frankel, A. D. & Williamson, J. R. (1996) *Science* **273,** 1547-51.
11. Coffin, J. M., Hughes, S. H. & Varmus, H. (1997) *Retroviruses* (Cold Spring Harbor Laboratory Press, Plainview, N.Y.).
12. Cochrane, A. (2004) *Curr Drug Targets Immune Endocr Metabol Disord* **4,** 287-95.
13. Lee, J.-H., Murphy, S. C., Belshan, M., Wannemuehler, Y., Liu, S. Hope, T. J., Dobbs, D. & Carpenter, S. (2005) *Submitted.*
14. Chakrabarti, P. & Janin, J. (2002) *Proteins* **47,** 334-43.
15. Berman, H. M., Westbrook, J., Feng, Z., Gilliland, G., Bhat, T. N., Weissig, H., Shindyalov, I. N. & Bourne, P. E. (2000) *Nucleic Acids Res* **28,** 235-42.
16. Wang, G. & Dunbrack, R. L., Jr. (2003) *Bioinformatics* **19,** 1589-91.
17. Allers, J. & Shamoo, Y. (2001) *J Mol Biol* **311,** 75-86.
18. Yan, C., Dobbs, D. & Honavar, V. (2004) *Bioinformatics* **20 Suppl 1,** I371-I378.
19. Terribilini, M., Lee, J.-H., Yan, C., Jernigan, R. L., Honavar, V. & Dobbs, D. (2005) *Submitted.*
20. Witten, I. H. & Frank, E. (2000) *Data Mining: Practical machine learning tools with Java implementations* (Morgan Kaufmann.
21. Baldi, P., Brunak, S., Chauvin, Y., Andersen, C. A. & Nielsen, H. (2000) *Bioinformatics* **16,** 412-24.
22. Yan, C., Terribilini, M., Wu, F., Jernigan, R. L., Dobbs, D. & Honavar, V. (2005) *Submitted.*
23. Hakata, Y., Yamada, M., Mabuchi, N. & Shida, H. (2002) *J Virol* **76,** 8079-89.
24. Mancuso, V. A., Hope, T. J., Zhu, L., Derse, D., Phillips, T. & Parslow, T. G. (1994) *J Virol* **68,** 1998-2001.
25. Belshan, M., Harris, M. E., Shoemaker, A. E., Hope, T. J. & Carpenter, S. (1998) *J Virol* **72,** 4421-6.
26. Chung, H. & Derse, D. (2001) *J Biol Chem* **276,** 18960-7.
27. Fridell, R. A., Partin, K. M., Carpenter, S. & Cullen, B. R. (1993) *J Virol* **67,** 7317-23.
28. Harris, M. E., Gontarek, R. R., Derse, D. & Hope, T. J. (1998) *Mol Cell Biol* **18,** 3889-99.
29. Rost, B., Liu, J., Nair, R., Wrzeszczynski, K. O. & Ofran, Y. (2003) *Cell Mol Life Sci* **60,** 2637-50.

MODELING AND SIMULATION WITH HYBRID FUNCTIONAL PETRI NETS OF THE ROLE OF INTERLEUKIN-6 IN HUMAN EARLY HAEMATOPOIESIS

SYLVIE TRONCALE AND FARIZA TAHI
LaMI, CNRS-UMR 8042, Université d'Evry Val-d'Essonne, Genopole, *France*

DAVID CAMPARD AND JEAN-PIERRE VANNIER
Laboratoire M.E.R.C.I (EA2122), Université de Rouen, France

JANINE GUESPIN
Laboratoire de Microbiologie de Rouen, France

The regulation of human haematopoiesis is a complex biological system with numerous interdependent processes. *In vivo* Haematopoietic Stem Cells (HSCs) self-renew so as to maintain a constant pool of these cells. It would be very interesting to maintain these cells *in vitro*, in view of their therapeutical importance. Unfortunately, there is currently no known process to activate HSCs self-renewal *in vitro*.
Since the difficulties related to *in vitro* experiments, modeling and simulating this process is indispensable. Moreover, the complexity of haematopoiesis makes it necessary to integrate various functionalities: both discrete and continuous models as well as consumption and production of resources. We thus focus on the use of Hybrid Functional Petri Nets, which offer a number of features and flexibility. We begin by modeling and simulating the role of a specific cytokine, interleukin-6, in the regulation of early haematopoiesis. Results obtained *in silico* lead to the disappearence of HSCs, which is in agreement with *in vitro* results.

1 Introduction

Haematopoiesis is a complex phenomenon leading to the continuous production of all types of mature blood cells. This process is ensured by a population of haematopoietic stem cells (HSCs) which are able by a process called "self-renewal" to maintain a constant pool *in vivo*. Nevertheless, the prolonged renewal of stem cells is not reproducible *in vitro*.

We then concentrated our work on the modeling and the simulation of this biological process. For this purpose, we needed a formalism which integrates a number of features since the complexity of the biological system. Thus, the notions of production and consumption of resources as well as the notions of discrete and continuous time must necessarily be taken into account. Among the formalisms existing in the literature, we focused on Hybrid Functional Petri Nets (HFPN) [1,2,3,4]. These nets can accumulate the notions of interest and they also have the particularity of allowing the definition of functions.

We expressed a hypothesis concerning the role of a specific cytokine, interleukin-6, in the fate of HSCs [5]. The receptor of this interleukin (sIL-6R) is thought to be a candidate for a positive feedback loop. A consequence of this hypothetical feedback circuit would be bi-stability of the HSCs phenotype. Thus, depending on the concentration of sIL-6R, HSCs either commit to differentiation to form mature blood cells irreversibly or they self-renew. In this way, IL-6 activation can be considered as an epigenetic switch [6], responsible for bi-stability (self-renewal and differentiation). This hypothesis was modeled and simulated *in silico* before performing costly *in vitro* experiments. To test this hypothesis, a pulse of sIL-6R was used to generate a transient modification of the environment, thus modifying the stable state of HSCs.

This paper is organized as follows: the biological context is presented in section 2. Section 3 presents a brief state-of-the art on Hybrid Petri Net formalism. In section 4, we explain the different steps in the construction of our model. Finally, the last section presents the results.

2 Biological Context

The regulation of HSCs involves numerous growth factors. We focus on IL-6 and the molecules directly associated, since this signalling pathway is known to play a central role in stem cell biology [5]. Receptors involved in recognition of IL-6 are IL-6R and gp130. The assembly of the complete signalling receptor is sequential and hierarchically ordered. IL-6R binds IL-6 with a low affinity (Figure 1). IL-6R exists in two forms: a membrane-bound form, called mIL-6R, and a freely-soluble-form, called sIL-6R. The soluble form of the receptor, sIL-6R, is secreted in an autocrine fashion by HSCs.

IL-6/mIL-6R or IL-6/sIL-6R complexes are recruited by 2 gp130 subunit (catalytic receptor) and trigger activation of several intra-cellular pathways, particularly JAK *(Janus kinase)*, which subsequently activates the STAT proteins *(Signal transducer and Activators of Transcription)* [5]. gp130/IL-6/IL-6R complexes are endocytosed. The activation of JAK/STAT pathway leads to upregulation of gp130 [8] and mIL-6R [7] in order to replenish HSC in IL-6 receptors. JAK/STAT pathway plays also a major role in self-renewal of HSCs, and neo synthesis of receptors subunits to IL-6 should maintain sensitivity to this cytokine. We assumed that this up regulation of gp130 and mIL-6R is a significant mechanism supporting self-renewal. Production of sIL-6R is mediated not at a level of gene transcription but by a cleavage mechanism in-

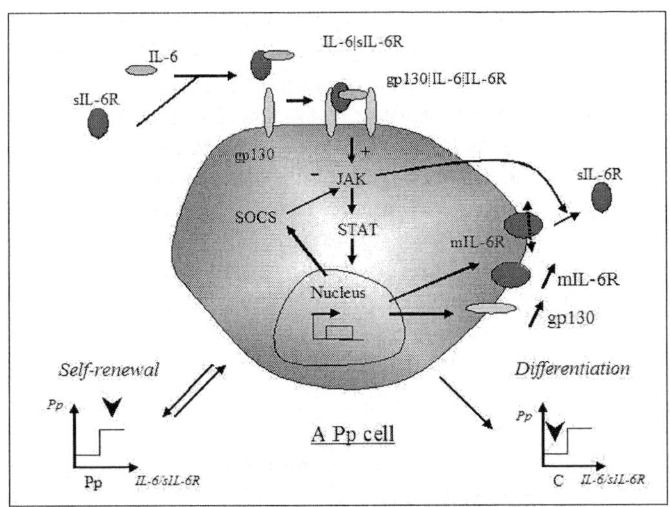

Figure 1. ***Signalling pathway of Interleukin-6.*** IL-6/IL-6R complexes and gp130/IL-6/IL-6R complexes form what activates the JAK/STAT pathway. The STAT act as transcription factors and enable gp130 and mIL-6R synthesis as well as expression of SOCS proteins which inhibit the JAKs. A part of the mIL-6R is cleaved in sIL-6R. Finally, according to the level of sIL-6R, the permissive HSC can self-renew or differentiate.

volving ADAM-protease [7]. Finally, the JAK/STAT pathway activates SOCS proteins *(Suppressor of Cytokine Signaling)* [5]. These SOCS act as inhibitors of the JAK/STAT pathway.

Gp130 stimulation is crucial for maintenance and proliferation of HSCs. We assume that the strength of gp130 activation determines cell fate, in accord with recent ligand/receptor signalling threshold theories [9]. Indeed, stem cell self-renewal is favored by maintenance of ligand/receptor complex level above a critical threshold; otherwise, below this threshold, the stem cells commit to differentiation. These data therefore suggest the presence of an autocrine loop but the functionality of the positive feedback circuit remains to be demonstrated. Before performing costly *in vitro* or *in vivo* experiments, it is reasonable to start with *in silico* experiments.

Moreover, the study of such regulation is difficult to carry out *in vitro*. Indeed, experimentally HSCs are usually isolated using a specific surface marker such as CD133. However, the ensuing "CD133 population" is heterogeneous, since it includes stem cells and progenitors. Progenitors committed in lineage differentiation represent the most important fraction: 95% of the population. We shall call this subpopulation C (committed)-cells. The remaining 5% are composed of quiescent HSCs residing in G_0 for 75% and recruited in the cell-cycle for 25% [10]. Quiescent HSCs will be referred to as *Pq*-cells and primitive cycling cells as *Pp*-cells. The *Pp* subpopulation is assumed to be permissive to signals from the micro-environment. This intermediate state permits us to link primitive- to committed HSCs and to integrate signals that determine fate of primitive HSCs (self-renewal or differentiation).

Thus, it is difficult to specifically study the evolution of HSCs, since they are not experimentally identifiable.

3 Hybrid Functional Petri Nets

To model the regulation of early haematopoiesis, we needed to take into account the molecular degradation as well as cellular evolution. Therefore, the notion of consumption and production of resources was important. Moreover, regulation of haematopoiesis has two time scales. We thus needed discrete time to model cellular evolution, since cells are enumerated, as well as continuous time to model molecular interactions. Consequently, the use of hybrid modeling conciliating discrete and continuous time was necessary. Hybrid Petri Nets represent a good tool to model and simulate such a process. There are currently two sorts of Hybrid Petri Nets in use: those which offer a maximum of functionalities so as to facilitate modeling [11,3,4,12] and those whose goal is to develop methods of analysis and model verification [13,14,15]. To enable formal analysis, the second class does not offer numerous functionalities.

Since the complexity of the regulation of haematopoiesis requires a maximum of functionalities, we used the first class of Hybrid nets. *Hybrid Petri Nets* (HPN) [11], defined by *David and Alla*, are the first Petri nets which integrate both discrete and continuous time:

- Discrete “places” represent entities which can be numbered (thanks to tokens).

- Continuous “places” represent entities which can not be numbered. A real number (representing a concentration for example) is associated with each continuous place.

- Discrete transitions fire after a delay of time *dt*.
- Continuous transitions continuously fire at a rate *v(t)*.
- Normal arcs activate transitions by consuming resources.
- Inhibitory arcs inhibit transitions and consume resources.
- Test arcs activate transitions without consuming resources.

The HPN model was improved by *Drath et al.* to obtain *Hybrid Dynamic Petri Nets* (HDN) [12]. The HDN model allows definition of functions on continuous arcs and continuous transitions but the consumed quantity has to be equivalent to the produced quantity. *Drath et al.* added the object oriented paradigm to HDN to obtain *Hybrid Object Nets* (HON) [12]. The main purpose was to encapsulate subnets within object frames.

Hybrid Functional Petri Nets (HFPN) have been recently developed [1,2,3,4] to integrate properties of HPN, HDN and HON. These nets allow the definition of functions on discrete and continuous arcs and transitions. The notions of functions on arcs and transitions seem very useful to model dynamic biological systems, since they enable establishment of links between places which are not directly related in the model. Another characteristic of this formalism is that the quantity of consumed resources can be different from the quantity of produced resources.

Since the HFPN model integrates a maximum of functionalities, they were chosen for our project. Throughout the article, we will use HFPN notations depicted in Figure 2.

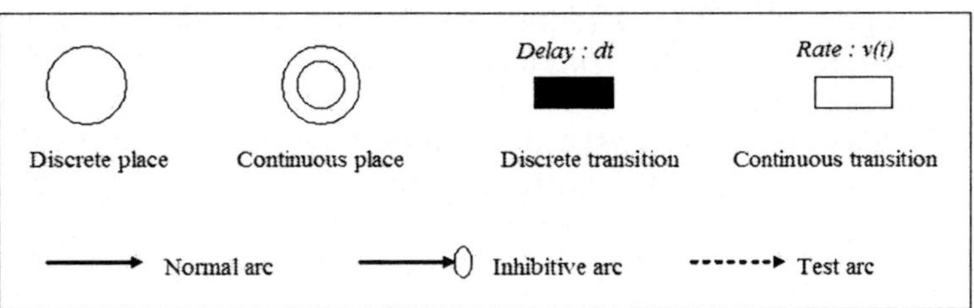

Figure 2. ***Notations used in Hybrid Functional Petri Nets (HFPN).*** Discrete transitions fire after a delay of time *Td*, whereas continuous transitions continuously fire at a rate *v(t)*. The HFPN model uses two kinds of places (discrete and continuous) and three kinds of arcs (normal arcs, inhibitive arcs and test arcs).

4 The use of HFPN to model the role of IL-6R in the regulation of early haematopoiesis

Haematopoiesis is a complex biological process comprising multiple and interdependent regulation networks. We therefore built the model in several steps. First, we built a sub-model for cellular dynamics and a sub-model for molecular dynamics. Then, we connected the two sub-models by modeling the autocrine secretion of sIL-6R by stem cells as well as the kinetics of the signalling pathway.

Experimental parameters were integrated in the final model. They came either from our experimental model or from literature. In the sequel, only few examples of these parameters will be done.

4.1 Cellular sub-model

The cellular model represents the evolution of each cell lineage as a function of time. This sub-model is built in discrete time, since cells can be numbered. The sub-model is shown in Figure 3. It contains three discrete places representing the three biological entities: quiescent stem cells *Pq*, permissive stem cells *Pp*, and cells committed to differentiation C. Discrete transitions represent all the processes that allow a cell to change its state:

- A *Pq* cell can enter the cell cycle to become a *Pp* cell, and inversely (respectively, transitions *T1* and *T2*). A quiescent cell (*Pq*) is consumed, whereas one permissive cell (*Pp*) is produced, and inversely.

- A *Pp* cell can self-renew with symmetric division when the two daughter cells remain primitive *(T3)*. After mitosis assumed a one day duration (delay of 1 on the transition *T3*), a permissive cell is consumed to produce two primitive daughter cells.

- A *Pp* cell can self-renew with asymmetric division when the two daughter cells are different; one stays primitive and the other differentiates *(T4)*.

- A *Pp* cell can differentiate without division *(T5)*.

- A *Pp* cell can divide *(T6)* to form two committed daughter cells.

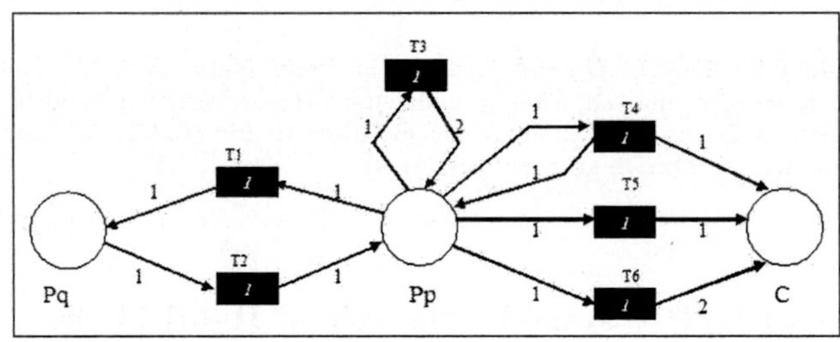

Figure 3. ***Cellular model.*** Discrete places (circles) represent the three types of cells: *Pq* are quiescent cells, *Pp* are permissive cells, and C are committed cells. Discrete transitions contain a time delay after which transitions fire. The number of consumed and produced cells is indicated on the arcs.

4.2 Molecular sub-model

The second sub-model is the molecular model representing interactions between cytokines. It represents the association and the dissociation of the complexes described above. Since we use growth factor concentrations, and since the formation and dissociation of complexes are continuous phenomena in cells, this sub-model was built in continuous time (Figure 4).

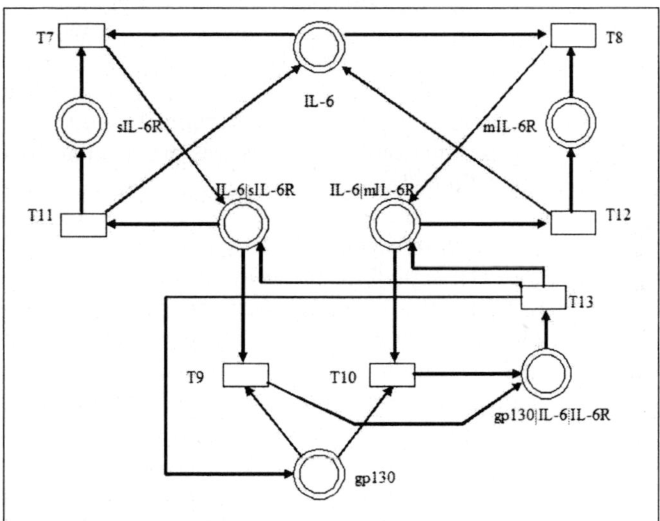

Figure 4. ***Molecular model.*** Continuous places model the molecules involved in the regulation of haematopoiesis by IL-6, and continuous transitions represent biological processes such as association of the complexes. The transitions *T7* and *T8* model the association of the IL-6/IL-6R complexes whatever the form of the receptor (sIL-6R or mIL-6R), whereas transitions *T11* and *T12* symbolize their dissociation. In the same way, transitions *T9* and *T10* model the formation of the gp130/IL-6/IL-6R complexes, whereas transition *T13* represents their dissociation.

Each cytokine as well as each complex of interest is modeled by a continuous place. We have places for the molecules IL-6, sIL-6R, mIL-6R, gp130 and places for all complexes: IL-6/IL-6R and gp130/IL-6/IL-6R with the soluble or membrane receptor. Continuous transitions used in the model represent biological processes such as formation and dissociation of all the complexes.

4.3 Autocrine secretion of the receptor of interleukin-6 and signalling pathway

Once both sub-models were built (the cellular sub-model and the molecular sub-model), we modeled the influence of IL-6R on cells. We thus numbered the biological links between molecular interactions and the cellular evolution. The model is presented on Figure 5.

As explained above, HSCs constitutively secrete the soluble receptor of IL-6, sIL-6R. In order to take this into account, we added a continuous transition (since *Pp* cells secrete a certain amount of molecules) where the entering arc is a test arc (since permissive cells are not consumed when secreting the receptors) (transition *T14* on Figure 5). The *in silico* quantity of sIL-6R secreted by permissive cells was determined thanks to [7] and [16], where the sIL-6R basal production by *Pp* cells was experimentally estimated to 43pg/mL/24h/150000 *Pp*-cells.

The most important link between cellular evolution and molecular interaction is the kinetics of the signalling pathway previously described. When the complex gp130/IL-6/IL-6R is formed, it activates the signal for the activation of JAK/STAT pathway. After one hour, all the gp130 are endocytosed, which interrupts signal activation. We therefore used a discrete transition with a delay of one hour to represent this process (*T17* on Figure 5). During this delay, gp130/IL-6/IL-6R complexes are not consumed, which is why we used a test arc. At the end of this delay, the virtual place *Pv_1* (without biological meaning) is filled. Consequently, the *T18* transition can fire triggering consumption of all the complexes. According to the quantity of formed complexes, either self-renewal is activated (*Pv_2* and *Pv_3*) or cells commit to differentiation *(Pv_4)*. For example, if the quantity of gp130/IL-6/IL-6R is sufficient, the virtual place *Pv_2* is filled; the transition *T3* can then fire and *Pp*-cells self-renew.

The JAK/STAT pathway is activated as soon as the concentration of gp130/IL-6/IL-6R is above a certain threshold. Then, the SOCS proteins inhibit this pathway. This negative loop introduces a refractory period, chosen as 8 hours, in analogy with [17]. After this delay (modeled by the discrete transition *T19*), the gp130 and newly synthesized mIL-6R return to the membrane. So the cell becomes once again responsive to the formation of complexes. This receptor accessibility is symbolized by the *T20* transition, which increases membrane receptor concentration. Finally, thanks to the JAK pathway, a part of the synthesized mIL-6R is cleaved into soluble receptor sIL-6R. The cleavage is represented by the *T21* transition.

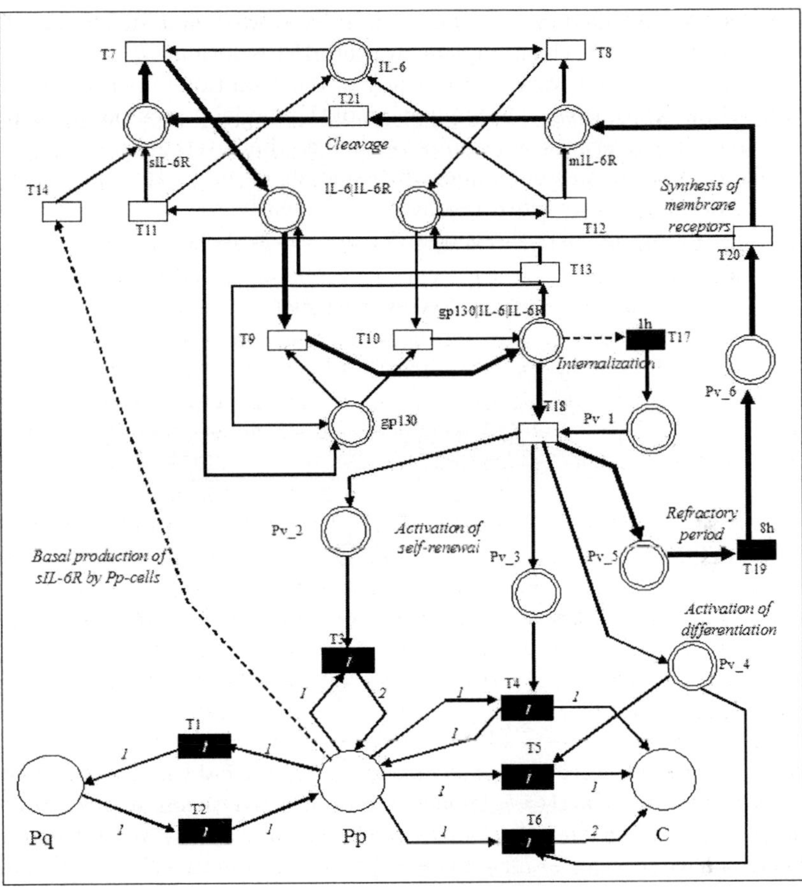

Figure 5. ***Global model.*** The cellular and molecular sub-models were joined thanks to biological links: the autocrine secretion of sIL-6R by permissive cells *(T14)* and the kinetics of signalling pathway *(Pv_1-Pv_6, T17-T20)*. The hypothetical retroaction loop is indicated in boldface arcs.

5 Simulation results and discussion

5.1 Results obtained with the model

Once the appropriate model of the IL-6-mediated regulation of the early haematopoiesis was built, a set of simulations was carried out with experimental values (detailed in [7]). The simulations were done using the *Cell Illustrator* software [3,4] which implements the HFPN formalism.

We concentrate our work on the study of the evolution of permissive cells (*Pp*) as a function of time. The simulation of the model presented in Figure 6 leads to the disappearance of permissive cells in about ten days.

This experiment was subsequently tested *in vitro* [18], and the obtained results conform to the *in silico* results. Consequently, our model is adequate for further testing our hypothesis of an epigenetic switch.

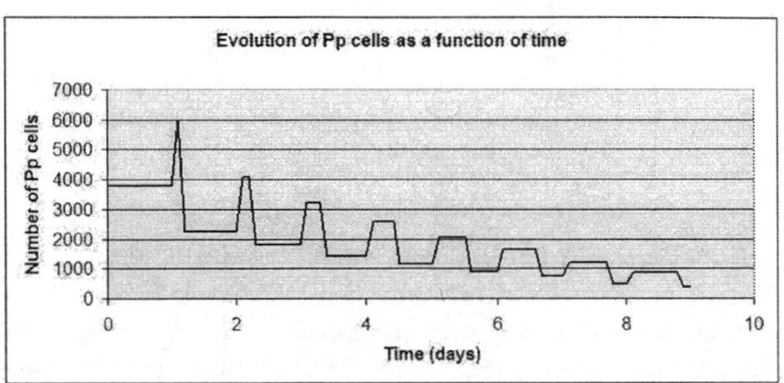

Figure 6. ***Evolution of Pp cells as a function of time.***

5.2 *Test of an epigenetic modification*

To test the functionality of the feedback circuit (in bold in Figure 5), it is necessary to verify whether a transient modification (using a pulse) of the sIL-6R (the candidate for the loop) is sufficient to activate the retrocontrol loop, and then to change the phenotype of permissive cells (self-renewal of stem cells). A pulse consists of adding a large quantity of a molecule, and after a delay, this molecule is completely removed. If haematopoietic stem cells are subject to an epigenetic modification, the feedback circuit would stay activated thanks to the inductor (sIL-6R), which exerts a positive regulation on its own synthesis. Two kinds of pulse were modeled: a pulse of sIL-6R and a pulse of HIL-6 (a fusion protein equivalent to the complex IL-6/IL-6R). Figure 7 presents the evolution of the number of *Pp* cells as a function of time, when a two hour pulse was applied. The symmetric self-renewal was activated, but only for one cycle after the pulse. Indeed, we observed a doubling of the totality of *Pp* cells. However, self-renewal was not activated longer and permissive cells became committed to differentiation.

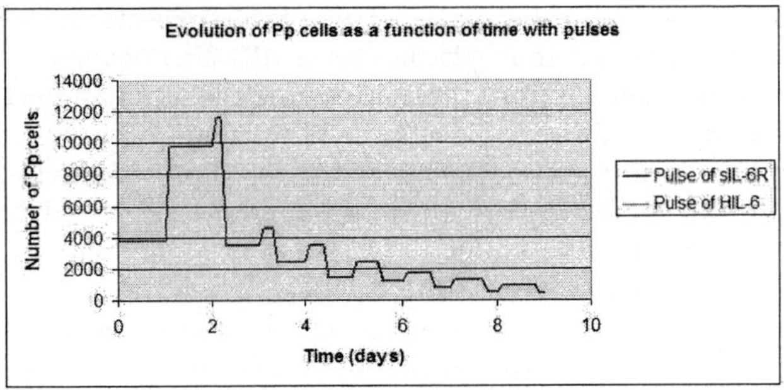

Figure 7. ***Evolution of Pp cells as a function of time, when sIL-6R and HIL-6 are pulsed (confounded curves).***

Simulation results revealed that there is no significant difference between the two kinds of pulses (sIL-6R and HIL-6). We thus conclude that the strong dissociation coefficient ($Kd = 10^{-9}$) of the complex IL-6/IL-6R do not influence the activation of the loop. Also, all permissive cells committed to differentiation despite the pulses. These results suggest that a pulse of sIL-6R (or HIL-6) with the experimental parameters is not sufficient to maintain stem cells *in silico*. Finally, the results of the simulations led us to suggest that activated cells produce sIL-6R in a quantity smaller than during the precedent cycle. Thus, each day a smaller fraction of *Pp* cells can self-renew (peaks observed every day). Consequently, a progressively larger part of *Pp* cells commits to differentiation.

Note that these experiments were also performed *in vitro* [18] and the results confirm the results obtained *in silico*.

6 Conclusion and future work

Haematopoiesis is a complex process involving numerous interdependent circuits in its regulation. A formalism which integrates a maximum of functionalities is therefore needed. The Hybrid Functional Petri Nets present a high level of integration. They enable the modeling of production and consumption of resources at a time scale which can be discrete or continuous. They also allow the definition of functions on arcs and transitions. Thanks to the numerous functionalities available with this formalism, we successed in modeling the entire process of early haematopoiesis by interleukin-6.

The simulation results predicted the disappearance of primitive HSC subpopulation whatever the tested pulse. This result suggests that this positive loop is not functional and that production of sIL-6R does not constitute an epigenetic modification determining the fate of HSCs, in the actual experimental conditions.

The agreement of *in silico* and *in vitro* results provides informal validation of our model. Nevertheless, it would be interesting to validate a model in a more formal manner. At present, the HFPN model only enables simulations. We are therefore interested in developing a system of validation and verification for hybrid petri nets considering a maximum of functionalities.

Finally, on a practical level, it would be useful in future work to vary those parameters whose values were only estimated (from the literature), i.e., those that could not be measured in our experimental model [7]. Also, we plan to add other factors, such as TGF-β (Transforming Growth Factor) which maintains primitive cells in quiescent state.

References

1. Nagasaki M *et al*, *Applied Bioinformatics* **2**, 181-184 (2003)
2. Doi A *et al*, *Applied Bioinformatics* **2(3)**, 185-188 (2003)
3. Matsuno H *et al* , *In Silico Biology* **3**, 389-404 (2003)
4. Doi A *et al*, *In Silico Biology* **4**, 271-291 (2004)
5. Heinrich PC *et al*, *Biochem J* **374**, 1-20 (2003)
6. Thomas R and d'Ari R, *CRC Press* , (1990)
7. Campard D *et al*, *Stem Cells (submitted)* , (2005)
8. O'Brien CA and Manolagas SC, **272**, 15003-15010 (1997)
9. Viswanathan S *et al*, *Stem Cells* **20**, 119-138 (2002)
10. Cheshier SH *et al*, *Proc Natl Acad Sci USA* **96**, 3120-3125 (1999)
11. David R, Alla H and Bail J, *Proc. 1st Int. ECC European Control Conference, Grenoble* , 1472-1477 (1991)
12. Drath R, *3rd International Conference on Automation of Mixed Processes, ADPM'98* , (1998)
13. Wolter K and Zisowsky A, *In Proceedings of the 4th International Computer Performance and Dependability Symposium IPDS'00* , (2000)
14. Müller Ch and Rake H, *Symposium on Mathematical Modelling* , 457-460 (2000)
15. Jörns C and Litz L, *In Proceeding of the 3rd European Control Conference ECC95* , 2035-2040 (1995)
16. Grafte-Faure S *et al*, *Br J Haematol* **105**, 33-39 (1999)
17. Gerhartz C *et al*, *Eur. J. Biochem* **223**, 265-274 (1994)
18. Götze K.S *et al*, *Experimental Heamtology* **29**, 822-832 (2002)

MODELING AND ANALYZING THREE-DIMENSIONAL STRUCTURES OF HUMAN DISEASE PROTEINS

YUZHEN YE, ZHANWEN LI and ADAM GODZIK

Bioinformatics and Systems Biology Program, The Burnham Institute, La Jolla, CA 92037, USA

Three-dimensional structures of proteins, experimental or predicted, show us how these molecular machines actually work. With the help of information on disease-related mutations, they can also show us how they malfunction in diseases. Such understanding, currently lacking for most human diseases, is an important first step before designing drugs or therapies to cure specific diseases. Here we used homology modeling to model human disease-related proteins, and studied structural characteristics of disease related mutations and compared them with non synonymous SNPs. 1484 domains from 874 proteins were modeled, and together with experimentally determined structures of 369 domains they provided the structural coverage of 48% of total residues in 1237 human disease proteins. We found that disease-related mutations have statistically significantly preference to form clusters on protein surfaces. In contrast, the non-synonymous SNPs appear to be randomly distributed on the surface. We interpret these results as an indication that disease mutations affect protein-protein interaction interfaces. This interpretation is supported by the analysis of 8 experimentally determined complexes between disease proteins, where disease-related mutations are clearly located in the binding interface of proteins, while SNPs are not. The non-uniform distribution of disease mutations indicates that we can use this feature as guidance in modeling and evaluating human disease proteins and their complexes. We set up a resource for Disease Protein Models (DPM at http://ffas.burnham.org/DPM), which can be used for studying the relation between disease and mutation / polymorphism sites in the context of protein 3D structures and complexes.

1. Introduction

Disease-related proteins are of great research interest for both experimental and computational scientists. Their high value in medicine and human health stems from the fact that they provide molecular picture of disease processes, a necessary prerequisite to rational drug development. As of today, thousands of genes (proteins) have been identified to be associated with various diseases in humans. Most often, mutations in these proteins have been identified in patients suffering from a particular disease. Mutation data is often the first source allowing us to study these diseases on the molecular level. Independently, technological advances in large-scale genome sequencing has allowed us to study human genomic variation and identify large numbers of SNPs (Single Nucleotide Polymorphisms), some of which cause changes of amino acids in the protein product of a gene (i.e., non-synonymous SNPs, nsSNPs) [1]. SNPs databases become an easy but valuable resource for studying genetic variations in human population. A vast majority of the nsSNPs has not been studied experimentally, and it is generally assumed that since nsSNPs are present in large sections of the population they are not strongly associated with diseases.

Many computational methods have been applied to study the effects of mutations (such as in P53 proteins [2]) and to predict the effects of the nsSNPs based on protein sequences, amino acid conservation and protein structures [3-6]. Structural information has been extensively used for studying the effects of mutations and nsSNPs [7] and a number of resources have been developed for mapping the SNPs onto the structures, such as MutDB (http://mutdb.org/) [8], SNPs3D (http://www.snps3d.org/), PolyPhen (http://www.bork.embl-heidelberg.de/PolyPhen/) [5] and SAAP (http://acrmwww.biochem.ucl.ac.uk/saap/) [2]. These resources are largely limited to human proteins with experimentally determined structures and there are still only a very small number of such proteins. Despite tremendous advances in recent years, experimental determination of protein structure is still time-consuming and expensive, especially for Eukaryotic proteins. It is possible to circumvent this limitation by use of comparative modeling, and this approach has been applied to disease proteins and used for SNPs annotation, including LS-SNP [9] and ModSNP [10].

In this work, we used distant homology recognition in conjunction with comparative modeling to build models of three-dimensional structures of human disease proteins. This strategy, validated in fold recognition test, can greatly increase the structural coverage of these proteins. Using the predicted structures, we further analyzed and compared the distribution of disease mutations and nsSNPs in the 3D space. The observation that spatial distribution of disease-related mutations is significantly different from that of usually benign nsSNPs suggests a possible explanation of different effects of the two groups. We hypothesize that disease mutations affect protein-protein interaction interfaces and thus disrupt functional networks within the cell. Detailed analyses of several available structures of experimentally determined complexes between disease proteins support this hypothesis.

2. Methods

2.1. *Data collection*

We focused on 1,237 human disease proteins and the corresponding mutation and SNPs information (discarding the variant sites marked as "unclassified") as identified by SwissProt database (http://us.expasy.org/sprot/). Structures of experimentally determined human disease proteins and their complexes were identified and downloaded from the Protein Database web site (http://www.rcsb.org).

2.2. *Homology modeling and model quality assessment*

We used FFAS [11], a profile–profile alignment and fold-recognition tool, for identifying the templates for modeling and generating the alignments. The alignments from FFAS were used as inputs for modeling packages Jackal [12] and Modeller [13], which were used to build three-dimensional models using the default options (Modeller models were used when no models can be produced by Jackal). All the models can be found at the DPM Web site at http://ffas.burnham.org/DPM and can be downloaded or viewed with a MDL Chime (http://www.mdl.com/products/framework/chime/) enabled browser.

Model quality was evaluated by the PSQS (Protein Structure Quality Score, http://www1.jcsg.org/psqs/psqs.cgi), an energy-like measure of quality of protein structures, calculated based on the statistical potentials of mean force describing interactions between residue pairs and between single residues and solvent, shown before to correlate well with model quality and accuracy [14]. Similar protocol is used for building molecular replacement templates in the Joint Center for Structural Genomics [15] and in several other large scale modeling projects (manuscript in preparation).

2.3. *Spatial distribution of disease-related mutants on the protein structures*

We define a residue as being in the core if its solvent accessible area is less than 5% of its maximum possible surface area in a fully extended conformation. By this definition about 25% of residues in an average protein are in the core. We used Lee & Richard method [16] to compute the atomic solvent accessible area of proteins.

We use the size of the *largest connected component* of the mutations graph as a measure of clustering of mutations, and its significance is calculated by a permutation test. For example, suppose a protein structure has N residues and M mutation sites. We first generate a graph of M nodes and link an edge between two nodes if they are within contact distance (i.e., minimum distance between any two atoms of the residues is $\leq$5.0Å). The graph is then partitioned into connected components [17]; and the size of the largest component is used as the clustering index of mutation sites. To compute the significance, we randomly select M residues out of N and do the same computation for R times. The significance of the clustering of the mutation positions is then defined as (the number of permutations with clustering index ≥ the clustering index of mutation sites) / *R*. See Figure 1 for a schematic illustration.

2.4. *Analysis of protein complex structures*

In addition to the models of disease proteins, we analyzed experimental structures of complexes of disease proteins, if available. Residues were considered as being on the complex interface if their burial/exposed status between the complex and individual structures changed, here defined as the difference of their solvent accessible area in complex and in individual is less than a cutoff (e.g, 5Å^2).

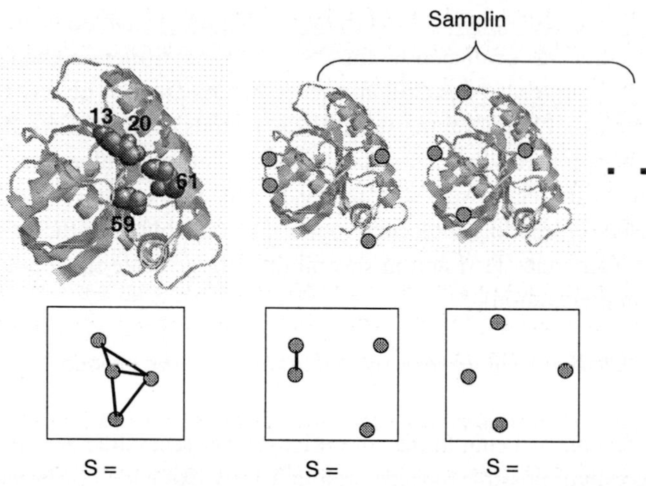

Figure 1. A schematic illustration of computing the significance of clustering of residues by a permutation test. Here 4 residues of position 13, 20, 59 and 61 highlighted in CPK are clustered, and the size of largest component of the graph with 4 nodes, S, is 4. The significance of the clustering P(S = 4) is (the number of samplings with S ≥4) / (total number of samplings).

3. Results

3.1. *Statistics of models and quality evaluation*

A total of 1,484 models were built for 874 proteins (many proteins are multi-domain and models were built for individual domains (if possible)). Together with 369 experimentally determined structures (only one structure from all that cover the same or very similar region in a target protein was counted), 1859 structures in total cover at least partially 1,064 out of the 1,237 human disease proteins. The structural coverage counted as the percentage of residues that could be mapped to the structures is 9% by experimentally determined structures, 39% by models, and 48% in total. For example, alpha-2-macroglobulin receptor-associated protein (SwissProt accession number P30533) is 357 aa long, and only a 82 aa N-terminal fragment was determined experimentally (PDB code 1op1); a model covering 199 aa at its C-terminal domain can be built using 1bf5 chain A as a template. While much higher structural coverage than 48% was reported for bacterial genomes [18], eukaryotic genomes, such as human were expected to have lower coverage.

The quality of the homology model is determined by a combination of the performance of the modeling algorithm and the quality of the alignments. Therefore, the quality of our models is largely determined by the performance of FFAS used to detect templates and produce alignment for homology modeling. FFAS benchmarks have shown that predictions with scores lower than –9.5 (the cutoff used in this work) should have less than 3% of false positives [11], and that its alignment quality is significantly higher than PSI-BLAST alignments. An independent measure of a model quality can be provided by empirical energy parameters, such as for instance calculated by a PSQS server (http://www1.jcsg.org/psqs/psqs.cgi) [14]. 81% of the models have good overall PSQSs (Protein Structure Quality Score < 0) (see the DPM website for the detailed results). In addition we emphasize that features such as relative position of a residue on the surface or in the core of the protein tend to be well conserved even in relatively inaccurate models.

3.2. *Distribution of mutations (disease-related mutations and SNPs)*

A total of 6,352 mutation and 954 nsSNP sites could be mapped onto the structures. As compared to the average residues and nsSNPs, more disease mutations are found in protein cores (all residues: core/total = 24.5%; nsSNPs: core/total = 20.1%; disease mutations: core/total = 34.9%).

Disease-related mutations tend to be clustered (as measured by the clustering index described in the methods section), in contrast to nsSNPs, which are not. In 97 out of 667 (14%) structures in which at least 2 mutations can be mapped onto the structure, disease-related mutations are significantly (0.05% significance) clustered; in comparison, in only 4 out of 205 structures with at least 2 nsSNPs mapped, nsSNPs are clustered together at the same significance threshold. In experimentally determined structures, disease-related mutations are significantly clustered in 27 out of 145 structures (19%), while nsSNPs are significantly clustered in only 2 out of 36 structures (6%). Both results show that no matter if predicted models were included for statistics or not, disease-related mutations are more significantly clustered together than nsSNPs.

As an example, Figure 2 shows the mapping of mutations and nsSNPs of Glutamate dehydrogenase 1 protein (SwissProt accession number P00367) onto its X-ray structure (PDB code 1l1f, chain A). This protein has 10 disease related mutations (associated with hyperinsulinism-hyperammonemia syndrome, HHS and highlighted red in Figure 2); all are closely located together, and the largest component of the mutation graph has 7 residues (with P-value = 0).

Figure 3 shows a model of a SWI/SNF-related matrix-associated actin-dependent regulator of chromatin subfamily A-like protein 1 (SwissProt accession number Q9NZC9) and the distribution of disease-related mutations and SNPs on this model. This protein has 10 mutations related to SIOD (Schimke immuno-osseous dysplasia) disease and 3 SNPs (collected in dbSNP) that can be mapped to the model. Analysis of the figure, supported by our statistical analysis, clearly shows that the disease mutations are clustered together (the largest component has 4 residues, P-value = 0.005), while the three SNPs do not (the largest component has 1 residue). Interestingly, one of the SNPs is located in the interface with many disease mutations; it suggests the possibility that this SNP may be deleterious as well. This example shows a possible way to study the effects of nsSNPs by comparing their spatial distribution with that of known disease mutation sites.

3.3. *Analysis of complexes*

The results from the modeling of human disease proteins and the analysis of the relative positions of nsSNPs and disease mutations on the structures strongly suggest that clusters of such mutations form specific patches on the surface and prompt the speculation that these patches are involved in protein-protein interactions. While a statistically rigorous evaluation of this hypothesis is not possible at present, we looked at details of a few available examples of experimentally determined complexes between human disease proteins.

We analyzed eight experimentally determined structures of complexes of disease proteins. This number seems rather small as compared to the number of potential protein-protein interactions collected in databases such as OPHID (http://ophid.utoronto.ca) [19] (its 5/2005 version collected 8836 protein-protein interactions involving at least one disease protein, and 1012 involving two disease proteins). This discrepancy illustrates the experimental difficulties involved in experimental studies of protein complex formations.

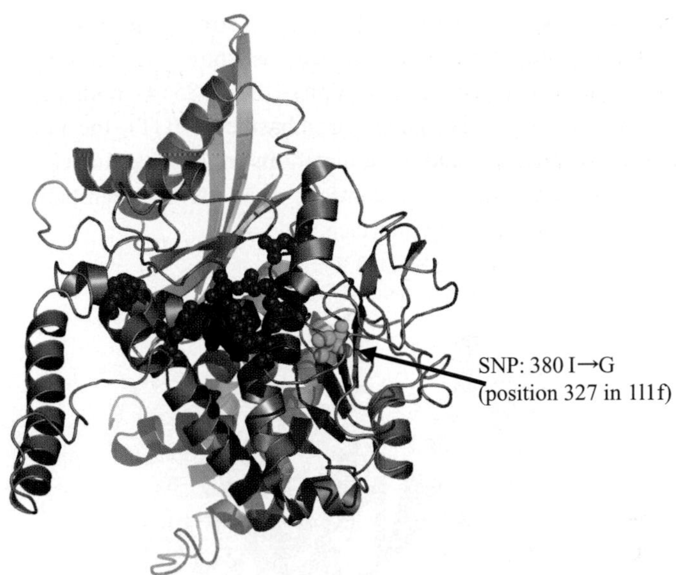

Figure 2. The mapping of mutations (in red ball-and-sticks) and nsSNPs (in green ball-and-stick) on the X-ray structure of SwissProt protein P00367 (PDB code 1l1fA)

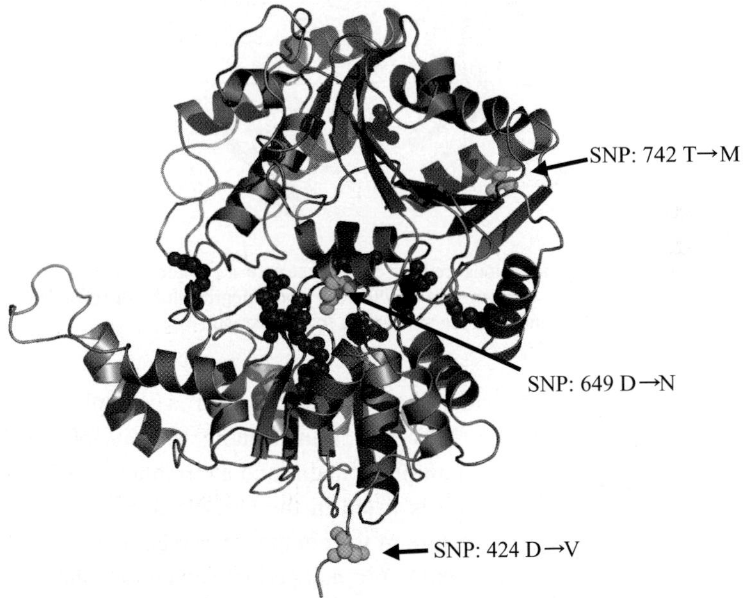

Figure 3. The mapping of mutations (in red ball-and-sticks) and nsSNPs (in green ball-and-stick) on the model of matrix-associated actin-dependent regulator of chromatin (SwissProt accession number Q9NZC9).

We found that most of such complexes show strong clustering of mutations around the binding interface. For instance, as shown in Figure 4, proteins integrin beta-3 (P05106) and integrin alpha-IIb (P08514) both have a lot of mutations associated with Glanzmann thrombasthenia (GT), the most common inherited disease of platelets, and these mutations (red ball-and-sticks) tend to be located in their binding interface; in contrast, three SNPs (green ball-and-sticks) are farther away from the binding interface.

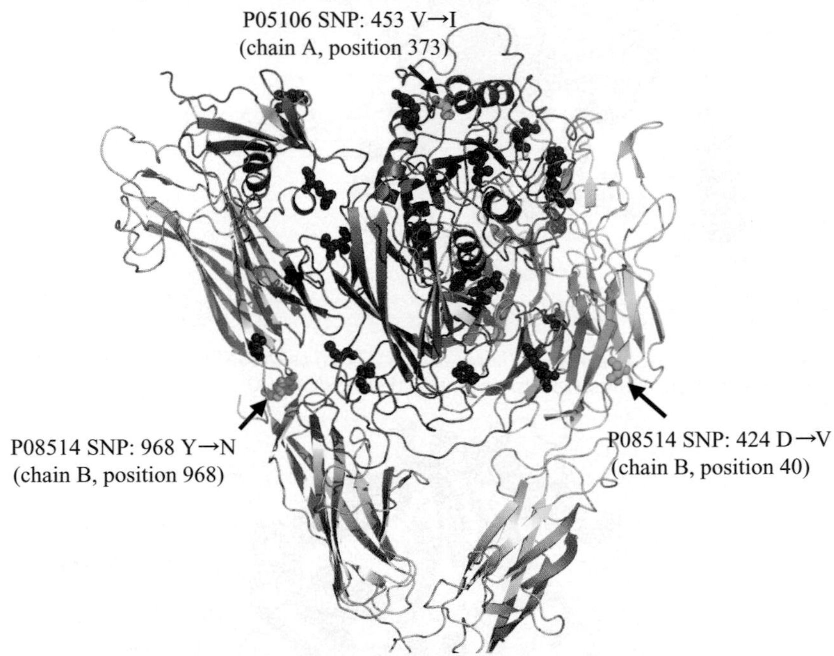

Figure 4. Clusters of disease mutations (red) on the structure of a complex (PDB code 1m1x) of two disease proteins, proteins: integrin beta-3 (P05106) and integrin alpha-IIb (P08514). Three SNPs highlighted in green ball-and-stick are also shown for comparison.

Another example is provided by the structure of a complex between GTPase-activating protein (P20936) and transforming protein p21/H-Ras-1 (P01112). Both proteins contain multiple domains and are implicated in a variety of human tumors. Their interaction is listed in the OPHID database, and more importantly, there is an X-ray structure of the complex between two domains one from each proteins (PDB code 1wq1). We mapped the mutations and SNPs of P20936 and P01112 onto structure 1wq1. Three mutations of P01112 are located on the domains present in the complex, forming two unique sites located exactly at the center of the interaction interface between protein P20936 and P01112 (see Figure 5). There are no experimentally determined structures for P09619

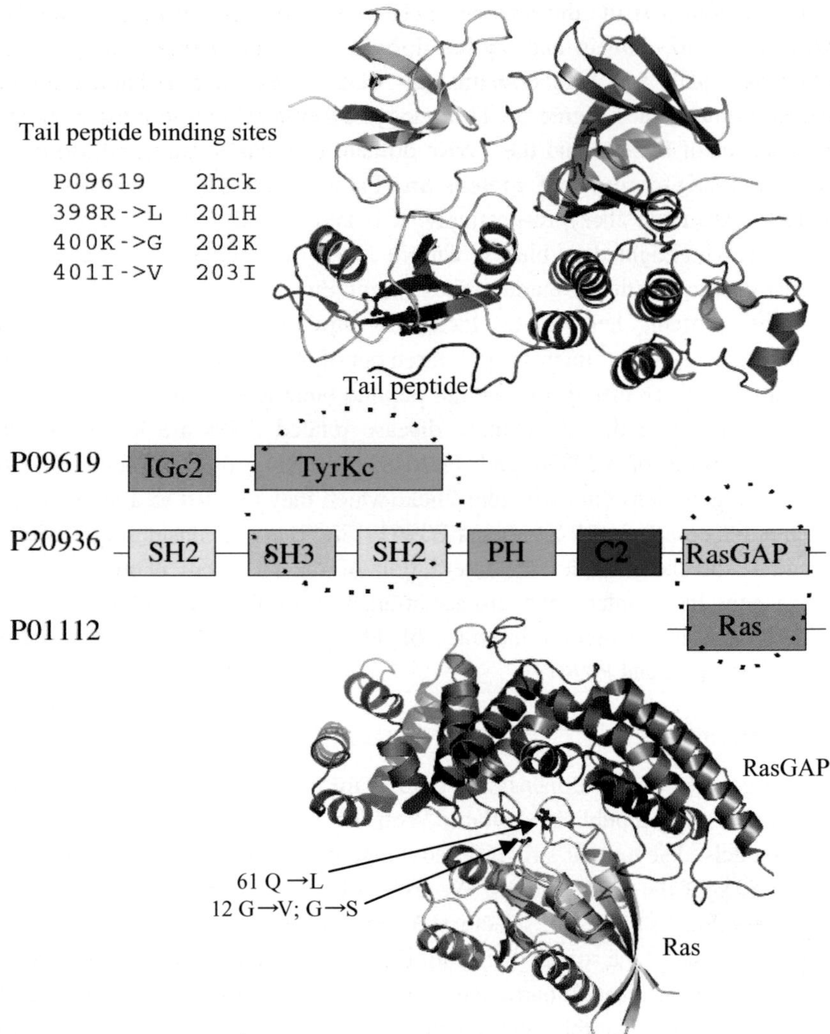

Figure 5. The interaction network of GTPase-activating protein (P20936), transforming protein p21/H-Ras-1 (P01112) and alpha-hemolysin (P09619). Two complexes are involved: one is the complex between RasGAP domain of protein P20936 and Ras domain of protein P01112 (PDB code 1w1q), with disease-related mutations highlighted in red ball-and-stick in the graph (shown in the bottom of this figure), and the other one is the complex between SH3 and SH2 domains of protein P20936 and the TyrKc domain of P09619 (PDB code 2hck, shown in the top of this figure).

and for P20936 (outside the region of 718-1037). Domain analysis shows that P09619 has Igc2 domain and TyrKc domain (Tyrosine kinase domain), and P20936 has SH3 and SH2 domains (as well as several other domains and RasGAP domain) (see Figure 5). Homology searching shows that the SH3 and SH2 domains of P20936 and the TyrKc domain of P09619 can match the SH3, SH2 and TyrKc domains of protein Src Family Kinase (with known X-ray structure, PDB code 2hck), respectively. It is known that in 2hck, the so-called tail peptide (highlighted in blue in Figure 5) of the catalytic domain (TyrKc domain) interacts with the binding sites (highlighted in red) in SH2 domain of the same protein, helping to lock the structure in an inactive form (autoinhibition) [20,21]. Interestingly, three out of four remaining disease-related SNPs in P20936 are matched to the tail peptide binding sites in the SH2 domain of 2hck, suggesting that these three disease related SNPs are located in the binding interface of P20936 and P09619 (involving the tail peptide). We mapped the mutations onto structure 2hck (which may be used as a template for modeling the complex of P20936 and P09619) in Figure 5, considering it will be difficult to precisely model the interaction of the short tail peptide with its binding site. In summary, the disease mutations of P20936 and P01112 are mainly located in the binding interface of the interaction network composed of these two proteins and P09619.

4. Conclusion

We have generated three-dimensional models for over a thousand human disease proteins and set up a publicly available Web site, showing annotated pictures for all the models. The current structural coverage of the human disease proteins is close to 50% of the total residues. We expect that the coverage would increase with the continuous growth of structural databases.

Our analysis of the spatial distribution of disease mutations shows their non-uniform distribution, and in particular forming patches on surfaces of proteins. It is tempting to speculate that such patches are located at or near protein-protein interaction interfaces.

To test this hypothesis we evaluated a number of structures of complexes, initially focusing on experimentally determined structures. The number of such complexes is rather small, but the examples are very suggestive. Indeed, in most cases disease mutations cluster at binding interfaces.

Recently, tremendous advances have been achieved in identifying the protein-protein interactions, both from large scale experiments and computational approach [22-24]. Thousands of protein-protein interactions involving disease proteins have been identified or predicted (as collected in

OPHID database). The huge gap between the number of potential interactions and the number of experimentally determined structures of such complexes suggests that modeling would play an important role in filling the gap. One possibility is to use the structure of a complex as a template [25,26] (as suggested by the example shown in Figure 5). But this method is limited because only a relatively small number of complexes are available for modeling. Large scale *ab initio* modeling of protein complexes would be necessary to further evaluate our hypothesis. We plan to extend the current study by using *ab initio* docking methods, such as GRAMM [27], to predict the structures of complexes of disease proteins. Another possibility is to use the non-uniform distribution of disease mutations and SNPs in this process as an additional guidance. Also we will search for other alternative ways of building models for protein-protein interactions [28], such as fitting models to the low resolution complex structure from electron microscopy (EM) when data is available. The resulting models of complexes would become an important resource for studying the functions of disease proteins and the mechanism of diseases.

Acknowledgments

We thank Dr. Lukasz Jaroszewski for his help with the protein modeling. This project was supported by NIH grant P01 GM63208.

References

1. M. Cargill, D. Altshuler, J. Ireland, P. Sklar, K. Ardlie, N. Patil, N. Shaw, C.R. Lane, E.P. Lim, N. Kalyanaraman, J. Nemesh, L. Ziaugra, L. Friedland, A. Rolfe, J. Warrington, R. Lipshutz, G.Q. Daley, and E.S. Lander, *Nature Genetics.* **22**(3), 231-238 (1999)
2. A.C. Martin, A.M. Facchiano, A.L. Cuff, T. Hernandez-Boussard, M. Olivier, P. Hainaut, and J.M. Thornton, *Hum Mutat.* **19**(2), 149-164 (2002)
3. M.A. Fleming, J.D. Potter, C.J. Ramirez, G.K. Ostrander, and E.A. Ostrander, *Proc Natl Acad Sci U S A.* **100**(3), 1151-1156 (2003)
4. P.C. Ng and S. Henikoff, *Nucleic Acids Res.* **31**(13), 3812-3814 (2003)
5. S. Sunyaev, V. Ramensky, and P. Bork, *Trends Genet.* **16**(5), 198-200 (2000)
6. Z. Wang and J. Moult, *Hum Mutat.* **17**(4), 263-270 (2001)
7. A. Cavallo and A.C. Martin, *Bioinformatics.* **21**(8), 1443-1450 (2005)
8. S.D. Mooney and R.B. Altman, *Bioinformatics.* **19**(14), 1858-1860 (2003)
9. R. Karchin, M. Diekhans, L. Kelly, D.J. Thomas, U. Pieper, N. Eswar, D. Haussler, and A. Sali, *Bioinformatics.* **21**(12), 2814-2820 (2005)
10. Y.L. Yip, H. Scheib, A.V. Diemand, A. Gattiker, L.M. Famiglietti, E. Gasteiger, and A. Bairoch, *Hum Mutat.* **23**(5), 464-470 (2004)

11. L. Rychlewski, L. Jaroszewski, W. Li, and A. Godzik, *Protein Science*. **9**, 232-241 (2000)
12. Z. Xiang and B. Honig, *J Mol Biol*. **311**(2), 421-430 (2001)
13. A. Sali and T.L. Blundell, *J Mol Biol*. **234**, 779-815 (1993)
14. L. Jaroszewski, K. Pawlowski, and A. Godzik, *J Mol Model*. **4**, 294 - 309 (1998)
15. R. Schwarzenbacher, A. Godzik, S.K. Grzechnik, and L. Jaroszewski, *Acta Crystallogr D Biol Crystallogr*. **60**(Pt 7), 1229-1236 (2004)
16. B. Lee and F.M. Richards, *J Mol Biol*. **55**, 379-400 (1971)
17. E. Minieka, *Optimization algorithms for networks and graphs*. New York: Marcel Dekker (1978)
18. I. Friedberg, L. Jaroszewski, Y. Ye, and A. Godzik, *Curr Opin Struct Biol*. **14**(3), 307-312 (2004)
19. K.R. Brown and I. Jurisica, *Bioinformatics*. **21**(9), 2076-2082 (2005)
20. T. Pawson, *Nature*. **385**(6617), 582-583 (1997)
21. W. Xu, S.C. Harrison, and M.J. Eck, *Nature*. **385**(6617), 595-602 (1997)
22. P. Uetz, L. Giot, G. Cagney, T.A. Mansfield, R.S. Judson, J.R. Knight, D. Lockshon, V. Narayan, M. Srinivasan, P. Pochart, A. Qureshi-Emili, Y. Li, B. Godwin, D. Conover, T. Kalbfleisch, G. Vijayadamodar, M. Yang, M. Johnston, S. Fields, and J.M. Rothberg, *Nature*. **403**(6770), 623-627 (2000)
23. C. von Mering, R. Krause, B. Snel, M. Cornell, S.G. Oliver, S. Fields, and P. Bork, *Nature*. **417**(6887), 399-403 (2002)
24. H. Yu, N.M. Luscombe, H.X. Lu, X. Zhu, Y. Xia, J.D. Han, N. Bertin, S. Chung, M. Vidal, and M. Gerstein, *Genome Res*. **14**(6), 1107-1118 (2004)
25. P. Aloy and R.B. Russell, *Proc Natl Acad Sci U S A*. **99**(9), 5896-5901 (2002)
26. L. Lu, A.K. Arakaki, H. Lu, and J. Skolnick, *Genome Res*. **16**(6A), 1146-1354 (2003)
27. E. Katchalski-Katzir, I. Shariv, M. Eisenstein, A.A. Friesem, C. Aflalo, and I.A. Vakser, *Proc Natl Acad Sci U S A*. **89**(6), 2195-2199 (1992)
28. R.B. Russell, F. Alber, P. Aloy, F.P. Davis, D. Korkin, M. Pichaud, M. Topf, and A. Sali, *Curr Opin Struct Biol*. **14**(3), 313-324 (2004)

DESIGN AND ANALYSIS OF GENETIC STUDIES AFTER THE HAPMAP PROJECT

FRANCISCO M. DE LA VEGA

Applied Biosystems, 850 Lincoln Centre Dr.
Foster City, CA 94404, USA

ANDREW G. CLARK

Department of Molecular Biology and Genetics, Cornell University
Ithaca, NY 14853, USA

ANDREW COLLINS

Human Genetics, University of Southampton
Duthie Building (808), Tremona Road, Southampton, England

KENNETH K. KIDD

Department of Genetics, Yale University School of Medicine
333 Cedar Street, New Haven, CT 06520, USA

A large international effort to define the fine patterns of sequence variation along the human genome will be essentially completed when this article is printed. The aim was to generate a genome-wide validated SNP resource and survey of the patterns of allelic association and common haplotypes useful for designing association studies. The outcome of this "HapMap" project is the genotypes of about four million SNPs in DNA samples of individuals from Africa, China, Utah, and Japan. This project has provided an unprecedented amount of empirical data on the patterns of linkage disequilibrium across the human genome that is fueling a large number of population genetic analyses. Aspects of the effective representation and use of this vast data resource in the design and implementation of cost-effective and more powerful disease association studies are among the topics of the eight papers included in this volume.

Three papers deal with the issue of how many tagSNPs will be needed to perform a genome-wide association scan without excessive loss of power compared to doing the scan will all the HapMap SNPs. The paper by deBakker et al. examines SNPs in 61 genes involved in DNA repair typed in people from 7 different population groups. The investigators then select "tag" SNPs using TAGGER, and ask a crucial question – how well does this same set of tag SNPs work to perform association testing in each of the 7 population samples. This has been dubbed the "transferability" problem, and until there is empirical confidence that tag SNPs have this transferability property, application of LD mapping in populations outside the initial HapMap set will be on rather uncertain ground. The good news is that deBakker et al. find that the percentage of SNPs with an $r^2 > 0.8$ to the tag SNPs ranges from 50% to 85% across the population samples.

The question of how many tag SNPs will be necessary to be able to perform genome-wide association testing is tackled by Magi et al. which use the HapMap sample and run it through an r^2-binning method as performed by the REAPER algorithm to identify tag SNPs. Then, taking another approach to the problem of transferability, they ask how many additional SNPs would have to be genotyped to attain the same power of an association test, and they find that from 10-35% more SNPs, beyond the minimal tag SNP set from the first population, need to be genotyped in the second population to reach the same power. Gopalakrishnan et al. apply yet a third algorithm, called FESTA, to infer the number of tag SNPs needed for whole genome association testing, and they reach a figure of 294,000 for the European HapMap sample. By picking this subset of SNPs instead of the full HapMap set, they further calculate a lost of power of only 5-10% under a variety of single-gene models for the disease (including dominance, recessiveness and additivity).

Given the vast number of SNPs now available for association studies and the complexity of data analysis it is essential to have software tools that enable both the selection of cost-effective SNP panels that provide high power and tools that assist in determining the optimal approaches to analysis. The SNPbrowser software, described by De La Vega et al., provides a powerful and flexible interface to an embedded database including the HapMap results together with SNP and gene annotations. The software also integrates metric linkage disequilibrium unit (LDU) maps of the genome and step-by-step wizards that implement a number of algorithms for the selection of non-redundant subsets of tag SNPs. The considerable flexibility offered by this tool makes it amenable to the design and implementation of the vast majority of association studies. The paper by Dudek et al. considers in particular how to optimize the analysis of genome-wide association data through the use of their software package genomeSIM. This program is designed to simulate large-scale datasets of population based case-control samples that allow thorough evaluation of alternative analytical methods since the parameters of the simulated disease model are known. While there are several simulation packages available for family-based study designs, the population-based simulation packages are limited to coalescent models which do not accommodate multiple penetrance functions that allow gene-gene interaction to be modeled. As genome-wide association data sets are now beginning to appear, the importance of modeling gene-gene interactions may soon be revealed and genomeSIM is likely to become an important tool in evaluating the effectiveness of competing approaches to data analysis.

The design and analysis of association studies is still challenging even as high throughput technologies allow the typing of thousands, or even hundreds of thousands of SNPs, since the cost of pursuing false leads after an initial scan needs to be contained. Peter Kraft presents a multi-stage approach where a portion of the samples are genotyped first with a high-throughput genotyping method, and a small number of the most promising variants are then genotyped in the remaining samples with a lower throughput method. The samples sizes in the first and subsequent stages and the corresponding significance levels are chosen to limit the False Positive Report Probability, while maximizing the number of Expected True Positives. Kraft shows that for a fixed budget, the multi-stage strategy has greater power than the single-stage strategy. The expected number of false positives does not change if the true number and effects of causal loci differs from the specified prior, thus limiting the amount of resources spent chasing false leads. On the other hand, Castellana et al. investigate the value of relaxing the rigidity of haplotype block models through a method called "haplotype motifs," which retains the notion of representing haploid sequences as concatenations of conserved haplotypes but abandons the assumption of population-wide block boundaries. They conclude that the benefits of haplotype models are modest, but that haplotype models in general and block-free models in particular are useful in picking up correlations near the boundaries of the detectable level.

Finally, the question of how easy would be to generalize the results of association studies between different populations around the world is addressed by Chen et al. In their paper, a candidate gene association study of the PPAR3 gene on body-mass-index (BMI) of an epidemiological cohort of public school students from Mexico is analyzed. Tag SNPs selected using the HapMap data were used to genotype 1200 subjects. While the present study confirms association between a set of SNPs in LD in the gene and BMI, and the results are promising, the paper discuss the requirements for replication and to discard population stratification effects that may complicate the analysis of associations found in admix populations.

Acknowledgements

We would like to acknowledge the generous help of the anonymous reviewers that supported the peer-review process for the manuscripts of this session.

RELAXING HAPLOTYPE BLOCK MODELS FOR ASSOCIATION TESTING

NATALIE CASTELLANA, KEDAR DHAMDHERE, SRINATH SRIDHAR, AND RUSSELL SCHWARTZ*

Computer Science Department and Biological Sciences Department
Carnegie Mellon University
4400 Fifth Avenue
Pittsburgh, PA 15213 USA
Email: russells@andrew.cmu.edu

The arrival of publicly available genome-wide variation data is creating new opportunities for reconciling model-based methods for associating genotypes and phenotypes with the complexities of real genome data. Such data is particularly valuable for testing the utility of models of conserved haplotype structure to association studies. While there is much interest in "haplotype block" models that assume population-wide regions of low diversity, there is also evidence that such models eliminate correlations potentially useful to association studies. We investigate the value of relaxing the rigidity of block models by developing an association testing method using the previously developed "haplotype motif" model, which retains the notion of representing haploid sequences as concatenations of conserved haplotypes but abandons the assumption of population-wide block boundaries. We compare the effectiveness of motif, block, and single-variant models at finding association with simulated phenotypes using real and simulated data. We conclude that the benefits of haplotype models in any form are modest, but that haplotype models in general and block-free models in particular are useful in picking up correlations near the boundaries of the detectable level.

1. Introduction

Searches for correlations between human genetic variations and disease phenotypes have often been fruitful for strongly hereditary diseases, but have had limited success at finding genetic risk factors for complex diseases. This failure is likely due at least in part to the challenges of distinguishing many relatively weak correlations from the noise produced by chance associations with the millions of known sites of common variation. One approach to address this problem involves identifying segments of correlated variations

*to whom correspondence should be addressed

known as haplotypes. By finding these co-associating sets of variations, one can in principle reduce the amount of data to be collected and analyzed in an association study and avoid some of the confounding effects of testing many variant sites. The prospects of such methods may be greatly facilitated by the recent construction of the HapMap[0], a publicly available collection of genome-wide single nucleotide polymorphism (SNP) variations separated by donor to allow for haplotype inference.

Studies of simulation models[22] and limited amounts of real data[1] have suggested potentially large advantages to haplotype-based association methods. Many such methods are based on the haplotype block model[3], which proposed that the genome consists of discrete regions of strongly correlated variations separated by recombination hotspots, across which correlations have been eliminated by frequent historical recombination. Numerous block construction criteria have since been proposed, generally based either on haplotype diversity or similar metrics[13] or on linkage disequilibrium statistics[5]. A haplotype-based association test may be conducted by directly testing for differences in frequencies of common haplotypes in individuals affected (cases) or unaffected (controls) by a disease[2,1]. Or they may use haplotypes to identify "haplotype tagging SNPs" (htSNPs), a subset of SNPs that contain most of the information contained in the full SNP set[9], which can reduce the cost of genotyping and the difficulty of finding meaningful associations in the resulting data.

While haplotype block models appear useful in facilitating association studies by reducing data complexity, there is evidence that they do not robustly capture true underlying haplotype conservation patterns[15]. In prior work, we developed the "haplotype motif" model[16], which explains individual genomes as concatenations of conserved haplotypes (or isolated variant sites). This model relaxes some of the rigidity of the block models, while still maintaining enough structure to allow for robust fitting[16] and efficient application to various computational analyses[17]. Figure 1 illustrates the difference between block and motif models of haplotype structure on a small hypothetical set of sequences. Other groups have since also developed "haplotype motif" models using various optimization metrics[18,10].

Here, we focus on the ultimate test of such relaxed conserved haplotype models: Does the extra information they preserve relative to block models provide an advantage in association testing? We approach that question with an empirical study of single-SNP, haplotype block, and haplotype motif methods for finding associations between genotype and phenotype, using simulated and HapMap data to understand how idealized models

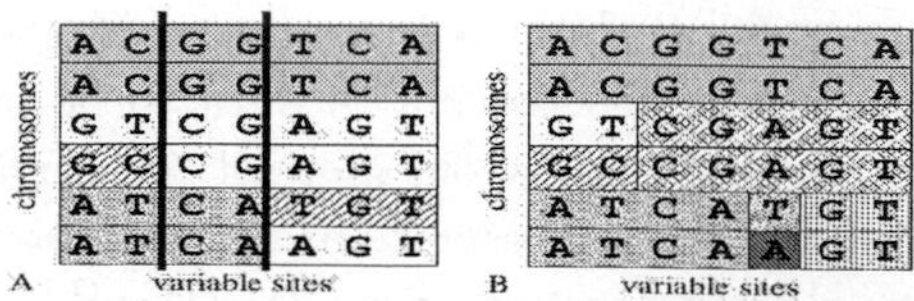

Figure 1. Illustration of possible block and motif partitions of a hypothetical set of sequences, each corresponding to variable sites in a given region of one chromosome from one individual. A: A block model, in which each chromosome is explained as a choice of one haplotype in each of three blocks. B: A motif model, in which each chromosome is explained as a concatenation of a set of "haplotype motifs" of varying length.

might mislead us with regard to real data. While we focus on the haplotype motif model, our interest is not specifically in that model *per se*, but rather in whether exploiting correlation information across haplotype block boundaries can lead to improvements in association study effectiveness.

2. Methods

2.1. *Haplotype Structure and Tagging SNP Inference*

Haplotype motif structure was inferred as described in Schwartz[16] with significance level 0.001 and maximum motif length 10. The method first identifies candidate motifs by finding each subsequence whose population frequency is significantly higher than would be predicted from the frequencies of substrings from which it might be assembled. It then uses an iterative algorithm to repeatedly explain the training sequences as maximum-likelihood concatenations of individual motifs then use these explanations to improve estimates of the motif frequencies. The reader is referred to Schwartz[16] for algorithm details. Tag SNPs were determined from the motif structure by a dynamic programming algorithm[17] to give an estimated maximum prediction error of 5% for each hidden SNP site on the training data. The maximum motif length was dictated by the prohibitive computational cost of htSNP selection using long motifs. The significance level is a program default selected to provide high confidence that almost all motifs represent truly conserved haplotypes.

We used two haplotype block methods, both implemented with the dynamic programming algorithm of Zhang et al.[21] One method we call bounded blocks finds block partitions by minimizing the total number of observed haplotypes over all blocks, subject to a maximum block length. The results reported here used a maximum length of 5, although we found nearly identical results with maximum length 10 (data not shown). We also used a simplified version of LD-based testing called four-gamete blocks, which minimizes the number of blocks given that sequences in each block must be consistent with a perfect phylogeny, or, equivalently, cannot have all four possible gametes for any pair of SNP sites. While many block metrics have been proposed, we chose these two as representatives because they tend to stress two different criteria that should make for a "good" model for association testing: few blocks and thus few distinct association tests (four gamete) or few haplotypes per block and thus less confounding from unassociated haplotypes (bounded block). Minimal tag SNPs were selected within each block by exhaustive enumeration.

2.2. *Association Testing Methods*

For each SNP in a data set, we counted occurrences of all motifs overlapping that SNP for which the total population frequency of the motif exceeded 10%. All other motifs were grouped into a common "other" class. We then performed a chi-square test of association on the contingency table of motif classes and case/control status. As the motif method does not separate the genome into putatively uncorrelated regions, we established p-values by a permutation test. We randomly reassigned case and control labels while preserving the size of each class and computed chi-square values as above, recording the maximum statistic value for each degree of freedom. P-values were estimated from one thousand permutations per data set.

With the two block methods, one class was developed for each of the three most common haplotypes in each block. All other haplotypes were assigned to a fourth class. A chi-square test of significance was applied to the two-by-four table of the haplotype classes and case/control statuses. As our concern in this study is whether correlations across block boundaries are useful to association studies, we exclude such cross-boundary correlation information by assuming no such correlations and using Bonferroni correction for multiple hypothesis control.

Association tests were also performed on individual SNPs using a chi-square test of the two-by-two contingency table of SNP allele and case-control status. Control for multiple hypotheses was again performed by Bonferroni correction assuming no correlations between SNPs. The same protocol was used to test association with tag SNPs.

To assess the influence of cross-block correlations, we repeated all Bonferroni-corrected tests with permutation tests using 1,000 permutations.

2.3. *Data Processing*

We evaluated the methods using two real and three simulated data sets. We downloaded phased data from a high-density 500 kb region of 7q21.13 from the ENCODE resequencing project[4] and the full chromosome 22 HapMap data set[6]. We believe the 7q21 data is a good approximation of the data to be expected in a candidate gene study while the larger but sparser chromosome 22 data provides a better approximation to the challenges involved in whole-genome studies. For each real data set, we removed all SNPs that were not variant in all four HapMap population groups: CEPH (Utah Residents with Northern and Western European ancestry); Han Chinese in Beijing, China; Japanese in Tokyo, Japan; and Yoruba in Ibadan, Nigeria. We were left with 548 such universal SNPs out of 1,523 total for the 7q21 data, an average marker distance of 912 bases, and 11,900 universal SNPs out of 19,250 for chromosome 22, an average marker distance of 4.7 kb.

Simulated data was generated by coalescent simulation under a Wright-Fisher neutral model using the ms program[7]. We followed a protocol developed for a prior empirical study of the utility of block and motif models for information compression[19]. We used a mutation rate of 2.5×10^{-8} per nucleotide per generation, a recombination rate of 10^{-8} per pair of sites per generation, and an effective population size of 10,000 based on estimated values of the human mutation[11] and recombination[8] rates and effective population size[14]. Each simulated data set consisted of 2,000 chromosomes in a region of 100,000 segregating sites, representative of a 100 kb genomic region. The resulting sequences were screened to remove any SNPs with population frequency below 10%. Pairs of sequences were combined at random assuming Hardy-Weinberg equilibrium to assign chromosomes to individuals. A total of 220 simulated population samples were created.

Simulated disease phenotypes were artificially imposed on all data sets. A disease SNP was assigned for each sample from among SNPs having population frequency between 40% and 60%. The disease SNP was also required to be within a region between 45% and 55% of the distance along the chromosome for real data or between 40% and 60% for simulated data. For simplicity, an additive model with a single disease penetrance parameter, p, was used to determine disease risk. Individuals homozygous for the disease allele had probability p of having the disease, those homozygous for the non-disease allele had probability $1-p$ of the disease, and others were assumed to have equal probability of having the disease or not. Individuals (pairs of chromosomes) were assigned to case and control sets accordingly. For the real data, cases and controls were assigned independently for each of the four population groups and any excess of cases over controls or vice-versa was discarded for each before pooling the four ethnicities for association testing. This protocol ensures an equal number of members of each group in the cases and controls in order to better simulate the demographically matched cases and controls to be expected in real association study data sets. For each simulated sample, a single case and a single control set were assigned and any excess of cases over controls or vice-versa was discarded.

For the 7q21 data, five case/control sets were constructed for each penetrance value from 55% to 100% in increments of 5%. For the chromosome 22 data, five case/control sets were constructed for each penetrance value from 60% to 100% in increments of 10%. One simulated data set was constructed by creating ten case/control sets for each penetrance from 55% to 100% in increments of 5% and a second by creating ten case/control sets for each penetrance from 51% to 60% in increments of 1%. A final set was developed for specificity testing using twenty sets of individuals assigned randomly to cases and controls independent of genotype.

Each association method was applied to all data sets. Success was evaluated by testing the fraction of associations detected at LOD cutoff values of 3 (p-value 0.001), 2.5 (p-value 0.0032), 2 (p-value 0.01), and 1.5 (p-value 0.032). Sensitivity was tested on the randomly assigned cases and controls at LOD cutoffs 3, 2.5, 2, 1.5, and 1 (p-value 0.1).

3. Results

We derived haplotype motif structures for all data sets and performed a visual inspection of the motif patterns for the real data. Figure 2 depicts the motif patterns assigned to the 7q21 region and a representative subregion of chromosome 22 selected for illustrative purposes. Both datasets are overwhelmingly assigned to motifs of the maximum allowed length (10 SNPs), suggesting considerable conserved structure at both marker densities. Common motifs are nearly identical between the two Asian samples, often shared between the Asian and European-ancestry samples, occasionally shared between the Asian and Yoruba samples, and very rarely shared by Yoruba and European but not Asian samples. These results provide an informal check on the method as they are consistent with recent reconstruction of the likely human evolutionary tree[20].

We began our quantitative analysis using the 7q21 data set. Figure 3 shows the results. All methods consistently fail for penetrance below 70%

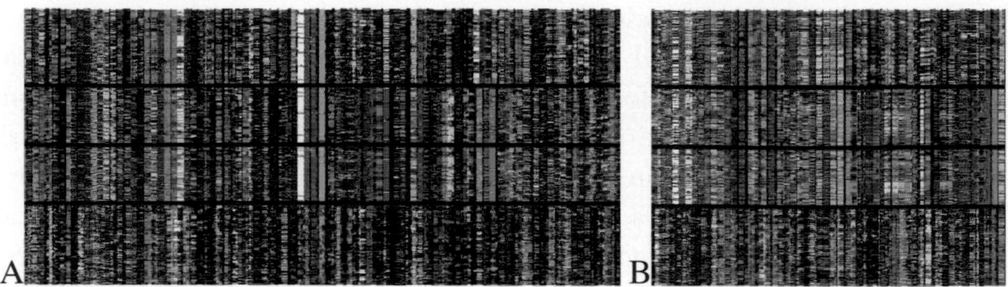

Figure 2. Visualization of haplotype motif assignments. Rows correspond to different chromosomes and columns to different SNP sites. Contiguous bars of a single color represent a single motif, with bars of the same color above and below them representing copies of that motif in other individuals. Each image shows motifs for the four HapMap populations (European-ancestry, Han Chinese, Japanese, and Yoruba) in order from top to bottom, separated by solid black rows. A: Motif assignments for a representative region of chromosome 22. B: Motif assignments for the 7q21 region.

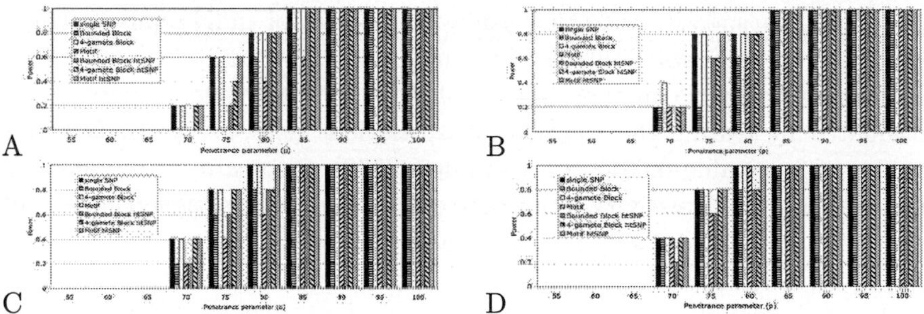

Figure 3. Power in identifying associations in 7q21 data. A: LOD cutoff 3; B: LOD cutoff 2.5; C: LOD cutoff 2; D: LOD cutoff 1.5

and succeed for penetrance values of at least 90% for LOD 3 or 85% for LOD 2.5 and below. The methods are therefore distinguishable only on a relatively narrow set of penetrances. Within that range, the straightforward motif method generally showed the least power. The single-SNP, 4-gamete block, and motif-based htSNP methods were most successful, with the 4-gamete block method outperforming the other two in one case.

We next examined the chromosome 22 data set. Preliminary visual inspection of the results confirms that the methods are finding significant associations only in the region of the disease SNP. While space does not permit us to present all of the detailed SNP-by-SNP scans, Fig. 4 shows a representative set of images from a sample with 90% penetrance. At the full-chromosome resolution, all methods show a single significant spike at

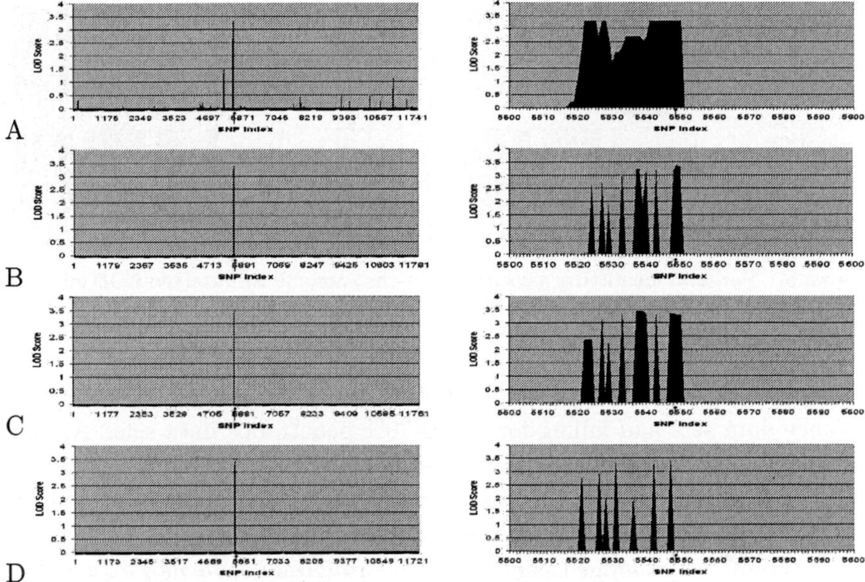

Figure 4. Representative selection of association scans for chromosome 22 with a simulated 90% penetrance disease SNP. Each line shows SNP-by-SNP LOD scores for the full chromosome (left) and a one hundred SNP region (right) around the disease site (marked by arrows). A: Motifs; B: Individual SNPs; C: Four-gamete blocks; D: Motif-selected htSNPs. Values below zero after Bonferroni correction are truncated to zero. Motif LOD scores above 3 (p-value 0.001) cannot be accurately estimated due to the use of a 1,000 trial permutation test and are therefore arbitrarily set to 3.3 (p-value 0.0005).

the location of the disease SNP (SNP position 5549). The motif method shows many smaller distant spikes, while the other methods do not. This is attributable to the fact that we assumed independence between tests for the other methods when correcting for multiple hypotheses and the resulting correction appears overly conservative. In the close-up view, all methods show a region of significant association centered slightly to the left of the disease SNP. The motif method shows significant associations across this entire region, while the others all show isolated spikes of association separated by regions of no association. This suggests that there are conserved haplotypes associated with the disease SNP spanning the entire region that are found by the motif method but often lost to the block methods.

Figure 5 shows the sensitivity of the methods on the chromosome 22 data. As with the 7q21 data, the methods behave identically for most

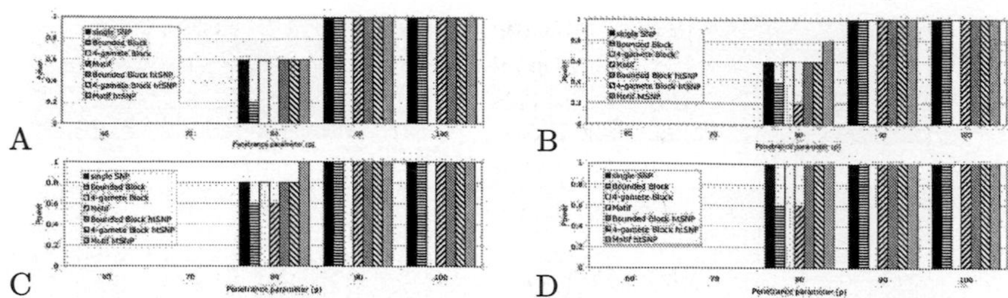

Figure 5. Power in identifying associations in chromosome 22 data. A: LOD cutoff 3; B: LOD cutoff 2.5; C: LOD cutoff 2; D: LOD cutoff 1.5

parameter values, successfully identifying association for all 90-100% penetrance data sets and failing for all 60-70% penetrance data sets. At 80% penetrance, the motif-based htSNP method is the most powerful, while the standard motif method is the least powerful. Bounded blocks also perform poorly, while all other methods perform equally well.

Although we examined here only one chromosome, we can extrapolate our results to a full-genome scan. Chromosome 22 is approximately 56 Mb, or about 1.8% of the human genome, which allows us to estimate that LOD scores detected by our methods would be approximately 1.8 lower if corrected for analysis to the full data set. Thus, the LOD 3 cutoff results would correspond to approximately a 94% confidence in a full genome scan.

We then analyzed the simulated data sets. We began by considering a broad range of penetrance parameters, 55% to 100% in increments of 5%. All methods are consistently successful for penetrance values of at least 75%. Occasional successes are observed even as low as 55%, suggesting that the larger population size used in the simulated test allows even relatively weak effects to be detected. The motif method appears most successful at detecting the weakest effects (penetrance 55%) but is the least successful on high-penetrance data sets. Overall, the motif-based htSNP method appears marginally the best at detecting associations in these data.

We then focused on the most difficult cases, examining simulated data with penetrances for each integer value from 51% to 60%. Figure 7 shows the results of these trials. No method was consistently dominant. The motif method appeared most successful on the hardest examples (penetrance 51%-55%). This success may be attributable to its better ability to find some conserved haplotype correlating with the disease SNP if any exists or it may be because its permutation test allows it to exploit correlations across block boundaries, a capability not permitted for the other methods.

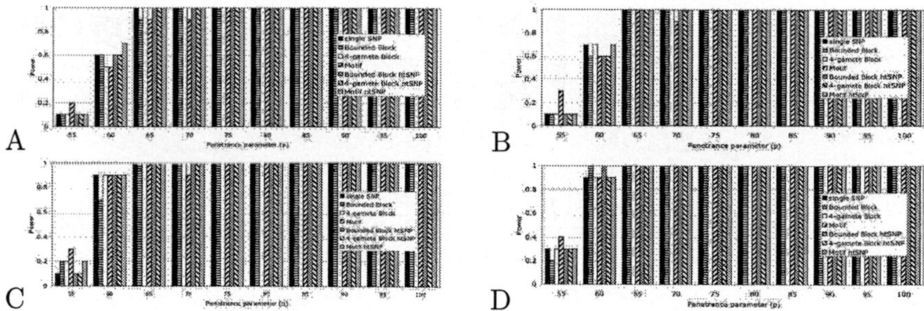

Figure 6. Power for penetrances 55%-100% using coalescent simulated data. A: LOD cutoff 3; ; B: LOD cutoff 2.5; C: LOD cutoff 2; D: LOD cutoff 1.5

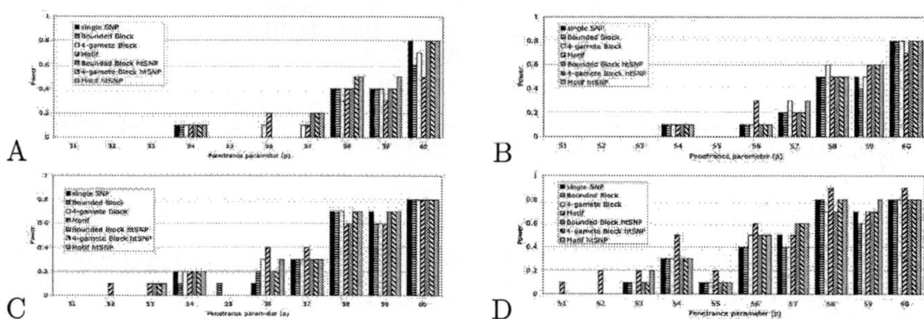

Figure 7. Power for penetrances 51%-60% using coalescent simulated data. A: LOD cutoff 3; B: LOD cutoff 2.5; C: LOD cutoff 2; D: LOD cutoff 1.5

Motif htSNPs become the most successful toward higher penetrance values.

Given indications that there is information useful to association tests lost by the assumption of independent blocks, we further asked whether dropping this assumption could recover some of that information. We therefore repeated all block and single-variant tests, replacing Bonferroni correction with permutation tests. Figure 8 shows power for chromosome 22 and coalescent simulated data at LOD 3. Compared to the prior Bonferroni-corrected graphs, the permutation tests yield a noticeable improvement in the block methods and a slight improvement for the block htSNPs. The motif htSNP method does not improve, suggesting that the assumption of independence is more nearly true of motif-selected htSNPs than block-selected htSNPs. Additional tests at other LOD values (data not shown) confirm that permutation tests lead to improved sensitivity of block, block htSNP, and single SNP tests, but not to motif htSNP tests.

We finally assessed the specificity of the methods using twenty samples

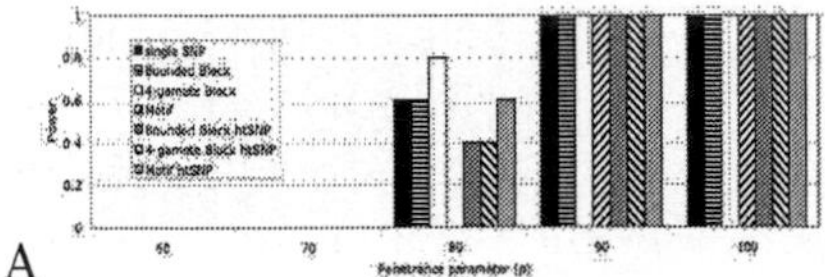
A

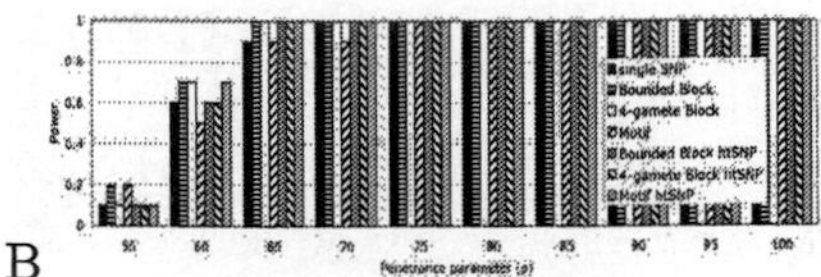
B

Figure 8. Power for permutation-test variants of all methods. A: chromosome 22 with LOD cutoff 3; B: simulated data with LOD cutoff 3.

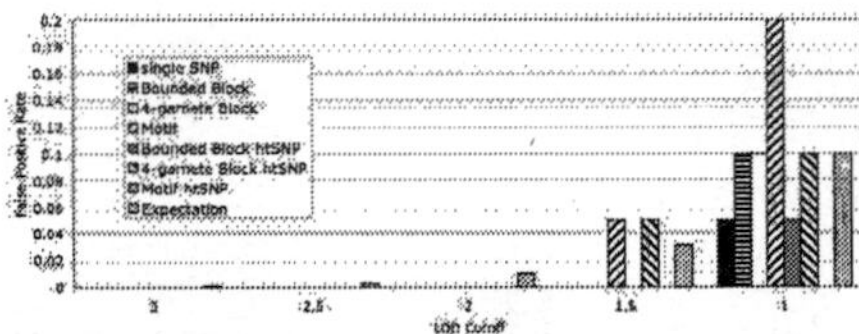

Figure 9. False positive identifications on twenty trials of randomly assorted simulated cases and controls as a function of LOD cutoff.

of randomly assorted simulated cases and controls. Figure 9 shows false positive rates as functions of the LOD cutoff. All values are within what can reasonably be expected by chance for each LOD score. All but motif htSNPs produce at least one false positive on 20 trials at LOD 1, while no methods produce any false positive values at LOD 2 or higher. Half of the methods using Bonferroni correction achieved exactly the expected number of errors at LOD 1 (2 errors); this appears to undermine our prior indication from chromosome 22 data that the Bonferroni correction is overconservative, suggesting that it is not excessively so.

4. Discussion

This study was intended to determine whether relaxing rigid block boundaries in models of haplotype structure would improve their utility for association tests. Our results provide an ambiguous answer to that question. Only a narrow range of parameter values discriminates between the methods for any data set, suggesting that the benefits of any one method over any other are relatively modest. The motif-based htSNP test appears to be marginally the best overall for both real and simulated data, consistent with prior work showing that motifs provide a clear advantage over comparable block methods at robustly selecting small htSNP sets[17]. The pure motif method does generally poorly, which we conjecture occurs because it tends to produce too many motifs covering each site, confounding the chi-square statistic. This could be an inherent problem of the motif approach or might be resolved by a different motif inference method or test statistic. Motifs do, however, appear to be the best for detecting the weakest correlations.

Both motif-based methods work comparatively better on simulated than on real data, suggesting that block-like patterns are more pronounced in real than in simulated data, even if they do not fully describe either. This conclusion is consistent with an emerging consensus in the field that inferred block patterns do not entirely reflect an inherent "blockiness" due to recombination hotspots, as was first proposed[3], but neither are they are fully explicable from uniform recombination rate models of human population history[12]. The conclusion is further supported by the improvement exhibited in block tests when using a permutation test rather than Bonferroni correction. The assumption of the block model that correlation information is captured within blocks is only partially valid, and methods for recovering that information either at the stage of model construction or application of the association test can lead to improved power. Both motif methods work better on the chromosome 22 data than on the denser 7q21 data, possibly because computational resource constraints limit us to short motifs (maximum of 10 SNPs in these tests), preventing them from taking advantage of long-range correlations in a dense marker set.

None of the methods considered — SNP, block, or motif — consistently dominates the others in all conditions; each can be expected to find some associations that would be missed by others. It would be self-defeating in practice to apply many methods to every data set, as the correction for multiple hypotheses would likely eliminate the small advantages of different methods for different cases. However, using a small number of very different methods may have advantages over applying only one "best" method. If we were to recommend one method from among those we examined, it would be an htSNP method. But if we were to recommend two, the second would be the motif method, as it is most likely to find associations the first misses. The field of association testing may benefit most by seeking a diversity of approaches, with particular emphasis on finding a few niche methods, like motifs, that are strongest in cases where others methods are weakest.

Acknowledgments

We thank R. Ravi and G. Blelloch for helpful discussions and comments on this manuscript. This work was supported in part by NSF awards #0122581 and #0346981 and by the Merck Program for Computational Biology and Chemistry at Carnegie Mellon University.

References

1. J. Akey, L. Jin, and M. Xiong. *Eur. J. Hum. Genet.* **9**, 291, (2001).
2. N.H. Chapman and E.M. Wijsman. *Am. J. Hum. Genet.* **63**, 1872 (1998).
3. M.J. Daly, J.D. Rioux, S.F. Schaffner, and T.J. Hudson, *Nat. Genet.* **29**, 229 (2001).
4. The ENCODE Project Consortium. *Science.* **306**, 636 (2004).
5. S. Gabriel, S. Schaffner, H. Nguyen, J. Moore, J. Roy, B. Blumenstiel, J. Higgens, M. DeFelice, A. Lochner, M. Faggart, S.N. Liu-Cordero, C. Rotimi, A. Adeyemo, R. Cooper, R. Ward, E.S. Lander, M.J. Daly, and D. Altschuler. *Science.* **296**, 2225 (2002).
6. The International HapMap Consortium. *Nature.* **426**, 789 (2003).
7. R.H. Hudson. *Bioinform.* **18**, 337 (2002).
8. M.I. Jensen-Seaman, T.S. Furey, B.A. Payseur, Y. Lu, K.M. Roskin, C.-F. Chen, M.A. Thomas, D. Haussler, and H.J. Jacob. *Genome Res.* **14**, 528 (2004).
9. G.C. Johnson, L. Esposito, B.J. Barret, A.N. Smith, J. Heward, G. Di Genova, H. Ueda, H.J. Cordell, I.A. Eaves, F. Dudbrigde, R.C. Twells, F. Payne, W. Hughes, S. Nutland, H. Stevens, P. Carr, E. Tuomilehto-Wolf, J. Tuomilehto, S.C. Gough, D.G. Clayton, and J.A. Todd. *Nat. Genet.* **29**, 233 (2001).
10. M. Koivisto, P. Rastas, and E. Ukkonen. *Lect. Notes Comp. Sci.* **3113**, 159 (2004).
11. M.W. Nachman and S.L. Crowell. *Genetics.* **156**, 297 (2000).
12. M. Nordborg and S. Tavaré, *Trends Genet.* **18**, 83 (2002).
13. N. Patil, A.J. Berno, D.A. Hinds, W.A. Barrett, J.M. Doshi, C.R. Hacker, C.R. Kautzer, D.H. Lee, C. Marjoribanks, D.P. McDonough, B.T. Nguyen, M.C. Norris, J.B. Sheehan, N. Shen, D. Stern, R.P. Stokowski, D.J. Thomas, M.O. Trulson, K.R. Vyas, K.A. Frazer, S.P. Fodor, and D.R. Cox. *Science.* **294**, 1719 (2001).
14. B. Rannala and Z. Yang. *Genetics*, **164**, 1645, (2003).
15. R. Schwartz, B. Halldórsson, V. Bafna, A.G. Clark, and S. Istrail. *J. Comp. Biol.* **10**, 13 (2003).
16. R. Schwartz, *Proc. IEEE Comp. Sys. Biotech. Conf.*, 306 (2003).
17. R. Schwartz, *Proc. IEEE Comp. Sys. Biotech. Conf.*, 90 (2004).
18. J. Sheffi. *MIT Comp. Sci. M.Eng. Thesis*, 2004.
19. S. Sridhar, K. Dhamdhere, G. E. Blelloch, R. Ravi and R. Schwartz. *Carnegie Mellon Comp. Sci. Tech Report,* **CMU-CS-040166** (2004).
20. S. Tishkoff and K.K. Kidd. *Nat. Genet.* **36**, S21 (2004).
21. K. Zhang, M. Deng, T. Chen, M. S. Waterman and F. Sun. *Proc. Natl. Acad. Sci. USA.* **99**, 7335 (2002).
22. K. Zhang, P. Calabrese, M. Nordborg, and F. Sun. *Am. J. Hum. Genet.* **71**, 1386 (2002).

EFFECT OF THE PEROXISOME PROLIFERATORS-ACTIVATED RECEPTOR (PPAR) GAMMA 3 GENE ON BMI IN 1,210 SCHOOL STUDENTS FROM MORELOS, MEXICO*

LINA CHEN†

Department of Social Medicine, University of Bristol, Bristol, UK

H. EDUARDO VELASCO MONDRAGÓN

National Institute of Public Health, Cuernavaca, Morelos, Mexico

EDUARDO LAZCANO-PONCE

National Institute of Public Health, Cuernavaca, Morelos, Mexico

ANDREW COLLINS

Human Genetics Research Division, University of Southampton, UK

YIN YAO SHUGART

Department of Epidemiology, Johns Hopkins Bloomberg School of Public Health, Baltimore, USA yyao@jhsph.edu

Department of Social Medicine, University of Bristol, Bristol, UK

Little research has been undertaken on risk factors for obesity in young people in Latin America, including Mexico, despite the fact that obesity constitutes the number one public health problem in Mexico. Our objective was to investigate the effect of the Peroxisome proliferators-activated receptor (PPAR)_3 gene on BMI measured among adolescents collected from a cohort study originally designed for epidemiological studies. METHODS: Blood samples and anthropometric measurements were collected from 1,210 out of 13,294 public school students of both sexes, aged 11-24 years in Morelos, Mexico. In this study, we genotyped 7 selected SNPs of the PPAR_ transcript variant 3 (including Pro12Ala) in a group of unrelated 717 males and 493 females (age range 11-24), including 3 SNPs located in the 5' untranslated region. These 7 SNPs were selected by the tagging algorithm implemented in the program haploview to scan the whole gene. We tested each of the 7 SNPs individually for association with the body mass index (BMI), and two SNPs (rs2938392 and rs1175542) revealed significant associations with BMI (p-value=0.008 and 0.029, respectively). The SNP rs2938392 is roughly 41.5 Kb from rs1801282 (Pro12Ala in PPAR_2). Furthermore, we examined the association between haplotypes built from 7 SNPs and BMI using a score statistic implemented in the program haplo.stats. While the permutation based global p-value was 0.544, one individual haplotype with a frequency of 0.279 gave a p-value of 0.089

* This work is supported by Glaxo-Smith-Kline Epidemiology

† This work is partially supported by University of Bristol

(permutation based). However, when the analyses were conducted in males only, the permutation based global p-value was 0.055 and one individual haplotype with a frequency of 0.28 gave a significant p-value of 0.013.

1. Introduction

The goal of this study was to investigate the effect of the Peroxisome proliferators-activated receptor (PPAR)_3 gene on Body Mass Index (BMI) measured among Mexican adolescents. Although it has been established that both PPAR_2 and PPAR_3 are expressed uniquely in colon and adipocyte tissue and their potential role in the metabolic disorders such as obesity and type 2 diabetes has been suggested [1], the reported effects of the Pro12Ala polymorphism on susceptibility for obesity have been inconsistent [2]. The detailed findings from all positive or negative studies are given in Table 1.

Table 1 A summary of Pro12Ala findings in different studies

Ref	Sample size	Ethnicity	P-value
[3]	517 being lean-to-moderately obese	Caucasian	0.01
[3]	169 very obese	Caucasian	<0.001
[4]	333 non-diabetic	Scandinavian	0.027
[4]	973 non-diabetic	Scandinavian	0.015
[5]	215 men	Asian	0.65
[6]	296 extremely obese	Caucasian	>0.05
[6]	130 underweight	Caucasian	>0.05
[7]	752 obese	Caucasian	0.008
[7]	869 non-obese	Caucasian	0.005
[8]	141 obese	Scandinavian	0.011
[9]	131 diabetic	Caucasian	0.8
[9]	312 normoglycemic	Caucasian	0.9
[10]	1025 diabetic	Caucasian	>0.05
[10]	310 with normal BMI	Caucasian	>0.05
[11]	108 non-diabetic	Caucasian	0.67
[11]	19 overweight	Caucasian	0.71
[12]	295 non-diabetic non-obese	Caucasian	>0.05
[12]	372 morbidly obese	Caucasian	>0.05
[12]	402 diabetic	Caucasian	>0.05
[13]	541 non-diabetic	Asian	0.15
[13]	415 diabetic subjects	Asian	0.10
[14]	165 obese	Caucasian	0.017
[15]	229	Asian	>0.05
[16]	921	Caucasian	0.011
[17]	476	Scandinavian	0.3
[18]	675 men	Caucasian	0.64

ID	Sample size	Ethnicity	P-value
[19]	228 with normal BMI	Scandinavian	0.070
[19]	217 with dyslipidemia	Scandinavian	0.034
[19]	649 without dyslipidemia	Scandinavian	0.080
[20]	280 with normal BMI	Caucasian	>0.05
[20]	95 obese	Caucasian	0.32
[20]	42 young obese	Caucasian	<0.05
[21]	619	Caucasian	0.035
[22]	453 from 10 families	Mexican	>0.05
[23]	2201 diabetics	Asian	0.881
[23]	1212 with normal BMI	Asian	0.846
[24]	292 obese	Caucasian	0.89
[24]	371 lean	Caucasian	0.47
[25]	259 men	Caucasian	0.554
[25]	333 women	Caucasian	0.678
[26]	124 non-diabetics	Caucasian	0.31
[27]	2245 non-diabetics	Scandinavian	>0.05
[28]	1107 diabetics	Caucasian	0.3
[29]	438	Caucasian	>0.05
[30]	478 men	Asian	>0.05
[30]	117 women	Asian	>0.05
[31]	145 obese	Caucasian	>0.05
[31]	317 non-obese	Caucasian	>0.05
[32]	210 monozygotic twins	Scandinavian	0.09
[32]	344 dizygotic twins	Scandinavian	>0.05
[33]	720	Caucasian	0.005
[34]	253 with low physical activity level	Caucasian	>0.05
[34]	253 with high physical activity level	Caucasian	p<0.05
[35]	420 diabetic	Asian	0.566
[35]	538 with impaired glucose tolerance	Asian	0.875
[35]	3080 with normal BMI	Asian	0.037

Few studies have investigated the association between PPAR_ gene and BMI in the Mexican population. The most recent report was from Hsueh *et al.* 2001 [22], based on the study of 453 subjects comprising of 10 pedigrees. However, no significant effect was found in their study.

While most of previous studies have focused on the effect of Pro12Ala or PPAR_2, we have chosen to study PPAR_3, which contains the region of PPAR_2, but also includes a 5' untranslated region of the PPAR_ gene. Interestingly, the expression of PPAR_3 is directed by an independent promoter, and to date, at least three promoters in the upstream region of PPAR_3 have been identified through molecular studies [1]. We analyzed seven tagging SNPs spanning 89.5kb of PPAR_3 gene region, and four of them were located in the long 5'-end untranslated region. Our motivation is to examine the effect of the haplotypes that represent the PPAR_3 genes, rather than focusing on Pro12Ala itself.

2. Methods

2.1. *Study Population*

This study builds on a parent cohort study carried out by Lazcano-Ponce et al. (2003). The parent study was the 1998-1999 baseline measurement of a cohort study of 13,293 students at public junior high and high schools and a state university in the central Mexican State of Morelos. The Research Ethics committee of the National Institute of Public Health approved the study Protocol for epidemiological studies. Further, The IRB review board of Johns Hopkins Bloomberg School of Public Health approved the current study which uses 1,270 anonymous DNA samples out of 13,293 Mexican students to test for association between BMI and SNPs residing in the PPARγ3 gene. Table 2 gives information on age and gender in this study population.

Table 2 Characteristics of the study group subjects

Variable	Male	Female	Whole population
No. of Subjects	717	493	1210
Age	15.56±2.67	15.84±2.93	15.67±2.78
BMI	23.76±5.76	23.45±5.61	23.63±5.7

2.2. *DNA extraction and Genotyping*

DNA extraction was performed using the Gentra DNA extraction kit following the manufacturer's suggestion. Out of 1,270 buffy coat samples, 1,210 samples gave sufficient DNA for this study. TaqMan, developed by Applied Biosystems (ABI), is an efficient system and genotyping was performed at the core genotyping facility at Johns Hopkins University.

The PCRs were conducted with both primers and probes added and only end point products were read. The Hydra and Biomek FX was used to dispense DNA samples and set up PCRs, respectively. The PCR was conducted in two 9700 thermocyclers each equipped with a dual 384-well blocks. Then the end point products were scored using the 7900HT. In each 384-well plate, two reference samples were included for quality control. The primers and probes were designed using Primer Express (ABI). The probes were labeled with two fluorescent dyes, one as an indicator and the other as a quencher. Two probes were synthesized for each locus, each labeled with two different dyes as indicators, respectively.

2.3. *SNP tagging*

We used the haploview program, http://www.broad.mit.edu/mpg/haploview/ [37] to compute the pair-wise LD measure D'. Haploview estimates the maximum-likelihood values of the 4 gamete frequencies, from which D' can be calculated. For these calculations we included members of our melanoma families as well as 350 individuals with BMI less than 25. The SNP haplotype tagging strategy (*htSNP*) allowed us to identify regions of strong LD using the Gabriel *et al* block definition [38], and kept every inter-block SNP plus the single SNP within each block with the highest minor allele frequency (MAF). The LD structure of these 7 selected SNPs is presented in Figure 1. The marker rs1801282 represents Pro12Ala in PPAR_2.

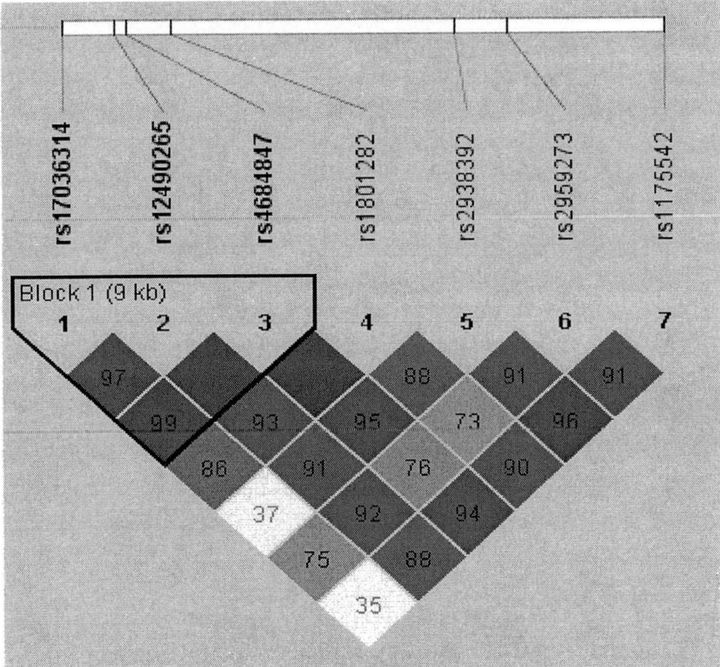

Figure 1 LD structure of the selected SNPs in PPAR_3. The numbers within each block indicate the pair –wise D' as a percentage. The higher LD linkages are shown in the darker color, while the first three SNPs constitute an LD block spanning 9kb of gene region. Moreover, all of these three SNPs are located in the 5'-untranslated region of PPAR_3.

2.4. *Statistical analysis*

Hardy-Weinberg proportions were tested among individuals with BMI less than 25 using the Pearson chi-square test. Pair-wise linkage disequilibrium was evaluated by two linkage disequilibrium parameters, Lewontin's D′ [39] and r^2 [40], which were both calculated in haploview (data not shown).

2.4.1. *Single SNP analysis*

Single SNP analyses were conducted using the linear regression procedure implemented in STATA version 9. The SNP effects were analyzed using three models which were, respectively: additive, dominant and recessive models. In the additive model, the wild type genotype, the homozygote consisting of the most common allele in the population, was treated as the baseline and coded as 1, while the heterozygote was coded as 2 and the other homozygote was coded 3. In the dominant model, the wild type homozygote was coded by 1 as baseline and the heterozygote and the other homozygote were both coded 2. Moreover, in the recessive model, both of the wild type homozygote and heterozygote were treated as baseline with the rare homozygote coded 2.

2.4.2. *Haplotype analysis*

All haplotype frequencies were estimated using the expectation maximization (EM) algorithm in haplo.stats in the R programming language (http://mayoresearch.mayo.edu/mayo/research/biostat/schaid.cfm). Schaid et al [41] developed a score statistic that can be used to test the statistical association between haplotypes and different types of human traits, including binary and quantitative traits. This method also allows adjustment for non-genetic factors. In this analysis, we used haplo.stats to compute the global score statistic (which allows us to test the significance of association considering all haplotypes) and obtain a global p-value and the haplotype specific statistic (which allows us to compare each haplotype with a selected common haplotype). All haplotypes with a frequency less than 1% were dropped from the score test in order to reduce the degrees of freedom. Another advantage of Schaid's method is to compute empirical p-values using simulation when the haplotype data is sparse. The empirical p-values are computed by repeatedly first permuting the genotypes among the subjects and then computing the score statistics.

3. Result

Hardy-Weinberg proportions were tested among individuals with BMI less than 25 using the Pearson chi-square test and no SNPs showed deviation from Hardy Weinberg equilibrium. The minor allele frequencies are 0.479, 0.130, 0.134,

0.143, 0.445, 0.151 and 0.441 for SNPs rs17036314, rs12490265, rs4684847, rs1801282, rs2938392, rs2959273, rs1175542, respectively.

Linear regression analysis of single SNP revealed two SNPs associated with BMI with significant p-values reported in Table 3 (non-adjusted for multiple testing).

Table 3. Significant associations revealed by single SNP analysis and the body mass index (BMI)

SNP ID	Location	Population	Model	Coefficient	P-value
rs2938392	Second intron	Whole	Dominant	1.0866	0.008
rs2938392	Second intron	Male	Dominant	1.3818	0.009
rs1175542	Fifth intron	Whole	Recessive	0.898	0.029
rs1175542	Fifth intron	Male	Recessive	1.171	0.029

In addition, we used a sliding window technique to fine map any potential signals within the larger haplotype using haplo.stats [41]. The size of the window varied from 3-4 SNPs in order to provide a comprehensive assessment of the haplotype subsets within the gene. Permutation based p-values were summarized in Table 4 for significant SNP combinations.

Table 4. Haplotypes revealing significant associations with BMI using 1,210 Mexican samples, p-value(sim) indicated the p-values based on permutation.

Haplotypes	Frequency	P-value	P-value (sim)
rs17036314*1/rs12490265*2/rs1801282*1	0.38952	0.05224	0.0495
rs17036314*1/rs4684847*1/rs1801282*1	0.38808	0.05637	0.058
rs17036314*2/rs2938392*1/rs1175542*2	0.28722	0.05607	0.0555
rs12490265*2 /rs1801282*1/ rs2938392*2	0.42052	0.05121	0.0512
rs12490265*2 /rs1801282*1/ rs1175542*1	0.42607	0.05451	0.0558
rs17036314*1/rs12490265*2/rs4684847*1/rs1801282*1	0.38607	0.03466	0.034
rs17036314*2/rs4684847*1/rs2938392*1/rs1175542*2	0.2869	0.05644	0.0564
rs12490265*2/rs4684847*1/rs1801282*1/ rs2938392*2	0.42129	0.0388	0.0387
rs12490265*2/rs4684847*1/rs1801282*1/ rs1175542*1	0.42648	0.04124	0.043

Furthermore, we examined the association between haplotypes built from 7 SNPs and BMI using a global score statistic. While the permutation based global p-value was 0.544 (degree of freedom=11), one individual haplotype gave a p-value of 0.086 (permutation based) in all samples for a haplotype frequency of 0.279. Interestingly, when the analyses were conducted in males only, the permutation based global p-value was 0.055, and one relatively frequent haplotype (frequency =0.28) gave a significant signal of 0.013.

3. Discussion

To summarize, we genotyped the 7 selected SNPs of the PPAR_ transcript variant 3 (including pro12Ala) in a group of unrelated 717 males and 493 females (ages ranging from 11-24), among which 3 SNPs are located in the 5' untranslated region. These 7 SNPs were selected by the tagging algorithm implemented in haploview to cover the whole gene and all SNPs conformed to Hardy-Weinberg equilibrium. We tested each of the 7 SNPs individually for association with body mass index (BMI). Although no association between Pro12Ala and BMI was observed, two SNPs rs2938392 and rs1175542 revealed significant associations with BMI (p-values=0.008 and 0.029, respectively). It is worth noting that the SNP rs2938392 is roughly 41.5 Kb away from Pro12Ala (rs1801282), a polymorphism known to be associated with BMI and diabetes type II. Furthermore, when all 7 SNPs were analyzed together, the permutation based global p-value was 0.544, and one individual haplotype (haplotype frequency =0.279) gave a significant signal of 0.086 (permutation based). Moreover, when the analyses were conducted in males only, the permutation based global p-value was 0.055, and one relatively frequent haplotype (frequency =0.28) gave a significant signal of 0.013.
Several speculations may explain our findings.

1. There is no association between Pro12Ala polymorphism and BMI in the 1,210 school students from Morelos, Mexico. The positive signals we detected are false positives, which may be due to the high-level admixture structure in the Mexican population.
2. There may not be a direct association between Pro12Ala and BMI, however, SNPs which are in high LD with Pro12Ala are associated with BMI in Mexican population, but this hypothesis will need to be further investigated.
3. Environmental factors may play a role to increase the genetic effect and the lack of data on dietary factors and physical activities in this data set may have limited the power to detect an association.

To summarize, our results indicate that SNPs in high linkage disequilibrium with Pro12Ala are associated with BMI in 1,210 students in Morelos, Mexico. However, due to limited access to the epidemiological factors such as dietary factors and physical activities, the results we reported here need to be interpreted with caution. Future study will explore causal associations between other genetic and non-genetic risk factors and obesity in this population. Further studies will consider increased sample sizes, and we will genotype 40 random SNPs to assess the level of population admixture this population.

Acknowledgments

Professor Andrew Collins would like to acknowledge funding from the Biotechnology and Biological Sciences Research Council (UK).

References

1. L. Fajas, J-C. Fruchart and J. Auwerx, *FEBS Lett.* **438(1-2),** 55 (1998).
2. M. Stumvoll and H.Haring, *Diabetes.* **51(8),** 2341 (2002).
3. B.A.Beamer, C.J.Yen, R.E.Andersen, D.Muller, D.Elahi, L.J.Cheskin, R.Andres, J.Roth, and A.R.Shuldiner, *Diabetes* **47(11)**, 1806 (1998).
4. S.S.Deeb, L.Fajas, M.Nemoto, J.Pihlajamaki, L.Mykkanen, J.Kuusisto, M.Laakso, W.Fujimoto, and J.Auwerx, *Nat. Gen.* **20(3)**, 284 (1998).
5. Y.Mori, H.Kim-Motoyama, T.Katakura, K.Yasuda, H.Kadowaki, B.A.Beamer, A.R.Shuldiner, Y.Akanuma, Y.Yazaki, and T.Kadowaki, *Biochem. Biophys. Res. Commun.* **251(1)**, 195 (1998).
6. A.Hamann, H.Munzberg, P.Buttron, B.Busing, A.Hinney, H.Mayer, W.Siegfried, J.Hebebrand, and H.Greten, *Eur J Endocrinol* **141(1)**, 90 (1999).
7. J.Ek, S.A.Urhammer, T.I.A.Sorensen, T.Andersen, J.Auwerx, and O.Pedersen, *Diabetologia* **42(7)**, 892 (1999).
8. R.Valve, K.Sivenius, R.Miettinen, J.Pihlajamaki, A.Rissanen, S.S.Deeb, J.Auwerx, M.Uusitupa, and M.Laakso, *J. Clin. Endocrinol. Metab.* **84(10)**, 3708 (1999).
9. F.P.Mancini, O.Vaccaro, L.Sabatino, A.Tufano, A.A.Rivellese, G.Riccardi, and V.Colantuoni, *Diabetes* **48(7)**, 1466 (1999).
10. J.Ringel, S.Engeli, A.Distler, and A.M.Sharma, *Biochem. Biophys. Res. Commun.* **254(2)**, 450 (1999).
11. M.Koch, K.Rett, E.Maerker, A.Volk, K.Haist, M.Deninger, W.Renn, and H.U.Haring, *Diabetologia* **42(6)**, 758 (1999).
12. K.Clement, S.Hercberg, B.Passinge, P.Galan, M.Varroud-Vial, A.R.Shuldiner, B.A.Beamer, G.Charpentier, B.Guy-Grand, P.Froguel, and C.Vaisse, *Int. J. Obes.* **24(3)**, 391 (2000).
13. K.Hara, T.Okada, K.Tobe, K.Yasuda, Y.Mori, H.Kadowaki, R.Hagura, Y.Akanuma, S.Kimura, C.Ito, and T.Kadowaki, *Biochem. Biophys. Res. Commun.* **271(1)**, 212 (2000).
14. W.D.Li, J.H.Lee, and R.A.Price, *Mol. Genet. Metab.* **70(2)**, 159 (2000).
15. E.Y.Oh, K.M.Min, J.H.Chung, Y.K.Min, M.S.Lee, K.W.Kim, and M.K.Lee, *J. Clin. Endocrinol. Metab.* **85(5)**, 1801 (2000).
16. S.A.Cole, B.D.Mitchell, W.C.Hsueh, P.Pineda, B.A.Beamer, A.R.Shuldiner, A.G.Comuzzie, J.Blangero, and J.E.Hixson, *Int. J. Obes.* **24(4)**, 522 (2000).
17. J.G.Eriksson, V.Lindi, M.Uusitupa, T.J.Forsen, M.Laakso, C.Osmond, and D.J.P.Barker, *Diabetes* **51(7)**, 2321 (2002).
18. O.Poirier, V.Nicaud, F.Cambien, and L.Tiret, J. Mol. Med. **78(6)**, 346 (2000).

19. J.Pihlajamaki, R.Miettinen, R.Valve, L.Karjalainen, L.Mykkanen, J.Kuusisto, S.Deeb, J.Auwerx, and M.Laakso, *Atherosclerosis* **151(2)**, 567 (2000).
20. O.Vaccaro, F.P.Mancini, G.Ruffa, L.Sabatino, V.Colantuoni, and G.Riccardi, *Int. J. Obes.* **24(9)**, 1195 (2000).
21. S.J.Hasstedt, Q.F.Ren, K.Teng, and S.C.Elbein, *J. Clin. Endocrinol. Metab.* **86(2)**, 536 (2001).
22. W.C.Hsueh, S.A.Cole, A.R.Shuldiner, B.A.Beamer, J.Blangero, J.E.Hixson, J.W.MacCluer, and B.D.Mitchell, *Diabetes Care* **24(4)**, 672 (2001).
23. H.Mori, H.Ikegami, Y.Kawaguchi, S.Seino, N.Yokoi, J.Takeda, I.Inoue, Y.Seino, K.Yasuda, T.Hanafusa, K.Yamagata, T.Awata, T.Kadowaki, K.Hara, N.Yamada, T.Gotoda, N.Iwasaki, Y.Iwamoto, T.Sanke, K.Nanjo, Y.Oka, A.Matsutani, E.Maeda, and M.Kasuga, *Diabetes* **50(4)**, 891 (2001).
24. M.M.Swarbrick, C.M.L.Chapman, B.M.McQuillan, J.Hung, P.L.Thompson, and J.P.Beilby, *Eur J Endocrinol* **144(3)**, 277 (2001).
25. J.Luan, P.O.Browne, A.H.Harding, D.J.Halsall, S.O'Rahilly, V.K.K.Chatterjee, and N.J.Wareham, *Diabetes* **50(3)**, 686 (2001).
26. M.Hara, S.Y.Alcoser, A.Qaadir, K.K.Beiswenger, N.J.Cox, and D.A.Ehrmann, *J. Clin. Endocrinol. Metab.* **87(2)**, 772 (2002).
27. L.Frederiksen, K.Brodbaek, M.Fenger, T.Jorgensen, K.Borch-Johnsen, S.Madsbad, and S.A.Urhammer, *J. Clin. Endocrinol. Metab.* **87(8)**, 3989 (2002).
28. A.Doney, B.Fischer, D.Frew, A.Cumming, D.M.Flavell, M.World, H.E.Montgomery, D.Boyle, A.Morris, and a.Palmer et, *BMC Genet* **3(1)**, 21 (2002).
29. O.Vaccaro, F.P.Mancini, G.Ruffa, L.Sabatino, C.Iovine, M.Masulli, V.Colantuoni, and G.Riccardi, *Clinical Endocrinology* **57(4)**, 481 (2002).
30. Y.Yamamoto, H.Hirose, K.Miyashita, K.Nishikai, I.Saito, M.Taniyama, M.Tomita, and T.Saruta, *Metabolism* **51(11)**, 1407 (2002).
31. J.L.Gonzalez Sanchez, M.Serrano Rios, C.Fernandez Perez, M.Laakso, and M.T.Martinez Larrad, *Eur. J. Endocrinol.* **147(4)**, 495 (2002).
32. P.Poulsen, G.Andersen, M.Fenger, T.Hansen, S.M.Echwald, A.Volund, H.Beck-Nielsen, O.Pedersen, and A.Vaag, *Diabetes* **52(1)**, 194 (2003).
33. J.Robitaille, J.P.Despres, L.Perusse, and M.C.Vohl, *Clin. Genet.* **63(2)**, 109 (2003).
34. P.W.Franks, J.Luan, P.O.Browne, A.-H.Harding, S.O'Rahilly, V.K.K.Chatterjee, and N.J.Wareham, *Metabolism* **53(1)**, 11 (2004).
35. E.S.Tai, D.Corella, M.urenberg-Yap, X.Adiconis, S.K.Chew, C.E.Tan, and J.M.Ordovas, *J.Lipid Res.* **45(4)**, 674 (2004).
36. E. C. Lazcano-Ponce , B. Hernandez, A. Cruz-Valdez, B. Allen, R. Diaz, C. Hernandez, R. Anaya and M. Hernández-Avila, *Arch. Med. Res.* **34(3),** 222 (2003).
37. J. C. Barrett, B. Fry, J. Maller and M. J. Daly, *Bioinformatics*, **21(2)**, 263 (2005).
38. S. B. Gabriel, S. F. Schaffner, H. Nguyen, J. M. Moore, J. Roy, B. Blumenstiel, J. Higgins, M. DeFelice, A. Lochner, M. Faggart, S. N. Liu-

Cordero, C. Rotimi, A. Adeyemo, R. Cooper, R. Ward, E. S. Lander, M. J. Daly and D. Altshuler, *Science* **296(5576)** 2225 (2002).
39. R. C. Lewontin and M. W. Feldman, *Theor Popul Biol* **34(2)** 177 (1988).
40. B. Devlin and N. A. Risch, *Genomics* **29(2),** 311 (1995).
41. D. J. Schaid, C. M. Rowland, D. E. Tines, R. M. Jacobson and G. A. Poland *Am. J. Hum. Genet.* **70(2),** 425 (2002).

TRANSFERABILITY OF TAG SNPS TO CAPTURE COMMON GENETIC VARIATION IN DNA REPAIR GENES ACROSS MULTIPLE POPULATIONS

PAUL I.W. DE BAKKER, ROBERT R. GRAHAM, DAVID ALTSHULER

Broad Institute of Harvard and Massachusetts Institute of Technology, Cambridge, MA

BRIAN E. HENDERSON, CHRISTOPHER A. HAIMAN

Department of Preventive Medicine, Keck School of Medicine, University of Southern California, Los Angeles, CA

Genetic association studies can be made more cost-effective by exploiting linkage disequilibrium patterns between nearby single-nucleotide polymorphisms (SNPs). The International HapMap Project now offers a dense SNP map across the human genome in four population samples. One question is how well tag SNPs chosen from a resource like HapMap can capture common variation in independent disease samples. To address the issue of tag SNP transferability, we genotyped 2,783 SNPs across 61 genes (with a total span of 6 Mb) involved in DNA repair in 466 individuals from multiple populations. We picked tag SNPs in samples with European ancestry from the Centre d'Etude du Polymorphisme Humain, and evaluated coverage of common variation in the other samples. Our comparative analysis shows that common variation in non-African samples can be captured robustly with only marginal loss in terms of the maximum r^2. We also evaluated the transferability of specified multi-marker haplotypes as predictors for untyped SNPs, and demonstrate that they provide equivalent coverage compared to single-marker tests (pairwise tags) while requiring fewer SNPs for genotyping. The efficacy of a tagging-based approach in studying genotype-phenotype correlations in complex traits is strongly supported by our empirical results.

1. Introduction

A significant fraction of the risk of developing common diseases such as cancer is due to genetic variation. Knowledge of the genetic basis of these complex traits may lead to new insights into disease pathogenesis, the identification of novel drug targets, and ultimately contribute to human health.

Family-based linkage analysis has been very successful in localizing causal variants for monogenic, Mendelian diseases. However, success has been rather limited for common diseases, where multiple loci are likely to act in concert and contribute only probabilistically [1]. Testing genetic variants for association between cases and appropriate controls offers a more powerful approach to detect putative causal variants, but require large sample sizes to achieve adequate power [2].

Complete ascertainment of genetic variation by resequencing is the only comprehensive approach to test all variants (both common and rare) directly for association. For the foreseeable future, routine resequencing in thousands of individuals will not be practical. But high-throughput technology to type large numbers of SNPs in thousands of people is rapidly improving, making it possible to probe the vast majority of human heterozygosity due to common variations. In addition, public databases of SNP variation have swelled to 10 million variants. The International HapMap Project provides genome-wide data in 269 individuals from four different population groups [3], and supports the selection of informative markers ("tag SNPs") by exploiting redundancies among nearby polymorphisms due to linkage disequilibrium (LD) [4]. Tagging approaches may substantially improve the cost-effectiveness of association studies by delivering greater power and better genotyping efficiency through the selection of tag SNPs and definition of statistical tests based on the empirical LD patterns in HapMap (and similar resources such as [5]).

An important outstanding question is whether tag SNPs picked from HapMap will be transferable across independent disease samples, and how this varies for different testing strategies (especially methods based on haplotypes). The precise LD patterns are likely to differ between population groups, given the many forces that determine the patterns of genetic variation. Thus, empirical evidence is required to study the efficiency and coverage of tag SNPs across different populations, and to validate LD-based tagging approaches in general. The work described here builds upon early studies that have begun to address this issue [6-8].

We report a large data set of genes involved in DNA repair within which we have performed dense genotyping in multiple population samples. By picking tag SNPs in one population sample, we can perform a blind assessment as to how well these tag SNPs—and allelic tests for association based on them—capture the variation in any of the other population samples.

2. Methods and Materials

2.1. *DNA samples*

We have collected genotype data from seven population samples. The CEPH (Centre d'Etude du Polymorphisme Humain) samples are a subset (20 trios, 60 individuals in total) of the 30 trios used in HapMap (designated as CEU samples) [3]. The African American (AA, n = 70), Native Hawaiian (NH, n = 67), Latino (LA, n = 70), Japanese (JA, n = 70) and White (WH, n = 70) samples were selected from the Multiethnic Cohort (MEC) conducted in Hawaii and California (mainly Los Angeles) [9]. The Chinese samples (CH, n = 59) were selected from an ongoing study in Shanghai and from the Singapore Chinese Health Study [10].

2.2. *Genotyped SNPs*

SNPs for genotyping were selected from dbSNP in 61 DNA repair genes (Table 1) with a total span of 5.7 Mb. This resulted in a working set of 2,783 successfully genotyped SNPs from all samples with an average marker density of 1 SNP every ~2 kb. Criteria for successful conversion are: Hardy-Weinberg $P > 0.01$ (for five of the six ethnic groups), genotyping percentage >75%, no more than one discordant blinded replicate (9 total) or Mendel inconsistency in parent-offspring trios (CEPH only).

As expected, we observe more common SNPs in AA than in the other samples, reflecting greater genetic diversity (heterozygosity) in African-derived populations.

The data sets were phased using the program EMPHASE (written by Nick Patterson) to give 140 unrelated chromosomes (haplotypes) for AA, LA, JA, WH; 134 for NH; 118 for CH and 80 for CEPH. EMPHASE is based on the expectation-maximization algorithm [11].

Table 1. List of selected DNA repair genes and number of successfully genotyped SNPs in all population samples.

Locus	# SNPs	Locus	# SNPs	Locus	# SNPs
APE1	30	Ku80	58	POLE	74
ATM	57	LIG1	55	POLI	34
ATR	33	LIG3	25	POLK	32
Artemis	45	LIG4	28	RAD50	42
BLM	71	MGMT	114	RAD51	20
BRCA1	32	MLH1	41	RAD52	45
BRCA2	59	MLH3	26	RPA1	68
CHEK1	44	MRE11	47	RPA2	33
CHEK2	39	MSH2	38	RPA3	73
CSA	52	MSH3	133	TP53	19
CSB	69	MSH6	25	XPA	40
DNA-PK	43	NEIL1	13	XPB	41
ERCC1	26	NEIL2	46	XPC	55
FANCA	64	OGG1	39	XPD	28
FANCC	50	PARP1	66	XPF	55
FANCD2	38	PCNA	27	XPG	61
FANCE	34	PMS1	49	XRCC1	42
FANCF	17	PMS2	29	XRCC2	38
FANCG	22	POLB	31	XRCC3	39
FEN1	19	POLD	43	XRCC4	83
Ku70	22				
				Total	**2,783**

2.3. *Selection of tag SNPs*

Many different methods have been proposed for selecting tag SNPs [12-16]. Pairwise methods offer straightforward analysis, but fail to exploit long-range haplotype structure. We have developed a tagging approach—called Tagger—that combines the simplicity of pairwise methods with the potential efficiency gains of multi-marker approaches [17].

In this study, we focus specifically on *common* variants with a frequency of ≥ 5%, given the limited ascertainment of less common SNPs in this data set. We picked tags from the CEPH samples as the reference panel so that all observed common variants are captured with $r^2 \geq 0.8$. Use of this threshold has become common practice in the field [15].

Tagging was performed in two modes: (*a*) by a greedy pairwise approach, in which every common allele is captured by a single tag at the prescribed r^2 threshold [15], and (*b*) by aggressively searching for specific multi-marker (haplotype) tests to improve tagging efficiency. We achieve the latter by first picking pairwise tags, and then iteratively dropping tags, one by one, and replacing them with a specific multi-marker predictor (using any of the remaining tag SNPs). That predictor is accepted only if it can capture the alleles originally captured by the discarded tag at the required r^2; otherwise, that provisionally dropped tag is considered indispensable and kept. This multi-marker approach essentially finds an identical set of 1 d.f. tests of association, only now using certain specific haplotypes as effective surrogates for single tag SNPs, thereby requiring fewer tag SNPs for genotyping. To minimize risk of overfitting, tag SNPs within a specified multi-marker test are forced to be in strong LD (defined as LOD > 3) with one another and with the predicted allele.

Tagger thus outputs (1) a list of tag SNPs, and (2) a list of allelic tests, both central for the evaluation of tag SNP transferability.

Tagger is available in the stand-alone application Haploview [18] and as a web server at http://www.broad.mit.edu/mpg/tagger/.

2.4. *Evaluation of tag SNPs*

Given the lists of tag SNPs, we evaluated the coverage of the common variants in the population samples (other than CEPH) by computing the maximum r^2 between the common variants observed in those samples and the specified allelic tests. For pairwise tagging, these tests simply correspond to the genotypes of every tag SNP (as single-marker tests). For multi-marker tagging, tests were specified during tag SNPs selection from the reference panel. (Importantly, in the evaluation of tag SNPs, we do not allow ourselves to derive better allelic tests by looking at LD patterns in the population sample under evaluation.)

3. Results

3.1. *Selection of tag SNPs*

To mimic how investigators will be using the HapMap resource, we used the CEPH samples as the reference panel for picking tag SNPs. For all 61 loci, we required all common variants (≥5%) observed in the reference panel to be captured at $r^2 \geq 0.8$.

We picked a total of 718 tag SNPs by pairwise tagging, and 631 tag SNPs when we allowed Tagger to form multi-marker predictors in place of single-marker tests (Table 2). For both tagging approaches, the mean r^2 for all common alleles (in the reference panel) was 0.97, and the minimum r^2 was 0.86 (these are averages over all 61 loci).

Table 2. Tag SNPs picked from CEPH as the reference panel. The mean and minimal r^2 are averages over all 61 loci studied.

Method	Number of tag SNPs	Mean r^2	Minimum r^2
Pairwise tagging	718	0.97	0.86
Multi-marker tagging (specified haplotype tests)	631	0.97	0.86

This suggests that a nontrivial boost in genotyping efficiency can be achieved by multi-marker tagging, exploiting the underlying haplotype structure, in contrast to pairwise tagging which relies solely on single-marker relationships between SNPs.

We note that the efficiency gain between pairwise and multi-marker tagging observed here (~12%) is significantly lower than that typically obtained in broader genomic regions such as the data from the HapMap-ENCODE project [17]. It is not uncommon for distant (> 100 kb) markers to be in strong LD and to form haplotypes that proxy for other SNPs. Since this study was performed on multiple genes (with an average span of 94 kb), overall efficiency was reduced compared to tagging in large contiguous regions of the genome.

3.2. *Evaluation of tag SNPs*

Having picked tag SNPs and defined statistical tests from the CEPH reference panel, we evaluated the performance of pairwise tagging in terms of the r^2 at which common variants are captured in each of the other six population samples (AA, HA, LA, JA, CH and WH). For every locus, we computed the percentage of common SNPs captured at $r^2 \geq 0.2$, 0.5 and 0.8 as well as the mean r^2 and minimum r^2. We present these metrics as averages over all 61 loci (Table 3).

Table 3. Coverage of common (≥5%) SNPs in six population samples by pairwise tag SNPs picked in CEPH as the reference panel. Values are averages over all 61 loci studied.

Population sample	Number of common (≥5%) SNPs	Percentage of common SNPs captured at $r^2 \geq$ 0.2	0.5	0.8	Mean r^2	Minimum r^2
AA	2347	88.8%	69.3%	50.4%	0.68	0.06
HA	2196	97.2%	92.7%	85.3%	0.90	0.45
LA	2273	97.9%	93.9%	80.5%	0.88	0.40
JA	2028	95.9%	92.4%	82.3%	0.88	0.33
CH	2030	97.0%	91.7%	79.2%	0.87	0.37
WH	2191	98.6%	95.8%	87.3%	0.92	0.51

Most importantly, coverage of common alleles in the HA, LA, JA, CH and WH samples appears to be robust. In the WH samples (which is most "similar" from a population-genetic standpoint), we observe a marginal drop in mean r^2 from 0.96 (in the CEPH reference panel) to 0.92. Between 80% and 87% of common variants are captured at $r^2 \geq 0.8$ in the non-African samples, and the overwhelming majority (> 92%) are captured at $r^2 \geq 0.5$. Of course, not all alleles are captured equally well: a small fraction (3-4%) of the common alleles in the non-African samples are not captured at all ($r^2 < 0.2$) by any of the allelic tests.

Not surprisingly, fewer common variants are captured in the AA samples: only 50% of the common alleles are captured with $r^2 \geq 0.8$; and the mean r^2 dropped down to 0.68. This can be attributed to the significantly lower extent of LD in African populations [19]. We emphasize that in practice, however, investigators will likely pick tag SNPs from a reference panel that is more representative (such as the HapMap samples of Yoruba from Ibadan, Nigeria). Due to greater genetic diversity and less LD, more tag SNPs will be required for capturing common variation in African-derived samples.

We next evaluated the performance of the multi-marker predictors on the basis of the 631 tag SNPs picked by Tagger. Again, we computed the percentage of common alleles captured at $r^2 \geq 0.2$, 0.5 and 0.8 as well as the mean r^2 and minimum r^2 (Table 4). The coverage with our haplotype-based approach is roughly equivalent to that of pairwise tagging but require fewer tag SNPs. Thus, the multi-marker approach in Tagger is not only more efficient than a pairwise tagging method, but the specified haplotype predictors capture common variation in the other (non-African) population samples almost as well as the single-marker tests (Table 3).

Table 4. Coverage of common (≥5%) SNPs in six population samples by tag SNPs picked and specified multi-marker tests defined in CEPH as the reference panel. Values are averages over all 61 loci studied.

Population sample	Number of common (≥5%) SNPs	Percentage of common SNPs captured at $r^2 \geq$ 0.2	0.5	0.8	Mean r^2	Minimum r^2
AA	2347	88.8%	66.6%	46.7%	0.66	0.06
HA	2196	97.3%	92.3%	83.7%	0.89	0.45
LA	2273	97.8%	92.8%	78.8%	0.86	0.38
JA	2028	95.3%	90.9%	79.1%	0.86	0.32
CH	2030	96.9%	90.5%	76.7%	0.85	0.35
WH	2191	98.6%	95.6%	87.0%	0.91	0.51

4. Discussion

Using empirical genotype data in genes of medical relevance, we find that (*a*) tag SNPs picked in the CEPH samples provide good coverage of common variants in the non-African population samples studied here; and (*b*) specified haplotype tests can improve overall tagging efficiency with minimal loss of coverage.

Even though the fine details of LD patterns are known to differ between population samples, these results demonstrate that tag SNPs chosen from the CEPH reference panel (used in HapMap) are able to effectively capture the majority of common alleles in other (non-African) samples in a cost-effective manner.

Although this work focuses only on a limited set of parameters, we believe that the results presented here are fairly representative of the practical decisions that investigators face in the design of tag SNP sets.

Our tagging approach, like that of others, is explicitly not based on haplotype "blocks," hotspots of recombination, or other features of empirical data. We agree with commentators who have noted that while blocks may be a convenient descriptor of genotype data, a block-by-block approach ignores the sometimes substantial correlations between blocks, and as not all SNPs are contained within blocks, block-based selection of tag SNPs is likely to give inadequate coverage [20].

While many different approaches exist for selecting tag SNPs from a reference panel and for performing tests, these concepts are sufficiently intertwined and should be considered as a unit. Tag SNPs may perform well under the particular analytical strategy for which they were designed, but not under another. We do not address in this study the tradeoff between the amount of required genotyping and statistical power to detect an association in an actual disease study. We have addressed these issues elsewhere [17].

Acknowledgments

We would like to thank Melissa A. Frasco for laboratory assistance and Xin Sheng, John T. Casagrande and David Van Den Berg for technical support. We would also like to acknowledge Laurence N. Kolonel, Loïc Le Marchand, Ronald K. Ross, Mimi C. Yu and Juan-Min Yuan for providing the samples for this study.

References

1. Altmuller, J., et al., *Genomewide scans of complex human diseases: true linkage is hard to find.* Am J Hum Genet, 2001. **69**(5): p. 936-50.
2. Lohmueller, K.E., et al., *Meta-analysis of genetic association studies supports a contribution of common variants to susceptibility to common disease.* Nat Genet, 2003. **33**(2): p. 177-82.
3. The International HapMap Consortium, *The International HapMap Project.* Nature, 2003. **426**(6968): p. 789-96.
4. Johnson, G.C., et al., *Haplotype tagging for the identification of common disease genes.* Nat Genet, 2001. **29**(2): p. 233-7.
5. Hinds, D.A., et al., *Whole-genome patterns of common DNA variation in three human populations.* Science, 2005. **307**(5712): p. 1072-9.
6. Nejentsev, S., et al., *Comparative high-resolution analysis of linkage disequilibrium and tag single nucleotide polymorphisms between populations in the vitamin D receptor gene.* Hum Mol Genet, 2004. **13**(15): p. 1633-9.
7. Ahmadi, K.R., et al., *A single-nucleotide polymorphism tagging set for human drug metabolism and transport.* Nat Genet, 2005. **37**(1): p. 84-9.
8. Mueller, J.C., et al., *Linkage disequilibrium patterns and tagSNP transferability among European populations.* Am J Hum Genet, 2005. **76**(3): p. 387-98.
9. Kolonel, L.N., et al., *A multiethnic cohort in Hawaii and Los Angeles: baseline characteristics.* Am J Epidemiol, 2000. **151**(4): p. 346-57.
10. Hankin, J.H., et al., *Singapore Chinese Health Study: development, validation, and calibration of the quantitative food frequency questionnaire.* Nutr Cancer, 2001. **39**(2): p. 187-95.

11. Excoffier, L. and M. Slatkin, *Maximum-likelihood estimation of molecular haplotype frequencies in a diploid population.* Mol Biol Evol, 1995. **12**(5): p. 921-7.
12. Stram, D.O., et al., *Choosing haplotype-tagging SNPs based on unphased genotype data using a preliminary sample of unrelated subjects with an example from the Multiethnic Cohort Study.* Hum Hered, 2003. **55**(1): p. 27-36.
13. Weale, M.E., et al., *Selection and evaluation of tagging SNPs in the neuronal-sodium-channel gene SCN1A: implications for linkage-disequilibrium gene mapping.* Am J Hum Genet, 2003. **73**(3): p. 551-65.
14. Meng, Z., et al., *Selection of genetic markers for association analyses, using linkage disequilibrium and haplotypes.* Am J Hum Genet, 2003. **73**(1): p. 115-30.
15. Carlson, C.S., et al., *Selecting a maximally informative set of single-nucleotide polymorphisms for association analyses using linkage disequilibrium.* Am J Hum Genet, 2004. **74**(1): p. 106-20.
16. Lin, Z. and R.B. Altman, *Finding haplotype tagging SNPs by use of principal components analysis.* Am J Hum Genet, 2004. **75**(5): p. 850-61.
17. de Bakker, P.I.W., et al., *Efficiency and power in genetic association studies.* Nat Genet, 2005. **In the press.**
18. Barrett, J.C., et al., *Haploview: analysis and visualization of LD and haplotype maps.* Bioinformatics, 2005. **21**(2): p. 263-5.
19. Reich, D.E., et al., *Linkage disequilibrium in the human genome.* Nature, 2001. **411**(6834): p. 199-204.
20. Wall, J.D. and J.K. Pritchard, *Haplotype blocks and linkage disequilibrium in the human genome.* Nat Rev Genet, 2003. **4**(8): p. 587-97.

A TOOL FOR SELECTING SNPS FOR ASSOCIATION STUDIES BASED ON OBSERVED LINKAGE DISEQUILIBRIUM PATTERNS

FRANCISCO M. DE LA VEGA[†], HADAR I. ISAAC, CHARLES R. SCAFE
Applied Biosystems, 850 Lincoln Centre Dr., Foster City, CA 94404, USA

The design of genetic association studies using single-nucleotide polymorphisms (SNPs) requires the selection of subsets of the variants providing high statistical power at a reasonable cost. SNPs must be selected to maximize the probability that a causative mutation is in linkage disequilibrium (LD) with at least one marker genotyped in the study. The HapMap project performed a genome-wide survey of genetic variation with about a million SNPs typed in four populations, providing a rich resource to inform the design of association studies. A number of strategies have been proposed for the selection of SNPs based on observed LD, including construction of metric LD maps and the selection of haplotype tagging SNPs. Power calculations are important at the study design stage to ensure successful results. Integrating these methods and annotations can be challenging: the algorithms required to implement these methods are complex to deploy, and all the necessary data and annotations are deposited in disparate databases. Here, we present the SNPbrowser™ Software, a freely available tool to assist in the LD-based selection of markers for association studies. This stand-alone application provides fast query capabilities and swift visualization of SNPs, gene annotations, power, haplotype blocks, and LD map coordinates. Wizards implement several common SNP selection workflows including the selection of optimal subsets of SNPs (e.g. tagging SNPs). Selected SNPs are screened for their conversion potential to either TaqMan® SNP Genotyping Assays or the SNPlex™ Genotyping System, two commercially available genotyping platforms, expediting the set-up of genetic studies with an increased probability of success.

1. Introduction

One problem researchers face when designing and executing human genetic studies with single nucleotide polymorphisms (SNPs) is the difficult task of selecting the most suitable set of the variants for the goal at hand in a cost-effective manner. This task is time-consuming and overwhelming due to the millions of SNPs currently listed on the public databases and the fact that relevant information is often distributed among multiple repositories which sometimes are difficult to access or the access is slow due to bandwidth and server load issues. Often this requires advanced algorithm development.

[†] To whom correspondence should be addressed. Email: delavefm@appliedbiosystems.com.

Furthermore, once a set of SNPs is selected, researchers lack a rapid way to obtain reliable, predictable assays for multiple SNPs that work together under the same experimental conditions.

To overcome these barriers, we developed the SNPbrowser Software, a freely available tool providing an intuitive interface to search a stand alone, embedded database that contains detailed information on millions of validated SNPs. Included in this SNP collection are over a million genome-wide distributed SNPs that the International HapMap Project recently genotyped,[2] as well as 160,000 intragenic SNPs previously validated by us in four populations using TaqMan SNP Genotyping Assays.[3,5] The depth of SNP and genomic information in the database together with the swift visual interface and embedded selection algorithms provides researchers greater flexibility when designing associations studies with an increased probability of success.

2. Methods

2.1. *SNP genotype and annotation data*

We previously genotyped DNA samples from 45 African-Americans, 46 Caucasians, 45 Chinese, and 45 Japanese, all unrelated individuals.[4] Over 160,000 TaqMan® SNP Genotyping Assays were used to genotype these samples. For the HapMap project dataset, we utilized genotypes from the public release 16 (Phase I data freeze) ignoring the children on the CEU and YRB trios[2]. Only SNPs having a unique mapping location on the NCBI b35 assembly and a minor allele frequency (MAF) of >5% were considered for further analysis. Gene annotation including HUGO names, exon and intron boundaries of all reported (RefSeq NM) and predicted (XM) transcripts were obtained from NCBI Entrez. Transcripts were coalesced into "supertranscript" constructs with boundaries delimited by the coordinates of the first and last base transcribed.

2.2. *SNP screening for genotyping assay development*

All the SNPs in our database were passed through the high-throughput design pipelines for both TaqMan SNP Genotyping Assays, and the SNPlex Genotyping System[6]. SNPs that passed the design rules of either platform and thus are candidates for the development of good assays where flagged and subsequent analyses were performed separately for each subset. In the case of TaqMan, assay designs (primers and probes) were uploaded into our TaqMan Predesigned database for immediate commercial availability. In the case of the

SNPlex System, since it is a multiplexed assay format, we perform a pre-screen for the "single-plex" part of the pipeline; when users submit an actual design request, a few SNPs may still be lost at the final design due to multiplexing rules.[6]

2.3. *Analysis of linkage disequilibrium*

We constructed metric maps scaled to the strength of LD that can guide the selection of SNPs for association studies. Linkage disequilibrium units (LDUs) define a metric coordinate system where locations are additive and distances are proportional to the allelic association between markers[10]. The LDMAP software v0.9 (available at: http://cedar.genetics.soton.ac.uk/public_html/helpld.html) was applied separately to each chromosome and population to construct the corresponding LDU maps. Haplotype blocks were estimated by a rule-based algorithm which uses the D' confidence interval,[7] optimized through a dynamic programming algorithm,[11] or dynamically by LDUs, as user-defined intervals with a very small distance in this coordinate system (the default value is 0.3, which returns similar blocks to previous methods[7]).

2.4. *Selection of minimum informative subsets of SNPs*

We utilized three algorithms[9] to select minimum informative subsets of SNPs or tag-SNPs: (i) simple genotype correlation between samples (allowing for one item of missing data); (ii) pair-wise r^2; and (iii) haplotype R^2. First, minimum sets of tag SNPs were selected on a chromosome-wide basis at three thresholds of pair-wise r^2 or haplotype R^2 through the use of a block-free dynamic programming algorithm framework.[8] The output of these calculations was included in the software database. Alternatively, we implemented an on-the-fly selection of tagging SNPs where the user has a greater choice of parameters which is suitable for smaller regions.

2.5. *Power calculations for case/control studies*

We calculated power for a fixed sample size of cases and controls on a per gene basis. For each gene, power is calculated using a haplotype based test, for each of the common haplotypes in the window, and entering in the calculation the empirically observed average LD on the gene region. Using a multiplicative genetic model with relative risk ratio of 3 and prevalence of 1.5%, power is calculated for each haplotype and a frequency weighted average is provided as the summary. This is repeated separately for each population, for three settings of sample sizes of cases and controls (250/250, 500/500, 1000/1000), and assuming a disease allele frequency of either 10 or 20%. The resulting estimated power is visualized using a color scale ranging from 0.5 to 1.0 displayed as a background to each gene region.[4]

2.6. *Downloading, installing, and updating SNPbrowser*

SNPbrowser is developed using the Microsoft® Visual C++ IDE and compiler, and currently is available only as a native Windows application requiring a system with 512 Mbytes of RAM. However, the software can be readily used with the MacOS platform with Microsoft Virtual PC, a commercially available emulation environment. The latest version of SNPbrowser is always freely available for download at http://www.allsnps.com/snpbrowser/. Once installed, the software checks for updated versions either automatically or manually.

3. Results

3.1. *SNPbrowser Software embedded database*

When SNPbrowser is launched, the user has the option to select which reference database they want to utilize: either the maps and data derived from the HapMap Project[2], or the gene-centric maps obtained by Applied Biosystems (AB) by typing 160,000 SNPs in four populations[3,5]. After the user selection, SNP and gene annotations, their physical coordinates on the NCBI b35 assembly, genotypes on the corresponding reference populations, and the results of a series of LD analysis, tagging SNP, and power calculations pipelines performed offline are loaded from a set of binary files distributed with the application into a highly compressed and indexed embedded database maintained in memory. The SNPbrowser database also includes a set of *metric* LD maps[10], which are empirically derived from the patterns of allelic association observed on the hundreds of millions of genotypes analyzed, and provide information on how to best position SNPs across the genes or regions of interest in a study[1].

3.2. *Visualization and query tools*

The SNPbrowser main interface is a visualization panel consisting of a chromosome map viewer representing the location in the physical map of SNPs, and their relationship to annotated human genes and exons. Researchers studying a particular gene or a set of genes can easily pan and zoom to the region of the genome of interest. For example, investigators studying a candidate gene for type 2 diabetes, calpain 10, can type the gene HUGO name (CAPN10) into the search box and quickly see that the gene spans about 31 kb and just 0.5 LDUs in the Caucasian population. The intron/exon structure of the gene is readily apparent, as is the haplotype block structure and the location of SNPs along the chromosomal axis (Fig. 1).

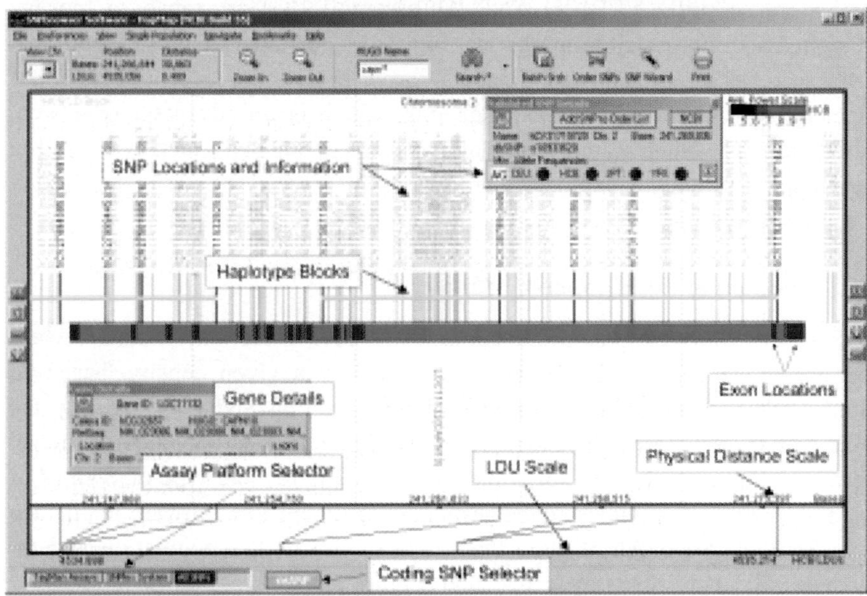

Figure 1. The SNPbrowser Software allows visualization of extensive gene and genomic information, including the physical and LD maps, intron/exon structure, the locations and allele frequencies of SNPs, and putative haplotype blocks in four different populations.

Vertical blue lines represent SNPs validated, either by the HapMap project or AB and for which genotypes are available, while grey lines represent SNPs corresponding to over 5 million putative SNP deposited at public databases. The user can select between by clicking the bottom tabs to display the SNPs which can be developed as assays for the SNPlex Genotyping System or as TaqMan SNP Genotyping assays.[6] By clicking the "All SNPs" tab, the union of both sets is displayed.

The vertical lines representing SNPs connect to their locations on the LDU coordinates shown on the bottom horizontal axis, in many cases coalescing together into a single position when LD is extensive (i.e. a haplotype block[7]). By clicking and dragging with the mouse any interval can quickly be measured in both base pairs or interpolated LDUs (see distance box upper left, Fig 1).

Finally, overall statistical power of the full SNP map, estimated per gene for a pre-selected genetic model, assumed disease allele frequency, and sample size,[4] is shown color coded within the intronic regions (scale is visible at the upper right corner).

Searches can be performed with a variety of terms, including gene name, RefSeq transcript ID, NCBI ID, SNP ID, assembly base-pair range, or Linkage Mapping Set microsatellite marker set intervals. For most of these identifiers, batch searches are also allowed. Since SNPbrowser database is loaded into RAM memory, searches are almost instantaneous, which is an advantage over web-based tools. The batch search feature allows users to quickly search genes in big candidate lists and to explore interactively the results of various selection scenarios.

3.3. *Selection of Evenly Spaced Markers*

SNPbrowser Software provides a number of SNP selection "wizards" where researchers can define a region and select SNPs at a given density, based on either LDU or kilobase (kb) distances. When selecting SNPs by spacing, the wizards also allow researchers to prioritize the SNPs that are included in the set based on criteria such as minor allele frequency (MAF) and type of SNP. For example, with a few clicks, researchers can configure the software to include only SNPs with a MAF of more than 10% in the CEPH population and for which a validated SNP assay is available (Figure 2).

Another typical use case for study design is the candidate region study, where the researchers already performed a linkage study and the goal is to perform fine mapping of an implicated chromosomal region to find the disease gene. For example, choosing an arbitrary region on chromosome 4, and searching for validated SNPs spaced at least 20 kb across the region, the SNPbrowser Software identified 33 appropriate SNPs and indicated that it was possible to achieve this spacing across the entire region (Figure 2). If validated SNPs had not been available, a red indicator bar would replace the green indicator bar in the bottom right-hand corner of the read-out window. The slider allows researchers to modify the spacing or MAF parameters to quickly visualize the level of coverage that is possible in the region given their other requirements. Alternatively, SNPs can be selected to try to achieve an even spacing of 0.5 LDUs on the metric LD map for CEPH by simply going back and changing the density parameters. In this case the wizard selects only 6 SNPs due to the extensive LD, although there is one interval at the 3'-end of the gene (red bar) where the LDU distance was greater than this value, suggesting the presence of a recombination hotspot around this location. If desirable, the user can request the wizard to fill this "gap" with as many non validated SNPs (gray vertical lines) as required. All this process can be carried out in seconds. Since the selection process is carried out selecting a particular genotyping platform (selected by the platform tabs), additional time is saved by not having to go back and refill gaps created by SNPs that cannot be converted to a given assay format. Furthermore, the SNPbrowser wizard can also take into account SNPs for which the user already developed assays, and fill gaps around them.

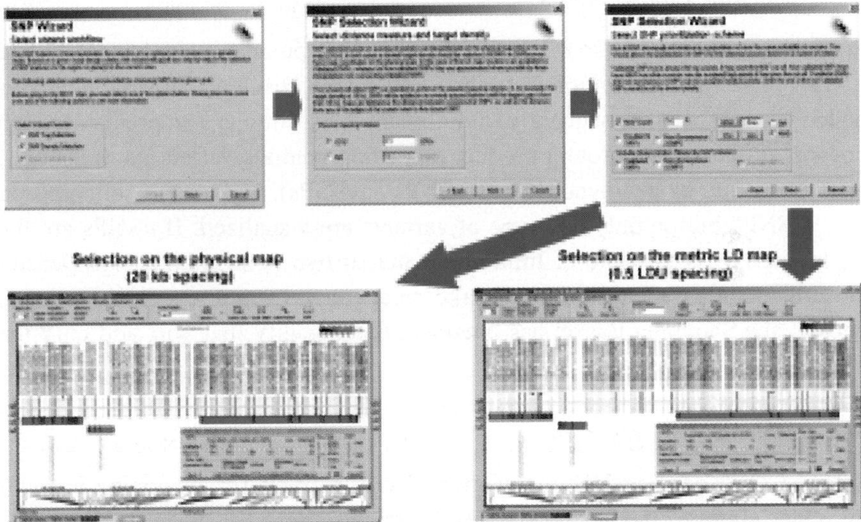

Figure 2. The SNP Selection wizard in density mode allows researchers to find SNPs at even LDU or kb intervals. In this example, a search on chromosome 4 for validated SNPs in the CEPH population with MAF of at least 10% and a gap no larger than either 20 kb or 0.5 LDU was carried out, and generated a set of either 37 (kb) or 6 (LDU) selected SNPs (shown in red).

3.4. *Selection of tagging SNPs*

Because SNPs that show high LD result in chromosomal segments in which a limited number of haplotypes are found in a population (i.e. a haplotype block[7]), it is possible to select a small subset of SNPs that distinguish, or "tag", the common haplotypes previously found in a gene or region. This eliminates a large number of SNPs from the study that would only provide redundant information. In principle, this reduction in markers brings down the cost and time necessary to conduct a study retaining good statistical power.[4] An example of this strategy is provided by inspecting the BRCA1 gene, which is covered by a single, continuous block of LD in all four populations studied (i.e., all SNPs within the gene fall in the same location on the LD map). Therefore, although there are a vast number of SNPs in the gene, and including 37 validated SNPs, the SNPbrowser Software Tagging SNP wizard reveals that only 8 SNPs are actually required to retain most of the haplotype diversity of the gene observed in the reference samples (using a haplotype r^2 metric threshold of 0.99; see Fig. 3).

3.5. *Selecting Coding SNPs*

SNPbrowser Software also makes it easy to include putatively functional coding SNPs (cSNPs) in association studies. SNPs that result in non-synonymous codon changes and consequently, amino-acid substitutions (or premature stop codons) in the gene's protein product that can potentially affect its function[12], also referred to as non-synonymous cSNPs (nsSNPs). By simply clicking on the "nsSNP" button only this type of variants are visualized. If cSNPs are the study focus, it is possible to limit the search at two points. First, the Density Wizard includes a checkbox to make selecting cSNPs the search priority. Second, the Shopping Basket has one-click functionality that will add only the cSNPs to the cart.

3.6. *Implementing the study*

Selected SNPs can be added to a working list of markers (or "shopping basket") by either simply clicking on the results bar of the SNP wizards, manually adding individual markers with the right click option, or by invoking the "shopping basket" window and adding markers from the current view in many forms. There are two separate shopping baskets: one for each TaqMan and SNPlex platforms. In the case of TaqMan, assay availability and previous performance validation is indicated. The contents of each basket can be exported and saved for use in a future session. Finally, once the researcher has identified the ideal set of SNPs for an association study, genotyping reagents can be easily be obtained. The user can also export the list of SNPs from the shopping basket to a text file, including a number of the annotations maintained in the software internal database.

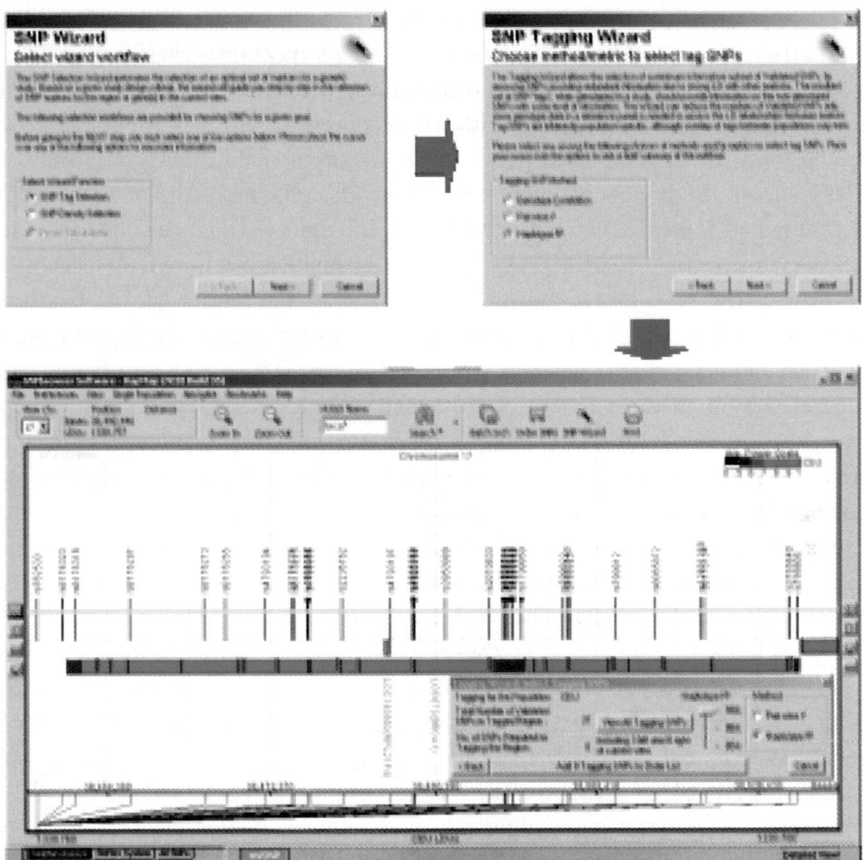

Figure 3. For genes in regions of strong LD, such as the BRCA1 gene, the SNP tagging wizard allows the selection of minimum subsets of SNPs (shown in red) retaining a threshold value of the selected quality metric (either pair-wise r^2 or haplotype R^2).

4. Discussion

Since there is no single SNP selection approach that can serve all the requirements of different types of studies, the SNPbrowser Software offers researchers a choice of methods for picking markers suited to a wide range of objectives and disease characteristics. Two basic paradigms for selecting SNP markers are supported: 1) selection of evenly spaced markers on the physical or metric linkage disequilibrium (LD) maps[10], and 2) selection of non-redundant subsets of haplotype "tagging" SNPs.[9] Furthermore, the tool immediately indicates the SNPs for which genotyping assays are viable and available from commercial sources. This means that researchers can get promptly started in their study after identifying an optimal SNP set. Previously, identifying the most efficient and highly informative SNP set for a multi-megabase region (e.g. a candidate region from a previous linkage study performed with microsatellites) was extremely time-consuming. With the SNPbrowser wizards it only takes a few seconds, for example, to get a list of evenly spaced, highly-informative SNPs across the region of interest either on the physical (kb) or metric LD (LDU) maps. A metric LD map, expressed in LD units (LDUs) calculated by the LDMAP software,[10] places SNPs on a coordinate system where distances between SNPs are additive and directly related to the degree of LD between them. For example, SNPs in perfect LD (completely correlated) have zero distance between them, whereas SNPs with no significant correlation are separated by over three LDUs in this map. Analogous to the genetic map expressed in centi-Morgans commonly used for selecting markers for linkage studies in families, the LD map can be used to efficiently position markers for population-based disease association studies.[1]

Normally, the HapMap database would be preferred due to the depth of coverage, but often the AB maps could be useful, for example, if the study involves African-Americans (The HapMap Project did not genotype samples of this population). Although it is always preferable to utilize validated SNPs when designing genetic studies, there may be circumstances when it would be desirable to include SNPs present in the public databases but that have not been validated, e.g. by the HapMap project. SNPbrowser allows displaying the complete SNP complement for the visible region that can be converted to commercially available genotyping assays, whether validated or not, making it easy to select additional SNPs that can be used to fill gaps left by the validation projects.

Sometimes nsSNPs are included because they are referenced in the literature, and other times adding nsSNPs to the study may increase its power because in some instances an nsSNP can be indeed a causative variant for the phenotype under investigation. It is important to note that non-coding SNPs such as those in regulatory regions or splice junctions can also influence the trait of interest and thus cannot be completely ignored, but these are difficult to identify or predict. Further, if their penetrance is high, cSNPs may not occur in sufficient frequency in the population to be informative in a study with a typical sample size. Ultimately, most researchers find that it is most productive to include a mix of nsSNPs and surrogate marker SNPs with high minor allele frequencies.

In summary, SNPbrowser is a free tool that allows researchers to easily select SNPs for genetic association or other types of studies involving human

SNPs. Its main advantages include: Ease of use; swift interaction and searches; informative visualization; intuitive wizards that automate the most common selection workflows; no need to be online to access the data; completeness in terms of data and selection algorithms, enabling rapid experimental cycles by considering an assay platform conversion potential from the beginning. The software also includes extensive online help describing in detail additional features and facilities that due to length limitations cannot be discussed in this manuscript. The extensive and detailed information available through the SNPbrowser Software solves many of the major challenges that researchers face when designing human association studies, including visualizing complete genomic information in their region or gene of interest, leveraging the extensive reference genotype datasets becoming available from the HapMap project, identifying the best set of SNPs for their studies, and easily obtaining reliable assays that correspond to those SNPs.

Acknowledgments

We are very grateful to Andrew Collins, for providing the LDMAP software, Bjarni Halldórsson and Ross Lippert, who provided the block-free tagging SNP selection pipeline and haplotype phasing code, and Derek Gordon, who provided power calculation algorithms. We acknowledge the valuable support and feedback provided by Joanna Curlee, Pius Brzoska, Dennis Gilbert, Toinette Hartshorne, Fiona Hyland, Michael Rhodes, Katherine Rogers, Leila Smith, Eugene Spier, Rob Tarbox, Fenton Williams, and Trevor Woodage.

References

1. Collins, A., Lau, W. and De La Vega, F.M. *Hum Hered* 2004; **58**:2-9.
2. Consortium, T.I.H. *Nature* 2003; **426**:789-796.
3. De La Vega, F.M., Dailey, D., Ziegle, J., Williams, J., Madden, D. and Gilbert, D.A. *Biotechniques* 2002; **Suppl**:48-50, 52, 54.
4. De La Vega, F.M., Gordon, D., Su, X., Scafe, C., Isaac, H., Gilbert, D. and Spier, E.G. *Hum Hered* 2005; **60**:In Press.
5. De La Vega, F.M., Isaac, H., Collins, A., Scafe, C.R., Halldorsson, B.V., Su, X., Lippert, R.A., Wang, Y., Laig-Webster, M., Koehler, R.T., et al. *Genome Res* 2005; **15**:454-462.
6. De la Vega, F.M., Lazaruk, K.D., Rhodes, M.D. and Wenz, M.H. *Mutat Res* 2005; **573**:111-135.
7. Gabriel, S.B., Schaffner, S.F., Nguyen, H., Moore, J.M., Roy, J., Blumenstiel, B., Higgins, J., DeFelice, M., Lochner, A., Faggart, M., et al. *Science* 2002; **296**:2225-2229.

8. Halldorsson, B.V., Bafna, V., Lippert, R., Schwartz, R., De La Vega, F.M., Clark, A.G. and Istrail, S. *Genome Res* 2004; **14**:1633-1640.
9. Halldorsson, B.V., Istrail, S. and De La Vega, F.M. *Hum Hered* 2004; **58**:190-202.
10. Maniatis, N., Collins, A., Xu, C.F., McCarthy, L.C., Hewett, D.R., Tapper, W., Ennis, S., Ke, X. and Morton, N.E. *Proc Natl Acad Sci U S A* 2002; **99**:2228-2233.
11. Schwartz, R., Halldorsson, B.V., Bafna, V., Clark, A.G. and Istrail, S. *J Comput Biol* 2003; **10**:13-19.
12. Thomas, P.D. and Kejariwal, A. *Proc Natl Acad Sci U S A* 2004; **101**:15398-15403.

DATA SIMULATION SOFTWARE FOR WHOLE-GENOME ASSOCIATION AND OTHER STUDIES IN HUMAN GENETICS

SCOTT M. DUDEK, ALISON A. MOTSINGER, DIGNA R. VELEZ, SCOTT M. WILLIAMS, MARYLYN D. RITCHIE

Center for Human Genetics Research, Vanderbilt University, 519 Light Hall, Nashville, TN 37232, USA

Genome-wide association studies have become a reality in the study of the genetics of complex disease. This technology provides a wealth of genomic information on patient samples, from which we hope to learn novel biology and detect important genetic and environmental factors for disease processes. Because strategies for analyzing these data have not kept pace with the laboratory methods that generate the data it is unlikely that these advances will immediately lead to an improved understanding of the genetic contribution to common human disease and drug response. Currently, no single analytical method will allow us to extract all information from a whole-genome association study. Thus, many novel methods are being proposed and developed. It will be vital for the success of these new methods, to have the ability to simulate datasets consisting of polymorphisms throughout the genome with realistic linkage disequilibrium patterns. Within these datasets, we can embed genetic models of disease whereby we can evaluate the ability of novel methods to detect these simulated effects. This paper describes a new software package, genomeSIM, for the simulation of large-scale genomic data in population based case-control samples. It allows for single SNP, as well as gene-gene interaction models to be associated with disease risk. We describe the algorithm and demonstrate its utility for future genetic studies of whole-genome association.

1. Introduction

The identification and characterization of susceptibility genes for common complex human diseases, such as cardiovascular disease, is a difficult challenge for genetic epidemiologists. This is because many disease susceptibility genes exhibit effects that are partially or solely dependent on interactions with other genes. In addition, selection of the appropriate candidate genes limits our ability to identify novel genetic factors associated with disease. Whole-genome association has been proposed as a solution to these problems; however, the appropriate analytical methods for this type of data are unknown. To deal with this issue, many groups, including our own, are in the process of developing new computational approaches for the analysis of whole-genome association studies, but without a priori knowledge of the genetic model underlying the phenotype it is unclear whether a given method is accurate.

Strategies for analyzing datasets on the scale of whole genome association studies data have not kept pace with the laboratory methods that generate the data. Because of this it is unlikely that technological advances will immediately lead to an improved understanding of the genetic contribution to common human disease and drug response. Currently, no single analytical method will allow us to extract all information from a whole-genome association study. In fact, no single method can be optimal for all datasets, especially if the genetic architecture for disease is substantially different.

One way to better design analytical protocols is to have datasets with known answers, but this is not possible using real data. When real data are used to test new methods, and significant results are found, it is impossible to know if they are false positives or true positives. Similarly, if nothing significant is detected, one cannot know if this is a lack of power, or the data had no true signal. Thus, it will be vital for the success of genome-wide association methods, to have the ability to simulate datasets consisting of polymorphisms throughout the genome on the scale of what is technically feasible. Having simulated data allows one to evaluate whether a methodology can detect known effects, and if the simulations are well-designed one can potentially embed a variety of genetic models of disease, making the evaluation of methods robust to genetic architecture.

Data simulations are often criticized because they are much cleaner than real data. However, simulating data remains an important component of most new methods development projects. To this end, any advances to improve the complexity of the data simulations will permit investigators to better assess new analytical methods. The present study was motivated by this lack of appropriately complex simulated data for association studies.

Several data simulation packages are currently available for family based study designs. SIMLINK[1,2], SIMULATE, and SLINK[3] will simulate pedigrees from an existing dataset. SIMLA[4] is a very nice software package for simulating both linkage and association in pedigree data. However, it does not allow for epistasis models or population-based simulations. Coalescent-based methods[5] have been used for population based simulation in genetic studies, however they do not allow for the tracking of ancestral information. In recent years, forward-time population simulations have been developed including easyPOP[6], FPG[7], and simuPOP[8]. simuPOP is the newest simulation package. It performs forward-time population simulations and allows the user to manipulate the evolutionary features. simuPOP is implemented in Python and provides flexibility for the user to run interactively using a Python shell or writing batch files[8]. The main weakness of simuPOP is the inability to simulate data based on complex gene-gene interaction penetrance functions. In addition, the programming environment is specific to Python, therefore, may not be user-friendly for all users. This paper describes a new software package, genomeSIM, for the simulation of large-scale genomic data in population based case-control samples. It is a forward-time population simulation algorithm that allows the user to specify many evolutionary parameters and control evolutionary processes. It allows for single SNP, as well as gene-gene interaction models to be associated with disease risk. We describe the algorithm and demonstrate its utility for future genetic studies of whole-genome association.

2. Methods

2.1. *Algorithm*

genomeSIM utilizes two different methods to generate datasets. An initial population can be generated on the basis of allele frequencies of the SNPs and then further generations are created by crossing the members of successive generations. The simulator assigns affection status only after a specified number of generations. Alternatively, the simulator can construct a case-control dataset by generating individuals as above, assigning affection status, and selecting cases and controls until the dataset is complete.

Fig. 1 illustrates the general steps involved in producing a simulated dataset utilizing successive generations. As a first step, genomeSIM establishes the genome based on the parameters passed to it. The total number of SNPs is not limited except by hardware considerations. The user specifies the number of SNPs per gene and the total number of genes in the genome. The simulator randomly determines the number of SNPs per gene based on the minimum and maximum parameters. The simulator then randomly determines the recombination fraction between adjacent SNPs within each gene based on maximum and minimum recombination fraction parameters. The recombination fraction between any pair of SNPs is independent of the recombination fraction between other pairs of SNPs within a given gene. Similarly, recombination fractions between genes are independent. Thus, all recombination fractions are random and independent. SNPs are unlinked across genes. Finally, the allele frequencies are randomly set for each SNP based on preset maximum and minimum allele frequency parameters. For all these parameters, when the minimum is set equal to the maximum, the values across the simulated genome will be identical. Specific SNPs can also be set so that the disease SNPs allele frequencies will match the expected frequencies for the model used.

genomeSIM then generates an initial population based on the genome established in the previous step. Each individual in the population has two binary chromosomes. For each SNP in the genome, the simulator randomly assigns an allele to each chromosome based on the allele frequencies of the SNP. The dual chromosome representation allows for an efficient representation of the genome and for crossover between chromosomes during the mating process. The genotype at any SNP can be determined simply by adding the values of the two chromosomes at that position. As a result, the genotypes range from 0 to 2 at any SNP.

The initial population forms the basis for the second generation in the simulation. For each cross two individuals are randomly selected with replacement to be the parents for a member of the new generation. Each parent contributes one haploid genome to the child. genomeSIM creates the gametic genotype by recombining the parent's chromosomes. The total number of individuals in each population is constant so the number of crosses conducted equals the number of individuals in the population for each generation.

A crossover is conducted as follows. genomeSIM selects one chromosome to be the start chromosome and begins copying allele values from that chromosome into the new chromosome. At every interval between SNPs, the simulator checks the recombination fraction against a randomly generated number. When the number is less than or equal to the fraction, the simulator switches chromosomes (assuming independent assortment) and begins taking allele values from the second chromosome. The simulator continues to check each interval and copies the allele values for the current chromosome until it reaches the end of the genome or another crossover takes place.

genomeSIM continues producing generations for the number specified and then assigns affection status to the final generation. Affection status is determined by the penetrance table for the simulation. To determine status, the simulator determines the genotype of the individual at the disease SNPs. The simulation then determines the penetrance for that genotype and generates a random number to determine if this individual is affected.

Alternatively, genomeSIM can produce the final dataset by producing individuals using the allele frequencies. The simulator's goal in this case is to generate the desired number of cases and controls. Each individual is checked against the penetrance table and then kept if there are not enough individuals with that affection status in the dataset. Additional individuals with that status are discarded. For example, if 500 cases and controls are needed, the simulator will take the first 500 controls that are generated but will then ignore any more while continuing to select the cases as needed. The simulator initially only generates the disease SNPs for each individual. If the simulator then needs to keep the individual based on its status, the rest of the alleles for the individual are generated.

genomeSIM can produce genetic heterogeneity by utilizing multiple penetrance tables. Each table is used for a portion of the final population. Datasets can also be produced with no disease model. If no penetrance table is used, then the individual has an equal chance of being a case or control. In addition, the simulator can generate phenocopies by assigning a fraction of the unaffected population to be affected at random. Finally, the simulator can introduce genotyping errors into the final population. The rate determines the expected number of errors per SNP. For each SNP, individuals are randomly selected and their genotypes are adjusted in a direction specified by the user if possible. For example, a selected individual may have a genotype of 2 and the error direction is specified as -1. In this case, the reported genotype for the individual will be 1. If the individual had a genotype of 0, no error would occur and another individual would be selected.

2.2. *Implementation*

genomeSIM is written in **ANSI-C++** and compiled using the GNU compiler into a library that can be linked to programs to generate datasets without the need for intermediate files. For the analyses done in this paper, the library was linked to a simple driver program that created input files for the Multifactor Dimensionality Reduction (MDR)[9] analysis software. The library provides simulation classes to be accessed by the main program for simulating both generational-based and frequency-based datasets.

The analysis can be run using functions in the library classes or the library can accept a configuration file as input for easy linkage with existing programs. The simulator accepts keywords and values as the configuration format. Table 1 displays the keywords that control the dataset production. Some keywords (POPSIZE, GENES, NUMGENS, MAXSNP, MINSNP, MAXRECOMB, MINRECOMB) are only used when simulating multiple generations to produce the final population. Other keywords (AFFECTED, UNAFFECTED, SIMLOCI) are only used when simulating a case-control set based on allele frequencies without crossing individuals. The differences arise from optimization of the process in the two cases. When simulating a case-control dataset without generations, the individuals can be set without regard to recombination rates between SNPs. In addition, the final dataset can be set to produce the desired number of cases and controls and the simulator will continue until it generates those numbers. The simulator only produces bi-allelic SNPs. This limitation allows the simulator to represent each chromosome as a series of bits and reduces the memory requirements.

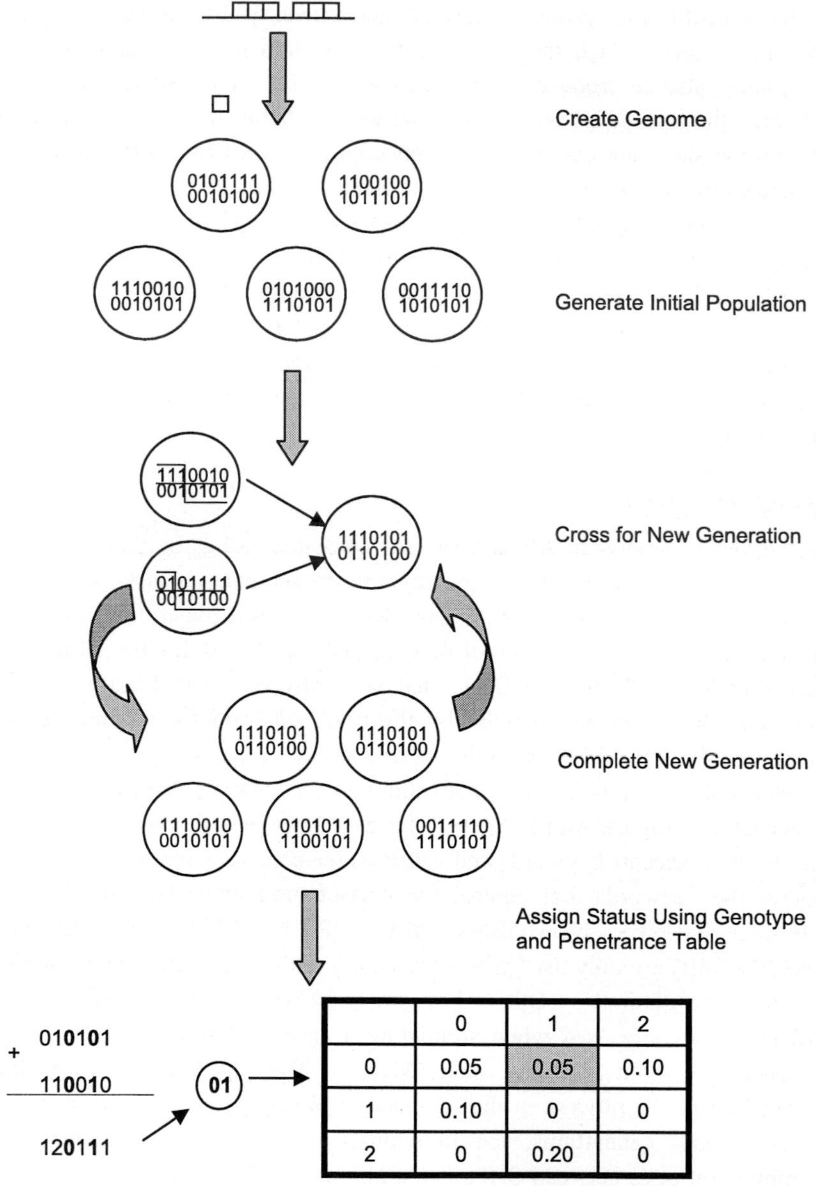

Figure 1. Summary of process involved in producing a simulated dataset. After the genome is constructed, an initial population of individuals is created and individuals cross by contributing one chromosome each to the offspring. These crosses create the next generation and the process repeats until the specified number of generations has occurred. In the last generation, the genotypes for the individual are produced by summing the chromosomes at each position. The genotype at the disease SNPs is used to find the penetrance value in the penetrance table.

Memory requirements vary with both the number of individuals and number of SNPs in the set being produced. For example, 10,000 individuals and 100,000 SNPs require slightly less than 400 MB of RAM. Total numbers of individuals and SNPs are only limited by the memory of the system running the software. We have successfully simulated 10,000 individuals and 400,000 SNPs on a system with 2 GB of RAM.

The library outputs data in a simple text format. Each line consists of one individual with the first column being the case or control status of the individual. Each additional column lists the genotype of the individual (0, 1, 2). This information is available through accessor functions of the library so that the output can be easily formatted to meet the needs of multiple software packages.

Table 1. Descriptive list of simulator parameters

Parameter	Example	Description
RAND	712	Sets random seed for creating dataset
MODELFILES	Model1.smod 0.7 Model2.smod 0.3	Lists model files that detail the penetrance table. Also indicates fraction of population that uses indicated model.
GENOTYPEERROR	.02	Per SNP error rate
PHENOCOPY	.05	Phenocopy rate in final population
AFFECTED	200	Number of cases in final population when only generating case-control set without crossing
UNAFFECTED	200	Number of controls in final population when only generating case-control set without crossing
SIMLOCI	500	Number of SNPs to simulate in a case-control set without crossing
ALLELELIMITS	0.05 0.50	Sets the range for the minor allele frequency of the SNPs in the simulation
ALLELEFREQS	1 0.7 0.3	Specifies allele frequencies for specific SNPs (overrides ALLELELIMITS for the SNP)
POPSIZE	1000	Size of simulated population
NUMGENS	100	Number of generations to simulate
GENES	100	Number of genes to simulate
MINSNP	5	Minimum number of SNPs per gene
MAXSNP	10	Maximum number of SNPs per gene
MINRECOMB	0.005	Minimum recombination rate between adjacent SNPs
MAXRECOMB	0.05	Maximum recombination rate between adjacent SNPs

2.3. *Benchmarks*

To test the genomeSIM's performance we simulated a dataset with 10,000 individuals and varying numbers of SNPs. The population underwent 100 generations of mating. We ran the tests on a PC with Intel Xeon 3.06 GHz CPUs and 2GB of RAM running Red Hat Enterprise Linux WS release 3 (Taroon Update 5). The simulator produces the dataset in 2 hours 48 minutes when simulating 100,000 SNPs and 12 hours 17 minutes when simulating 400,000 SNPs.

We also tested the data simulator's performance in producing a set of 500 cases and 500 controls without mating generations. For 100,000 SNPs, the simulator produced the dataset in 11.7 seconds on the system listed above. For 400,000 SNPs the simulator produced the set in 48.8 seconds.

2.4. *Data Simulations*

For this paper, we performed several data simulations to demonstrate the utility of our new data simulation software. First, we simulated a single SNP recessive model with the penetrance table shown in Table 2. The allele frequency of the functional SNP was p=0.7, q=0.3, where p is the frequency of the A allele. Next, we simulated the two SNP gene-gene interaction model shown in Table 3. For this model, the allele frequencies of both functional SNPs were p=0.6, q=0.4.

Table 2. Single SNP recessive model with reduced penetrance

Genotype	Probability(disease|genotype)
AA	0.0
Aa	0.0
aa	0.9

Table 3. Two SNP gene-gene interaction model

	BB	Bb	bb
AA	0.177	0.080	0.005
Aa	0.074	0.150	0.017
aa	0.014	0.013	0.569

Table 4. Parameters for data simulations for MDR analysis

Parameter	Value
Population size	10,000
Total SNPs	50,000
Genes	5,000
SNPs per gene	10
Generations	150
Minimum recombination between SNPs	0.0
Maximum recombination between SNPs	0.10
Minimum minor allele frequency	0.05
Maximum minor allele frequency	0.5

We used one set of simulation parameters for these simulations (shown in Table 4). A total of 500 cases and 500 controls were extracted from the simulated population.

3. Results

To validate that the data simulations are indeed functioning as expected, we analyzed the datasets simulated to determine if statistical methodologies are able to detect the effects simulated. We applied the Multifactor Dimensionality Reduction (MDR)[9-11]approach to detect all single SNP and two-SNP models. We performed the MDR analysis without cross-validation due to the computation time required to analyze 50,000 SNPs. We selected the model with the minimum classification error and calculated a chi-square test for association. The results of the analyses are shown in Table 5. These results show uncorrected chi-square p-values. In the dataset with a recessive model simulated, MDR identified the correct model, SNP 5, as the optimal model. In the dataset with a two-SNP model simulated, MDR identified the optimal model as the two SNP model (SNP 5 and SNP 10). The best single SNP model in that dataset was not as significant as the two-SNP model. Thus, we would select the two-SNP model as the best model.

Table 5. Results of MDR analysis on simulated data

Model	SNPs	Classification error	Chi Square p-value
Recessive	5	1.10	0.00000000
Two SNP	7792	42.9	0.00001822
Two SNP	5 10	30.50	0.00000000

In addition to demonstrating the ability to simulate known effects, we also wanted to determine if our simulation algorithm was able to simulate linkage disequilibrium across the genome. Figure 2 shows a Haploview plot generated on one dataset simulated with our new software[12]. The data simulation parameters used for this particular dataset are shown in Table 6. There are several blocks of strong LD across this particular area of the genome. This indicates that this software is able to simulate LD in addition to specified genetic models.

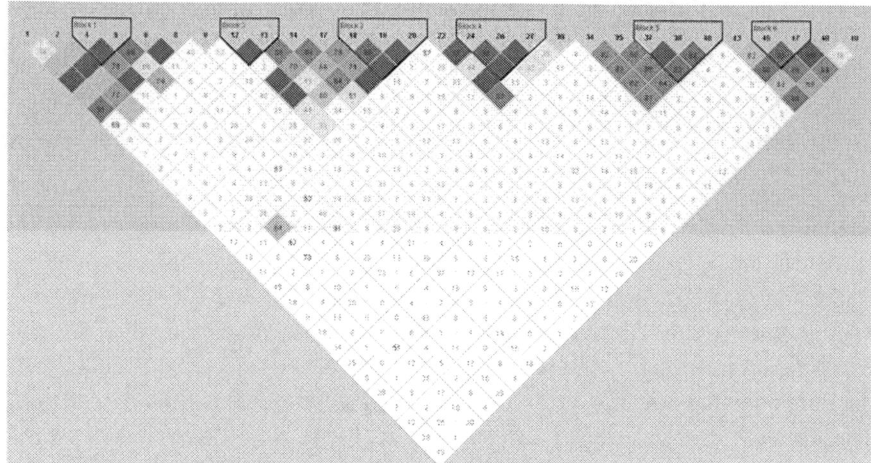

Figure 2. Plot generated by Haploview on one simulated dataset.

Table 6. Parameters for data simulations for Haploview plot

Population size	500
Total SNPs	50
Genes	5
SNPs per gene	10
Generations	500
Minimum recombination between SNPs	0.001
Maximum recombination between SNPs	0.001
Minimum minor allele frequency	0.05
Maximum minor allele frequency	0.5

4. Discussion

Detecting disease susceptibility genes for common disease is a major focus of study in human genetics. The ability to achieve success in this endeavor is dependent upon intelligent study design, accurate genotyping, and efficient algorithms for analysis. Methodology development in statistical and computational genetics continues to advance the field, and novel approaches are being developed in an attempt to keep pace with the development of genotyping technology. Evaluating and comparing these methods requires the ability to perform complex data simulations to efficiently test the new algorithms. Several methods currently exist for the simulation of family-based data including SIMLINK, SIMULATE, and SIMLA. Coalescent-based and forward-time population based algorithms have also been developed, however, to our knowledge, none have the flexibility of genomeSIM. genomeSIM is a new data simulation package that uses forward-time population based simulations, user-specified evolutionary features, and the ability to specify simple or complex penetrance functions to assign disease status, including gene-gene interaction models. We believe that since interactions are likely to be an important component of complex disease[13,14] having the capability of evaluating new methods in this type of data will be a true test of the method's success.

While we believe that genomeSIM is an advance over current data simulation methods, we will continue to add additional features. There are currently no family based simulation algorithms that allow for the simulation of complex gene-gene interaction models. We are in the process of allowing genomeSIM to generate pedigree data under such penetrance functions. We plan on simulating larger sets more quickly by parallelizing the algorithm. In addition, there are many evolutionary features that could be parameter options in the algorithm including random genetic drift, population bottlenecks, and selection that we plan to implement. genomeSIM is freely available from the authors upon request. It will also be available via the internet at http://chgr.mc.vanderbilt.edu/ritchielab.

5. Acknowledgements

This work was supported by National Institutes of Health grants GM31304, AG20135, and in part by HL65962, the Pharmacogenomics of Arrhythmia Therapy U01 site of the Pharmacogenetics Research Network.

6. References

1. Boehnke,M. Estimating the power of a proposed linkage study: a practical computer simulation approach. *Am. J. Hum. Genet.* **39**, 513-527, (1986)
2. Ploughman,L.M. and Boehnke,M. Estimating the power of a proposed linkage study for a complex genetic trait. *Am. J. Hum. Genet.* **44**, 543-551, (1989)
3. Weeks, D. E, Ott, J, and Lathrop G.M. SLINK: A general simulation program for linkage analysis. *American Journal of Human Genetics* **47**, A204. (1990)
4. Bass,M.P. et al. Pedigree generation for analysis of genetic linkage and association. Pac. Symp. *Biocomput.*, 93-103, (2004)
5. Kingman, J. The coalescent. *Stochastic Processes Appl* **13**, 235-248. (1982)
6. Balloux,F. EASYPOP (version 1.7): a computer program for population genetics simulations. *J. Hered.* **92**, 301-302, (2001)

7. Hey, J. Nielsen, R. Multilocus methods for estimating population sizes, migration rates and divergence time, with applications to the divergence of Drosophila pseudoobscura and D. persimilis. *Genetics* **167**, 747-60. (2004)
8. Peng,B. and Kimmel,M. simuPOP: a forward-time population genetics simulation environment. *Bioinformatics*. **21**,3686-7 (2005)
9. Ritchie,M.D. et al. Multifactor-dimensionality reduction reveals high-order interactions among estrogen-metabolism genes in sporadic breast cancer. *Am J Hum. Genet* **69**, 138-147, (2001)
10. Hahn,L.W. et al. Multifactor dimensionality reduction software for detecting gene-gene and gene-environment interactions. *Bioinformatics*. **19**, 376-382, (2003)
11. Ritchie,M.D. et al. Power of multifactor dimensionality reduction for detecting gene-gene interactions in the presence of genotyping error, missing data, phenocopy, and genetic heterogeneity. *Genet. Epidemiol.* **24**, 150-157, (2003)
12. Barrett,J.C. et al. Haploview: analysis and visualization of LD and haplotype maps. *Bioinformatics*. **21**, 263-265, (2005)
13. Moore,J.H. The ubiquitous nature of epistasis in determining susceptibility to common human diseases. *Hum. Hered.* **56**, 73-82, (2003)
14. Sing,C.F. et al. Genes, environment, and cardiovascular disease. *Arterioscler. Thromb. Vasc. Biol.* **23**, 1190-1196, (2003)

TAGSNP SELECTION BASED ON PAIRWISE LD CRITERIA AND POWER ANALYSIS IN ASSOCIATION STUDIES

SHYAM GOPALAKRISHNAN, ZHAOHUI S QIN

Center for Statistical Genetics, Department of Biostatistics, School of Public Health, University of Michigan,
Ann Arbor, MI 48109-2029, USA
E-mail: gopalakr@umich.edu

TagSNP selection is an important step in designing case control association studies. Among selection methods that have proliferated, the ones based on pairwise LD measurement are attractive for the purpose of designing association studies. The goal is to minimize the number of markers selected for genotyping in a particular platform and therefore reduce genotyping cost while simultaneously representing information provided by all other markers. Depending on the platform, it is also important to select sets that are robust against occasional genotyping failure. An array of methods has been proposed to effectively select these tagSNPs using various criteria. In this study, we extend the algorithms used in FESTA, a computer program we previously developed for picking tagSNPs using r^2 criteria. We applied FESTA to the HapMap whole chromosome data in two different populations, and we also performed a power analysis for case-control association studies using simulated data. FESTA chooses 294322 tagSNPs in the autosomes in the CEPH samples. The YORUBA samples require 61.5% more tagSNPs than the CEPH samples. The power study showed that limiting ourselves to only tagSNPs, instead of choosing all SNPs in the interval for an association study, results in a power loss of only about 5-10%.

1. Introduction

Rapid advancement in genotyping technologies together with the successful deployment of the International HapMap project [1,2] further popularized the genome-wide association studies, where a dense set of SNP markers across the whole genome is assayed to locate a susceptible chromosomal region that potentially harbors disease predisposing genetic variants. An important initial step in designing an association study is to choose a set of SNPs to represent all variants in the genomic regions of interest. Various algorithms have been proposed for selecting these so called tagSNPs[4,5,6,7,8,9,10,11,12,13]. Most of these strategies aim at choosing "haplotype tagging" SNPs, which are able to capture most of the haplotype diversity, and therefore, could potentially capture most of the information for association between a trait and the marker loci[14]. Recently, Zhang and Jin[15] and Carlson et al.[4] introduced a simpler criterion for choosing tagSNPs which is based on the pairwise LD measure[16]. These methods search for a small set of SNPs that are in strong LD (measured through pairwise r^2) with all the other SNPs that are not selected for genotyping. Pairwise r^2 is an attractive criterion for tagSNP selection since it is closely related to statistical power for case control association studies, where a directly associated SNP is replaced with an indirectly associated tagSNP[17].

In studies conducted in this manuscript, we adopted the newly developed pairwise LD-based algorithm named **F**ragmented **E**xhaustive **S**earch for **TA**gSNPs[18]. FESTA implements a novel partition step to allow comprehensive search to be carried out. Therefore, it produces fewer tagSNPs than the greedy approach. FESTA also incorporates alternative solution picking according to additional criteria; it can force certain markers in or out of the tagSNP set; and find double coverage tagSNPs. FESTA readily identifies equivalent tagSNP sets, so that additional selection criteria can be incorporated.

We extended the FESTA algorithms by adding a new user-defined criterion, which can be used to pick among the alternative tagSNP sets identified by FESTA. This added flexibility can be quite useful under some situations. We also applied FESTA to whole chromosome HapMap data to identify tagSNPs genome wide. Next, we conducted a simulation study for power analysis; comparing power of detecting association to the disease causing variant using tagSNPs chosen by FESTA. We use two benchmarks to compare the performance of the tagSNPs, (a) all the SNPs in the interval and (b) the same number of random SNPs in the interval.

2. Methods

2.1. *FESTA: Algorithms*

In this section, for the sake of completeness, we briefly review the algorithm implemented in FESTA. The basic idea is to replace a greedy search, where the most connected markers are added sequentially to the tagSNP set, with an exhaustive search where all marker combinations are evaluated. In most settings, our method is guaranteed to find the optimal tagSNP set(s) defined by the r^2 criterion. The details of the FESTA program and the results of comparison with the greedy approach can be found in Qin et al[18].

Define $\mathbb{S}$ to be the set of all SNPs in the precinct under consideration.

Our aim is to find a tagSNP set, denoted by T, a subset of $\mathbb{S}$ such that for all a_i not in T, there exists a_j in T such that $r^2(a_i, a_j) \geq r_0$. In our explanation of the algorithm, we introduce two intermediate SNP sets, P and Q. The candidate set P contains all the markers that are eligible to be chosen as tagSNPs and the target set Q contains all the markers that are yet to be tagged, i.e. no marker in Q is in LD with any tagSNP in T. Typically, the candidate set P is the complement of the tagSNP set T, and $P = Q$. We describe several different algorithms for updating P, Q and T starting with a greedy approach[4]. We then outline successive refinements of a partition and exhaustive search algorithm, designed to allow processing of very large number of markers. Finally, we discuss enhancements to our algorithm.

2.1.1. *Greedy Approach*

The greedy algorithm[4] constructs a tagSNP set by adding the most connected marker to the tagSNP set. It then removes the chosen marker and all connected markers from consideration. This is repeated till there are no markers to be considered. Though the greedy approach is efficient, it does not always find the optimal solution[18].

2.1.2. *Exhaustive Search and Partitioning*

An exhaustive search guarantees the minimum tagSNP set. Genome-wide tagSNP selection requires considering thousands of SNP markers. In these cases, exhaustive searches can not be directly applied due to prohibitive computation costs. Here we use the spatial locality property of LD, i.e. high LD can only be maintained over short distances; therefore we can decompose the set of markers into disjoint precincts such that no marker in a precinct is in high LD with any marker outside the precinct.

After the partitioning step, we perform the tagSNP selection within each precinct using exhaustive search. The result of the greedy algorithm can be used as an upper bound on the number of tagSNPs required in the precinct. The detailed algorithm follows;

(1) Apply BFS[20] to decompose the entire set of markers into precincts $\mathbb{S}_i$ such that strong LD can only be observed within precincts. $\mathbb{S} = \bigcup_{i=1}^{n} \mathbb{S}_i$, and $\mathbb{S}_i \bigcap \mathbb{S}_j = \varnothing \forall i \neq j$;
(2) Within each precinct $\mathbb{S}_i$, set $k_i = 1$,
    a Enumerate all possible k_i-marker combinations. $P_i = Q_i =$

$\mathbb{S}_i$. If no such combination can cover the entire precinct, set $k_i = k_i + 1$ and repeat this step;

b Record all tagSNP sets that can cover the precinct. These form the complete minimum tagSNP sets $T_i^j : j = 1, ..., J_i$, where J_i is the number of such minimum tagSNP sets.

(3) Any combination of tagSNP sets identified from all disjoint subsets forms a tagSNP set for the whole set $\mathbb{S}$, the overall size of such minimum tagSNP sets is $\sum_{i=1}^{n} k_i$, and the total number of minimum tagSNP sets is $\prod_{i=1}^{n} J_i$.

FESTA uses a hybrid of greedy and exhaustive algorithms to solve precincts that are not computationally feasible.

In addition to the basic tagSNP selection, we have implemented the following additional features to assist in tagSNP selection.

(1) Include/Exclude tagSNP markers: As discussed earlier, it may be important to include/exclude some SNPs in the tagSNP set to reduce genotyping cost or ensure genotyping success. Specific SNPs may be included/excluded from the tagSNP set using the mandatory/exclude option in FESTA respectively.
(2) Choosing between alternate solutions based on LD: Exhaustive search may return more than one tagSNP solution for a given precinct. All these sets contain the same number of tagSNPs. Three additional criteria were implemented in FESTA to select one set, (a) Maximize the average r^2 between tagSNPs and the untagged SNPs they represent; (b) Maximize the lowest r^2 between tagSNPs and the untagged SNPs they represent; (c) Minimize the average r^2 among all pairs of tagSNPs;
(3) Double coverage: Current pairwise LD based tagSNP picking algorithms aim to find a tagSNP set such that each SNP marker is either a tagSNP itself or is in LD with at least one of the tagSNPs. Random genotyping failure or error on these tagSNPs can result in loss of power. To be more robust, FESTA implemented a more stringent criterion requiring that, if possible, every untyped marker should be in LD with at least two tagSNPs.

2.1.3. *Extensions to FESTA*

How to choose an optimal set of tagSNPs for genotyping is a practical problem. Specific issues may arise in various scenarios, therefore it is imperative that the tagSNP selection tool is flexible enough to let the user impose different optimization rules or apply certain restrictions by themselves. One idea is to introduce an additional criteria to constrain all the available results.

It is common to obtain a large number of tagSNP sets of the same size using the pairwise LD criterion based tagging tools such as FESTA. To select a particular tagSNP set for a particular study, additional criteria need to be introduced to narrow down to the ultimate optimal tagSNP set according to the study requirement. In addition to the additional criteria described above which are already implemented in FESTA, we added a new feature to the FESTA program to allow optimization based on user-specified *ad hoc* variables. An example of such a variable is the quality or design scores of some genotyping platforms such as Illumina. The design score is a continuous variable, which ranges from 0 to 100, where high scores indicate higher genotyping success rate. By entering such scores for all the candidate SNPs, FESTA will to identify the tagSNP set that is optimized in the sense of high design scores among all the tagSNP sets that have the same size according to the pairwise r^2 criteria alone. By adding this constraint, genotyping failure rate will be reduced. Another variable that can be assigned to each SNP is the minor allele frequency (MAF). In this case FESTA can produce a tagSNP set that maximizes the average MAF of all tagSNPs.

This additional variable can also be discrete. For example, whether or not this SNP is in the coding region (cSNP), missense SNP or double hit SNP can be indicated using this variable. FESTA can then report which of the tagSNP sets contained the largest number of such desired SNPs. Another example, some of the markers are "preferred" because they may have already been typed in earlier rounds of studies. To minimize the cost of retyping them, an additional indicator variable can be added showing whether this marker is preferred. Then the tagSNP set containing the most number of preferred tagSNPs can be selected among all tagSNP sets picked by pairwise LD criterion alone. In association studies, these practical constraints can be quite valuable.

2.2. *Power Analysis*

Similar to the power comparison study[21] by Zhang et al., we conducted a simulation study to assess the power of performing case control association studies using tagSNPs identified by FESTA.

2.2.1. *Simulation scheme*

We first simulated a large number of chromosomes consisting of many consecutive SNPs across a 500 kb genomic region using the 'ms' program[22]. It assumes the standard coalescent approximation to the Wright-Fisher model. We assume a constant population size, without subpopulation or gene conversion. We further assume a constant mutation rate throughout this region. The mutation parameter, $\theta = 4N_0\mu$, was chosen to be 200, where N_0 is the effective diploid population size, and $\mu = 10^{-8}$ is the neutral mutation rate per site for this segment. The recombination parameter, $\rho = 4N_0r$ was set to 20. Here, $r = 10^{-9}$ is the probability of recombination in this interval. A hundred populations each containing 2200 chromosomes were generated.

After generating the haplotypes, we randomly chose a marker locus as the disease locus as long as its minor allele frequency was greater than 0.05. The remaining marker loci that had minor allele frequency greater than 0.05 were also retained.

The tagSNP sets were selected using FESTA. The pairwise LD measurement, r^2, was calculated from two marker haplotype frequencies and allele frequencies calculated from the first 200 chromosomes in each population. The case control samples were generated using the remaining 2000 chromosomes, where a hypothetical individual was formed by randomly picking two chromosomes from the pool. The disease status for each hypothetical individual was determined by the penetrance of the individual's disease locus genotype. We assumed the four common disease models: additive, multiplicative, dominant and recessive. In order to mimic a common disease, common variant situation, we specified population disease prevalence to be $P = 0.05$ and $P = 0.1$, and the sibling recurrence risk ratio λ_s[23] was fixed at 1.02.

The association test is based on the difference in allele frequency between case individuals and control individuals[24]. Suppose N is the number of case/control individuals, n_i and m_i are the number of allele A_i in case and control individuals, and p_i and q_i are the frequency of allele A_i in case and control individuals respectively. The test statistic is

$$\chi^2 = \sum_{i=1}^{2} \frac{(n_i - m_i)^2}{n_i + m_i} = 2n \sum_{i=1}^{2} \frac{(p_i - q_i)^2}{p_i + q_i} \tag{1}$$

The above test statistic approximately has a χ^2 distribution with 1 degree of freedom under the null hypothesis of no association. The use of this test statistic assumes Hardy-Weinberg Equilibrium[25], as shown by Sasieni.

Since there are a large number of marker loci involved, multiple testing is a critical issue for the performance of association test. Simple adjustment approaches such as Bonferroni correction do not perform well. An alternative approach is the permutation test using a Monte Carlo strategy[26]. The maximum value of the test statistic from all markers, denoted as χ^2_{max}, was taken as the test statistic for the association test of the interval. The same test statistic is also calculated for each of the permuted case control samples (generated by switching case control labels for randomly picked individuals). The overall p-value is calculated as the proportion of permuted case control samples that have higher χ^2_{max} value than the one observed from the original case control sample.

The following procedure illustrates our simulation scheme:

(1) Generate 100 populations of 2200 chromosomes using ms program.
(2) In each population, use the first 200 chromosomes to calculate the pairwise r^2, and select tagSNPs using the FESTA program. Subsequently, generate 500 case and 500 controls by sampling from the remaining 2000 chromosomes.
(3) Calculate the test statistic χ^2_{max} under three different cases:

    a. all SNPs in this segment,
    b. only the tagSNPs,
    c. the same number of randomly chosen SNPs.

(4) Perform random permutation 100 times within each case control sample, and calculate the same test statistics χ^2_{max}.
(5) Calculate the overall p-value by determining the proportion of the permuted samples that have higher test statistics than the ones observed from the original case control sample.

Since FESTA provides multiple tagSNP sets for each precinct, and thus for the entire set of markers as well, we chose 3 tagSNP sets, which we used to calculate the power of the tagSNPs to associate the interval to a disease. We used the 3 in-built criteria based on LD, described in 2.1.2, to choose from the alternative solutions. The power of the tagSNPs was reported as the average power of the 3 chosen solutions.

3. Results

3.1. *Genome wide tagSNP selection*

We applied FESTA to pick tagSNPs in the entire human genome using the HapMap data. Two different populations with African and European ancestry were used. A minor allele frequency (maf) threshold of 0.05 was used to prune the SNP map. The total number of SNPs is 742180 in the CEPH samples (European ancestry) and it is 775420 in the YORUBA samples (African ancestry). The total number of tagSNPs in autosomes, using a threshold of $r^2 = 0.8$, identified in the CEPH samples is 294322 and it is 475307 in the YORUBA samples. The YORUBA samples contain almost 61.5% more tagSNPs compared to the CEPH samples. The summary of percentage of tagSNPs in each chromosome is summarized in table 1. It is interesting to note that chromosome 19, the most gene-rich of all human chromosomes, has the largest proportion of tagSNPs in three out of four cases.

Table 1. Proportions of tagSNPs in 22 human autosomes in 2 populations.

	$r^2 > 0.5$			$r^2 > 0.8$		
	Mean	Min	Max	Mean	Min	Max
CEPH	0.251	0.178(8)	0.352(19)	0.410	0.312(8)	0.517(19)
YORUBA	0.434	0.335(8)	0.534(19)	0.624	0.514(9)	0.705(16)

The average computation time, in seconds, in the CEPH samples is 603.54 with a minimum of 169.7 (chr 20) and a maximum of 1313.25 (chr 3), whereas in the YORUBA samples, the average computation time is 628.05, with a minimum of 183.87 (chr 19) and a maximum of 1421.15 (chr 8).

Figure 1 shows the proportion of tagSNps selected in the 22 autosomes in the two populations using the thresholds of $r^2 = 0.5$ and $r^2 = 0.8$.

3.2. *Power results*

We analyzed the power of the tagSNPs to detect association of a disease to the interval as mentioned in the previous section. We simulated a 100 populations of 2200 haplotypes each. The number of selected SNPs (MAF > 0.05) in a population ranges from 236 to 902 with an average of about 584 SNPs per population. The number of tagSNP in a population ranges from 66 to 346, and the average number of tagSNPs selected in a population was about 162 markers. On average, 28% of SNPs were selected as tagSNPs,

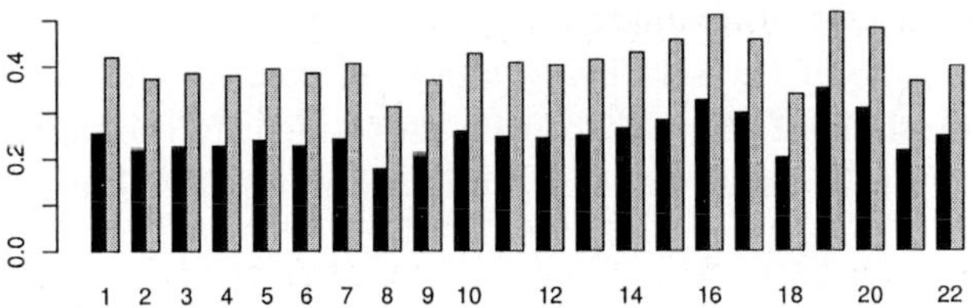

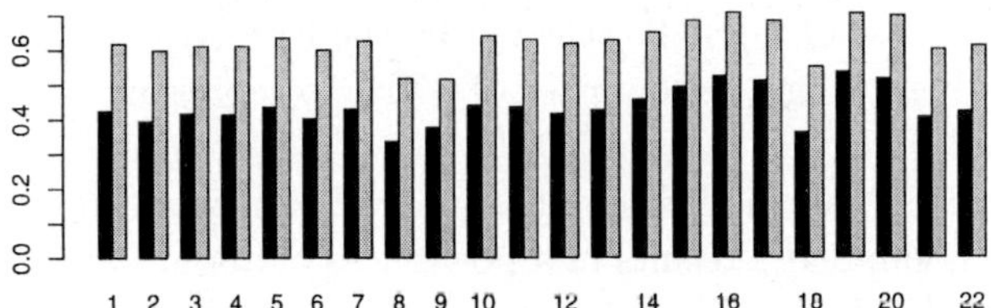

Figure 1. Proportion of tagSNPs in human autosomes using $r^2 = 0.5$ (black) and $r^2 = 0.8$ (gray) thresholds.

across the 100 populations. The disease marker had an average MAF of about 0.216.

Table 2. Power analysis results with disease marker included.

Disease Model	*Prevalence* = 0.05			*Prevalence* = 0.1		
	All SNPs	TagSNPs	Random	All SNPs	TagSNPs	Random
Additive	0.49	0.45	0.407	0.63	0.603	0.563
Multiplicative	0.11	0.11	0.087	0.145	0.1367	0.13
Dominant	0.45	0.42	0.417	0.54	0.547	0.513
Recessive	0.42	0.427	0.407	0.47	0.47	0.437

Table 3. Power analysis results with disease marker excluded.

Disease Model	*Prevalence* = 0.05			*Prevalence* = 0.1		
	All SNPs	TagSNPs	Random	All SNPs	TagSNPs	Random
Additive	0.61	0.603	0.586	0.6	0.6	0.6
Multiplicative	0.1	0.115	0.087	0.15	0.13	0.123
Dominant	0.41	0.39	0.37	0.49	0.4833	0.453
Recessive	0.51	0.49	0.487	0.59	0.58	0.58

We also conducted the power analysis in two ways, by: (i) including the disease marker in the set of simulated SNPs and (ii) excluding the disease marker from the set of simulated SNPs. Exclusion of the disease marker ensures that none of the sets being analyzed contain the disease marker. We used a threshold of $r^2 = 0.8$ for FESTA to identify the tagSNPs in both cases. The results of the simulation study are summarized in tables 2 and 3 given above.

4. Discussion

As can be observed from the results in the above tables, the loss of power is minimal in the case of tagSNPs selected by the FESTA program. However, it is higher when using a random set of SNPs to represent the information in the interval. There is about a 5-10% loss of power when we choose only tagSNPs instead of all the SNPs in the interval, whereas choosing the same number of random SNPs results in a power loss of about 20%. With higher prevalence of the disease, we get better power to associate the interval to the disease.

We also compared the performance of the algorithm under two other situations, viz., when the disease marker is central to the interval, i.e. if we represent the interval as (0, 1), the disease marker lies in the region [0.4 0.6], and when the disease marker is not central to the interval. We find that a central location of the disease marker favors tagSNPs more heavily than it does random SNPs. If, however the marker is not centrally located, random SNPs perform almost as well as tagSNPs; e.g., in the multiplicative model with $P = 0.1$, the power of the tagSNPs is about 0.17 whereas the random SNPs show a power of 0.13 when the disease marker is central; when the disease marker is not central both the tagSNPs and the random SNPs exhibit a power of about 0.14.

Our current simulation study is still very limited. More comprehensive comparison is needed for us to better understand the effect of tagSNPs under different scenarios.

Pairwise LD is just one criterion for choosing tagSNPs. An interesting alternative is to consider multipoint LD instead. Since a marker may not be in high LD with any single marker, but may be correlated well with haplotypes consisting of multiple linked markers. Therefore, typically multipoint LD[12,27] based tagSNP selection algorithms such as Tagger[28] produce fewer tagSNPs compared to pairwise LD based approaches. However, when conducting association studies using single markers, tagSNPs picked based on pairwise LD criterion are likely to show better power.

The extended FESTA program is freely available at http://www.sph.umich.edu/csg/qin/FESTA.

Acknowledgements

This work is partially supported by NIH RO1-HG002651-01. We would like to thank Dr. Gonçalo Abecasis for insightful discussion about this project, and the four anonymous reviewers for their constructive comments and suggestions.

References

1. The International HapMap Consortium, The International HapMap Project. *Nature* **426**, 789-796, (2003).
2. Sachidanandam R, International SNP Map Working Group A map of human genome sequence variation containing 1.42 million single nucleotide polymorphisms. *Nature* **409**, 928-933, (2001).
3. Avi-Itzhak HI, Su X, De La Vega FM Selection of minimum subsets of single nucleotide polymorphisms to capture haplotype block diversity. *Pac Symp Biocomputing* 466-477, (2003).
4. Carlson CS, Eberle MA, Rieder MJ, Yi Q, Kruglyak L and Nickerson DA, Selecting a maximally informative set of single-nucleotide polymorphisms for association analysis using linkage disequilibrium. *Am. J. Hum. Genet.* **74**, 106-120, (2004).
5. Hampe J, Schreiber S, Krawczak M Entropy-based SNP selection for genetic association studies. *Hum Genet.* **114**, 36-43, (2003).
6. Halldórsson BV, Bafna V, Lippert R, Schwartz R, De La Vega FM, Clark AG, Istrail S. Optimal haplotype block-free selection of tagging SNPs for genome-wide association studies. *Genome Res.* **14**, 1633-1640, (2004).
7. Johnson GC, Esposito L, Barratt BJ, Smith AN, Heward J, Di Genova G, Ueda H, Cordell HJ, Eaves IA, Dudbridge F Haplotype tagging for the identification of common disease genes. *Nat. Genet.* **29**, 233-237, (2001).
8. Ke X, Cardon LR Efficient selective screening of haplotype tag SNPs. *Bioinformatics* **19**, 287-288, (2003).
9. Lin Z, Altman RB Finding haplotype tagging SNPs by use of principal components analysis. *Am. J. Hum. Genet.* **75**, 850-861, (2004).
10. Meng Z, Zaykin DV, Xu CF, Wagner M, Ehm MG. Selection of genetic markers for association analyses, using linkage disequilibrium and haplotypes. *Am. J. Hum. Genet.* **73**, 115-130, (2004).
11. Sebastiani P, Lazarus R, Weiss ST, Lunkel LM, Kohane IS and Romani MF, Minimal haplotype tagging *Proc. Natl. Acad. Sci. USA* **100**, 9900-9905, (2003).
12. Stram DO, Haiman CA, Hirschhorn JN, Altshuler D, Kolonel LN, Henderson BE and Pike MC Choosing haplotype-tagging SNPs based on unphased

genotype data using preliminary sample of unrelated subjects with an example from the multiethnic cohort study. *Hum. Hered.* **55**, 27-36 (2003).
13. Zhang K, Deng M, Chen T, Waterman MS and Sun F A dynamic programming algorithm for haplotype partitioning. *Proc. Natl. Acad. Sci. USA* **99**, 7335-7339, (2002).
14. Chapman JM, Cooper JD, Todd JA, Clayton DG, Detecting disease associations due to linkage disequilibrium using haplotype tags: a class of tests and the determinants of statistical power. *Hum Hered* **56**, 1831, (2003).
15. Zhang, K. and Jin, L. HaploBlockFinder: Haplotype block analysis. *Bioinformatics* **19**, 1300-1301, (2003).
16. Delvin B, Risch N, A comparison of linkage disequilibrium measures for fine-scale mapping. *Genomics* **29**, 311-322, (1995).
17. Pritchard JK, Przeworski M Linkage disequilibrium in humans: models and data. *Am. J. Hum. Genet.* **69**, 1-14, (2001).
18. Qin ZS, Gopalakrishnan S, Abecasis G, An efficient comprehensive search algorithm for tagSNP selection using linkage disequilibrium criteria. *Unpublished manuscript.* http://www.sph.umich.edu/csg/qin/FESTA. (2005).
19. Patil N, Berno AJ, Hinds DA, Barrett WA, Doshi JM, Hacker CR, Kautzer CR, Lee D H, Marjoribanks C, McDonough DP, et al. Blocks of limited haplotype diversity revealed by high-resolution scanning of human chromosome 21. *Science* **294**, 1719-1723, (2001).
20. Cormen TH, Leiserson CE, Rivest RL, Introduction to algorithms.*McGraw-Hill Publications*, (2001).
21. Zhang K, Calabrese P, Nordborg M, Sun F, Haplotype Block Structure and Its Applications to Association Studies: Power and Study Designs *Am J Hum Genet.* **71(6)**. 13861394, 2002.
22. Hudson RR, Generating samples under a Wright-Fisher neutral model. *Bioinformatics*, **18**, 337-338, (2002).
23. Risch N, Linkage strategies for genetically complex traits II: The power of affected relative pairs. *Am. J. Hum. Genet.* **46**, 229-41, (1990).
24. Olson JM, Wijsman EM, Design and sample size considerations in the detection of linkage disequilibrium with a disease locus. *Am J Hum Genet* **55**, 574-580, (1994).
25. Sasieni PD, From genotypes to genes: doubling the sample size. *Biometrics* **53**, 1253-1261, (1997).
26. McIntyre LM, Martin ER, Simonsen KL, Kaplan NL, Circumventing multiple testing: a multilocus Monte Carlo approach to testing for association. *Genet Epidemiol* **19**, 18-29, (2000).
27. Stram DO, Tag SNP selection for association studies. *Genet Epidemiol.* **27**, 365-374 (2005).
28. Paul de Bakker, Tagger http://www.broad.mit.edu/mpg/tagger

EFFICIENT TWO-STAGE GENOME-WIDE ASSOCIATION DESIGNS BASED ON FALSE POSITIVE REPORT PROBABILITIES

PETER KRAFT

Program in Molecular and Genetic Epidemiology, Harvard School of Public Health, 655 Huntington Avenue, Boston , MA 02112, United States of America

Despite recent advances, very-high-throughput (VHT) technologies capable of genotyping hundreds of thousands of SNPs in individual samples remain prohibitively expensive for the large studies necessary to screen substantial sections of the genome for variants with modest effects on disease risk. This paper presents a two-stage strategy, where a portion of available samples are genotyped with VHT technology, and a small number of the most promising variants are genotyped with standard high-throughput techniques in the remaining samples as an independent replication study. The sample sizes in the first and second stages and the corresponding significance levels are chosen to limit False Positive Report Probability (FPRP), while maximizing the number of Expected True Positives (ETPs). (The FPRP is the conditional probability that a marker is not truly associated with disease, given the a significant test for disease-marker association.) For a fixed budget, the two-stage strategy has greater power (a larger number of ETPs) than the single-stage strategy (where all subjects are genotyped using expensive VHT technology). Furthermore, concentrating on the FPRP leads to considerable savings relative to strategies designed to control the family-wise error (e.g. Bonferonni correction). The FPRP and number of ETPs can also accommodate researchers' prior beliefs about the number of causal loci and the magnitude of their effects. The expected number of false positives does not change if the true number and effects of causal loci differs from the specified prior (although the false discovery rate will vary), thus limiting the absolute amount of resources spent chasing "false leads."

1. Introduction

Genome-wide linkage scans have successfully located the genes underlying simple Mendelian disorders, including rare, high-risk hereditary forms of cancer. However, studies based on genetic cosegregation have been less successful in finding susceptibility loci for complex disease; for example, high-risk cancer genes only account small percentage of the familial aggregation of cancer.[*1*, *2*] Genome-wide association (GWA) studies are likely to have more power to detect the common, low- to moderate-risk genes that have the greatest impact on morbidity and mortality due to complex disease at the population level.

Advances in our knowledge of the architecture of the human genome—e.g. studies that examine linkage disequilibrium (LD) patterns among dense sets of Single Nucleotide Polymorphisms (SNPs)[*3*, *4*]—and advances in high-throughput genotyping technology have made GWA studies feasible. Several GWA studies are currently underway, including the NCI's Cancer Genetic Markers of Susceptibility study, which aims to identify susceptibility genes for breast and prostate cancer using a series of nested case-control studies.[*5*]

GWA studies still face a number of design and analysis challenges. Even using the fine-scale correlation structure of the genome to choose a subset of maximally informative SNPs, theoretical and empirical studies suggest hundreds of thousands of SNPs will be needed to cover the genome.[*3, 4, 6*] Furthermore, to reliably detect low- to moderate-risk genes while controlling the number of false positives, large sample sizes will be necessary.[*7-9*] Despite rapidly decreasing genotyping costs, over the next few years it will remain prohibitively expensive to genotype all the SNPs needed for a genome-wide scan in all available subjects.

A multi-stage approach may provide a cost-efficient alternative. In the first stage, the full panel of markers is genotyped on a subset of subjects; in the second and subsequent stages, the most promising markers are followed up in the remaining subjects. Given a fixed budget and a fixed sample size, the number of subjects in the first stage and the number of markers to follow up in the second stage can be chosen so as to maximize power while controlling the number of false positives. Satogopan et al. have explored this design in the context of controlling the family-wise error rate (FWER), that is, limiting the number of false positives to zero with high probability.[*10-12*] This paper introduces a multi-stage framework that limits the False Positive Report Probability (FPRP) recently introduced by Wacholder et al.[*13*]

The FPRP provides a weaker form of control that is useful in the context of genome-wide association scans. Rather than definitively proving the causality of a locus, the goal of genome-wide association scans will be to suggest a list of candidate genes or regions with high probability of causality for further study. Strong control of the FWER can lead to reduced power or an impractical increase in required sample size. Researchers will be willing to accept a limited number of false positives results if that ensures causal loci will be detected using available resources, especially if any positive results from a genome-wide association scan will be quickly followed up—tested in other populations, studied in vitro, etc.

The next section discusses general technical and logistical constraints on genome wide association studies. It also reviews the FPRP and extends it to two-stage designs. Since the FPRP is a quasi-Bayesian tool, I discuss the choice of priors for genome-wide association scans. The third section presents a hypothetical example involving 100,000 markers and compares the performance of one- and two-stage designs aimed at controlling the FWER and the FPRP.

This example shows that for a fixed budget the two-stage and FPRP approaches can be considerably more powerful (in terms of expected number of causal loci detected) than one-stage and FWER designs. The final discussion includes a comparison of the multi-stage FPRP design and analysis with related designs, such as group sequential sampling and multi-stage designs aimed at controlling the False Discovery Rate.[*14-16*]

2. Materials and methods

For simplicity I assume researchers have access to two classes of genotyping technology: "very high throughput technology" and "high throughput technology." Very high throughput technology (VHT) is good at measuring many genotypes simultaneously on each sample, at a low per-genotype cost. But because of the sheer number of genotypes, each sample is expensive to genotype. Furthermore, the set of SNPs genotyped is relatively inflexible, as developing new arrays or multiplexes is expensive and time consuming. High throughput (HT) technology is somewhat more expensive per sample, but more flexible in terms of choosing the SNPs to be genotyped. It is also currently more widely available.

In the example presented below, I assume that the "very high throughput" technology is used at the first stage of the study, but the more flexible "high throughput" technology is used in subsequent stages. This assumption is not intrinsic to the statistical methods, however, as they simply allow for per-genotype costs to vary across the stages. The distinction between "very high throughput" and "high throughput" technology will likely soon fade, as the former becomes more flexible and start-up costs decline. It may currently be useful in practice to consider three tiers of technology, with per-genotype price breaks occurring between hundreds, tens of thousands, and hundreds of thousands of markers typed per sample.

Notation for various study parameters is presented in Table I. All subsequent discussion is limited to the context of two-stage designs; the calculations should easily extend to designs with three or more stages (where the number of stages is fixed ahead of time).

I assume that researchers have fixed numbers of cases and controls available for study. This would be the case when using DNA samples from existing cohort or case-control studies for genome-wide association. Furthermore, for many rare diseases there is effectively a rather low upper limit on the number of cases that can be studied in a reasonable time frame. I assume that there are an equal number of cases and controls. This assumption is not essential to the method; only minor modifications to power calculations would be needed to account for case:control ratios different than 1:1. Note that although I focus on case-control studies using unrelated individuals, the design concepts can be easily applied to family-based studies, e.g. where researchers have a fixed number N of case-parent trios. I also assume researchers have a fixed budget B for genotyping.

Table 1. Notation

Parameter	Notation	Example value
Total number of subjects	N	2000
Number of subjects, stages 1 and 2	N_1, N_2	*
Set of markers studied	$\mathcal{M}$	
Total number of marker studied	M	100,000
Allele frequency for marker $m \in \mathcal{M}$	q_m	≡0.10
Target FPRP	φ	0.50
Significance threshold at first stage, marker m	α_{1m}	*
Significance threshold at second stage	α_{2m}	*
Possible (non-null) relative risks of disease	$RR_1,\ldots,RR_A$	2,1.5,1.3
Prior probability that a marker has relative risk RR_j	π_j	1,2,4***
Type II error rates at stage i for locus m with RR_j	β_{mji}	**
VHT genotyping cost per subject	K	1
HT:VHT genotyping cost ratio	κ	10

* To be solved for given fixed budget. ** Fixed given N, α_{1m}, α_{2m}. *** Value × 10^{-5}

In the first stage, $N_1 \leq N$ cases and controls are genotyped at M independent markers using very high throughput technology. (I assume these markers are diallelic, but they could also be made up of several correlated SNPs, as would be the case in haplotype-tagging studies.) Each marker m is tested for association with disease; each marker m that is significant at the α_{1m} level is then genotyped using high throughput technology in the remaining N_2 cases and controls. Each marker m that is significantly associated with disease in this second sample at the α_{2m} level is declared "overall significant." (As discussed below, the FPRP depends on both the significance level and the power of the test for association between marker m and disease. As power will depend on the allele frequency q_{m} which may differ across markers, the significance thresholds at the first and second stage can vary across markers.)

The goal is to find a two-stage design that maximizes the expected number of true associations detected while controlling the FPRP $\leq \varphi$, given the number of subjects N and the budget B. This involves maximizing the expected number of true positives over a grid of designs parameterized by the first stage sample size N_1 and the first stage significance level $\boldsymbol{\alpha}_1 = (\alpha_{11},\ldots,\alpha_{1M})$ (N_2 is fixed given N_1; $\boldsymbol{\alpha}_2$ is fixed given $\boldsymbol{\alpha}_1$, φ, and N_1) such that the expected overall cost remains below B. Assuming that the expected number of truly associated variants is very small relative to the total number M genotyped, the expected cost is

$$\Sigma_m K (N_1 + N2\, \alpha_{1m}\ \kappa),$$

where K is the per-genotype cost of the very high throughput technology and κ is the ratio of high-throughput to very-high-throughput genotyping costs. The calculations developed here could also be used to examine the impact of increasing sample size and budget on the expected number of true positives. This would provide a guide for researchers designing a de novo study or contemplating the cost-benefit ratio for enrolling more subjects or increasing the genotyping budget.

2.1. *Two-stage false positive report probability*

The FPRP is defined as the probability that a variant that has been found statistically significant is actually not associated with disease, but appears statistically significant merely by chance. The FPRP depends on the Type I error rate α of the applied test, the Type II error rate β, and the prior probability that a given locus is truly associated with disease π:

$$\frac{\alpha(1-\pi)}{\alpha(1-\pi)+(1-\beta)\pi}.$$

As originally proposed, the prior density on the strength (relative risk) of the variant-disease association put point masses at unity (no association) and a single non-unity value (which was used to compute the power 1-β).[*13, 17*] This assumption is easily relaxed, allowing for a range of possible non-null relative risks (and hence a range of βs = β_1,.., β_A) with a range of prior probabilities $\pi_1,\ldots,\pi_A$, leading to the following expression for the FPRP:

$$\frac{\alpha(1-\sum_{j=1\ldots A}\pi_j)}{\alpha(1-\sum_{j=1\ldots A}\pi_j)+\sum_{j=1\ldots A}(1-\beta_j)\pi_j}.$$

To calculate the FPRP for the two-stage design, I assume that only second stage subjects are used to test the most promising markers, so that the tests in the first and second stage are independent. (This approach differs from that of Satogopan et al., who use first and second stage subjects to test the most promising markers.[*12*] Although limiting second-stage tests to the second-stage sample may reduce power somewhat, it may also be most appropriate if the second-stage sample is a separate study.) The two-stage FPRP for marker m then has a simple form:

$$\frac{\alpha_{2m}\alpha_{1m}(1-\sum\pi_j)}{\alpha_{2m}\alpha_{1m}(1-\sum\pi_j)+\sum(1-\beta_{mj2})(1-\beta_{mj1})\pi_j}.$$

Here the Type II error probabilities β_{mjk} are indexed by the marker $m \in \mathcal{M}$, the relative risks j=1,...,A, and stage k; the power 1-β_{mjk} depends on marker allele frequency, relative risk, number of subjects in stage k. Note this expression assumes the priors $\pi_1,\ldots,\pi_A$ are identical for all markers; this could be easily modified to incorporate prior beliefs about probability a particular SNP plays a causal role. For example, non-synonymous coding SNPs could be upweighted relative to intergenic "tag SNPs."

Assuming further that the priors and tests at multiple loci are independent, the expected number of true positives in a two-stage study can be calculated as:

$$\sum_{m,j}(1-\beta_{mj2})(1-\beta_{mj_1})\pi_j \cdot$$

This expression is used to solve for first stage sample size and first-and second stage significance level that maximize the ETP while controlling the minimum FPRP and remaining under budget.

2.2. *The FPRP and marker choice*

Marker choice is a key factor in genome-wide association studies. Several overlapping paradigms have been suggested.[*4, 8, 18*] An advantage of the FPRP is that the prior probability that a given marker is associated with disease can account for the fact that the set $\mathcal{M}$ of markers measured does not contain all causal loci, and some causal loci may not even be in linkage disequilibrium with markers in $\mathcal{M}$.

2.3. *The FPRP and prior choice*

Choosing a prior for a given candidate gene can be quite difficult and rather subjective. Often the best researchers can do is set a range of priors that spans several orders of magnitude.[*13*] On the other hand, priors on the number of loci with a detectable marginal effect on a disease (and to a lesser extent priors on the sizes of those effects) are somewhat easier to specify, as the number of such loci is believed to be quite small relative to the number of loci screened. At most there will be several score of such loci; perhaps more realistically, there will be less than a dozen. For example, several authors have argued that the prior probability a randomly chosen marker is associated with disease should be on the order of 1 in 10,000.[*13, 19*] For the example presented below, I assume prior probabilities that a marker has a genetic relative risk of 2.0, 1.5 or 1.3 are 1, 2 and 4 in 100,000, respectively. Thus on average seven of the 100,000 markers tested will be truly associated with disease. More realistic priors could be developed using what is known about the distribution of the size of genetic effects in general, characteristics of the disease under study (such as sibling relative risks), and plausible distributions for the allele frequencies of susceptibility loci.[*9*]

3. Results

I calculated the expected number of true positives given a fixed budget for four designs: a single stage case-control study that aims to maximize the number of expected true positives (ETP) while holding the FWER below 5%; a two-stage study that maximizes ETP while also holding the FWER below 5%; a one-stage study that aims to maximize the number of ETP while holding the minimum FPRP below 50%; and a two-stage study that maximizes ETP while holding the minimum FPRP below 50%. Parameter values underlying these calculations are summarized in Table 1. Power was calculated for the standard Pearson's chi-squared statistic for 2×3 tables assuming the risk allele had a multiplicative effect on the risk of disease. For the two-stage studies aimed at controlling the overall FWER, the first- and second-stage samples are analyzed independently, so that the overall Type I error rate is $\alpha_1\alpha_2$. The FWER for one- and two-stage studies is controlled by ensuring the overall Type I error is below $1\text{-}(1\text{-}\alpha_*)^{1/M}$, where α_* is the target family-wise error rate (e.g. 5%).

Figure 1 shows the maximum expected true positives for the four designs for a range of budgets. The two stage designs are always more powerful than the analogous one-stage designs, although for large enough budgets the one- and two-stage designs have equivalent power. This reflects the fact that if we could afford to genotype all available subjects using the very high-throughput technology, we would. The power advantage of the two-stage designs comes from the ability to genotype more subjects at the second stage; simply splitting the sample and testing the same set of markers in each sub-sample always results in less power. [*20*] Note that if there were an unlimited number of cases available for enrollment, the two-stage designs would remain more powerful than the one-stage designs as budget increased.

The designs that control the FPRP also have greater power than the analogous designs that control the family-wise error. The expected number of false positives for the FPRP-based designs is larger, as in this case (FPRP=50%) the expected number of false positives for the FPRP is equal to the expected number of false negatives, while for FWER designs it is fixed at ≈ 0.05. However, the expected number of false positives is fixed by design, regardless of the true number of associated markers and their relative risks. This is because limiting the FPRP at φ requires that on average the number of false positives should not exceed $[\varphi / (1\text{-}\varphi)] \times$ the expected number of true positives. Thus, when using an FPRP-based design the expected resources spent following false leads is limited, while the chance of detecting a true association is increased.

Figure 2 shows the number of expected true positives when $N_1 = N_2 = 1000$ as a function of α_1, the first stage significance level and roughly equal to the proportion of markers taken to the second stage. These sample sizes were chosen because the number of expected true positives for the two-stage FPRP design begins to plateau when N_1= 1000, at a budget of 1082 (in units of cost to genotype all M markers on one subject using the very-high-throughput technology). The parameters that maximize the expected true positives for that budget are $\alpha_1 = 0.0067$ (on average roughly 670 markers are taken to the second stage) and α_2=0.0055. The sharp initial increase in Figure 2 suggests that the power of the two stage design is driven in large part by the Type II error rate of the first stage: if a truly associated marker does not make out of the first stage, no association can be found at the second. On the other hand, the eventual slow decline in expected number of true positives with increasing α_1 is due to the increasingly stringent α_2 level necessary to control the two-stage FPRP.

Finally, Figure 3 shows the number of expected true positives for two-stage designs as a function of α_1 and N_1 for the FPRP and FWER designs. The power surface has similar shape for both designs, although the precise allocation of samples and number of markers to carry to the second stage that maximizes power differs somewhat. For fixed N_1, the FPRP is maximized at higher α_1 than the FWER; for fixed α_1, the FPRP is maximized at a higher N_1.

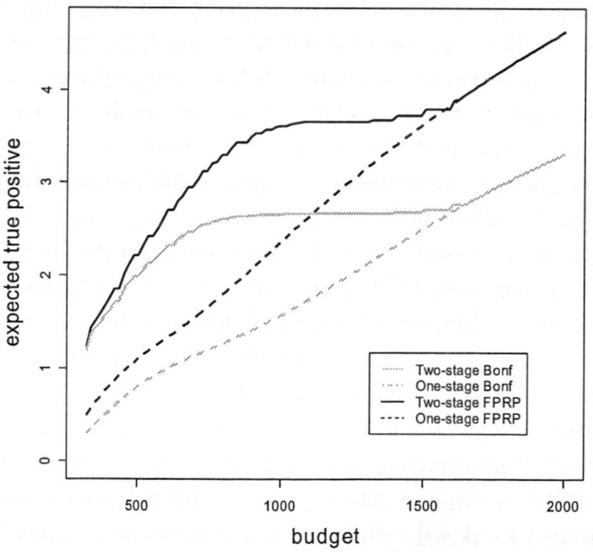

Figure 1. Expected true positives for four study designs as a function of budget, when the total available sample size is fixed at 2000 (see Table 1 for other study parameters). The budget is given in terms of the cost to genotype one subject using the very-high-throughput technology. Thus for a budget of 2000, all subjects could be genotyped using the very-high-throughput technology.

4. Discussion

Although I presented the two-stage FPRP design in the context of case-control studies using unrelated individuals, it can be easily adapted to other contexts, including continuous traits and family-based designs. Other methods for power calculations can also be used. For example, van den Oord and Sullivan proposed a two-stage design very similar to ours, except they used a liability-threshold model in their power calculations.[*16*] They also assumed marker allele frequencies were randomly distributed similar to Schork[*21*], although in a genome-wide association analysis using a set of markers chosen from a screening set like the International HapMap it is likely researchers will have accurate estimates of marker allele frequencies beforehand and can use them in power calculations as we did here.

I have also presented this two-stage design in the context of genotyping technologies that allow for subject-specific genotypes, but DNA pooling techniques could also used at either or both stages. This may lead to a further reduction in genotyping costs, although the modest genetic effects anticipated in common complex disease may lie below the signal detection threshold without increasing the number of pools to the point where pooling loses much of its efficiency relative to individual genotyping.

One concern with large genetic association studies using unrelated individuals is that markers may be associated with a trait not because the marker is near a gene on the causal pathway, but because of population stratification or cryptic relatedness.[*22-24*] In a well-designed study in a relatively homogeneous population such as non-Hispanic European Americans, such effects should be small and could in principle be accounted for using a multinomial analog of the FPRP that calculates the posterior probability of a positive test being a false positive, a true positive association due to population stratification, or a true positive due to linkage disequilibrium with a causal locus. It should be possible to differentiate between the latter two results, because the prior distribution for population stratification effect sizes will differ from that for effect sizes due to linkage disequilibrium.

Sequential designs are an alternative that may be more efficient than the two stage procedure presented here.[*14, 15*] Instead of treating the sample size at the first stage as fixed, sequential designs consider it to be random and keep adding subjects and recalculating test statistics until one of two significance thresholds is crossed. This allows researchers to stop genotyping early if a significant result is observed or stop early for futility if early returns look "really null." Many such designs are based on controlling the Type I error rate for a single independent variable (here: marker), but it should be straightforward to adapt them to control the FPRP over multiple markers. Sequential multiple decision procedures[*15*] are intriguingly analogous the FPRP procedure presented here in that instead of controlling the experiment-wide Type I error rate they seek to partition the markers into a set that is enriched for truly associated with disease and a set that is overwhelmingly not associated with disease.

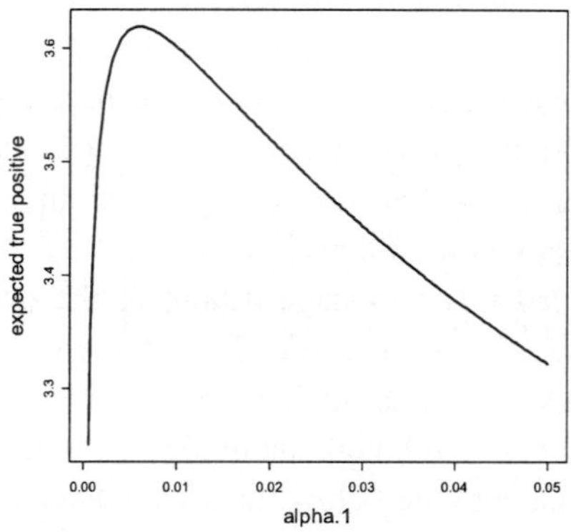

Figure 2. Number of expected true positive results as a function of the first stage significance level α_1 when the sample is split between first and second stage samples, N_1=N_2=1000.

However, there are logistical barriers in genetic association studies that may keep such sequential designs from achieving their theoretical gains in efficiency. High- and very-high throughput genotyping technologies require certain economies of scale: many subjects must be genotyped simultaneously, often at many markers simultaneously. This may limit the number of sequential tests. Two- or three-stage studies are likely feasible, but a ten-stage study may be too complex logistically. In particular, it may be infeasible or overly costly to change the set of markers typed more than two or three times. Going from 100,000 markers to ca. 700 is a sufficiently large decrease to justify redesigning genotyping protocols; going from 100,000 to 65,000 may not be.

The False Positive Report Probability is closely related to the False Discovery Rate, defined as the expected fraction of positive tests that are false positives. Procedures that control the FDR provide a weaker form of false-positive control than those which control the family-wise error. The FDR is widely used in hypothesis-generating microarray studies, and its use has been proposed in the context of genetic association studies where researchers are testing many hypotheses.[*16, 25*] When the number of true positives is small relative to the total number of tests (as it is here), the FDR ≈ FPRP.[*26*] Although we do not claim the FPRP procedure presented here controls the false discovery rate, it should be accurate when the number of truly associated markers with clinically relevant effect sizes is very small relative to the total number tested, and the power to detect the truly associated markers is high. Further, the FPRP has a practical advantage over standard FDR procedures that require all test p-values to be ranked: if markers are genotyped in batches, researchers can analyze each batch as it comes in, rather than waiting for data on all markers before moving on.

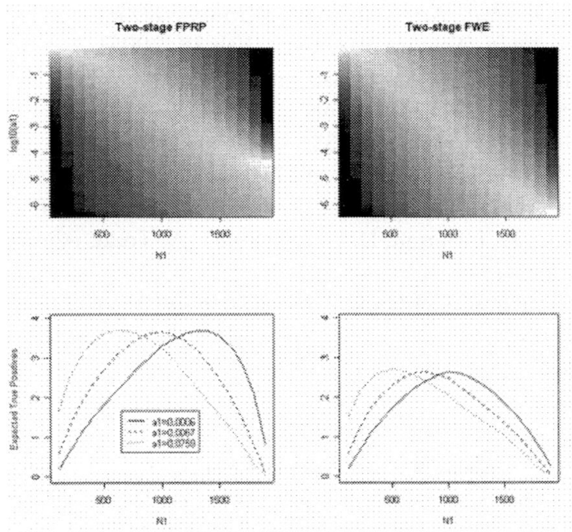

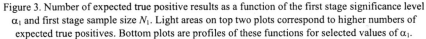

Figure 3. Number of expected true positive results as a function of the first stage significance level α_1 and first stage sample size N_1. Light areas on top two plots correspond to higher numbers of expected true positives. Bottom plots are profiles of these functions for selected values of α_1.

In summary: two-stage designs that limit the number of markers studied in all available subjects can lead to considerable savings. Furthermore, designs that control the two-stage False Positive Report Probability can be more powerful than designs that control the family-wise error rate. The expected number of false positives is higher using FPRP designs, but this should be acceptable in the context of genome-wide association studies, where any positive results would be followed up in other epidemiological and laboratory studies. Although the choice of priors for the FPRP is subjective, this is true of all power calculations, and the quasi-Bayesian framework of the FPRP allows uncertainty in the number of causal loci and their strength to be built in to design calculations.

Acknowledgments

This work was supported by NIH grants 5 R01 MH59532 and U01CA098233. The author thanks Drs. David Hunter, Gilles Thomas and Stephen Chanock for helpful discussion.

References

1. A. Balmain, J. Gray, B. Ponder, *Nat Genet* **33 Suppl**, 238 (Mar, 2003).
2. P. D. Pharoah *et al.*, *Nat Genet* **31**, 33 (May, 2002).
3. D. A. Hinds *et al.*, *Science* **307**, 1072 (Feb 18, 2005).
4. The International HapMap Consortium *Nature* **426**, 789 (2003).
5. *NCI Cancer Bulletin*. (2005), vol. 2, pp. 7.
6. C. S. Carlson *et al.*, *Nat Genet* **33**, 518 (Apr, 2003).
7. N. Risch, K. Merikangas, *Science* **273**, 1616 (1996).
8. J. N. Hirschhorn, M. J. Daly, *Nat Rev Genet* **6**, 95 (Feb, 2005).
9. W. Y. Wang, B. J. Barratt, D. G. Clayton, J. A. Todd, *Nat Rev Genet* **6**, 109 (Feb, 2005).
10. J. M. Satagopan, R. C. Elston, *Genet Epidemiol* **25**, 149 (Sep, 2003).
11. J. M. Satagopan, E. S. Venkatraman, C. B. Begg, *Biometrics* **60**, 589 (Sep, 2004).
12. J. M. Satagopan, D. A. Verbel, E. S. Venkatraman, K. E. Offit, C. B. Begg, *Biometrics* **58**, 163 (Mar, 2002).
13. S. Wacholder, S. Chanock, M. Garcia-Closas, L. El Ghormli, N. Rothman, *J Natl Cancer Inst* **96**, 434 (Mar 17, 2004).
14. I. R. Konig, A. Ziegler, *Hum Hered* **56**, 63 (2003).
15. M. A. Province, *Genet Epidemiol* **19**, 301 (Dec, 2000).
16. E. J. van den Oord, P. F. Sullivan, *Hum Hered* **56**, 188 (2003).
17. D. C. Thomas, D. G. Clayton, *J Natl Cancer Inst* **96**, 421 (Mar 17, 2004).
18. D. Botstein, N. Risch, *Nat Genet* **33 Suppl**, 228 (Mar, 2003).
19. J. N. Hirschhorn, D. Altshuler, *J Clin Endocrinol Metab* **87**, 4438 (Oct, 2002).
20. D. C. Thomas *et al.*, *Am J Epidemiol* **122**, 1080 (1985).
21. N. Schork, B. Thiel, P. St. Jean, *J Exper Zool* **282**, 133 (1999).
22. M. L. Freedman *et al.*, *Nat Genet* **36**, 388 (Apr, 2004).
23. D. Thomas, J. Witte, *Cancer Epidemiol Prev Biom* **11**, 505 (2002).
24. S. Wacholder, N. Rothman, N. Caporaso, *Cancer Epidemiol Prev Biomarkers* **11**, 513 (2002).
25. C. Sabatti, S. Service, N. Freimer, *Genetics* **164**, 829 (Jun, 2003).
26. J. D. Storey, R. Tibshirani, *Proc Natl Acad Sci U S A* **100**, 9440 (Aug 5, 2003).

THE WHOLE GENOME TAGSNP SELECTION AND TRANSFERABILITY AMONG HAPMAP POPULATIONS

REEDIK MÄGI, LAURIS KAPLINSKI, MAIDO REMM

Department of Bioinformatics, University of Tartu, Riia str. 23
Tartu, 51010, Estonia

One of the crucial issues of association studies is the selection of markers. One possible approach would be to select tagging SNPs (tSNPs) according the HapMap information. In this study we present the number of tSNPs required for the association analysis of the entire human genome for all available HapMap population samples: CEPH, Nigerian, Chinese and Japanese. For future association studies, it is also important to know how well the tSNP set of one population sample can describe the markers of another population. Therefore, we have calculated the proportion of markers adequately described by tSNPs and how many additional tSNPs we need to describe all markers of another population.

1. Introduction

The selection of markers is a very important step for a successful association study. The amount of markers and thereby the cost of the study can be significantly reduced by selecting markers according to any popular tagging SNP (tSNP) selection method[1-3]. Studies have shown that it takes approximately 500,000 – 1,000,000 to cover the whole human genome[4-6]. Lately, haploblock independent methods have gathered more popularity among researchers, particularly the r^2-bin based method, which was introduced by Carlson et al.[7].

The HapMap Project started in 2001 with the goal of determining the common patterns in the human genome and to make that information freely available in the public domain[8]. The HapMap information gives us an opportunity to calculate the approximate number of tSNPs necessary to cover the whole human genome. This approach can be used for marker selection in national genome projects or in other large association studies.

An important aspect of marker selection in standardized populations is the transferability of chosen tSNP sets to other populations[2,9]. As the allele frequencies of many markers differ among populations, the selected markers of one population sample may be insufficient to describe the whole heterogeneity of another population. Ahmadi *et al.* found, in their study of European and Japanese population samples, that it is possible to identify tSNPs that work adequately in multiple population groups[10]. Also, it is known that despite of the role of a populations' demographic history, the linkage disequilibrium (LD) pattern of all four HapMap populations is remarkably similar[11]. However, the extension of LD and haplotype allele frequencies of areas with high LD, may be very different[12].

Two important goals were set for this work:

- To find the number of tSNPs necessary to describe the whole human genome, according to HapMap population samples;
- To evaluate the transferability of HapMap tSNPs among populations – the proportion of markers adequately described by tSNP sets and the number of additional tSNPs that have to be chosen from a population sample to describe all of its markers.

2. Methods

2.1. *Population samples*

In this study we used public data from HapMap release #16c.1.* Genotype information of the following four population samples has been used for tSNP selection:

- CEPH (Utah residents with ancestry from northern and western Europe) (abbreviation: CEU) 1,104,996 markers genotyped at 60 founders
- Japanese in Tokyo, Japan (abbreviation: JPT) 1,087,297 markers genotyped at 45 population samples
- Han Chinese in Beijing, China (abbreviation: CHB) 1,087,297 markers genotyped at 45 population samples
- Yoruba in Ibadan, Nigeria (abbreviation: YRI) 1,076,381 markers genotyped on 60 founders

SNP data of all chromosomes, except for Y-chromosome, were used in this study.

2.2. *r^2-bin based tSNP selection*

We created a software program, REAPER, which is optimized for fast tSNPs selection. The program implements the r^2-bin algorithm[7] for tSNP selection and is specifically designed for full genome scale analysis. It is written in C++ and is available for 32bit Linux and Windows operating systems.

---

* Publicly available final data freeze of Phase I (21. June 2005) from http://www.hapmap.org/. Redundant-filtered data was used.

The algorithm works as follows: REAPER uses the greedy approach, always starting from the biggest possible bin. In the first step, all r^2 values between markers less than N positions apart are calculated for the full genome. Lowering the distance threshold N increases the calculation speed and lowers memory consumption, but increases the risk of leaving some relevant SNPs out of bins. In the current analysis, a default distance threshold of 1024 positions was used, giving us reasonable calculation times with the assumption that no two markers more than 1024 positions apart show significant r^2. Only boolean values are stored for each analyzed marker pair, based on whether the calculated r^2 was above or equal to the linkage threshold or not. The default linkage threshold is 0.8. Lowering the threshold results in larger r^2-bins and smaller number of tSNPs. However, with a lower linkage threshold, the average prediction power of tSNPs decreases, thus requiring more individuals in further association studies. The program allows the use of any r^2 threshold between 0 and 1. For each bin, the marker with the largest number of "good" r^2 values (above or equal to the threshold) from the set of unbinned markers, is selected as the r^2-bin seed marker. All other markers, whose r^2 with the seed, is above or equal to the threshold are put into the bin, together with the seed marker. The markers in the bin are sorted by the number of "good" r^2 values with other bin members. All markers in the bin having, "good" r^2 values with all other bin members, are marked as alternative tSNPs. The candidate tSNP with the highest average r^2 will be reported for the given bin. The same algorithm is repeated, choosing a new seed marker from the remaining, ungrouped markers of the genome, until no SNP has a r^2 value with other ungrouped markers above or equal to the linkage threshold. The remaining SNPs are then added to the tSNP list as single-marker bins. REAPER can also be forced to use a pre-selected list of tSNPs from a file. In this case, the tSNP selection procedure is identical, except that in the first step bins are constructed by picking seeds from the tSNP list. Remaining markers are then distributed to bins using the abovementioned greedy algorithm.

The calculations described in this paper (ca 25 sets of whole-genome tSNPs) were made by parallel computing on 14 Pentium 4 2.200MHz computers with 1GB memory each. This calculation process took about 1 week to complete. REAPER is freely available for academic users at http://bioinfo.ebc.ee/download/.

2.3. *tSNP transferability calculation*

The HapMap data is likely to be used for the selection of tSNPs in the future. However, the cases for the association study may be collected from different populations. In order to evaluate the suitability of the HapMap based tSNPs are on other populations, we calculated the transferability of tSNP sets between HapMap populations. For each population, the tSNP set was calculated with the r^2-bin method ($r^2 \geq 0.8$). These tSNP sets were used in other populations' tSNP calculations as pre-selected seeds. For each population, the number of additional tSNPs to describe all markers in the dataset was calculated.

2.4. *Strategies for tSNP selection from other populations*

If the tSNP set is calculated from a different population, it might be possible to genotype some additional individuals from the observed population to increase the efficiency of the tSNP set. The tSNP sets that were defined in the CEPH trios were tested on all other populations. We added an additional, randomly selected, 10, 20 and 30 persons from the other datasets and tested how efficiently the tSNP sets of these mixed populations can describe Chinese, Japanese and Nigerian population samples. The efficiency of a given tSNP set in the other population samples is evaluated by the following criteria: an average r^2 among all typed SNPs and 25% lower quartile of r^2 among all typed SNPs.

3. Results

3.1. *The number of tSNPs required for the whole human genome*

The idea of tSNP based association analysis is currently very popular. Using this analysis method, it is therefore interesting to understand how many tSNPs does it take to cover the whole human genome. To answer this question, we have calculated the tSNP sets for all four HapMap populations using the r^2-bin method. The number of tSNPs needed to describe the population samples varied from 579,978 in the CEPH sample set to up to 716,617 in the Nigerian set (Table 1).

We have also analyzed the distribution of r^2-bins by size (Fig.1). The proportion of markers in single-marker-bins is varying from 75.9% in the CEPH samples to 79.6% in the Nigerian sets. The r2-bins over 30 markers were rarely found in any population. R^2-bins with more than 10 markers are most common in the CEPH population sample (1.9%) and are quite rare in the Nigerian samples (0.6%). Largest r^2-bins are in chromosome 12 of the CEPH population (257 markers) and in chromosome X of the Japanese population (255 markers).

Table 1. The number of tSNPs necessary to describe all markers in each HapMap population sample.[†] The numbers on the diagonal show how many tSNPs are necessary to describe each sample. Each populations sample tSNP sets were also tested on other samples to test the transferability of tSNPs. Additional tSNPs have been found for the markers that are not described by a tSNP set (added after '+' signs) of each corresponding population sample. tSNP sets were calculated according to the r^2-bin method with $r^2 \geq 0.8$ in bins.

		population sample of tSNP selection			
		CEU	JPT	CHB	YRI
studied population sample	CEU	579,978	+91,495 (15.5%)	+63,088 (9.9%)	+67,555 (9.4%)
	JPT	+92,840 (16.0%)	590,979	+50,993 (8.0%)	+82,165 (11.5%)
	CHB	+88,018 (15.2%)	+54,470 (9.2%)	639,459	+78,163 (10.9%)
	YRI	+200,358 (34.6%)	+209,398 (35.4%)	+122,019 (19.0%)	716,617

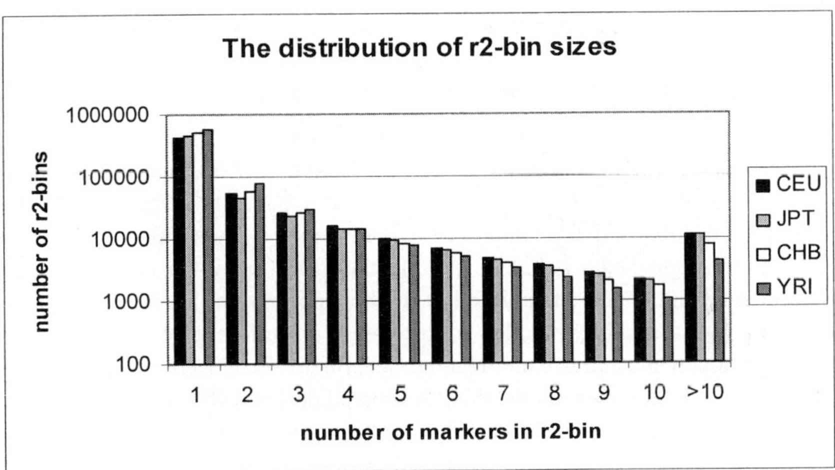

Fig 1. The distribution of r^2-bin sizes. The counts of r2-bins with all marker numbers have been determined for all four populations.

[†] Chromosome Y not included.

3.2. *tSNP transferability between populations*

To find the union and the intersection of different tSNP sets of population samples, we selected the tSNP list of one population sample and used this as a pre-selected seed list for REAPER, while calculating r^2-bins of another population. The percentage of additional tSNPs needed is indirectly reflecting the union between the tSNP sets of the two populations. We have also determined the proportion of tSNPs that is not used in another population sample by comparing a population's own tSNP count with the number of tSNPs required from another population (Table 1, outside of diagonals). The number of additional tSNPs required to cover the other populations is the smallest for the Nigerian population sample tSNP set (additional 9.4% - 11.5%) and the largest in the Japanese populations sample tSNPs (additional 35.4% of tSNPs to cover the Nigerian samples). The number of redundant tSNPs can be found by conducting a comparison of the number of tSNPs found from the population sample with the number of transferred tSNPs + additional tSNPs. For example, if we use the Japanese tSNP set on the Chinese population sample, we need 590,979 markers + 54,470 additional tSNPs. It takes 639,459 markers to describe the Chinese population sample with its own tSNPs. Therefore, we have genotyped 639,459 – 645,449 = 5990 tSNPs in excess.

3.3. *Strategies for tSNP selection from other populations*

In the future, the association studies will be performed on various populations, many of which are not covered by the current HapMap. If the standard HapMap population tSNPs were used for studying other population samples, then the performance of tSNP might be improved by genotyping a small amount of people from the same population before tSNP selection (local HapMap approach). We have used the CEPH tSNP set on the other HapMap populations in order to test how many individuals should be added to increase the performance of tSNPs. To understand how many extras were required, local individuals were mixed with the CEPH individuals' samples. As the tSNP set should describe all genotyped markers in the dataset, we determined the best r^2 score between each marker and any tSNP. For the whole population's dataset, we found the average of these r^2 scores and the lesser quartile of 25%. As expected, additional genotyped persons increase the r^2 values between tSNPs and other markers most efficiently by adding the Nigerian population samples (Fig. 2). Also note that the performance of some poor markers (indicated by lower quartile) increases significantly by using local population samples for the selection of tSNPs.

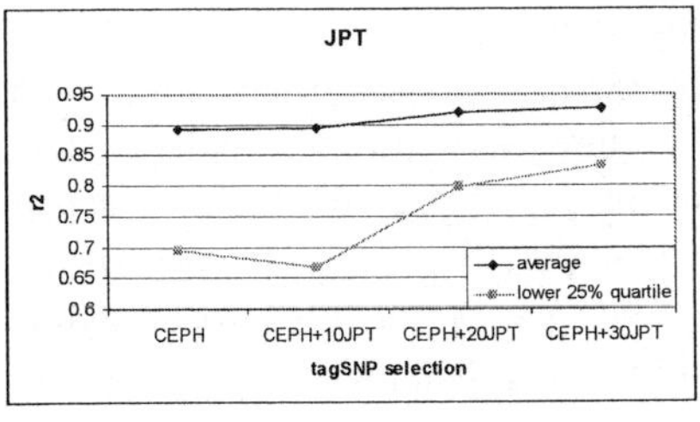

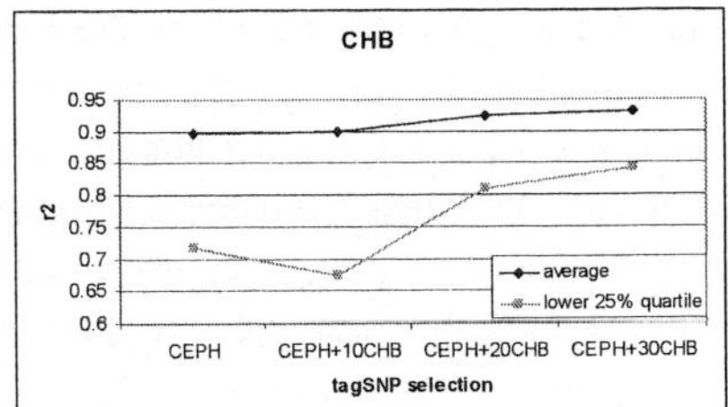

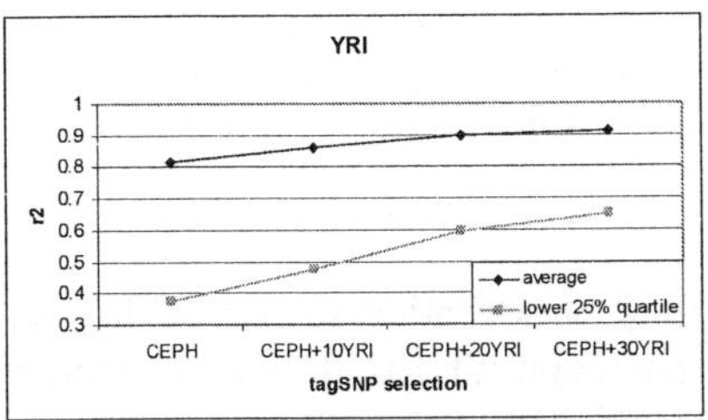

Fig 2A-C. Mean and 25% lower quartile of r^2 value between tSNPs and all the markers in the observed population sample. tSNPs were calculated according to the CEPH population sample; CEPH + 10 random persons from the observed population sample; CEPH + 20 random persons from the observed population sample; CEPH + 30 random persons from the observed population sample.

4. Discussion

The results of the association analysis are strongly determined by marker selection. Thereby the selection of tSNPs is one of the most important aspects while designing a new study. The HapMap Consortium has contributed greatly to this research by making four population samples, with more than million markers genotyped in each, publicly available. That data gives researchers an opportunity to use fine scale LD data for designing and using different marker selection algorithms.

The approximate number of tSNPs to cover the whole human genome has been predicted to be between 500,000 and 1,000,000 – one marker per 3 - 5kb[4,5]. Our results re-confirm the order of magnitude for HapMap populations. The large number of singleton markers indicates that the current HapMap marker density is still too sparse for adequate coverage of the whole human genome. Therefore, the number of tSNPs may grow, when new data is added to HapMap in Phase II.

As the idea of tSNP based association analysis is currently very popular, it is important to find out how well the tSNP set of HapMap population will work on other population samples. Our results indicate that tSNP sets from different populations can describe other populations reasonably well. This result is similar to the findings of Ahmadi *et al.*[10]. In their study of Japanese and European (CEPH) populations they found that the number of tSNPs that adequately cover both populations is only 19% higher than the one required for only one population. Our findings show that the percentage is somewhat lower – 15.5%-16% in the same pair of populations. However, we also found that the tSNP sets of non-African populations describe Nigerian data poorly. And *vice versa,* if we use Nigerian tSNPs on non-African populations, we found that we genotyped a large number of unnecessary markers.

The current study required numerous calculations of the whole-genome tSNP sets, which was possible thanks to the speed and low memory demand of REAPER.

Acknowledgments

This work is supported by the Estonian Ministry of Education and Research grant no. 0182649s04 and by the applied research grant EU19730 from Enterprise Estonia. We thank Ulvi Gerst-Talas, Jody Novakoski and Katre Palm for valuable help with English grammar.

References

1. Halldorsson, B. V., Istrail, S. & De La Vega, F. M. Optimal selection of SNP markers for disease association studies. *Hum Hered* **58**, 190-202 (2004).
2. Mueller, J. C. et al. Linkage disequilibrium patterns and tagSNP transferability among European populations. *Am J Hum Genet* **76**, 387-98 (2005).
3. Gabriel, S. B. et al. The structure of haplotype blocks in the human genome. *Science* **296**, 2225-9 (2002).
4. Kruglyak, L. Prospects for whole-genome linkage disequilibrium mapping of common disease genes. *Nat Genet* **22**, 139-44 (1999).
5. Dunning, A. M. et al. The extent of linkage disequilibrium in four populations with distinct demographic histories. *Am J Hum Genet* **67**, 1544-54 (2000).
6. Judson, R., Salisbury, B., Schneider, J., Windemuth, A. & Stephens, J. C. How many SNPs does a genome-wide haplotype map require? *Pharmacogenomics* **3**, 379-91 (2002).
7. Carlson, C. S. et al. Selecting a maximally informative set of single-nucleotide polymorphisms for association analyses using linkage disequilibrium. *Am J Hum Genet* **74**, 106-20 (2004).
8. The International HapMap Project. *Nature* **426**, 789-96 (2003).

9. Liu, N. et al. Haplotype block structures show significant variation among populations. *Genet Epidemiol* **27**, 385-400 (2004).
10. Ahmadi, K. R. et al. A single-nucleotide polymorphism tagging set for human drug metabolism and transport. *Nat Genet* **37**, 84-9 (2005).
11. De La Vega, F. M. et al. The linkage disequilibrium maps of three human chromosomes across four populations reflect their demographic history and a common underlying recombination pattern. *Genome Res* **15**, 454-62 (2005).
12. Sawyer, S. L. et al. Linkage disequilibrium patterns vary substantially among populations. *Eur J Hum Genet* **13**, 677-86 (2005).

COMPUTATIONAL APPROACHES FOR PHARMACOGENOMICS

MICHELLE W CARRILLO
Department of Genetics, Stanford University, 300 Pasteur Drive L331 Stanford, CA 94305 USA

RUSSELL A WILKE
Personalized Medicine Research Center, Marshfield Clinic Research Foundation, 1000 N Oak Avenue, Marshfield, WI 54449 USA

MARYLYN D RITCHIE
Center for Human Genetics Research, Department of Molecular Physiology & Biophysics, Vanderbilt University, 519 Light Hall Nashville, TN 37232 USA

Pharmacogenomics is an interdisciplinary field that combines genetics, classical pharmacology, and molecular biology to address individual variation in drug response. Coordinated research efforts are therefore multidisciplinary by necessity. Genetic and, maybe more importantly, genomic variation explains many differences in how people respond to medical treatments. Environmental factors including diet, age, and lifestyle influence response to medicines as well, but these also need to be placed in the context of genetics to fully understand their significance. Pharmacogenomics may ultimately lead to personalized medicine in which the most effective medicine is prescribed to individuals based on their genetic information, and adverse drug events could be virtually eliminated.

As the research emphasis on pharmacogenomics increases, there is a corresponding demand for computational and statistical methods and tools to analyze the data produced. Genome scans, gene expression arrays in the presence of drugs and large scale genotyping experiments have become common. These types of experiments yield enormous amounts of data that need to be organized, stored, analyzed and disseminated. There is an important need for computational approaches to address these issues in order to answer scientific questions such as: how do phenotypic results relate to genomic variation, and what is the significance of a particular drug on gene expression? Some of the current computational challenges in the field include database design and implementation, data sharing among pharmacogenomics centers, statistical analysis, statistical and computational method development, and real data applications.

The goal of The Pacific Symposium on Biocomputing is to explore current research in the theory and application of computational methods as they apply to problems of biological significance. The significance of pharmacogenomics is clear. Many population geneticists anticipate that this will be the first discipline wherein functional genomics translates into clinical application on a large scale. As such, PSB represents an ideal forum to further the analytical and computational approaches associated with such an endeavor. The biological and chemical technology is advancing, and a merging of pharmacogenomics with biocomputing is inevitable.

This session is dedicated to computational approaches for pharmacogenomics. The papers representing this session range from large-scale data analysis to applications of statistical methods to pharmacodynamic data models.

Micorarrays are increasingly utilized in pharmacogenomic studies. Because of the vast amount of data generated by these experiments, computational approaches are vital. Several papers in this session involve analysis of microarray data in a pharmacogenomics context. Imoto et al. describe a computational strategy for creating gene networks with regard to some chemical compound, such as a drug. They apply their approach to microarray data for human endothelial cells treated with the drug fenofibrate. The authors discovered a gene network involving a known target of fenofibrate, PPAR-α.

Borgwardt et al. developed a kernel-based approach to classify time series microarray expression data. As proof of concept, they use their method to predict drug response in Multiple Sclerosis patients from a published dataset.

As an alternative approach to microarray data analysis, Richter et al. investigate an opportunity to apply gene expression results to predict growth inhibition of human tumor cells. The authors used gene expression data to extend the classical Structure Activity Relationship (SAR) and Quantitative Structure Activity Relationship (QSAR) paradigm. They found that, in this particular case, the addition of transciptomic information worsened performance instead of enhancing it, unless the data were first aggregated in some manner.

Statistical analysis methods can also be successfully applied to pharmacogenomics data. Motsinger, et al. applied a statistical method called Multifactor Dimensionality Reduction (MDR) designed to detect gene-gene and gene-environment interactions in pharmacogenomics studies. The authors were able to identify associations between candidate genes and environmental factors and a common arrhythmia called postoperative atrial fibrillation (PoAF).

The last paper in this session focuses on a data model relating to pharmacodynamic reactions. Lin et al. applied a bivariate model created to detect the genetic determinants affecting drug response curves for systolic and diastolic blood pressures. The authors looked at the β2AR gene and discovered a haplotype associated with the response of SBP and DBP to the drug dobutamine.

CLASS PREDICTION FROM TIME SERIES GENE EXPRESSION PROFILES USING DYNAMICAL SYSTEMS KERNELS

KARSTEN M. BORGWARDT
Institute for Computer Science, Ludwig-Maximilians-University of Munich, Oettingenstr. 67, 80538 Munich, Germany
kb@dbs.ifi.lmu.de

S.V.N. VISHWANATHAN
Statistical Machine Learning Program, National ICT Australia, Canberra, 0200 ACT, Australia
SVN.Vishwanathan@nicta.com.au

HANS-PETER KRIEGEL
Institute for Computer Science, Ludwig-Maximilians-University of Munich, Oettingenstr. 67, 80538 Munich, Germany
kriegel@dbs.ifi.lmu.de

We present a kernel-based approach to the classification of time series of gene expression profiles. Our method takes into account the dynamic evolution over time as well as the temporal characteristics of the data. More specifically, we model the evolution of the gene expression profiles as a Linear Time Invariant (LTI) dynamical system and estimate its model parameters. A kernel on dynamical systems is then used to classify these time series. We successfully test our approach on a published dataset to predict response to drug therapy in Multiple Sclerosis patients. For pharmacogenomics, our method offers a huge potential for advanced computational tools in disease diagnosis, and disease and drug therapy outcome prognosis.

1 Introduction

Gene expression levels change over time, as proteins interfere with gene transcription. Proteins and DNA interact in a complex feedback system of gene expression control, in which some proteins foster gene expression as *transcription factors*, while others reduce transcription activity as *inhibitors* (for details see [1]). Furthermore, protein-protein interactions can increase or reduce the influence of certain proteins on transcription. These networks of gene expression control form the basis of essential cellular processes such as the cell cycle, development, and disease progression.

Single microarray profiles describe one current state of a cell only and may prove inadequate to study these complex interactions that steer biological processes. Therefore, it becomes necessary to view and analyze gene expression profiles as dynamical systems evolving with time. Such an approach may lead to a more expressive model for interpreting molecular processes within cells.

Over recent years, a growing number of time series of microarray data has become available in databases such as the Stanford Microarray Database [2] and GEO [5]. Whereas early studies used algorithms on microarray time series that had been developed for static data (for example [12]), the interest in algorithms that are able to handle and analyze time series of gene expression data in particular has grown tremendously since. The interested reader is referred to [3] for a full review of approaches and challenges in this field.

1.1 Classification Using Time Series Gene Expression Profiles

Most data mining algorithms that have been developed for gene expression time series (for example [10]) deal with the problem of clustering. Given a set of data points D, clustering algorithms try to decompose D into subsets $\{D_1, \ldots, D_n\}$ such that similarity between data points is maximized within each cluster, i.e. each subset, and minimized between distinct clusters. In short, clustering finds classes of data when classes are unknown. In medical applications, clustering is most required when exploring subtypes of the same disease or when searching groups of genes with similar expression profiles, i.e. to find classes in unorganized data.

On the other hand, classification deals with the problem of predicting class membership of unlabeled test data points after learning from a training set of data points with known class memberships. Predicting class labels can be regarded as equivalent to predicting unknown characteristics of data points. Central questions in pharmacogenomics constitute such classification problems: Will patient X respond well to a certain therapy or drug treatment? Has patient X been infected by a pathogen? Is patient X recovering from a disease? Pharmacogenomics could greatly benefit from computational tools that answer or at least help human experts to answer these questions. The growing number of time series microarray data provide the training data from which such advanced classifiers can learn.

Goal and outline of this article In this project, our aim was as follows: Define a novel approach to time series microarray classification which uses Support Vector Machine (SVM) classification and a kernel function which respects the temporally changing character of these expression profiles. In what follows, we will present our dynamical systems kernel for gene expression time series data. Modeling the time series as dynamical systems, we measure distances between these systems and then classify them using a SVM. In Section 2, we will briefly review kernel methods. In Section 3, we show how microarray time series data can be modeled as dynamical systems and how kernels can be defined on them. We show the feasibility of our approach in experiments in Section 4 on drug response prediction. We conclude with a discussion of our findings and an outlook to future extensions and refinements of our method.

2 Support Vector Machines and Kernels

In this section we give a brief overview of binary classification with SVMs and kernels. For a more extensive treatment we refer the reader to [11], and the references therein.

Given m observations (x_i, y_i) drawn iid (independently and identically distributed) from a distribution over $\mathcal{X} \times \{\pm 1\}$ our goal is to find a function $f : \mathcal{X} \to \{\pm 1\}$ which classifies observations $x \in \mathcal{X}$ into classes $+1$ and -1. In particular, SVMs assume that f is a linear function given by

$$f(x) = \mathrm{sign}(\langle w, x \rangle + b), \tag{1}$$

and maximize the margin of separation between the decision boundary and the points from opposite classes. We also need to take into account the slack when the two classes are not linearly separable. Without going into details (which can be found in [11]) this leads to the optimization problem:

$$\begin{aligned} \underset{w,b,\xi}{\text{minimize}} \quad & \frac{1}{2}\|w\|^2 + C\sum_{i=1}^{m} \xi_i \\ \text{subject to} \quad & y_i\left(\langle w, x_i \rangle + b\right) \geq 1 - \xi_i \;\; \forall 1 \leq i \leq m \\ & \xi_i \geq 0 \end{aligned} \tag{2}$$

Here, the constraint $y_i\left(\langle w, x_i \rangle + b\right) \geq 1$ ensures that each (x_i, y_i) pair is classified correctly. The slack variable ξ_i relaxes this condition at penalty $C\xi_i$. Finally, minimization of $\|w\|^2$ ensures maximization of the margin by seeking the smallest $\|w\|$ for which the condition $y_i\left(\langle w, x_i \rangle + b\right) \geq 1$ is still satisfied.

2.1 Kernel Expansion

To obtain a nonlinear classifier, one simply replaces the observations x_i by $\Phi(x_i)$. That is, we extract *features* $\Phi(x_i)$ from x_i and compute a linear classifier in terms of the features. Note that there is no need to compute $\Phi(x_i)$ explicitly, since Φ only appears in terms of dot products:

- $\langle \Phi(x), w \rangle$ can be computed by exploiting the linearity of the scalar product, which leads to $\sum_i \alpha_i y_i \langle \Phi(x), \Phi(x_i) \rangle$.

- Likewise $\|w\|^2$ can be expanded in terms of a linear combination scalar products by exploiting the linearity of scalar products twice to obtain $\sum_{i,j} \alpha_i \alpha_j y_i y_j \langle \Phi(x_i), \Phi(x_j) \rangle$.

Furthermore, if we define

$$k(x, x') := \langle \Phi(x), \Phi(x') \rangle, \tag{3}$$

we may use $k(x, x')$ wherever $\langle x, x' \rangle$ occurs. This is often referred to as the *kernel trick* and the resulting hyperplane (now in feature space) is written as

$$f(x) = \langle \Phi(x), w \rangle + b = \sum_{i=1}^{m} \alpha_i y_i k(x_i, x) + b. \tag{4}$$

The family of methods which relies on the kernel trick are popularly called *kernel methods*, and SVMs are one of the most prominent kernel methods.

2.2 Kernels and Microarrays

In evaluation studies comparing different classification techniques, SVMs outperformed all competitors such as Fisher's linear discriminant, Parzen windows and two decision tree learners [6, 13]. Kernel methods have been applied to classify microarray data in numerous studies such as [6], and [9]. An increasing number of application studies in medicine and pharmacogenomics utilize kernel methods to disease, especially cancer diagnosis (for example [8]). Although these studies dealt with time series of microarray measurements, they ignore the temporal character of the data; i.e. microarray data are compared without exploiting the fact that measurement i follows measurement $i - 1$. As these dynamics might reflect, for example, patients' reaction to drug therapy or infection with a pathogen, pharamacogenomics requires a classification method based on the temporal dynamics in gene expression profiles.

A key advantage of kernel methods is that meaningful classifiers can be constructed from non-vectorial data if a sufficiently expressive kernel function can be designed. We are therefore not restricted to vectorial representations when classifying time series of microarray expression data. We exploit this property, in the sequel, by defining a kernel on dynamical systems for time series gene expression profiles.

3 Kernels on Time Series Microarray Data

In this section, we discuss how time series microarray data can be modeled as a Linear Time Invariant (LTI) dynamical system. We then present sub-optimal, but fast, methods for identifying model parameters. Using these estimated parameters we define kernels on dynamical systems to compare different time series which, in turn, are used by a SVM for classification.

3.1 The Model

We begin by modeling time series microarray data as a partially observed discrete time LTI model. These models are also popular in control theory where they are often called Auto Regressive Moving Average (ARMA) models or Kalman Filters. The time-evolution of this model is described as

$$y_t = Px_t + w_t \text{ where } w_t \sim \mathcal{N}(0, R) \tag{5a}$$

$$x_t = Qx_{t-1} + v_t \text{ where } v_t \sim \mathcal{N}(0, S). \tag{5b}$$

Here, $y_t \in \mathbb{R}^m$ is observed, $x_t \in \mathbb{R}^n$ is the *hidden* or *latent* variable, and $P \in \mathbb{R}^{m \times n}, Q \in \mathbb{R}^{n \times n}, R \in \mathbb{R}^{m \times m}$ and, $S \in \mathbb{R}^{n \times n}$, moreover R and S are positive semi-definite matrices. Typically $m \gg n$, and we set $P^\top P = \mathbf{1}$ to fix the scaling (see e.g. Section 4 [7]). The way to understand these models is to assume that the actual dynamics of the system are guided by a very small number of latent variables while the output space might be very high-dimensional. For instance, gene expression profiles might contain many thousands of genes, while only a few factors may be responsible for the expression.

3.2 Estimating the Parameters

Given a sequence of τ observations the identification problem is to estimate the model parameters P, Q, R and S. Exact solutions like the `n4sid` method in the systems identification toolbox of MATLAB(TM) exist, but they are very expensive to compute when the output space is high dimensional. Instead, we use a sub-optimal closed form solution proposed by [7]. Set $Y := [y_1, \ldots, y_\tau]$, $X := [x_1, \ldots, x_\tau]$, and $W := [w_1, \ldots, w_\tau]$ and solve

$$\min ||Y - PX||_F = \min ||W||_F \text{ such that } P \in \mathbb{R}^{m \times n} \text{ and } P^\top P = \mathbf{1}.$$

The unique solution to the above problem is given by $\hat{P} = U_n$ and $\hat{X} = \Sigma_n V_n^\top$ where $U_n \Sigma_n V_n^\top$ is the best rank n approximation of Y. U_n, Σ_n and V_n can be estimated in a straightforward manner from the $Y = U \Sigma V^\top$, the Singular Value Decomposition (SVD) of Y.

In order to estimate Q we solve $\min ||Q\hat{X} - X'||_F$ where $X' = [x_2, \ldots, x_\tau]$ which again has a closed form solution using SVD. Now, we can compute

$\hat{v}_t = x_t - Qx_{t-1}$, and set

$$S = \frac{1}{\tau - 1} \sum_{i=1}^{\tau-1} \hat{v}_i \hat{v}_i^\top .$$

The covariance matrix R can also be computed from the columns of W in a similar manner. For more information including details of efficient implementation we refer the interested reader to [7].

3.3 Dynamical System Kernels

For the sake of defining kernels we use the behavioral framework of [15] and identify dynamical systems, $X := (P, Q, R, S, x_0)$, with their trajectories $\mathrm{Traj}(X) = \{y_t | t \in 1, \ldots, \infty\}$. These trajectories can now be interpreted as linear operators mapping from $\mathbb{R}^m$ (the space of observations y) into the time domain ($\mathbb{N}$ in discrete time systems).

By using a exponentially decaying weighting factor we can define the kernel between two dynamical systems X and X' as the dot product of the trajectories. In other words:

$$k(X, X') := \sum_{t=1}^{\infty} e^{-\lambda t} y_t^\top y_t' = \mathrm{tr}(\mathrm{Traj}(X)\ T\ \mathrm{Traj}(X')^\top), \qquad (6)$$

where T is a diagonal operator with entries $e^{-\lambda t}$. Since y_t, y_t' are random variables (5a), we also need to take expectations over w_t, v_t, w_t', v_t'. Some tedious yet straightforward algebra [14] allows us to compute (6) as follows:

$$k(X, X') = x_0^\top M_1 x_0' + \frac{1}{e^\lambda - 1} \mathrm{tr}\left[S M_2 + R \right], \qquad (7)$$

where x_0 and x_0' are the initial conditions, and M_1, M_2 satisfy the Sylvester equations:

$$M_1 = e^{-\lambda} Q^\top P^\top P' Q' + e^{-\lambda} Q^\top M_1 Q' \text{ and } M_2 = P^\top P' + e^{-\lambda} Q^\top M_2 Q'. \quad (8)$$

The important point to note is that even though the kernel involves summing over infinite terms, they can be computed in $O(m^3)$ time. These kernels can also be interpreted in a more general framework using the Binet-Cauchy formula. More details can be found in [14].

In our experiments we will use this kernel to compute similarities between gene expression profiles, after having encoded the latter as a dynamical system. This approach has the further advantage that it allows us to compare sequences of different lengths, as they are all mapped to dynamical systems in the first place.

4 Experiments

4.1 Drug response prediction

A central prognosis problem in pharmacogenomics is drug response prediction. We therefore tested our dynamical systems kernel classifier on a set of microarray expression time series data from [4], to predict whether Multiple Sclerosis patients will respond well to treatment with recombinant human interferon beta (rIFNβ).

The dataset contains time expression profiles of 52 multiple sclerosis patients, out of which 33 were good and 19 were poor responders to rIFNβ. Expression profiles of 70 genes were measured for up to seven times per patient; the first five observations were at a regular interval of 3 months each while the last two observations were spaced 6 months apart. 17 patients missed a test and hence have only 6 measurements, 8 patients missed two tests and hence have only 5 measurements.

On this data, Baranzini et al. aimed at determining higher-order predictive patterns associated with treatment outcome and tried to uncover key players (i.e. responsible genes) associated with a good or poor response. When searching for gene expression signatures associated with drug response, they conducted clustering of samples using normalized data for all 70 genes at each time point. Although they applied different similarity measures and clustering algorithms, they did not observe concomitant segregation of samples according to their responder status. They therefore applied a method for feature selection which allowed them to determine triplets of genes whose early expression correlates with responder status. We were interested in the question of predicting user response to the drug using our dynamical systems kernels. As a side effect, we also wanted to estimate the latent dimension of the system, i.e. the number of factors which actually leads to the observed behavior.

For our experiments we modeled the temporal microarray data from [4] as LTI dynamical systems. The parameters of our LTI model were estimated using the methods described in Section 3.2. λ was set to 10 by cross-validation, n was set to a default value of 1. We then computed the kernel matrix for all pairs of dynamical systems and used a SVM for classification. We took 100 random splits of the data into 4 folds of equal size and performed 4-fold cross-validation on each of these splits. We report our classification accuracies as averages over these 100 repetitions.

Our classifier achieved an average prediction accuracy of 87.05% which compares very favorably with the 87.8% accuracy reported by [4]. We must note here that our results are obtained without any specialized knowledge of the dataset or feature selection methods; Baranzini et al. obtained their result by selecting features, namely triplets of genes whose early expression levels correlate best with response outcome. Furthermore, standard state of the art clustering techniques such as two-way hierarchical clustering failed to satisfactorily separate these samples (see Figure 1 in [4]).

In our second experiment, we checked the influence of the number of measurements per patient on classification accuracy. From a pharmacogenomics point of view, it is interesting to know if we are already able to predict therapy outcome correctly after few measurements. We repeated our classification experiment considering the first k measurements with $k \in \{2, ..7\}$ (at least two measurements are needed to derive a dynamical system). If a patient's data contains less than k measurements, we derive the dynamical systems using all available measurements. As before, in Figure 1 we report results as average classification accuracy of 4-fold cross-validation repeated 100 times.

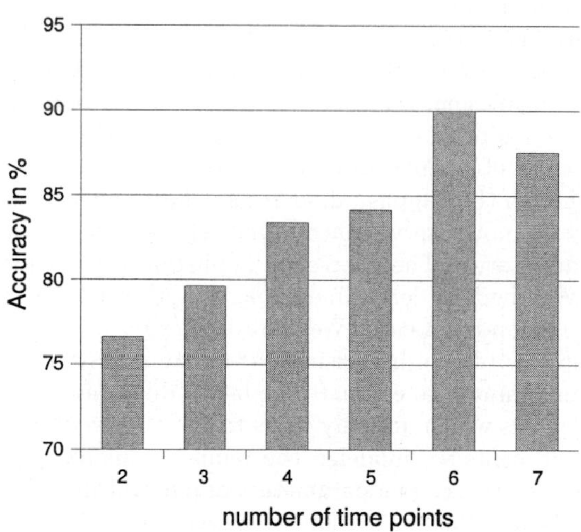

Figure 1: Mean classification accuracy in 100 repetitions versus number of measurements.

As expected, classification accuracy steadily increases as the number of measurements grows. After 3 measurements we already reach a prediction accuracy of more than 80%. But curiously enough, classification accuracy after 7 observations is less than the classification accuracy after 6 observations. This can be explained as follows: The last two observations are taken at a different time interval (6 months) than the first five observations (3 months). Therefore, the last two observations encode longer range interactions. But, for some patients only six or less observations are available. This means that we are not able to effectively model this long range interaction for those patients and this is reflected in the classification accuracy.

Also observe that by looking at only the first 6 measurements leads to an average classification accuracy of around 90%. This significant jump in classification accuracy (from around 83.5% for 5 measurements) can also be attributed to the modeling of long range effects between 5th and 6th measurement.

Observe that for 6 measurements, we now found a mean accuracy of 89.81% which is significantly better than that of the best-scoring gene triplet reported in [4] with 87.8% accuarcy (Yates' corrected $\chi^2 = 10.26$, $P = 0.0014$).

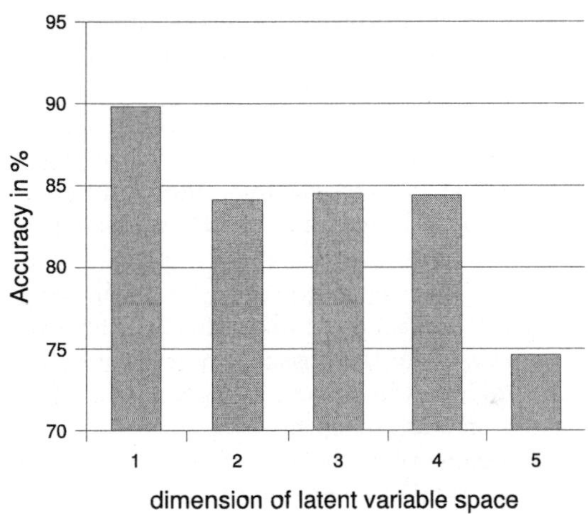

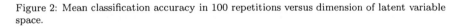

Figure 2: Mean classification accuracy in 100 repetitions versus dimension of latent variable space.

As a final experiment, we tested the impact of the latent dimension n of the system on classification accuracy. All results reported so far were with $n = 1$ which was chosen using cross-validation. We repeated our classification experiment once for each n in $\{2, \ldots, 5\}$. The best accuracy was achieved when we considered only 6 measurements. Results of 4-fold cross-validation repeated

100 times is reported in Figure 2.

Interestingly, a one-dimensional latent variable is the best choice for our dynamical systems model. As a consequence, one single factor, i.e., one group of genes with highly correlated expression levels, seems to be responsible for the microarray time series expression data which allow us to separate good and poor responders with high accuracy.

5 Discussion

In this paper, we modeled a series of gene expression profiles as LTI dynamical systems, and used a kernel on dynamical systems to classify them. The main advantage of our method is that it respects the temporal nature of the data, and is independent of the length of the time series being compared.

Our model can also be extended to predict future values of the time series. In other words, after building the LTI model, we can simulate it to predict the value of a gene expression profile at a future time step. This could have potential applications in disease progression prognosis or in simulations of the cellular gene expression control network.

Ideally, domain knowledge about the underlying dynamics of the data should be used to estimate the latent dimension of our model. But in some cases this knowledge might not be available, and we might have to resort to methods like Locally Linear Embedding (LLE) which estimate the *effective* data dimension.

Our model assumes that the time series has been sampled at constant discrete time intervals. If this assumption is violated then we can use Kalman Filtering to predict the value of the missing observations. In our experiments, the results are encouraging even though we ignore the missing observations. Future work will focus on addressing this issue.

Furthermore, we assume that gene expression level dynamics can be modeled as a linear process. While this appears to be a rather simplistic assumption, our experiments verify that it is rich enough to capture complex dynamics. But, if one examines gene expression dynamics that evolve nonlinearly, our approach might fail to describe the underlying biological process appropriately. In these cases we need to resort to more advanced methods described in [14].

We have shown the potential of our method in drug response prediction in our experiments, but our method offers far more possibilities for applications in pharmacogenomics and medicine, namely in disease diagnosis, and disease and therapy outcome prognosis. For example, one could predict if cancer patients should continue to receive chemotherapy, given their response to initial treatments. Besides, observing microarray time series data, one could try to predict if a person has been infected by a pathogen or has been exposed to extreme environmental conditions such as heat or stress at a certain point in time.

Furthermore, it will be interesting to combine our classification approach with two central topics of interest in microarray data analysis, namely methods to detect genes that are key players in biological processes and to derive gene regulatory networks from gene expression data.

Acknowledgments

This work was supported in part by National ICT Australia and by the German Ministry for Education, Science, Research and Technology (BMBF) under grant no. 031U112F within the BFAM (Bioinformatics for the Functional Analysis of Mammalian Genomes) project which is part of the German Genome Analysis Network (NGFN). National ICT Australia is funded through the Australian Government's *Backing Australia's Ability* initiative, in part through the Australian Research Council.

References

[1] Bruce Alberts, Alexander Johnson, Julian Lewis, Martin Raff, Keith Roberts, and Peter Walter. *Molecular Biology of the Cell, Fourth Edition.* Garland Science, New York, 2002.

[2] C. A. Ball, I. A. Awad, J. Demeter, J. Gollub, J. M. Hebert, T. Hernandez-Boussard, H. Jin, J. C. Matese, M. Nitzberg, F. Wymore, Z. K. Zachariah, P. O. Brown, and G. Sherlock. The stanford microarray database accommodates additional microarray platforms and data formats. *Nucleic Acids Res*, 33(Database issue):D580–D582, Jan 2005.

[3] Z. Bar-Joseph. Analyzing time series gene expression data. *Bioinformatics*, 20(16):2493–2503, Nov 2004.

[4] S. E. Baranzini, P. Mousavi, J. Rio, S. J. Caillier, A. Stillman, P. Villoslada, M. M. Wyatt, M. Comabella, L. D. Greller, R. Somogyi, X. Montalban, and J. R. Oksenberg. Transcription-based prediction of response to IFNbeta using supervised computational methods. *PLoS Biology*, 3(1):e2, Jan 2005.

[5] T. Barrett, T. O. Suzek, D. B. Troup, S. E. Wilhite, W. C. Ngau, P. Ledoux, D. Rudnev, A. E. Lash, W. Fujibuchi, and R. Edgar. NCBI

GEO: mining millions of expression profiles–database and tools. *Nucleic Acids Res*, 33(Database issue):D562–D566, Jan 2005.

[6] M. P. Brown, W. N. Grundy, D. Lin, N. Cristianini, C. W. Sugnet, T. S. Furey, J. r. Ares M, and D. Haussler. Knowledge-based analysis of microarray gene expression data by using support vector machines. *Proc Natl Acad Sci U S A*, 97(1):262–267, Jan 2000.

[7] G. Doretto, A. Chiuso, Y.N. Wu, and S. Soatto. Dynamic textures. *International Journal of Computer Vision*, 51(2):91–109, 2003.

[8] Y. Lee and C. K. Lee. Classification of multiple cancer types by multicategory support vector machines using gene expression data. *Bioinformatics*, 19(9):1132–1139, Jun 2003.

[9] G. Natsoulis, L. El Ghaoui, G. R. Lanckriet, A. M. Tolley, F. Leroy, S. Dunlea, B. P. Eynon, C. I. Pearson, S. Tugendreich, and K. Jarnagin. Classification of a large microarray data set: algorithm comparison and analysis of drug signatures. *Genome Research*, 15(5):724–736, May 2005.

[10] M. F. Ramoni, P. Sebastiani, and I. S. Kohane. Cluster analysis of gene expression dynamics. *Proceedings of the National Academy of Science*, 99(14):9121–9126, Jul 2002.

[11] B. Schölkopf and A. Smola. *Learning with Kernels.* MIT Press, Cambridge, MA, 2002.

[12] P. T. Spellman, G. Sherlock, M. Q. Zhang, V. R. Iyer, K. Anders, M. B. Eisen, P. O. Brown, D. Botstein, and B. Futcher. Comprehensive identification of cell cycle-regulated genes of the yeast saccharomyces cerevisiae by microarray hybridization. *Mol Biol Cell*, 9(12):3273–3297, Dec 1998.

[13] A. Statnikov, C. F. Aliferis, I. Tsamardinos, D. Hardin, and S. Levy. A comprehensive evaluation of multicategory classification methods for microarray gene expression cancer diagnosis. *Bioinformatics*, 21(5):631–643, Mar 2005.

[14] S. V. N. Vishwanathan, R. Vidal, and A. J. Smola. Kernels and dynamical systems. *Automatica*, 2005. submitted.

[15] J. C. Willems. From time series to linear system. I. Finite-dimensional linear time invariant systems. *Automatica J. IFAC*, 22(5):561–580, 1986.

COMPUTATIONAL STRATEGY FOR DISCOVERING DRUGGABLE GENE NETWORKS FROM GENOME-WIDE RNA EXPRESSION PROFILES

SEIYA IMOTO[1,*], YOSHINORI TAMADA[2,*], HIROMITSU ARAKI[3,*], KAORI YASUDA[3], CRISTIN G. PRINT[4,†], STEPHEN D. CHARNOCK-JONES[4], DEBORAH SANDERS[4], CHRISTOPHER J. SAVOIE[3], KOUSUKE TASHIRO[5], SATORU KUHARA[5], SATORU MIYANO[1]

[1]*Human Genome Center, Institute of Medical Science, University of Tokyo, 4-6-1, Shirokanedai, Minato-ku, Tokyo, 108-8639, Japan*

[2]*Bioinformatics Center, Institute for Chemical Research, Kyoto University, Gokasho, Uji, Kyoto, 611-0011, Japan*

[3]*Gene Networks International, 4-2-12, Toranomon, Minato-ku, Tokyo, 105-0001, Japan*

[4]*Department of Pathology, Cambridge University, Tennis Court Road, Cambridge, CB2 1QP, United Kingdom*

[5]*Graduate School of Genetic Resources Technology, Kyushu University, 6-10-1, Hakozaki, Higashi-ku, Fukuoka, 812-8581, Japan*

We propose a computational strategy for discovering gene networks affected by a chemical compound. Two kinds of DNA microarray data are assumed to be used: One dataset is short time-course data that measure responses of genes following an experimental treatment. The other dataset is obtained by several hundred single gene knock-downs. These two datasets provide three kinds of information; (i) A gene network is estimated from time-course data by the dynamic Bayesian network model, (ii) Relationships between the knocked-down genes and their regulatees are estimated directly from knock-down microarrays and (iii) A gene network can be estimated by gene knock-down data alone using the Bayesian network model. We propose a method that combines these three kinds of information to provide an accurate gene network that most strongly relates to the mode-of-action of the chemical compound in cells. This information plays an essential role in pharmacogenomics. We illustrate this method with an actual example where human endothelial cell gene networks were generated from a novel time course of gene expression following treatment with the drug fenofibrate, and from 270 novel gene knock-downs. Finally, we succeeded in inferring the gene network related to *PPAR-α*, which is a known target of fenofibrate.

*These authors contributed equally to this work.

†Current affiliation: Department of Molecular Medicine & Pathology, School of Medical Sciences, University of Auckland, Private Bag 92019, Auckland, New Zealand

1. Introduction

The microarray technology has produced a huge amount of gene expression data under various conditions such as gene knock-down, overexpression, experimental stressors, transformation, exposure to a chemical compound, and so on. Using a large volume of microarray gene expression data, a number of algorithms together with mathematical models[1,5,7,9,12,23] for estimating gene networks has been proposed and successfully applied to the gene network estimation of *S. cerevisiae*, *E. coli* etc. As a real application of gene network estimation techniques, computational drug target discovery[19] enhanced with gene network inference[6,14,20,22] has made tremendous impacts on pharmacogenomics.

In this paper, we propose a computational strategy for discovering the druggable gene networks, which are most strongly affected by a chemical compound. For this purpose, we use two types of microarray data: One is gene expression data obtained by measuring transcript abundance responses over time following treatment with the chemical compound. The other is gene knock-down expression data, where one gene is knocked-down for each microarray. Figure 1 is the conceptual view of our strategy. First, we estimate dynamic relationships denoted by G_T between genes based on time-course data by using dynamic Bayesian networks.[17] Second, in gene knock-down expression data, since we know the information of knocked-down genes, possible regulatory relationships between knocked-down gene and its regulatees can be obtained. We denote this information by R. Finally, the gene network G_K is estimated by gene knock-down data denoted by $\boldsymbol{X}_K$ together with G_T and R by using Bayesian networks based on multi-source biological information.[13] The key idea for estimating a gene network based on multi-source biological information is to use G_T and R as the Bayesian prior probability of G_K. The prior probability of the graph proposed by Imoto *et al.*[13] only uses binary prior information, i.e. known or unknown for each gene-gene relation. In this paper, we extend the prior probability of graph[13] in order to use prior information represented as continuous values. After estimating a gene network, for extracting biologically plausible information from the estimated gene network, we have also developed a gene network analysis tool called iNET that is an extended version of G.NET.[14] The iNet tool provides a computational environment for various path searches among genes with annotated gene network visualization.

As for related works, Basso *et al.*[2] estimated a gene network of human B cells as an undirected graph by their proposed algorithm. Our aim is to

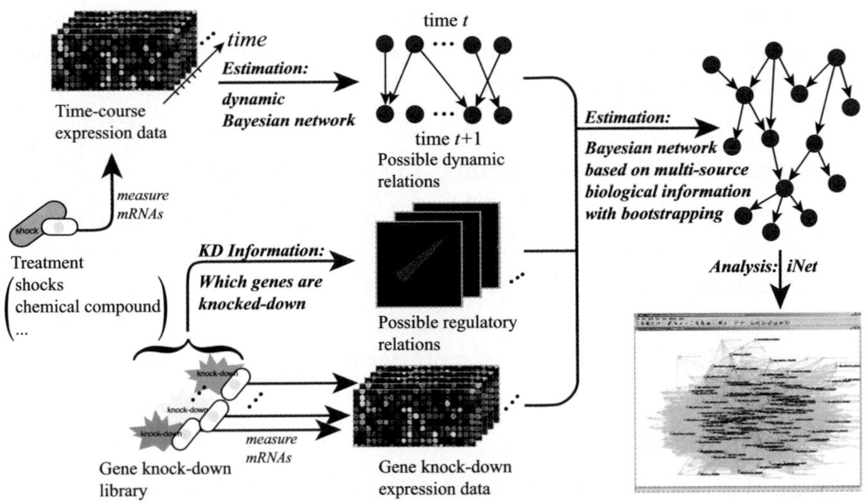

Figure 1. Conceptual view of the proposed method.

estimate druggable gene networks as directed graphs, that are sub-networks of the tissue-specific network. In this method, the edge direction is very important information and selection of compound-related genes is necessary. Therefore, the aim of this paper is clearly different from theirs. Di Bernardo *et al.*[6] proposed an interesting method for identifying mode-of-action of a chemical compound based on microarray gene expression data. Di Bernardo *et al.*[6] used statistical inference of a linear regression-based network model to find affected genes by a chemical compound. On the other hand, our interest is not only in the identification of affected genes, but also in the elucidation of their dependency as the network. In addition, since di Bernardo *et al.*[6] used examples of *S. cerevisiae* genes, more discussions might be needed in order to apply their method to human genes.

To demonstrate the whole process of the proposed method, we analyze expression data from human endothelial cells. We generate new time-course data that reveal the responses of human endothelial cell transcripts to treatment with the anti-hyperlipidaemia drug fenofibrate. We also generate new data from 270 gene knock-down experiments in human endothelial cells. The fenofibrate-related gene network is estimated based on fenofibrate time-course data and 270 gene knock-down expression data by the proposed method. The estimated gene network reveals gene regulatory relationships related to *PPAR*-α, which is known to be activated by fenofibrate. Our computational analysis suggests that this computational strategy based on gene knock-down and drug-dosed time-course microarrays will give a new way to druggable gene discovery.

2. Methods for Reverse-Engineering Gene Networks

In the proposed method, we use Bayesian networks and dynamic Bayesian networks for estimating gene networks from gene knock-down and time-course microarray data, respectively. In this section, we briefly describe these two network models and then elucidate how we combine multi-source biological information to estimate more accurate gene networks.

2.1. *Preliminary*

Suppose that we have the observational data $\boldsymbol{X}$ of the set of p random variables $\mathcal{X} = \{X_1, ..., X_p\}$ and that the dependency among p random variables, shown as a directed graph G, is unknown and we want to estimate it from $\boldsymbol{X}$. In gene network estimation based on microarray data, a gene is regarded as a random variable representing the abundance of a specific RNA species, and $\boldsymbol{X}$ is the microarray data. From a Bayes approach, the optimal graph is selected by maximizing the posterior probability of the graph conditional on the observed data. By the Bayes' theorem, the posterior probability of the graph can be represented as

$$p(G|\boldsymbol{X}) = \frac{p(G)p(\boldsymbol{X}|G)}{p(\boldsymbol{X})} \propto p(G)p(\boldsymbol{X}|G),$$

where $p(G)$ is the prior probability of the graph, $p(\boldsymbol{X}|G)$ is the likelihood of the data $\boldsymbol{X}$ conditional on G and $p(\boldsymbol{X})$ is the normalizing constant and does not depend on the selection of G. Therefore, we need to set $p(G)$ and compute $p(\boldsymbol{X}|G)$ for the graph selection based on $p(G|\boldsymbol{X})$.

The prior probability of the graph $p(G)$ enables us to use biological data other than microarray data to estimate gene networks and the likelihood $p(\boldsymbol{X}|G)$ can be computed by Bayesian networks and dynamic Bayesian networks from gene knock-down and time-course microarray data, respectively. We elucidate how we construct $p(G|\boldsymbol{X})$ in the following sections.

2.2. *Bayesian Networks*

Bayesian networks are a graphical model that represents the causal relationship in random variables. In the Bayesian networks, we use a directed acyclic graph encoding Markov relationship between connected nodes. Suppose that we have a set of random variables $\mathcal{X} = \{X_1, ..., X_p\}$ and that there is a causal relationship in $\mathcal{X}$ by representing a directed acyclic graph G_K. Bayesian networks then enable us to compute the joint probability by the product of conditional probabilities

$$\Pr(\mathcal{X}) = \prod_{j=1}^{p} \Pr(X_j | Pa_j), \tag{1}$$

where Pa_j is the set of random variables corresponding to the direct parents of X_j in G_K. In gene network estimation, we regard a gene as a random variable representing the abundance of a specific RNA species, shown as a node in a graph, and the interaction between genes is represented by the direct edge between nodes.

Let $\boldsymbol{X}_K$ be an $N \times p$ gene knock-down data matrix whose (i, j)-th element $x_{j|D_i}$ corresponds to the expression data of j-th gene when D_i-th gene is knocked down, where $j = 1, ..., p$ and $i = 1, ..., N$. Here we assume that i-th knock-down microarray is measured by knocking-down D_i-th gene. Since microarray data take continuous variables, we represent the decomposition (1) by using densities

$$f_{\mathrm{BN}}(\boldsymbol{X}_K | \boldsymbol{\Theta}, G_K) = \prod_{i=1}^{N} \prod_{j=1}^{p} f_j(x_{j|D_i} | \boldsymbol{pa}_{j|D_i}, \boldsymbol{\theta}_j),$$

where $\boldsymbol{\Theta} = (\boldsymbol{\theta}'_1, ..., \boldsymbol{\theta}'_p)'$ is a parameter vector, $\boldsymbol{pa}_{j|D_i}$ is the expression value vector of Pa_j measured by i-th knock-down microarray. Hence, the construction of the graph G_K is equivalent to model the conditional probabilities f_j $(j = 1, ..., p)$, that is essentially the same as the regression problem. For constructing $f_j(x_{j|D_i} | \boldsymbol{pa}_{j|D_i}, \boldsymbol{\theta}_j)$, we assume the nonparametric regression model with B-splines of the form

$$x_{j|D_i} = \sum_{k=1}^{|Pa_j|} m_{jk}(\boldsymbol{pa}^{(k)}_{j|D_i}) + \varepsilon_{j|D_i},$$

where $\boldsymbol{pa}^{(k)}_{j|D_i}$ is the k-th element of $\boldsymbol{pa}_{j|D_i}$, $\varepsilon_{j|D_i} \sim i.i.d. N(0, \sigma^2)$ for $i = 1, ..., N$, and m_{jk} $(k = 1, ..., |Pa_j|)$ are smooth functions constructed by B-splines as $m_{jk}(x) = \sum_{m=1}^{M_{jk}} \gamma_m^{(jk)} b_m^{(jk)}(x)$. Here $\gamma_m^{(jk)}$ and $b_m^{(jk)}(x)$ $(m = 1, ..., M_{jk})$ are parameters and B-splines, respectively.

The likelihood $p(\boldsymbol{X}_K | G_K)$ is then obtained by

$$p(\boldsymbol{X}_K | G_K) = \int f_{\mathrm{BN}}(\boldsymbol{X}_K | \boldsymbol{\Theta}, G_K) p(\boldsymbol{\Theta} | \boldsymbol{\lambda}, G_K) d\boldsymbol{\Theta}, \tag{2}$$

where $p(\boldsymbol{\Theta}|\boldsymbol{\lambda}, G_K)$ is the prior distribution on the parameter $\boldsymbol{\Theta}$ specified by the hyperparameter $\boldsymbol{\lambda}$. The high-dimensional integral can be asymptotically approximated with an analytical form by the Laplace approximation and Imoto *et al.*[12] defined a graph selection criterion, named BNRC, of the form

$$\begin{aligned}\mathrm{BNRC}(G_K) = -2\log\{p(G_K)\} - r\log(2\pi/N) \\ + \log|J_\lambda(\hat{\boldsymbol{\Theta}}|\boldsymbol{X}_K)| - 2Nl_\lambda(\hat{\boldsymbol{\Theta}}|\boldsymbol{X}_K),\end{aligned}$$

where

$$l_\lambda(\boldsymbol{\Theta}|\boldsymbol{X}_K) = \frac{1}{N}\left\{\log f_{\mathrm{BN}}(\boldsymbol{X}_K|\boldsymbol{\Theta}, G_K) + \log p(\boldsymbol{\Theta}|\boldsymbol{\lambda}, G_K)\right\},$$

$$J_\lambda(\boldsymbol{\Theta}|\boldsymbol{X}_K) = -\frac{\partial^2}{\partial\boldsymbol{\Theta}\partial\boldsymbol{\Theta}'} l_\lambda(\boldsymbol{\Theta}|\boldsymbol{X}_K),$$

r is the dimension of $\boldsymbol{\Theta}$, and $\hat{\boldsymbol{\Theta}}$ is the mode of $l_\lambda(\boldsymbol{\Theta}|\boldsymbol{X}_K)$. The network structure is learned so that $\mathrm{BNRC}(G_K)$ decreases by the greedy hill-climbing algorithm.[12] We should note that the solution obtained by the greedy hill-climbing algorithm cannot be guaranteed as the optimal. To find better solution, we repeat the greedy algorithm and choose the best one as $\hat{G}_K$. It happens quite often that the likelihood $p(\boldsymbol{X}_K|G_K)$ gives almost the same values for several network structures, construction an effective $p(G_K)$ based on various kinds of biological information is a key technique. We elucidate how we construct $p(G_K)$ in Section 2.4.

2.3. *Dynamic Bayesian Networks*

Dynamic Bayesian networks represent the dependency in random variables based on time-course data. Let $\mathcal{X}(t) = \{X_1(t), ..., X_p(t)\}$ be the set of p random variables at time t $(t = 1, ..., T)$. In the dynamic Bayesian networks, a directed graph that contains p nodes is rewritten as a complete bipartite graph that allows direct edges from $\mathcal{X}(t)$ to $\mathcal{X}(t+1)$, where $t = 1, ..., T-1$. The directed graph G_T of the causal relationship among p random variables is then constructed by estimating the bipartite graph defined above. Under G_T structure, we then have the decomposition

$$\Pr(\mathcal{X}(1), ..., \mathcal{X}(T)) = \prod_{t=1}^{T}\prod_{j=1}^{p} \Pr(X_j(t)|Pa_j(t-1)), \qquad (3)$$

where $Pa_j(t)$ is the set of random variables at time t corresponding to the direct parents of X_j in G_T.

Let $\boldsymbol{X}_T$ be a $T \times p$ time-course data matrix whose (t, j)-th element $x_j(t)$ corresponds to the expression data of j-th gene at time t, where $j = 1, ..., p$ and $t = 1, ..., T$. As we described in the Bayesian networks, the decomposition in (3) holds by using densities

$$f_{\text{DBN}}(\boldsymbol{X}_T | \boldsymbol{\Xi}, G_T) = \prod_{t=1}^{T} \prod_{j=1}^{p} f_j(x_j(t) | \boldsymbol{pa}_j(t-1), \boldsymbol{\xi}_j, G_T),$$

where $\boldsymbol{\Xi} = (\boldsymbol{\xi}_1', ..., \boldsymbol{\xi}_p')'$ is a parameter vector, $\boldsymbol{pa}_j(t)$ is the expression value vector of direct parents of X_j measured at time t. Here we set $\boldsymbol{pa}_j(0) = \emptyset$. We can construct f_{DBN} by using nonparametric regression with B-splines in the same way of the Bayesian networks. Therefore, by replacing f_{BN} by f_{DBN} in (2), Kim *et al.*[17] proposed a graph selection criterion for dynamic Bayesian networks, named $\text{BNRC}_{dynamic}$, with successful applications.

2.4. *Combining Multi-Source Biological Information for Gene Network Estimation*

Imoto *et al.*[13] proposed a general framework for combining biological knowledge with expression data aimed at estimating more accurate gene networks. In Imoto *et al.*[13], the biological knowledge is represented as the binary values, e.g. known or unknown, and is used for constructing $p(G)$. In reality, there are, however, various confidence in biological knowledge in practice. Bernard and Hartemink[3] constructed $p(G)$ using the binding location data[18] that is a collection of p-values (continuous information). In this paper, we construct $p(G)$ by using multi-source information including continuous and discrete prior information.

Let $\boldsymbol{Z}_k$ is the matrix representation of k-th prior information, where (i, j)-th element $z_{ij}^{(k)}$ represents the information of "gene $i \rightarrow$ gene j". For example, (1) If we use a prior network G_{prior} for $\boldsymbol{Z}_k$, $z_{ij}^{(k)}$ takes 1 if $e(i, j) \in G_{\text{prior}}$ or 0 if $e(i, j) \notin G_{\text{prior}}$. Here $e(i, j)$ denotes the direct edge from gene i to gene j. (2) By using the gene knock-down data for $\boldsymbol{Z}_k$, $z_{ij}^{(k)}$ represents the value that indicates how gene j changes by knocking down gene i. We can use the absolute value of the log-ratio of gene j for gene i knock-down data as $z_{ij}^{(k)}$. Using the adjacent matrix $E = (e_{ij})_{1 \le i,j \le p}$ of G, where $e_{ij} = 1$ for $e(i, j) \in G$ or 0 for otherwise, we assume the Bernoulli distribution on e_{ij} having probabilistic function

$$p(e_{ij}) = \pi_{ij}^{e_{ij}} (1 - \pi_{ij})^{1 - e_{ij}},$$

where $\pi_{ij} = \Pr(e_{ij} = 1)$. For constructing π_{ij}, we use the logistic model with linear predictor $\eta_{ij} = \sum_{k=1}^{K} w_k(z_{ij}^{(k)} - c_k)$ as $\pi_{ij} = \{1 + \exp(-\eta_{ij})\}^{-1}$, where w_k and c_k $(k = 1, ..., K)$ are weight and baseline parameters, respectively. We then define a prior probability of the graph based on prior information $\boldsymbol{Z}_k$ $(k = 1, ..., K)$ by

$$p(G) = \prod_i \prod_j p(e_{ij}).$$

This prior probability of the graph assumes that edges $e(i,j)$ $(i,j = 1, ..., p)$ are independent of each other. In reality, there are several dependencies among e_{ij}'s such as $p(e_{ij} = 1) < p(e_{ij} = 1|e_{ki} = 1)$, and so on, we consider adding such information into $p(G)$ is premature by the quality of such information.

3. Application to Human Endothelial Cells' Gene Network

3.1. *Fenofibrate Time-Course Data*

We measure the time-responses of human endothelial cell genes to 25μM fenofibrate. The expression levels of 20,469 probes are measured by CodeLink^{TM} Human Uniset I 20K at six time-points (0, 2, 4, 6, 8 and 18 hours). Here time 0 means the start point of this observation and just before exposure to the fenofibrate. In addition, we measure this time-course data as the duplicated data in order to confirm the quality of experiments.

Since our fenofibrate time-course data are duplicated data and contain six time-points, there are $2^6 = 64$ possible combinations to create a time-course dataset. We should fit the same regression function to a parent-child relationship in the 64 datasets. Under this constrain, we consider fitting nonparametric regression model to the connected data of 64 datasets. That is, if we consider gene $i \to$ gene j, we will fit the model $x_j^{(c)}(t) = m_j(x_i^{(c)}(t-1)) + \varepsilon_j(t)$, where $x_j^{(c)}(t)$ is the expression data of gene j at time t in the c-th dataset for $c = 1, ..., 64$. In the Bayesian networks, the reliability of estimated edges can be measured by using the bootstrap method. For time-course data, several modifications of the bootstrap method are proposed such as block resampling, but it is difficult to apply these methods to the small number of data points generated by short time-courses. However, by using above time-course modeling, we can define a method based on the bootstrap as follows: Let $D = \{D(1), ..., D(64)\}$ be the combinatorial time-course data of all genes. We randomly resample $D(c)$ with replacement and define a bootstrap sample $D^* = \{D^*(1), ..., D^*(64)\}$. We then re-estimate a gene network based on D^*. We repeat 1000 times bootstrap replications and obtain $\hat{G}_T^{*1}, ..., \hat{G}_T^{*1000}$, where $\hat{G}_T^{*B}$ is the estimated graph based on the B-th bootstrap sample. The estimated reliability of edge can be used as the matrix representation of the first prior information $\boldsymbol{Z}_1$ as $z_{ij}^{(1)} = \#\{B | e(i,j) \in \hat{G}_T^{*B}, B = 1, ..., 1000\}/1000$.

3.2. *Gene Knock-Down Data by siRNA*

For estimating gene networks, we newly created 270 gene knock-down data by using siRNA. We measure 20,469 probes by CodeLink^{TM} Human Uniset I 20K for each knock-down microarray after 24 hours of siRNA transfection. The knock-down genes are mainly transcription factors and signaling molecules. Let $\tilde{\boldsymbol{x}}_{D_i} = (\tilde{x}_{1|D_i}, ..., \tilde{x}_{p|D_i})'$ be the raw intensity vector of i-th knock-down microarray. For normalizing expression values of each microarray, we compute the median expression value vector $\boldsymbol{v} = (v_1, ..., v_p)'$ as the control data, where $v_j = \text{median}_i(\tilde{x}_{j|D_i})$. We apply the loess normalization method to the MA transformed data and the normalized intensity $x_{j|D_i}$ is obtained by applying the inverse transformation to the normalized $\log(\tilde{x}_{j|D_i}/v_j)$. We refer to the normalized $\log(\tilde{x}_{j|D_i}/v_j)$ as the log-ratio.

In 270 gene knock-down microarray data, we know which gene is knocked-down for each microarray. Thus, when we knock-down gene D_i, genes that significantly change their expression levels can be considered as the direct regulatees of gene D_i. We measure this information by computing corrected log-ratio as follows: The fluctuations of the log-ratios depend on their sum of sample's and control's intensities. From the normalized MA transformed data, we can obtain the conditional variance $s_j = \text{Var}[\log(x_{j|D_i}/v_j) | \log(x_{j|D_i} \cdot v_j)]$ and the log-ratios can be corrected $z_{ij}^{(2)} = \log(x_{j|D_i}/v_j)/s_j$ satisfying $\text{Var}(z_{ij}^{(2)}) = 1$.

3.3. *Results*

For estimating fenofibrate-related gene networks from fenofibrate time-course data and 270 gene knock-down data, we first define the set of genes that are possibly related to fenofibrate as follows: First, we extract the set of genes whose variance-corrected log-ratios, $|\log(x_{j|D_i}/v_j)/s_j|$, are greater than 1.5 from each time point. We then find significant clusters of selected genes using GO Term Finder. Table 1 shows the significant clusters of genes at 18 hours. The first column indicates how expression values are changed, i.e. "$\nearrow$" and "$\searrow$" mean "overexpressed" and "suppressed", respectively. The GO annotations of clusters with "$\searrow$" are mainly related to cell cycle,

Table 1. Significant GO annotations of selected fenofibrate-related genes from 18 hours microarray.

		GO Function	p-value	#genes
↘	GO:0007049	cell cycle	1.0E-08	35
↘	GO:0000278	mitotic cell cycle	3.7E-07	19
↘	GO:0000279	M phase	5.0E-06	17
↗	GO:0006629	lipid metabolism	1.3E-05	25
↘	GO:0007067	mitosis	1.3E-05	15
↘	GO:0000087	M phase of mitotic cell cycle	1.6E-05	15
↘	GO:0000074	regulation of cell cycle	2.7E-05	22
↗	GO:0044255	cellular lipid metabolism	4.4E-05	21
↗	GO:0016126	sterol biosynthesis	4.3E-04	6
↗	GO:0016125	sterol metabolism	4.5E-04	8
↗	GO:0008203	cholesterol metabolism	1.5E-03	7
↗	GO:0006695	cholesterol biosynthesis	2.4E-03	5
↗	GO:0008202	steroid metabolism	3.6E-03	10
↘	GO:0000375	RNA splicing, via transesterification reactions	4.1E-03	9
↘	GO:0000377	RNA splicing, via transesterification reactions with bulged adenosine as nucleophile	4.1E-03	9
↘	GO:0000398	nuclear mRNA splicing, via spliceosome	4.1E-03	9
↗	GO:0006694	steroid biosynthesis	6.0E-03	7
↘	GO:0016071	mRNA metabolism	6.3E-03	13

the genes in these clusters are expressed ubiquitously and this is a common biological function. On the other hand, the GO annotations of clusters with "↗" are mainly related to lipid metabolism. In biology, it is reported that the fenofibrate acts around 12 hours after exposure.[8,10] Our first analysis for gene selection suggests that fenofibrate affects genes related to lipid metabolism and this is consistent with biological facts. We also focus on the genes from the 8 hour time-point microarray. Unfortunately, no cluster with specific function could be found in the selected genes from the 8 hour time-point microarray However, there also exist some genes related to lipid metabolism. Therefore we use the genes from the 8 and 18 hour time-point microarrays. Finally we add the 267 knock-down genes (three genes are not spotted on our chips) to the selected genes above, total 1192 genes are defined as possible fenofibrate-related genes and used for the next network analysis.

By converting the estimated dynamic network and knock-down gene information into the matrix representations of the first and second prior information $\boldsymbol{Z}_1$ and $\boldsymbol{Z}_2$, respectively, we estimate the gene network $\hat{G}_K$ based on $\boldsymbol{Z}_1$, $\boldsymbol{Z}_2$ and the knock-down data matrix $\boldsymbol{X}_K$. For extracting biological information from the estimated gene network, we first focus on lipid metabolism-related genes, because the clusters related this func-

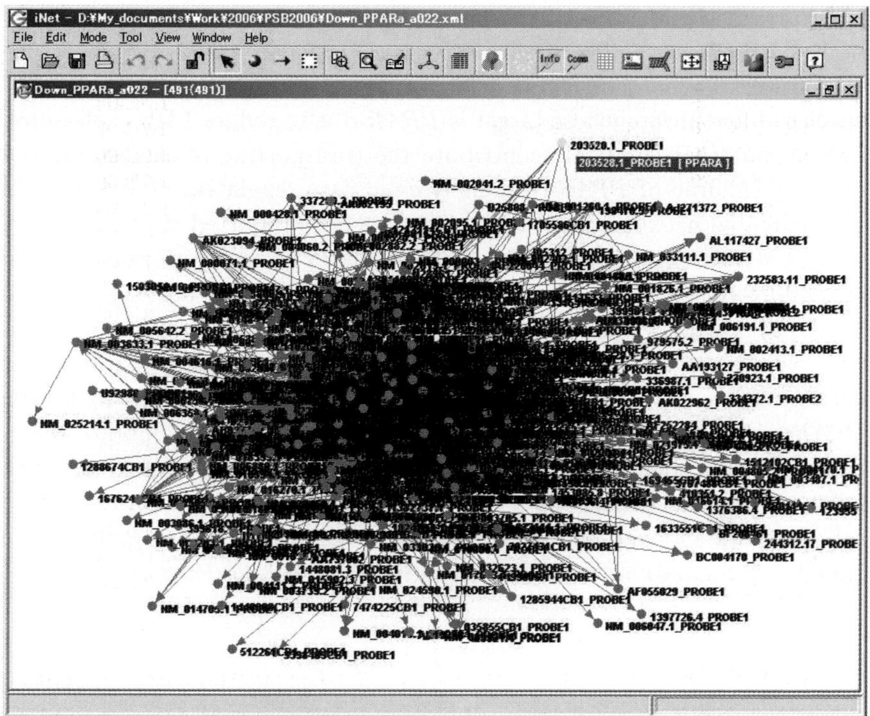

Figure 2. Down-stream of *PPAR-α*.

tion are significantly changed at 18 hours microarray. In the estimated gene network, there are 42 lipid metabolism-related genes and *PPAR-α* (*Homo sapiens* peroxisome proliferative activated receptor, alpha) is the only transcription factor among them. Actually, *PPAR-α* is a known target of fenofibrate. Therefore, we next focus on the node down-stream of *PPAR-α*. In Figure 2, the node down-stream of *PPAR-α* (491 genes). Here we consider that genes in the four steps down-stream of *PPAR-α* are candidate regulatees of *PPAR-α*. Among the candidate regulatees of *PPAR-α*, there are 21 lipid metabolism-related genes and 11 mole-

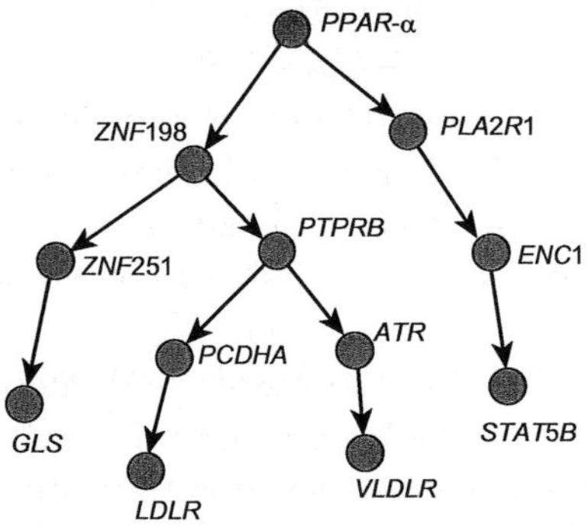

Figure 3. A sub-network related to *PPAR-α*.

cules previously identified experimentally to be related to *PPAR-α*. Actually, *PPAR-α* is known to be activated by fenofibrate. We show one sub-network having *PPAR-α* as a root node in Figure 3. One of the drug efficacies of fenofibrate whose target is *PPAR-α* is to reduce LDL cholesterol. *LDLR* and *VLDLR* mainly contribute the transporting of cholesterol and they are children of *PPAR-α*, namely candidate regulatees of *PPAR-α*, in our estimated network. As for *LDLR*, it has been reported the relationship with *PPAR-α*.[15] Moreover, several genes related to cholesterol metabolism are children of *PPAR-α* in our network. We also could extract *STAT5B* and *GLS* that are children of *PPAR-α* and have been reported their regulation-relationships with *PPAR-α*.[16,21] Therefore, it is not surprising that our network shows that many direct and indirect relationships involving known *PPAR-α* regulatees are triggered in endothelial cells by fenofibrate treatment. In the node up-stream of *PPAR-α*, *PPAR-α* and *RXR-α*, which form a heterodimer, share a parent. We could extract fenofibrate-related gene network and estimate that *PPAR-α* is the one of the key molecules of fenofibrate regulations without previous biological knowledge.

4. Discussion

From the point of view of pharmacogenomics, it is very important to know druggable gene networks. Our gene networks have the potential to predict the mode-of-action of a chemical compound, discover more effective drug target and predict side-effects. In this paper, we proposed a computational method to discover gene networks relating to a chemical compound. We use gene knock-down microarray data and time-course response microarray data for this purpose and combine multiple information obtained from observational data in order to estimate accurate gene networks under a Bayesian statistics framework. We illustrated the entire process of the proposed method using an actual example of gene network inference in human endothelial cells. Using fenofibrate time-course data and data from gene knock-downs in human endothelial cells, we successfully estimated a gene network related to the drug fenofibrate, which is a known agonist of *PPAR-α*. In the estimated gene network, *PPAR-α* has many direct and indirect regulatees including lipid metabolism related genes and this result indicates *PPAR-α* works as a trigger of the estimated fenofibrate-related network. There are many known relationships in the candidate regulatees of *PPAR-α* and we could find the relationship between *PPAR-α* and *RXR-α* in the estimated network. Peroxisome proliferator-activated receptors

(PPARs) are ligand-activated transcription factors expressed by endothelial cells and several other cell types. They are activated by ligands such as naturally occurring fatty acids and synthetic fibrates. Once activated, they heterodimerize with the retinoid-X-receptor (RXR) to activate the transcription of target genes. Many of these genes encode proteins that control carbohydrate and glucose metabolism and down-regulate inflammatory responses.[4] The further details on the relation between *PPAR-α* and *RXR-α* and their common parent will be discussed in another paper with biological evidences.

Acknowledgements

We wish to acknowledge Ben Dunmore, Sally Humphries, Muna Affara and Yuki Tomiyasu for assistance with endothelial cell culture and gene array analysis. Computation time was provided by the Super Computer System, Human Genome Center, Institute of Medical Science, University of Tokyo.

References

1. T. Akutsu *et al.*, *Pac. Symp. Biocomput.*, **4**:17–28, 1999.
2. K. Basso *et al.*, *Nat. Genet.*, **37**:382–390, 2005.
3. A. Bernard and A.J. Hartemink, *Pac. Symp. Biocomput.*, **10**:459–470, 2005
4. A. Cabrero *et al.*, *Curr. Drug Targets Inflamm. Allergy*, **1**:243–248, 2002.
5. T. Chen *et al.*, *Pac. Symp. Biocomput.*, **4**:29–40, 1999.
6. D. di Bernardo *et al.*, *Nat. Genet.*, **37**:382–390, 2005.
7. N. Friedman *et al.*, *J. Comp. Biol.*, **7**:601–620, 2000.
8. K. Goya *et al.*, *Arterioscler. Thromb. Vasc. Biol.*, **24**:658–663, 2004.
9. A.J. Hartemink *et al.*, *Pac. Symp. Biocomput.*, **7**:437–449, 2002.
10. K. Hayashida *et al.*, *Biochem. Biophys. Res. Commun.*, **323**:1116–1123, 2004.
11. D. Heckerman *et al.*, *Machine Learning*, **20**:197–243, 1995.
12. S. Imoto *et al.*, *Pac. Symp. Biocomput.*, **7**:175–186, 2002.
13. S. Imoto *et al.*, *J. Bioinform. Comp. Biol.*, **2**:77–98, 2004.
14. S. Imoto *et al.*, *J. Bioinform. Comp. Biol.*, **1**:459–474, 2003.
15. K.K. Islam *et al.*, *Biochim. Biophys. Acta.*, **1734**:259–268, 2005.
16. S. Kersten *et al.*, *FASEB J.*, **15**:1971–1978, 2001.
17. S. Kim *et al.*, *Biosystems*, **75**:57–65, 2004.
18. T.I. Lee *et al.*, *Science*, **298**:799–804, 2002.
19. M.J. Marton *et al.*, *Nat. Med.*, **4**:1293–1301, 1998.
20. C.J. Savoie *et al.*, *DNA Res.*, **10**:19–25, 2003.
21. J.M. Shipley and D.J. Waxman, *Mol. Pharmacol.*, **64**:355–364, 2003.
22. Y. Tamada *et al.*, *Genome Informatics*, **16**:182–191, 2005.
23. E.P. van Someren *et al.*, *Pharmacogenomics*, **3**:507–525, 2002.

A BIVARIATE FUNCTIONAL MAPPING MODEL FOR IDENTIFYING HAPLOTYPES THAT CONTROL DRUG RESPONSE FOR SYSTOLIC AND DIASTOLIC BLOOD PRESSURES

MIN LIN*

Duke University, Department of Biostatistics and Bioinformatics,
Duke Clinical Research Institute, P.O. Box 17969, Durham, NC 27715, USA
E-mail: annie.lin@duke.edu

RONGLING WU

University of Florida, Department of Statistics,
533 McCarty Hall C, Gainesville, FL 32611, USA
E-mail: rwu@stat.ufl.edu

JULIE JOHNSON

University of Florida, Department of Pharmacy Practice,
Box 100486, Health Sciences Center, Gainesville, FL 32610, USA
E-mail: Johnson@cop.ufl.edu

A bivariate functional mapping model has been proposed to detect haplotype-based DNA sequence variants that regulate the response curves of systolic and diastolic blood pressures (SBP and DBP) to a particular drug. This model capitalizes on the haplotype structure constructed by single nucleotide polymorphisms (SNPs) and incorporates the mathematical aspects of pharmacodynamic reactions into the estimation process, aimed to identify DNA sequence variants responsible for drug response. In this way, by estimating and testing the curve parameters that define drug response, many genetically and clinically meaningful hypotheses regarding the degree and pattern of the genetic control of SBP and DBP can be formulated, tested and disseminated. In a pharmacogenetic study composed of 107 subjects, our bivariate model has probed two haplotypes within the β2AR candidate gene that exert a significant effect on both SBP and DBP respond to dobutamine. With this candidate gene, two SNPs are genotyped, with allele Gly16 (G) and Arg16 (A) at codon 16 and alleles Glu27 (G) and Gln27 (C) at codon 27, respectively. The significant haplotypes are [AC] for SBP and [GG] for DBP. This model provides a powerful tool for elucidating the genetic variants of drug response and ultimately designing personalized medications based on each patient's genetic makeup.

*The corresponding author

1. Introduction

The question of whether genes control drug response has been recognized from simple association analysis between genetic ethnicity and aberrant drug response in the 1950s to more precise family and twin studies in the 1960s and 70s to biochemical studies in the 1970s and 80s to molecular genetic studies in the 1980s and 90s[1]. Today, a more challenging question is not whether there are genes involved in drug response rather than how genes control drug response. Tremendous efforts have been made to isolate genes or polymorphisms responsible for inherited differences in drug metabolism and disposition, drug effects and drug transporters and targets[1].

The identification of genes for drug response is difficult for two reasons. First, inter-individual variation in drug response is regulated by a multitude of genes each with a small effect and segregating in Mendelian laws. With the near completion of the human genome project, it will be possible to characterize fine-structured DNA variation in the human genome and further identify the chromosomal regions associated with the variation in drug response. Second, patients' response to a particular drug involves a series of sequential biochemical pathways and reactions, which are described by two different but related processes, known as pharmacokinetics (PK) and pharmacodynamics (PD)[2]. While PK concerns the change of drug concentration in the body with time, PD deals with different drug effects under changing concentrations. Because both PK and PD each presents a dynamic process, the effects of genes involved are supposed to display particular trajectories. Statistical models for analyzing such trajectory or longitudinal data with multiple measurements are qualitatively complicated, as compared to those for single measurements.

More recently, statistical models for detecting the genetic architecture of longitudinal traits by mapping the underlying quantitative trait loci (QTL) have been proposed in the literature[3–5]. These models, called *functional mapping*, approximate time-dependent genetic effects based on mathematical equations of biological relevance and have proven instrumental for the identification of QTL for growth traits[4–5]. Functional mapping has now been extended to map QTL that control drug response through the incorporation of the mathematical aspects of pharmacodynamic processes[6]. Taking advantages of sequence-based association studies, functional mapping has been modified to characterize the DNA sequence structure of drug response[7] and compare the genetic differences between efficacy and toxicity at the single DNA base level[8].

Congestive heart failure (CHF) is a pervasive and insidious clinical syndrome that most commonly results from ischemic heart disease and hypertension. It is estimated that almost 5 million Americans are affected by CHF. Dobutamine is primarily an agonist at β_1-adrenergic receptors (β1ARs) that predominate in the heart and also has some β_2-adrenergic receptors (β2ARs) agonist properties. It is used to relieve symptoms in patients with CHF by increasing stroke volume in a dose-dependent manner. The understanding of the association between genetic polymorphisms and the inter-patient variability in SBP and DBP responding to dobutamine will provide an objective genetic basis for individualization in treating CHF. In this article, we modify Lin and Wu's bivariate functional mapping model[8] to detect the haplotype-based DNA sequence variants associated with SBP and DBP.

2. Methods

2.1. *Likelihood functions*

Suppose there are two SNPs that are co-segregating with the linkage disequilibrium of D in a human population at Hardy-Weinberg equilibrium. Let $p_1^{(1)}$, $p_0^{(1)}$ and $p_1^{(2)}$, $p_0^{(2)}$ be the relative frequencies of two alleles, designated as 1 and 0, at each of the two SNPs, respectively, where the superscript stands for the identification of SNP and $p_1^{(1)} + p_0^{(1)} = 1$ and $p_1^{(2)} + p_0^{(2)} = 1$. These two SNPs form 4 possible haplotypes [11], [10], [01] and [00] whose frequencies are expressed as

$$p_{r_1r_2} = p_{r_1}^{(1)}p_{r_2}^{(2)} + (-1)^{r_1+r_2}D,$$

where $\sum_{r_1=0}^{1}\sum_{r_2=0}^{1} p_{r_1r_2} = 1$ and $r_1, r_2 = 1, 0$ denote the alleles of the two SNPs, respectively[9]. If the haplotype frequencies are known, then the allelic frequencies and linkage disequilibrium, arrayed by the population genetic parameter vector $\mathbf{\Omega}_p = (p_{r_1}, p_{r_2}, D)$, can be solved with the above equation.

We assume that a specific haplotype among the four ones may affect drug response which is called the reference or risk haplotype[7]. The four haplotypes form 10 indistinguishable diplotypes (i.e., a combination between maternally- and paternally-derived haplotypes) and 9 observable genotypes with respective frequencies expressed in terms of haplotype frequencies in a population[7]. Thus, although different diplotypes contribute to inter-individual variation in drug response, we can only construct the likelihoods based on observable genotypes of size expressed as $n_{r_1r_1'/r_2r_2'}$

$(r_1 \geq r'_1 = 1, 0; r_2 \geq r'_2 = 1, 0)$. In statistics, this is a classic mixture model problem and can be solved by implementing the EM algorithm.

Without loss of generality, we assume that haplotype [11] is the risk haplotype and the other haplotypes [10], [01] and [00], collectively designated as $[\overline{11}]$, are the non-risk haplotype. Thus, we have three possible composite genotypes [11][11] (2), $[11][\overline{11}]$ (1) and $[\overline{11}][\overline{11}]$ (0) whose concentration-dependent genotypic values are expressed as $\mathbf{u}_2$, $\mathbf{u}_1$ and $\mathbf{u}_0$, respectively. Let $\mathbf{\Sigma}$ be the residual covariance matrix within each composite genotype. We use $\mathbf{\Omega}_q = (\mathbf{u}_2, \mathbf{u}_1, \mathbf{u}_0, \mathbf{\Sigma})$ to denote the quantitative genetic parameters for the two SNPs under consideration.

For a total of n subjects, the measures of SBP, $\mathbf{y}_s$, and DBP, $\mathbf{y}_d$, are recorded at C different concentration levels of a drug. Let $\mathbf{y}_i = (\mathbf{y}_{si}, \mathbf{y}_{di})$ be the bivariate drug response for subject i. The log-likelihood functions of trait values and SNPs given observed genotypes ($\mathbf{G}$) are expressed as

$$\begin{aligned}
&\log L(\mathbf{\Omega}_p, \mathbf{\Omega}_q | \mathbf{y}, \mathbf{G}) \\
&= \sum_{i=1}^{n_{11/11}} \log f_2(\mathbf{y}_i) + \sum_{i=1}^{n_{11/10}} \log f_1(\mathbf{y}_i) + \sum_{i=1}^{n_{11/00}} \log f_0(\mathbf{y}_i) + \sum_{i=1}^{n_{10/11}} \log f_1(\mathbf{y}_i) \\
&+ \sum_{i=1}^{n_{10/10}} \log[\varpi f_1(\mathbf{y}_i) + (1-\varpi) f_0(\mathbf{y}_i)] + \sum_{i=1}^{n_{10/00}} \log f_0(\mathbf{y}_i) \\
&+ \sum_{i=1}^{n_{00/11}} \log f_0(\mathbf{y}_i) + \sum_{i=1}^{n_{00/10}} \log f_0(\mathbf{y}_i) + \sum_{i=1}^{n_{00/00}} \log f_0(\mathbf{y}_i) \qquad (1)
\end{aligned}$$

and

$$\begin{aligned}
\log L(\mathbf{\Omega}_p | \mathbf{G}) \propto\ & 2n_{11/11} \log p_{11} + n_{11/10} \log(2p_{11}p_{10}) + 2n_{11/00} \log p_{10} \\
& + n_{10/11} \log(2p_{11}p_{01}) + n_{10/10} \log(2p_{11}p_{00} + 2p_{10}p_{01}) \\
& + n_{10/00} \log(2p_{10}p_{00}) + 2n_{00/11} \log p_{01} \\
& + n_{00/10} \log(2p_{01}p_{00}) + 2n_{00/00} \log p_{00} \qquad (2)
\end{aligned}$$

where the double heterozygote is the mixture of two possible diplotypes weighted by $\varpi = \frac{p_{11}p_{00}}{p_{11}p_{00}+p_{10}p_{01}}$ and $1 - \varpi = \frac{p_{10}p_{01}}{p_{11}p_{00}+p_{10}p_{01}}$.

The multivariate normal distribution of SBP and DBP for composite genotype j, $f_j(\mathbf{y}_i)$ ($j = 2, 1, 0$), can be expressed as

$$f_j(\mathbf{y}_i; \mathbf{u}_j, \mathbf{\Sigma}) = \frac{1}{(2\pi)^C |\mathbf{\Sigma}|^{1/2}} \exp\left[-\frac{1}{2}(\mathbf{y}_i - \mathbf{u}_j)\mathbf{\Sigma}^{-1}(\mathbf{y}_i - \mathbf{u}_j)^{\mathrm{T}}\right],$$

with mean vector

$$\mathbf{u}_j = (\mathbf{u}_{sj}, \mathbf{u}_{dj}) = (u_{sj}(1), \ldots, u_{sj}(C), u_{dj}(1), \ldots, u_{dj}(C)),\ j = 2, 1, 0 \quad (3)$$

where $\mu_{sj}(c)$ and $\mu_{dj}(c)$ are the genotypic values of SBP and DBP of composite genotype j at concentration c, and covariance matrix

$$\mathbf{\Sigma} = \begin{pmatrix} \mathbf{\Sigma}_s & \mathbf{\Sigma}_{sd} \\ \mathbf{\Sigma}_{ds} & \mathbf{\Sigma}_d \end{pmatrix}, \tag{4}$$

where $\mathbf{\Sigma}_s$ and $\mathbf{\Sigma}_d$ are composed of $\sigma_s^2(c)$ and $\sigma_s(c_1, c_2)$, and $\sigma_d^2(c)$ and $\sigma_d(c_1, c_2)$ $(1 \leq c_1, c_2 \leq C)$, respectively; and $\mathbf{\Sigma}_{ds}$ and $\mathbf{\Sigma}_{ds}$ are composed of $\sigma_{sd}(c)$ and $\sigma_{sd}(c_{1s}, c_{2d})$ $(1 \leq c_{1s} \neq c_{2d} \leq C)$, and $\sigma_{ds}(c)$ and $\sigma_{sd}(c_{1d}, c_{2s})$ $(1 \leq c_{1d} \neq c_{2s} \leq C)$, respectively.

2.2. *Modelling the mean vector*

The concentration-dependent expected values of composite genotype j can be modelled for SBP and DBP by the sigmoid Emax model[2]. The $\mathrm{E}_{\max}$ model postulates the following relationship between drug concentration (c) and drug effect $(u(c))$ for composite genotype j

$$u_j(c) = E_{0j} + \frac{E_{\max j} c^{H_j}}{EC_{50j}^{H_j} + c^{H_j}}, \tag{5}$$

where E_0 is the constant or baseline value for the drug response parameter, $E_{\max}$ is the asymptotic (limiting) effect, EC_{50} is the drug concentration that results in 50% of the maximal effect, and H is the slope parameter that determines the slope of the concentration-response curve. Eq. (5) can be used to fit the responses of SBP and DBP. As a result, there are eight curve parameters together for these two blood pressures.

It is possible that SBP and DBP have different risk haplotypes, so their composite genotypes should be treated differently. Thus, eight curve parameters are defined for composite genotype j_1 for SBP and j_2 for DBP, which are arrayed by $\mathbf{\Omega}_{u_{j_1 j_2}} = (E_{0j_1}, E_{\max j_1}, EC_{50j_1}, H_{j_1}, E_{0j_2}, E_{\max j_2}, EC_{50j_2}, H_{j_2})$. If different composite genotypes have different combinations of these parameters, this implies that the DNA sequence under consideration plays a role in governing the differentiation of these two pressures. Thus, by testing for the difference of $\mathbf{\Omega}_{u_{j_1 j_2}}$ among different genotypes, we can determine whether there exists a specific sequence variant that confers an effect on these two pressures.

2.3. *Modelling the structure of the covariance matrix*

We use the first-order autoregressive [AR(1)] model to model the structure of the within-subject (co)variance matrix[10], expressed as

$$\sigma_s^2(1) = \cdots = \sigma_s^2(C) = \sigma_s^2, \quad \sigma_d^2(1) = \cdots = \sigma_d^2(C) = \sigma_d^2$$

for the variances, and

$$\sigma_s(c_1,c_2)=\sigma_s^2\rho_s^{|c_2-c_1|},\qquad \sigma_d(c_1,c_2)=\sigma_d^2\rho_d^{|c_2-c_1|}$$

for the covariances between any two concentration intervals c_1 and c_2, where $0<\rho_s,\rho_d<1$ are the proportion parameters with which the correlation decays with concentration lag. The covariances between two responses at the same concentration level of drug or different concentration levels are, respectively, modelled by

$$\sigma_{sd}(1)=\cdots=\sigma_{sd}(C)=\sigma_s\sigma_d\rho_{sd},\quad \sigma_{sd}(c_1,c_2)=\sigma_s\sigma_d\lambda^{|c_2-c_1|},\ 0<\lambda<1.$$

The parameters that model the structure of the (co)variance matrix $\mathbf{\Sigma}$ is arrayed by $\mathbf{\Omega}_v=(\sigma_s^2,\rho_s,\sigma_d^2,\rho_d,\rho_{sd},\lambda)$. Thus, instead of estimating all elements in matrix $\mathbf{\Sigma}$, we only need to estimate the parameters contained in $\mathbf{\Omega}_v$. This largely reduces the number of parameters to be estimated.

2.4. *Computational algorithm*

The EM algorithm is implemented to obtain the maximum likelihood estimates (MLEs) of the marker population parameters ($\mathbf{\Omega}_p$), the curve parameters ($\mathbf{\Omega}_{u_{j_1j_2}}$) that model the mean vector, and the parameters ($\mathbf{\Omega}_v$) that model the structure of the covariance matrix.

In the E step, we calculate the expected number (ϖ) of diplotype [11][00] contained in the double heterozygote 10/10. In the M step, we use the calculated ϖ to estimate the haplotype frequencies using a series of closed form[7], expressed as

$$\hat{p}_{11}=\frac{2n_{11/11}+n_{11/10}+n_{11/00}+\varpi n_{10/10}}{2n},$$

$$\hat{p}_{10}=\frac{2n_{11/00}+n_{11/10}+n_{10/00}+(1-\varpi)n_{10/10}}{2n},$$

$$\hat{p}_{01}=\frac{2n_{00/11}+n_{10/11}+n_{00/10}+(1-\varpi)n_{10/10}}{2n},$$

$$\hat{p}_{00}=\frac{2n_{00/00}+n_{00/10}+n_{00/11}+\varpi n_{10/10}}{2n},$$

But in this step, we encounter a considerable difficulty in deriving the log-likelihood equations for $\mathbf{\Omega}_{u_{j_1j_2}}$ and $\mathbf{\Omega}_v$ because they are contained in complex nonlinear equations. In this article, the simplex algorithm[11–12]is embedded in the EM algorithm above to provide simultaneous estimation of haplotype frequencies and curve parameters and matrix-structuring parameters.

2.5. *Hypothesis tests*

The existence of significant DNA sequence variants for drug response can be tested by formulating the hypothesis,

$$\begin{cases} H_0: \boldsymbol{\Omega}_{u_{j_1 j_2}} \equiv \boldsymbol{\Omega}_u, & j_1, j_2 = 2, 1, 0 \\ H_1: \text{at least one of the equalities above does not hold,} \end{cases} \tag{6}$$

where H_0 corresponds to the reduced model, in which the data can be fit by a single drug response curve, and H_1 corresponds to the full model, in which there exist different dynamic curves to fit the data. The test statistic for testing this hypothesis in Eq. (6) is calculated as the log-likelihood ratio (LR) of the reduced to the full model:

$$\mathrm{LR} = -2[\log L(\widetilde{\boldsymbol{\Omega}}|\mathbf{y}, \mathbf{G}) - \log L(\widehat{\boldsymbol{\Omega}}|\mathbf{y}, \mathbf{G})],$$

where $\widetilde{\boldsymbol{\Omega}}$ and $\widehat{\boldsymbol{\Omega}}$ denote the MLEs of the unknown parameters under H_0 and H_1, respectively. The LR is asymptotically χ^2-distributed with 16 degrees of freedom. An empirical approach for determining the critical threshold is based on permutation tests, as advocated by Churchill and Doerge[13]. By repeatedly shuffling the relationships between marker genotypes and phenotypes, a series of the maximum log-likelihood ratios are calculated, from the distribution of which the critical threshold is determined.

Table 1. Likelihood ratios for 16 possible combinations of assumed reference haplotypes for SBP and DBP within β2AR genes.

	DBP			
SBP	GC	GG	AC	AG
GC	15.13	17.31	11.28	8.63
GG	16.60	17.71	13.90	12.18
AC	16.60	**21.57**	13.04	10.58
AG	18.91	21.50	13.91	10.82

Note: The maximum likelihood ratio value is detected when [AC] and [GG] are used as the reference haplotypes for SBP and DBP, respectively.

If the same risk haplotype for the systolic and diastolic pressures can better explain the data, the next test is about the pleiotropic control of this risk haplotype on these two blood pressures. Such tests are

$$\begin{cases} H_0: \boldsymbol{\Omega}_{u_{sj_1 j_2}} \equiv \boldsymbol{\Omega}_{us}, & j_1, j_2 = 2, 1, 0 \\ H_1: \text{at least one of the equalities above does not hold,} \end{cases} \tag{7}$$

for the systolic blood pressure, and

$$\begin{cases} H_0: \boldsymbol{\Omega}_{u_{dj_1j_2}} \equiv \boldsymbol{\Omega}_{ud}, \quad j_1, j_2 = 2, 1, 0 \\ H_1: \text{ at least one of the equalities above does not hold,} \end{cases} \tag{8}$$

for the diastolic blood pressure. Only the null hypotheses of both Eq. (7) and Eq. (8) are rejected can we suggest the significance of the pleiotropic effect on the two pressures.

3. Subjects

A pharmacogenetic study of cardiovascular disease is used to demonstrate the usefulness of our model. Cardiovascular disease, principally heart disease and stroke, is the leading killer for both men and women among all racial and ethnic groups. Dobutamine is a heart-stimulating medication that is used to treat congestive heart failure by increasing heart rate and cardiac contractility through β-adrenergic receptors (βARs), with actions on the heart similar to the effect of exercise[14–15].

Table 2. MLEs of population genetic parameters (allele frequencies and linkage disequilibria) for SNPs as well as quantitative genetic parameters (drug response and matrix-structuring parameters) within β2AR gene.

Population genetic parameters		
p_1	p_2	D
0.62	0.60	0.05

Composite genotype	Curve parameters: SBP			
	E_0	E_{max}	EC_{50}	H
[AC][AC]	0.46	0.09	4.99	18.37
[AC][$\overline{AC}$]	0.40	0.17	5.09	16.40
[$\overline{AC}$][$\overline{AC}$]	0.44	0.09	6.14	3.43

Composite genotype	Curve parameters: DBP			
	E_0	E_{max}	EC_{50}	H
[GG][GG]	0.61	-0.10	5.11	16.69
[GG][$\overline{GG}$]	0.56	-0.05	18.91	8.01
[$\overline{GG}$][$\overline{GG}$]	0.57	-0.11	8.03	1.30

Matrix-structuring parameters			
σ^2_{SBP}	σ^2_{DBP}	ρ_{SBP}	ρ_{DBP}
0.03	0.01	0.83	0.84

Note: The risk haplotypes for SBP and DBP are [AC] and [GG], respectively.

Both the β1AR and β2AR genes have several polymorphisms that are common in the population. Two common polymorphisms are located at codons 49 (Ser49Gly) and 389 (Arg389Gly) for the β1AR gene and at codons 16 (Arg16Gly) and 27 (Gln27Glu) for the β2AR gene[14]. The polymorphisms in each of these two receptor genes are in linkage disequilibrium, which suggests the importance of taking into account haplotypes, rather than a single polymorphism, when defining biologic function. This study attempts to detect haplotype variants within these candidate genes which determine the response of SBP and DBP to varying concentrations of dobutamine.

A group of 163 men and women in ages from 32 to 86 years old participated in this study. Each of these subjects was genotyped for SNP markers at codons 49 and 389 within the β1AR gene and at codons 16 and 27 within the β2AR gene. Dobutamine was injected into these subjects to investigate their response in SBP and DBP to this drug. The subjects received increasing doses of dobutamine, until they achieved target SBP and DBP or predetermined maximum concentration. The concentration levels used were 0 (baseline), 5, 10, 20, 30 and 40 mcg—min, at each of which both SBP and DBP were measured. Raw data for SBP-concentration and DBP-concentration profiles are illustrated in Figure 1A and 1C, respectively. Only those (107) in whom there were SBP and DBP data at all the six concentration levels were included for data analyses.

4. Results

Statistical analysis and test suggested that different SNPs within each candidate gene have significant linkage disequilibria (results not shown). By assuming that one haplotype is the risk haplotype, we hope to detect a particular DNA sequence associated with the response of SBP and DBP to dobutamine. At the β1AR gene, we did not find any haplotype that contributed to inter-individual difference in the SBP and DBP. A significant effect was observed for haplotype Arg16(A)–Gln(C) for SBP and haplotype Gly16(G)–Glu27(G) for DBP within the β2AR gene. The log-likelihood ratio (LR) test statistics for the combination between these two risk haplotypes 21.57, which is statistically significant (P-value=0.05) based on the critical threshold determined from 1000 permutation tests and is also greater than the LR values for any other combinations (Table 1).

The MLEs of the population and quantitative genetic parameters were obtained by our bivariate model (Table 2). Using the estimated response parameters, we drew the profiles of SBP and DBP response to increasing concentration levels of dobutamine for three composite genotypes for these two types of blood pressures (Figure 1). As shown, the three composite genotypes displayed different curves across all concentration levels for each blood pressure (Figure 1B and 1D). The haplotype [AC] displayed over-dominant effect on SBP across all concentration levels (Figure 1B). In Figure 1D, the DBP curve of composite homozygote [GG][GG] showed rapid decreases when concentration reached 5 mcg. We used area under curve (AUC) to test in which gene action mode (additive or dominant) haplotypes affect drug response curves for blood pressures. The testing results suggest that both additive and dominant effects are important in determining the shape of the response curve (Table 3).

Since the pioneering simulation study was performed by Lin and Wu[7] to investigate the robustness and power of the method within a range of parameters, the simulation study, in this article, was conducted by mimicking the example used above in order to determine the reliability of our estimates in this real application. One haplotype was assumed to be different from the other three. The data simulated under this assumption were subject to statistical analysis, pretending that haplotype distinction is unknown. As expected, only under the correct haplotype distinction could the haplotype effect be detected and the parameters be accurately and precisely estimated (result not shown).

Table 3. Testing results for additive and dominant effects for SBP and DBP based on AUC in 107 subjects under the optimal haplotype model.

Test	Additive_{SBP}	Dominant_{SBP}	Additive_{DBP}	Dominant_{DBP}
LR	7.78	10.12	13.55	14.42
P value	<0.05	<0.05	<0.05	<0.05

5. Discussion

A growing body of data has shown that people differ in their response to the same medication. Although such variability in drug response among patients can result from nongenetic factors such as sex, age and race, genetic variants underlying pharmacological effect appear to receive more attention in revealing these inter-individual differences[16]. Increasing examples have shown that genetic variations lie in the encoding genes of drug metabolism enzymes, transporters, receptors, and other targets that can modulate drug response[17]. And the genetic differences can explain 20 to 95 percent of variability in drug effects[18]. Thus, pharmacogenetics or pharmacogenomics, the study of inherited variability in individuals' responses to drugs becomes

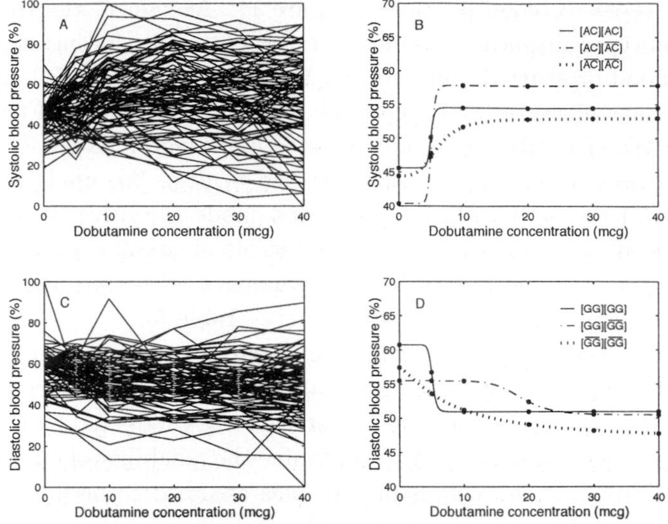

Figure 1. Response curves for systolic (SBP) (**A**) and diastolic blood pressure (DBP) (**C**) to dobutamine in a pharmacogenomic study composed of 107 patients. From these curves we have detected significant risk haplotypes [AC] (**B**) for SBP and [GG] for DBP (**D**) that form three composite genotypes for each type of blood pressure.

flourishing in biomedical science.

With the development of the haplotype map or HapMap project[19], more and more information of DNA sequence variation, such as "tag" single nucleotide polymorphisms (tag SNPs), a small fraction of SNPs required in distinguishing a large fraction of the haplotypes, gives the possibility to directly identify specific DNA sequence that influence drug response at a single DNA base[7]. In contrast to the traditional QTL mapping in which only hypothetical gene can be detected[20], the approach proposed by Lin et al.[7] can detect specific DNA sequence variants for a complex trait.

Although the intensity of drug effects resulting from varying drug concentrations at the effect site can be characterized by pharmacodynamics, such dynamic behavior of drug response provides a statistical problem involving longitudinal traits. In coupling with the advantages of functional mapping which maps dynamic QTL responsible for a biological process[3–5], Lin and Wu proposed a bivariate model for detecting specific DNA sequence variants that determine multiple processes of drug responses[8]. This model is incorporated by clinally meaningful mathematical functions into modelling concentration-dependent drug response and the statistical device used to model the correlated structure of the (co)variance matrix.

In this article, we adapt Lin and Wu' approach[8] to detect haplotype-based genetic variants that contribute to inter-individual variation in two response curves of SBP and DBP to a medication. The magnitudes of the SBP and DBP are used as an indicator of whether there is a hypertension for a patient. In a pharmacogenetic study composed of 107 subjects, we have detected two risk haplotypes, [AC] and [GG], within the β2AR candidate gene, that exert significant effects on the response profiles of the SBP and DBP to dobutamine, respectively. Biologically, such detected genetic variants provide a possible explanation for the variability of the inotropic effect among patients resulting from dobutamine. Previous study also indicates that haplotype [GG] has a significant impact on response in heart rate[7]. The results in this article suggest that a pleiotropic effect of haplotype [GG] on DBP and heart rate may exist. The genetic variants that regulate the response of SBP and DBP to a medication can therefore provide scientific guidance for designing individualized drugs and dosages based on a patient's genetic makeup.

References

1. W. E. Evans and M. V. Relling, *Nature* **429**, 464 (2004).
2. J. Giraldo, *Trends Pharmacolog. Sci.* **24**, 63 (2003).
3. C.-X. Ma, G. Casella and R. L. Wu, *Genetics* **161**, 1751 (2002).
4. R. L. Wu, C.-X. Ma, M. Lin and G. Casella, *Genetics* **166**, 1541 (2004).
5. R. L. Wu, C.-X. Ma, M. Lin, Z.-H. Wang and G. Casella, *Biometrics* **60**, 729 (2004).
6. Y. Gong, Z. H. Wang, T. Liu, W. Zhao, Y. Zhu, J.A. Johnson and R.L. Wu, *Pharmacogenomics J.* **4**, 315 (2004).
7. M. Lin, C. Aquilante, J. A. Johnson and R. L. Wu, *Pharmacogenomics J.* **5**, 149 (2005).
8. M. Lin and R. L. Wu, *Genetics* **170**, 919 (2005).
9. B.S. Weir, Sinauer, Sunderland, MA (1996).
10. P. J. Diggle, P. Heagerty, K. Y. Liang and S. L. Zeger, UK: Oxford University Press, (2002).
11. W. Zhao, R. L. Wu, C.-X. Ma and G. Casella, *Genetics* **167**, 2133 (2004).
12. J. A. Nelder and R. Mead, *Computer J.* **7**, 308 (1965).
13. G. A. Churchill and R. W. Doerge, *Genetics* **138**, 963 (1994).
14. E. G. Nabel, *N. Engl. J. Med.* **349**, 60 (2003).
15. J. A. Johnson and S. G. Terra, *Pharmaceutical Res.* **19**, 1779 (2002).
16. J. A. Johnson and W. E. Evans, *Trends Mol. Med.* **8**, 300 (2002).
17. H. L. McLeod and J. Yu, *Cancer Invest* **21**, 630 (2003).
18. W. Kalow, B. K. Tang and L. Endrenyi, *Pharmacogenetics* **8**, 283 (1998).
19. The International HapMap Consortium, *Nature* **426**, 789 (2003).
20. E. S. Lander and D. Botstein, *Genetics* **121**, 185 (1989).

RISK FACTOR INTERACTIONS AND GENETIC EFFECTS ASSOCIATED WITH POST-OPERATIVE ATRIAL FIBRILLATION

ALISON A. MOTSINGER[1], BRIAN S. DONAHUE[2], NANCY J. BROWN[3], DAN M. RODEN[3], MARYLYN D. RITCHIE[1]

[1]*Department of Molecular Physiology & Biophysics, Center for Human Genetics Research*

[2]*Department of Anesthesiology*

[3]*Departments of Medicine and Pharmacology*

Vanderbilt University, Nashville, TN 37232, USA

Postoperative Atrial Fibrillation (PoAF) is the most common arrhythmia after heart surgery, and continues to be a major cause of morbidity. Due to the complexity of this condition, many genes and/or environmental factors may play a role in susceptibility. Previous findings have shown several clinical and genetic risk factors for the development of PoAF. The goal of this study was to determine whether interactions among candidate genes and a variety of clinical factors are associated with PoAF. We applied the Multifactor Dimensionality Reduction (MDR) method to detect interactions in a sample of 940 adult subjects undergoing elective procedures of the heart or great vessels, requiring general anesthesia and sternotomy or thoracotomy, where 255 developed PoAF. We took a random sample of controls matched to the 255 AF cases for a total sample size of 510 individuals. MDR is a powerful statistical approach used to detect gene-gene or gene-environment interactions in the presence or absence of statistically detectable main effects in pharmacogenomics studies. We chose polymorphisms in three (IL-6, ACE, and ApoE) candidate genes, all previously implicated in PoAF risk, and a variety of environmental factors for analysis. We detected a single locus effect of IL-6 which is able to correctly predict disease status with 58.8% ($p<0.001$) accuracy. We also detected an interaction between history of AF and length of hospital stay that predicted disease status with 68.34% ($p<0.001$) accuracy. These findings demonstrate the utility of novel computational approaches for the detection of disease susceptibility genes. While each of these results looks interesting, they only explain part of PoAF susceptibility. It will be important to collect a larger set of candidate genes and environmental factors to better characterize the development of PoAF. Applying this approach, we were able to elucidate potential associations with postoperative atrial fibrillation.

1. Introduction

Atrial fibrillation (AF) is the most frequent complication after cardiac surgery, occurring in 25-40% of patients[1-3]. It is an abnormal irregular heart rhythm whereby electrical signals are generated apparently randomly throughout the upper chambers (atria) of the heart. Its onset leads to a significantly higher risk for stroke compared with patients in sinus rhythm and other adverse events[4]. In addition, patients who develop PoAF are more likely to have other postoperative complications such as peri-operative myocardial infarction, congestive heart failure and respiratory failure[5]. Post-op AF (PoAF) has been associated with increased frequency of inotropic and mechanical circulatory support, ventilation time[4], and increased length of hospital stay. Management strategies have focused on reducing PoAF mainly through antiarrhythmic drugs such as beta-blockers, sotalol, and amiodarone. These drugs have had some success in reducing risk, but are far from universally effective[6].

The mechanism of PoAF is complex and not fully understood, but almost certainly multifactorial, involving susceptibility and triggering factors. Genomic approaches offer one way of analyzing risk, but a significant challenge involves identifying sequence variations associated with increased risk. In the case of rare, Mendelian single-gene disorders such as sickle-cell anemia or cystic fibrosis, the genotype to phenotype relationship is often apparent, as disease phenotypes can be explicitly attributed to a mutant genotype. In the case of common complex diseases and pharmacological responses, such relationships are more difficult to characterize since the phenotype is likely the result of many genetic and environmental factors. In addition, epistasis, or gene-gene interaction, is increasingly assumed to play a crucial role in the genetic architecture of common diseases[7-9] and pharmacological responses[10].

There has been strong evidence for both genetic and environmental risk factors contributing to the development of AF[11]. The most frequently identified risk factors include increased age, valvular heart disease, atrial enlargement, preoperative atrial arrhythmias and chronic lung disease[12-14]. AF was first reported in a familial form in 1943[15]. More recent studies have indicated a genetic susceptibility to disease shown by the fact that parental AF increases the risk of AF in their offspring[16]. Linkage analysis has indicated a number of genetic loci in kindreds with a familial form of AF. Mutations in three potassium channel genes have been identified, each in a single kindred[17,18,19]. Other loci have been implicated, but no disease gene within these regions has yet been identified (10q22[20]; 6q14-16[21]; 5p13[22]). Though family studies have been successful in demonstrating a genetic component to AF, the familial form is uncommon. It is possible that this is largely a genetic disorder with highly variable penetrance. Association studies of acquired forms of AF have identified several candidate genes, but without much replication of results. One report form Japan identified a polymorphism in the ACE gene that confers disease risk[23]. Also, recently, a small study (110 patients) implicated the –174C/G polymorphism in the interleukin-6 (IL-6) gene as a risk factor for PoAF[24]. Earlier reports from our group confirm the association of the IL6 promoter polymorphism with PoAF[25].

The goal of this study was to determine whether interactions among candidate genes and a variety of clinical factors are associated in PoAF risk. We selected polymorphisms in six candidate genes, all chosen because of previous work implicating them in PoAF risk. We also chose a variety of recognized environmental factors to analyze.

For this study, we used Multifactor Dimensionality Reduction (MDR), a method for analyzing interactions designed to address many of the limitations of traditional methods. A key problem in traditional parametric methods is that the dimensionality involved in the evaluation of combinations of many genetic and environmental variables quickly diminishes their usefulness. Referred to as the curse of dimensionality[26], as the number of genetic or environmental factors increases and the number of possible interactions increases exponentially, many contingency table cells will be left with very few, if any, data points. This can result in increased type I errors and parameter estimates with very large standard errors[27]. Traditional approaches using logistic regression modeling are limited in their ability to deal with many factors and simultaneously fail to characterize epistasis models in the absence of main effects due to the hierarchical model building process[28]. This leads to an increase in type II errors and decreased power[29]. This is particularly a problem with relatively small sample sizes.

MDR reduces the dimensionality of multilocus data to improve the ability to detect genetic combinations that confer disease risk. MDR pools genotypes into "high-risk" and "low-risk" or "response" and "non-response" groups in order to reduce multidimensional data into only one dimension. It is a nonparametric method, so no hypothesis concerning the value of any statistical parameter is made. It is also a model free method, so no inheritance model is assumed [30].

MDR has been used to identify higher order interactions in the absence of any significant main effects in simulated data. In addition, MDR has demonstrated gene-gene interactions in a variety of different clinical datasets, including sporadic breast cancer[30], essential hypertension[28], type II diabetes[31], atrial fibrillation[32], amyloid polyneuropathy[33], and coronary artery calcification[34]. Studies with simulated data (of multiple models of different allele frequencies and heritability) have also shown that MDR has high power to identify interactions in the presence of many types of noise commonly found in real datasets (including missing data and genotyping error), while errors such as heterogeneity (genetic or locus), and phenocopy diminish the power of MDR [35]. Additionally, theoretical mathematical approaches strongly support the idea that MDR is an optimal method to discriminate between clinical endpoints using multi-locus genotype data more efficiently than any other method[36].

Using MDR, we identified both a genetic effect and an environmental interaction that confers increased risk of PoAF. These findings demonstrate the utility of novel computational approaches for the detection of disease susceptibility genes and risk factors.

2. Methods

2.1. *Sample Population*

Since 1999, our group has been enrolling elective cardiac surgery patients into a genetic registry to study genetic variables that impact clinical outcomes. Following IRB approval and informed consent, we evaluated 940 adult cardiac surgery patients in the registry, and determined the following polymorphisms: -174G/C of IL-6, angiotensin converting enzyme (ACE) intron 16 insertion/deletion (I/D) polymorphism, and the apolipoprotein E alleles 2, 3, and 4. These loci have been previously identified as genetic risk factors for cardiovascular disease[24,37-40]. PoAF was defined as having occurred if present on either a postoperative ECG or rhythm strip, or documented by at least two of the following: progress notes, nursing notes, discharge summary, consultation, or change in medication. Other clinical variables were determined by chart review, and are listed in Table 2. These include clinical variables which have previously been associated with PoAF[41]. Prophylactic beta blockade was defined as receiving beta blockers after surgery but before discharge or onset of atrial fibrillation, whichever happened first. Of the 940 subjects enrolled, 255 developed PoAF. We took a random sample of 255 controls along with the 255 AF cases for a total sample size of 510 individuals.

Table 1: Genetic Parameters Measured in PoAF sample

Number of IL-6 -174G alleles (0,1,2)
Number of ACE D alleles
Number of ApoE2 alleles
Number of ApoE3 alleles
Number of ApoE4 alleles

2.2. *Laboratory Techniques*

Genomic DNA was isolated from blood sampled at the time of surgery. DNA processing was performed by the Vanderbilt Center for Human Genetics DNA Core Laboratory using Puregene (Gentra Systems). ACE insertion/deletion polymorphism was determined by amplification of intron 16 and agarose gel fragment size determination, similar to that of Perticone et al[42]. The ABI Prism 7900HT Sequence Detection System (Applied Biosystems) was used for genotyping the IL-6 -174G/C and Apolipoprotein E (ApoE) alleles. This system utilized the 5' nuclease allelic discrimination *Taq*man assay in a 384-well format, a fluorescent method similar to that of MacLeod et al[43].

Table 2: Clinical Parameters:

Patient demographics			
Gender	Age	Ethnicity	
Surgical procedure			
Valve operation	Coronary bypass operation	Non-coronary, non-valve operation	Open chamber procedures
Duration of cardiopulmonary bypass	Offpump procedures	Repeat sternotomy	
Preoperative medications			
Aspirin	Corticosteroids	Nonsteroidal anti-inflammatory drugs	Alpha-2 antagonists
Beta-blockers	ACE inhibitors	Antilipid drugs	Calcium antagonists
Diuretics	Inotropes		
Medical history			
Hypertension	Diabetes	Preoperative tobacco history	Left ventricular ejection fraction
Preoperative use of intra-aortic balloon pump	History of congestive heart failure	Atrial fibrillation at time of surgery	History of atrial fibrillation
Postoperative events			
Reoperation for bleeding	Death during hospitalization	Use of intra-aortic balloon pump	New neurologic deficit
Use of prophylactic beta blockers	Blood loss during first 24 hours after surgery	Units of blood products transfused after surgery	Length of hospital stay

2.3. *Statistical Techniques*

To explore potential multifactor interactions, we applied the Multifactor Dimensionality Reduction (MDR) method. The details of the MDR algorithm have previously been described[30,35,44]. A diagram explaining the steps of the MDR algorithm is shown in Figure 1.

In the first step of MDR, the dataset is divided into multiple partitions for cross-validation. MDR can be performed without cross-validation; however, this is rarely done due to the potential for over-fitting[45]. Cross-validation[46] is an important part of the MDR method, as it tries to find a model that not only fits the given data, but can also predict on future, unseen data. Since attainment of a second dataset for testing is time-consuming and often cost-prohibitive, cross-validation produces a testing set from the given data to evaluate the predictive ability of the model produced. The training set is comprised of 9/10 of the data while the testing set is comprised of the remaining 1/10 of the data.

Second, a set of *n* genetic and/or environmental factors is selected for analysis, and a list of all possible combinations of factors is created. In the third step these *n* factors are arranged in contingency tables in *n*-dimensional space with all possible multifactorial combinations as individual cells in the table. The cases and controls for each locus combination are counted and in the fourth step the ratio of cases to controls within each cell is calculated. Each multilocus genotype combination is then labeled as "high risk" or "low risk" based on a threshold set at 1: if the ratio within a multifactor combination is >1, it is labeled as "high risk" for disease and if it is <1, it is labeled as "low risk" for disease. This step compresses multidimensional data into one dimension with two classes. For pharmacogenomic endpoints, each genotype combination could be labeled "response" or "non-response" based on the ratio of responders to non-responders.

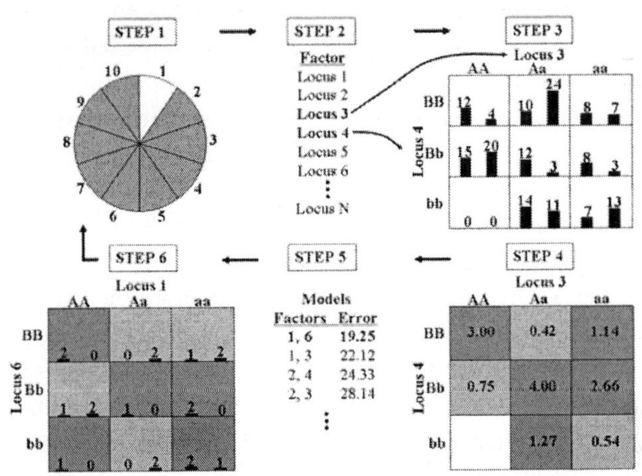

Figure 1. Summary of the general steps to implement the MDR method (adapted from [30]).

The disease risk classifications from each of the multifactorial combinations represent the MDR models for a particular combination of multilocus genotypes. The classification error for each model is calculated based on the number of individuals within the model that are actually cases in genotype combinations classified as "low risk" and the number of individuals that are actually controls in the genotype combination classified as "high risk." The best *n* locus model is selected and the model is evaluated against the testing group and prediction error is calculated. Prediction error is based on the number of misclassified individuals in the testing set, based on the model developed in the training set. This is repeated for each training set and the average classification error and prediction error are calculated. Among all of the models created, the one model with the lowest prediction error is chosen. This process is completed for each number of loci combinations that is computationally feasible. For this analysis, single-locus through four-way interactions were evaluated. A model is chosen for each number of loci considered; so a one-locus model, two-locus model, three-locus model, etc will each comprise a set.

Once this set of models is completed, a final model is chosen. The final model is selected based on minimization of prediction error and maximization of cross-validation consistency. Prediction error is how well the model predicts risk/disease status in independent testing sets - generated through cross-validation. The error for the model is calculated by taking the average of the prediction errors in each of the ten testing sets. Cross-validation consistency is the number of times a model is identified across the cross-validation sets. Therefore, for ten-fold cross-validation, the consistency can range from one to ten. The higher the cross-validation consistency is, the stronger the support for the model. When prediction error and cross-validation indicate different models, the rule of parsimony, or the simpler model, is used to choose between them.

Once a best/final model is chosen, permutation testing is used to test the significance of the hypothesis generated. Permutation testing involves creating multiple permuted datasets by randomizing the disease status labels. One thousand randomized datasets are generated. The entire MDR procedure is repeated for each randomized dataset. The best model is extracted for each random data set as described above which generates a distribution of one thousand prediction errors and cross-validation consistencies that could be expected by chance alone. The significance of the final model is determined by comparing the prediction error of the final model to the distribution. A p-value is extracted for the model by its location in this empirical distribution. A p-value < 0.05 is considered statistically significant.

3. Results

Table 3 shows the results of the MDR analysis of the genotype data. We detected a single locus effect of IL-6 which is able to correctly predict disease status with 58.8% accuracy (Figure 2). This model was significant at the $p<0.001$ level.

Table 3. Results of MDR Analysis in Genotype Sample

Number of Loci	Polymorphism in Model	Cross Validation Consistency	Prediction Error
1	*IL6*	10	41.2*
2	*IL6, APOE4*	5	46.8
3	*IL6, ACE, APOE3*	9	46.01
4	*IL6, ACE, APOE2, APOE3*	10	42.77

* p=<0.001

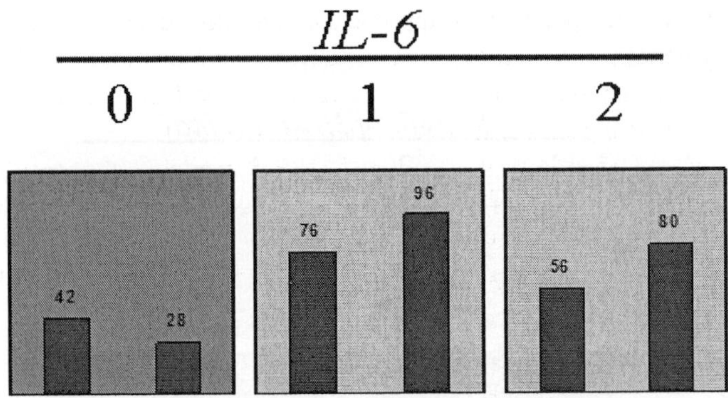

Figure 2. Single-locus MDR Model demonstrating effect of IL-6 which is able to correctly predict disease status with 58.8% accuracy. Light grey cells are low-risk, while dark grey cells are high risk. The number of cases is shown in the histogram on the left in each cell, while controls are shown by the histogram on the right. The genotype labels indicate number of IL-6 promoter -174 G alleles.

Table 4 shows the results of the clinical risk factor analysis. All four MDR models were found to be statistically significant at the $p<0.001$ level, but we focus on the two-locus interaction model because of the low prediction error and high cross-validation consistency. We detected an interaction between history of AF and length of hospital stay that predicted disease status with 68.34% accuracy (Figure 3).

Table 4. Results of MDR Analysis in Clinical Risk Factor Sample

Number of Loci	Variable in Model	Cross Validation Consistency	Prediction Error
1	*Length of stay*	10	33.06*
2	*History of AF, length of stay*	8	31.46*
3	*AF at time of surgery, age, length of stay*	3	38.64*
4	*AF at time of surgery, age, coronary bypass operation, length of stay*	10	33.4*

* p=<0.001

4. Discussion

Post-operative atrial fibrillation is likely the result of multiple genetic and environmental factors. In this study, we investigated potential associations between both candidate genes and risk of PoAF, and clinical risk variables and PoAF. A case-control study design was used, with a large population size and randomly selected controls. We detected an interesting single locus effect of IL-6 which is able to correctly predict disease status with 58.8% accuracy. PoAF is known to prolong length of hospital stay, and preoperative history of AF is a risk factor for postoperative AF[41,47]. Consistent with these findings, we also detected an interaction between history of AF and length of hospital stay that predicted disease status with 68.34% accuracy. These findings demonstrate the utility of novel computational approaches for the detection of disease susceptibility genes.

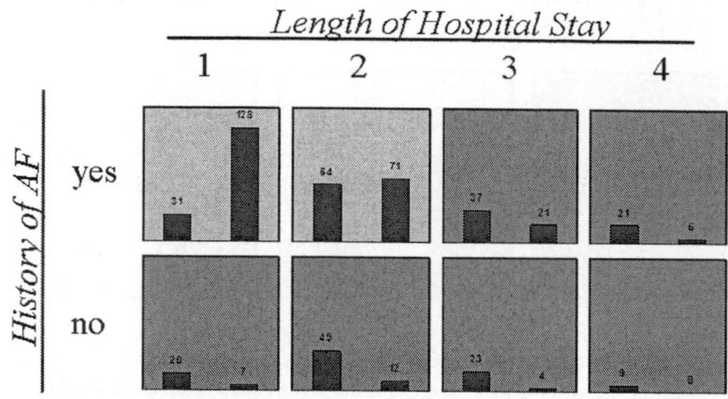

Figure 3. MDR model demonstrating an interaction between history of AF and length of hospital stay that predicted disease status with 68.34% (p<0.001) accuracy. Light grey cells are low-risk, while dark grey cells are high risk. The number of cases are shown in the histogram on the left in each cell, while controls are shown by the histogram on the right. Length of stay was defined as the date difference between surgery date and discharge date. It is coded as follows: 1=0-4 days, 2=5-7 days, 3= 8-13 days, 4=14 or more days. These cutoffs correspond to the 40th, 80th and 95th percentile cutoffs.

Part of the challenge in exploring epistatic interactions in pharmacogenomics or genetic epidemiology is the interpretation of results. Two interesting associations were found – both a main effect and an interactive effect. That IL-6 was shown to have an association with disease risk replicates the findings of[24] providing support for a postulated role for activation of inflammatory pathways in this[48] and perhaps other forms of AF[49]. This underscores a possible role for anti-inflammatory approaches for the prevention of this common complication.

The interaction model demonstrates the importance of genetic and environmental interactions, and represents a possible approach for detection of at-risk subgroups. The occurrence of multiple significant models also demonstrates the extreme complexity of the phenotype and could imply the importance of complicating issues such as heterogeneity and phenocopy.

Future studies may focus on a larger set of candidate genes and environmental factors to better characterize the development of post-operative AF. In addition, cases and controls could be matched on a number of clinical factors to control for confounding. This study demonstrated the importance of looking for both main effects and interactive effects, as well as demonstrating the utility of MDR in analyzing multiple gene-gene and gene-environment interactions.

5. Acknowledgments

This work was supported by National Institutes of Health grants GM31304, AG20135, and in part by HL65962, the Pharmacogenomics of Arrhythmia Therapy U01 site of the Pharmacogenetics Research Network

6. References

1. Ommen,S.R. *et al.* Atrial arrhythmias after cardiothoracic surgery. *N. Engl. J. Med.* **336**, 1429-1434, (1997)
2. Hravnak,M. *et al.* Predictors and impact of atrial fibrillation after isolated coronary artery bypass grafting. *Crit Care Med.* **30**, 330-337, (2002)
3. Aranki,S.F. *et al.* Predictors of atrial fibrillation after coronary artery surgery. Current trends and impact on hospital resources. *Circulation* **94**, 390-397, (1996)
4. Murphy,G.J. *et al.* Operative factors that contribute to post-operative atrial fibrillation: insights from a prospective randomized trial. *Card Electrophysiol. Rev.* **7**, 136-139, (2003)
5. Almassi,G.H. *et al.* Atrial fibrillation after cardiac surgery: a major morbid event? *Ann. Surg.* **226**, 501-511, (1997)
6. Crystal,E. *et al.* Atrial fibrillation after cardiac surgery: update on the evidence on the available prophylactic interventions. *Card Electrophysiol. Rev.* **7**, 189-192, (2003)
7. Moore,J.H. The ubiquitous nature of epistasis in determining susceptibility to common human diseases. *Hum. Hered.* **56**, 73-82, (2003)
8. Sing,C.F. *et al.* Genes, environment, and cardiovascular disease. *Arterioscler. Thromb. Vasc. Biol.* **23**, 1190-1196, (2003)
9. Thornton-Wells,T.A. *et al.* Genetics, statistics and human disease: analytical retooling for complexity. *Trends Genet.* **20**, 640-647, (2004)
10. Wilke, R. A, Reif, D. M, and Moore, J. H. Combinatorial pharmacogenetics. Nature Reviews Drug Discovery in press. (2005).
11. Darbar,D. *et al.* Familial atrial fibrillation is a genetically heterogeneous disorder. *J. Am. Coll. Cardiol.* **41**, 2185-2192, (2003)
12. Creswell,L.L. *et al.* Hazards of postoperative atrial arrhythmias. *Ann. Thorac. Surg.* **56**, 539-549, (1993)
13. Fuster,V. *et al.* ACC/AHA/ESC guidelines for the management of patients with atrial fibrillation: executive summary. A Report of the American College of Cardiology/ American Heart Association Task Force on Practice Guidelines and the European Society of Cardiology Committee for Practice Guidelines and Policy Conferences (Committee to Develop Guidelines for the Management of Patients With Atrial Fibrillation): developed in Collaboration With the North American Society of Pacing and Electrophysiology. *J. Am. Coll. Cardiol.* **38**, 1231-1266, (2001)
14. Maisel,W.H. *et al.* Atrial fibrillation after cardiac surgery. *Ann. Intern. Med.* **135**, 1061-1073, (2001)
15. Wolff, L. Familiar auricular fibrillation. New England Journal of Medicine **229**, 396. (1943)

16. Fox,C.S. *et al.* Parental atrial fibrillation as a risk factor for atrial fibrillation in offspring. *JAMA* **291**, 2851-2855, (2004)
17. Chen,Y.H. *et al.* KCNQ1 gain-of-function mutation in familial atrial fibrillation. *Science* **299**, 251-254, (2003)
18. Yang,Y. *et al.* Identification of a KCNE2 gain-of-function mutation in patients with familial atrial fibrillation. *Am. J. Hum. Genet.* **75**, 899-905, (2004)
19. Xia,M. *et al.* A Kir2.1 gain-of-function mutation underlies familial atrial fibrillation. *Biochem. Biophys. Res. Commun.* **332**, 1012-1019, (2005)
20. Brugada,R. *et al.* Identification of a genetic locus for familial atrial fibrillation. *N. Engl. J. Med.* **336**, 905-911, (1997)
21. Ellinor,P.T. *et al.* Locus for atrial fibrillation maps to chromosome 6q14-16. *Circulation* **107**, 2880-2883, (2003)
22. Oberti,C. *et al.* Genome-wide linkage scan identifies a novel genetic locus on chromosome 5p13 for neonatal atrial fibrillation associated with sudden death and variable cardiomyopathy. *Circulation* **110**, 3753-3759, (2004)
23. Yamashita,T. *et al.* Is ACE gene polymorphism associated with lone atrial fibrillation? *Jpn. Heart J.* **38**, 637-641, (1997)
24. Gaudino,M. *et al.* The -174G/C interleukin-6 polymorphism influences postoperative interleukin-6 levels and postoperative atrial fibrillation. Is atrial fibrillation an inflammatory complication? *Circulation* **108** Suppl 1, II195-II199, (2003)
25. Donahue, B, Darbar, D, George, AL, Li, C., Brown NJ., Ritchie, MD., and Roden, DM. Pharmacogenomics of Atrial Fibrillation. Clin Pharmacol Ther 77[2], P7. (2005)
26. Bellman, R. Adaptive Control Processes. Princeton, Princeton University Press. (1961)
27. Hosmer DW and Lemeshow S. Applied Logistic Regression. New York, John Wiley & Sons Inc. (2000)
28. Moore,J.H. and Williams,S.M. New strategies for identifying gene-gene interactions in hypertension. *Ann. Med.* **34**, 88-95, (2002)
29. Moore,J.H. Computational analysis of gene-gene interactions using multifactor dimensionality reduction. *Expert. Rev. Mol. Diagn.* **4**, 795-803, (2004)
30. Ritchie,M.D. *et al.* Multifactor-dimensionality reduction reveals high-order interactions among estrogen-metabolism genes in sporadic breast cancer. *Am J Hum. Genet* **69**, 138-147, (2001)
31. Cho,Y.M. *et al.* Multifactor-dimensionality reduction shows a two-locus interaction associated with Type 2 diabetes mellitus. *Diabetologia* **47**, 549-554, (2004)
32. Tsai,C.T. *et al.* Renin-angiotensin system gene polymorphisms and atrial fibrillation. *Circulation* **109**, 1640-1646, (2004)
33. Soares,M.L. *et al.* Susceptibility and modifier genes in Portuguese transthyretin V30M amyloid polyneuropathy: complexity in a single-gene disease. *Hum. Mol. Genet.* **14**, 543-553, (2005)
34. Bastone,L. *et al.* MDR and PRP: a comparison of methods for high-order genotype-phenotype associations. *Hum. Hered.* **58**, 82-92, (2004)
35. Ritchie,M.D. *et al.* Power of multifactor dimensionality reduction for detecting gene-gene interactions in the presence of genotyping error, missing data, phenocopy, and genetic heterogeneity. *Genet. Epidemiol.* **24**, 150-157, (2003)

36. Hahn,L.W. and Moore,J.H. Ideal discrimination of discrete clinical endpoints using multilocus genotypes. *In Silico. Biol.* **4**, 183-194, (2004)
37. Eichner,J.E. *et al.* Apolipoprotein E polymorphism and cardiovascular disease: a HuGE review. *Am. J. Epidemiol.* **155**, 487-495, (2002)
38. Sayed-Tabatabaei,F.A. *et al.* Angiotensin converting enzyme gene polymorphism and cardiovascular morbidity and mortality: the Rotterdam Study. *J. Med. Genet.* **42**, 26-30, (2005)
39. Stengard,J.H. *et al.* An ecological study of association between coronary heart disease mortality rates in men and the relative frequencies of common allelic variations in the gene coding for apolipoprotein E. *Hum. Genet.* **103**, 234-241, (1998)
40. Wang,J.G. *et al.* Family-based associations between the angiotensin- converting enzyme insertion/deletion polymorphism and multiple cardiovascular risk factors in Chinese. *J. Hypertens.* **22**, 487-491, (2004)
41. Mathew,J.P. *et al.* A multicenter risk index for atrial fibrillation after cardiac surgery. *JAMA* **291**, 1720-1729, (2004)
42. Perticone,F. *et al.* Hypertensive left ventricular remodeling and ACE-gene polymorphism. *Cardiovasc. Res.* **43**, 192-199, (1999)
43. MacLeod,M.J. *et al.* Lack of association between apolipoprotein E genoype and ischaemic stroke in a Scottish population. *Eur. J. Clin. Invest* **31**, 570-573, (2001)
44. Hahn,L.W. *et al.* Multifactor dimensionality reduction software for detecting gene-gene and gene-environment interactions. *Bioinformatics.* **19**, 376-382, (2003)
45. Coffey,C.S. *et al.* An application of conditional logistic regression and multifactor dimensionality reduction for detecting gene-gene interactions on risk of myocardial infarction: the importance of model validation. *BMC. Bioinformatics.* **5**, 49, (2004)
46. Hastie, T, Tibshirani, R, and Friedman, J. H. The elements of statistical learning. Springer Series in Statistics. Basel, Springer Verlag. (2001)
47. DiDomenico,R.J. and Massad,M.G. Pharmacologic strategies for prevention of atrial fibrillation after open heart surgery. *Ann. Thorac. Surg.* **79**, 728-740, (2005)
48. Aviles,R.J. *et al.* Inflammation as a risk factor for atrial fibrillation. *Circulation* **108**, 3006-3010, (2003)
49. Chung,M.K. *et al.* C-reactive protein elevation in patients with atrial arrhythmias: inflammatory mechanisms and persistence of atrial fibrillation. *Circulation* **104,** 2886-2891, (2001)

LEARNING A PREDICTIVE MODEL FOR GROWTH INHIBITION FROM THE NCI DTP HUMAN TUMOR CELL LINE SCREENING DATA: DOES GENE EXPRESSION MAKE A DIFFERENCE?

LOTHAR RICHTER, ULRICH RÜCKERT AND STEFAN KRAMER*

Institut für Informatik I12, Technische Universität München, Bolzmannstr. 3, Garching b. München, Germany

We address the problem of learning a predictive model for growth inhibition from the NCI DTP human tumor cell line screening data. Extending the classical Quantitative Structure Activity Relationship paradigm, we investigate whether including gene expression data leads to a statistically significant improvement of prediction quality. Our analysis shows that the straightforward approach of including individual gene expression as features does not necessarily improve, but on the contrary, may degrade performance significantly. When gene expression information is aggregated, for instance by features representing the correlation with reference cell lines, performance can be improved significantly. Further improvements may be expected if the learning task is structured by grouping features and instances.

1. Introduction

Pharmacogenomics is concerned with linking drug response with genomic as well as transcriptomic and proteomic data. Whereas many studies deal with genomic variation and single nucleotide polymorphisms (SNPs), we aim at extending the classical Quantitative Structure Activity Relationship (QSAR) paradigm by transcriptomic information. That is, we use not only the structural properties of the compounds for predicting pharmacological activity (as in classical QSAR), but also data about the biological environment, such as gene expression measurements of the involved cell lines. In this way we hope to improve the predictive accuracy of the induced models.

The study is based on the NCI DTP human tumor cell line screening database[1]. This database consists, by and large, of three parts: the first part contains measurements of the growth inhibition of human tumor cell

*Corresponding author: kramer@in.tum.de

lines caused by chemical compounds. The second part contains gene expression measurements for those cell lines. We do not use the (potentially useful) third part of the database containing data on so-called molecular targets. The NCI database has been the subject of a lot of interest in the past few years. So far, work has focused mainly on descriptive mining (variants of clustering, co-clustering/biclustering, and linking clustering results with other information)[2,3,4,5,6,7,8] and predictive mining for individual compounds or small, selected subsets of compounds[9]. In this paper, we develop a model that not only connects these different types of data, but also can be used to make predictions for new, yet unseen cases. We present the results of experiments in learning predictive models on all cell lines (their gene expression) and all compounds. The envisaged usage of the induced model is: given data on a new compound and a new cell line, predict the growth inhibition of that compound on the cell line. The goal of this study is not to achieve a particularly low error, but to test the null hypothesis that the inclusion of gene expression does not change the error of the predictive models.

This paper is organized as follows: In Section 2, we present the materials and methods. Section 3 describes the experimental set-up and data flow in detail. In Section 4, experimental results are presented both quantitatively and qualitatively, before Section 5 concludes and gives an outlook on future work.

2. Materials and Methods

2.1. *Materials*

The NCI DTP has tested tens of thousands of chemical compounds for growth inhibition on human tumor cell lines. Structure information about the compounds is available in a standard file format. For each cell line in the database, gene expression data from Affymetrix chips[9] and cDNA chips[10] is given. The overall dataset is constantly updated by the NCI and complemented by additional information.

For each chemical compound and tumor cell line, the database specifies an indicator of the growth inhibition of the compound on the cell line, the GI_{50} value. The GI_{50} is based on the concept of test through control (T/C, see figure 1(a)). Consider the growth of a tumor cell line until a certain point in time in an untreated control. The measured intensity changes from C_0 to C. In contrast, the measured intensity in a treated cell line changes from T_0 to only T. Given the measured intensities C_0, C, T and T_0, we can

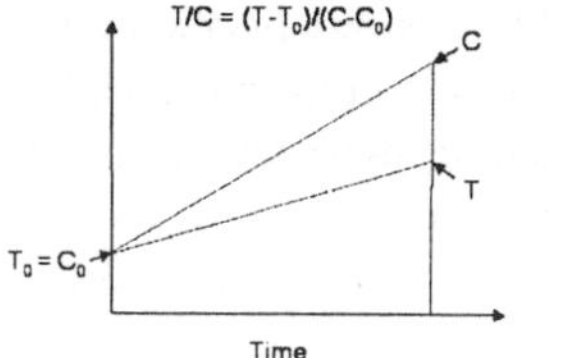

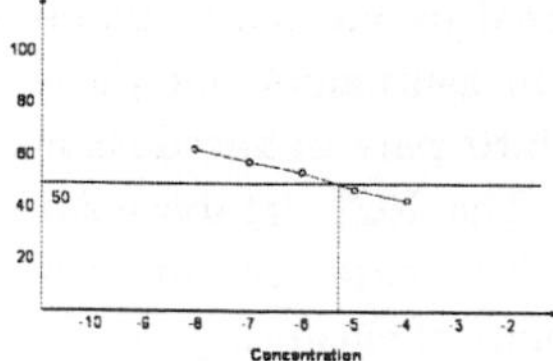

Figure 1. Left side (a): definition of T/C in general, right side (b): definition of GI_{50} from five measurements

calculate the growth (increase in measured intensity) relative to the control as $T/C = (T - T_0)/(C - C_0)$. If the T/C takes the value 0, then the growth of the tumor cell line has stopped completely. If it takes the value 1, the treatment did not make any difference compared to the untreated cell line. Thus, one (or in relative terms, 100%), implies low activity, and zero (or 0%) implies high activity[a].

Now, in general, the GI_{50} value is defined as the concentration of a compound for which the growth is reduced to 50% compared to the untreated control, that is for which the T/C is 50%. Since it would be too costly and time-consuming to determine this point exactly, it is usually determined by *interpolation*. For the given dataset, the first step in this process is the measurement of the T/C value for five concentrations, from the lowest concentration 10^{-8} M to the highest concentration 10^{-4} M (see Fig. 1(b)). Then, we connect all pairs of neighboring points of measurement by straight lines. Next, we determine the point where one of these lines intersects the horizontal line at $T/C = 50\%$. The value on the x-axis (the concentration) of this point is then defined as the GI_{50}.

Unfortunately, there are many complications in this process. First, the procedure might not be successful, because the connected lines might be all below or above $T/C = 50\%$. In this case, the procedure usually *extrapolates* to a range outside the interval between 10^{-8} M and 10^{-4} M. If the intersection is close to the interval boundaries, then the value is taken as it is. Otherwise, the extrapolation is considered too unreliable, in which case the value is rounded to, e.g., 10^{-4} exactly. Therefore, the rounded values in the dataset are indicators of unreliable extrapolations. For the purpose of this paper, we dropped rounded values altogether, since we might otherwise

[a]The explanation is slightly simplified for ease of presentation.

not predict growth inhibition, but rather other, more coarse effects such as solubility. Since some compounds were tested with one cell line at various concentrations we selected the GI_{50} value most distant from the upper and closest to the lower concentration border.

2.2. *Methods*

The main tools for the analysis of the data are the graph mining system *FreeTreeMiner*[12] and the regression rule learning system *Cubist*[11]. Due to space constraints we can not give a detailed description of the two systems and refer to the references for details.

Most learning systems expect the input data in an attribute-value representation, i.e. a table. Thus, we need a preprocessing step to extract features from the compound's structure. In our study we use frequently occurring substructures to characterize each molecule. Each attribute represents a substructure; it is set to 1, if the molecule contains the substructure and 0 otherwise. Of course, it is way to expensive to use all possibly occurring substructures. Instead, we use a graph mining tool to identify those substructures that occur frequently in the database.

In its most general form, graph mining is concerned with finding frequently occurring subgraphs in a database of graphs. More formally, for a database D of graphs let $f(s, D)$ denote the number of graphs containing the subgraph s. Then the goal of a graph mining system is to generate the set of all (connected) subgraphs that occur more often than a predefined threshold t: $S = \{s | f(s, D) \geq t\}$. As we represent a compound as a molecular graph, we can apply any general graph mining tool for the feature generation step. For our study we use FreeTreeMiner[12] to extract all acyclic subgraphs from the molecule database, whose relative frequency is greater than 3%[b]. The restriction on acyclic substructures is made mainly for efficiency purposes. However, unpublished experiments on standard SAR datasets shows no significant difference in performance between various graph mining approaches[13]. In the experiments we find that 5015 acyclic graphs occur in more than 3% of the database's compounds. Thus, we represent each compound by a list of Boolean values, where each value specifies whether the corresponding substructure occurs in the compound.

[b]Preliminary experiments indicate that a smaller threshold hardly makes a difference as we use only the most significant substructures (see below) and those tend to remain the same if one decreases the threshold.

Of course, many of the generated attributes might not contain any significant information about the GI_{50} value. Thus, we perform a second step to identify the meaningful attributes. We calculate a Wilcoxon rank sum statistic for each substructure with regard to the GI_{50} value and keep only those attributes that exceed a minimum significance level. By adjusting the significance threshold one can easily control the number of attributes in the dataset for performance tuning and overfitting avoidance. To calculate the statistic, we split the list of all compounds in two parts: those compounds which contain the substructure and those which do not.

We sort both lists according to the GI_{50} value for each cell line and compute the Wilcoxon rank sum statistic between the two lists. The resulting p-value is an indicator on how certain one can be that the compounds containing a particular substructure exhibit a different distribution of GI_{50} values than those compounds that do not contain the substructure. We calculate those p-values for each attribute and each cell line and use the minimum p-value over all cell lines as the overall p-value of the substructure. Thus, we assume a substructure is significant if it is significant for at least one cell line. We then sort the substructures according to these aggregate p-values to obtain the desired ranking of the generated features.

We chose the commercial tool Cubist for inducing a model that predicts the GI_{50} from the preprocessed data. Cubist generates regression rules with tests for attribute values in the body (antecedent) and linear models in the head (consequent). Table 2 contains typical rules generated by Cubist. Cubist implicitly selects features (for inclusion in the rule body or head), but is not designed to detect feature interactions. It is known to perform well in terms of predictive accuracy and to scale nicely to large datasets.

3. Experimental Set-Up and Data Flow

The goal of our experiments is to show that the inclusion of biological information makes a measurable difference in the predictive accuracy of the models. Note that it is hard to prove that including biological information is not useful at all, since we can only make statements about the particular approach we are following here. The results indicate that one can improve predictive accuracy if the problem representation is chosen carefully. The null hypothesis we want to put to the test in the following is that the inclusion of biological information does not improve performance at all.

We follow the hold-out procedure for evaluating the performance of the learned models: Two thirds of the examples are used for training the

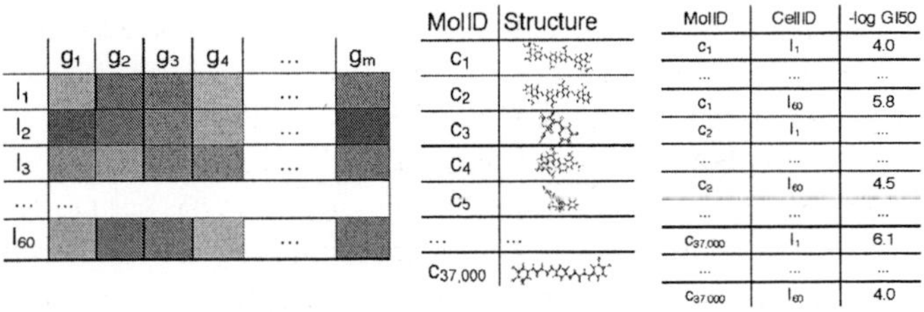

	g_1	g_2	g_3	g_4	...	g_m
l_1					...	
l_2					...	
l_3					...	
...	...					
l_{60}					...	

MolID	Structure
c_1	
c_2	
c_3	
c_4	
c_5	
...	...
$c_{37,000}$	

MolID	CellID	-log GI50
c_1	l_1	4.0
...	...	...
c_1	l_{60}	5.8
c_2	l_1	...
...	...	...
c_2	l_{60}	4.5
...	...	...
$c_{37,000}$	l_1	6.1
...	...	...
$c_{37,000}$	l_{60}	4.0

Figure 2. Input data for our approach. On the left, the table CellLines contains gene expression information for thousands of genes and around 60 cell lines. The Compounds table in the center gives 3D structure information for more than 37,000 chemical compounds. On the right, table GI_{50} contains the negative logarithm of the GI_{50} for combinations of compounds and cell lines.

model and one third is set aside to evaluate it. Since the dataset is very large, there is no need to perform the more time-consuming cross-validation procedure. Fig. 2 shows the input data for our approach. We are given a (target) table of $-\log GI_{50}$ values for combinations of molecules and cell lines. Thus, we have to deal with a multi-relational learning problem. The naïve approach would be to join the three tables and then predict the GI_{50} from the chemical and biological information. However, joining the full three tables would result in a table of approximately 40 GB of data, which makes this approach impractical. Therefore, we perform feature selection to reduce the dimensionality of the gene expression and molecular structure table before joining the tables[c]. Since we still join the tables, we accept a high degree of redundancy in the data as gene expression and compound information is duplicated many times in memory. This, however, only affects memory consumption and not predictive accuracy.

As explained above, we choose those substructures whose Wilcoxon p-values on the GI_{50} values exceeds a certain threshold. Feature selection of the gene expression data is done in a class-blind manner. For the first batch of experiments, we simply sort the genes according to their variance over the cell lines, and add the expression values in this order as features to the data. Thus, inclusion of genes can be parameterized in exactly the same way as the inclusion of substructures. For instance, we might run experiments with the first 100 genes and the first 500 substructures.

[c]Feature selection is necessary anyway, given the high dimensionality and the fact that the performance of most machine learning algorithms tends to degrade with increasing numbers of features.

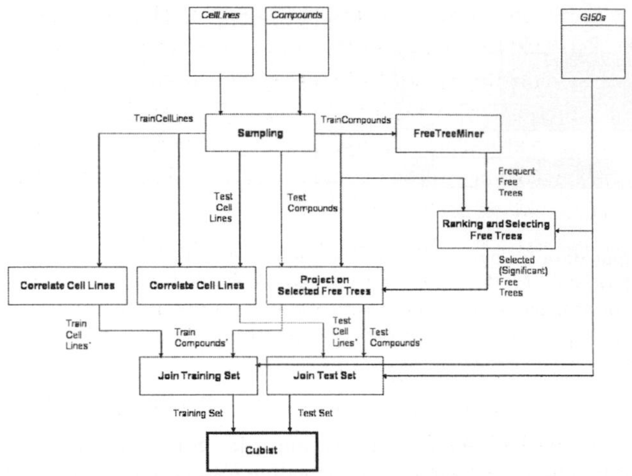

Figure 3. Data flow in the second batch of experiments (see also appendix A)

For the second batch of experiments, we do not use individual gene expression values as attributes. Instead, each attribute represents a reference cell line. For prediction, the new unseen cell line is compared to each reference cell line by calculating a correlation coefficient. This coefficient is then used as feature value. We conduct two experiments: In the first full-fledged variant, we use all 21 training cell lines as reference cell lines, and compute the Pearson correlation coefficient with these cell lines as features. In the second reduced variant, we use only 7 selected, representative cell lines from the training set as reference features. The reference cell lines are chosen to represent all 7 tissues: `A549/ATCC` for lung, `SW-620` for colon, `OVCAR-8` for ovary, `SF-295` for central nervous system, `MOLT-4` for blood, `SK-MEL-28` for skin and `UO-31` for kidney. The overall data flow for the second batch of experiments is shown in Fig. 3. A complete description of the data flow in both batches of experiments is given in appendix A.

4. Experimental Results

Given the above experimental set-up, we first state the results quantitatively (see table 1) and then qualitatively, that is, we present parts of the best model found (see table 2).

We test the influence of biological information describing the respective cell lines – represented in different forms – on prediction accuracy for various sets of substructures. The results are shown in table 1. As outlined in section 2.2 the sets of 100, 500 and 1000 substructures are selected according to their significance. As also outlined the biological information is provided in three different representations: Either as single genes, selected from the Scherf dataset[10] according to the variance over all cell lines ("single genes"), or as Pearson correlation coefficients with all training cell lines ("cell lines") or, further condensed, as correlation with only one cell line per originating tissue ("tissue"). We also obtain reference results from substructure occurrence information without any biological information ("none"). We would now like to accept or reject the null hypothesis that the addition of biological data reduces the prediction error. To do so we perform a paired t-test on the test set example predictions. The table denotes the t-statistic and the corresponding p-value.

The results indicate that the null hypothesis can not be accepted or rejected regardless of the chosen set-up. Instead, the improvement or deterioration of predictive accuracy depends very much on dataset size and the representation of the biological information. First of all, even if one does not include any biological information, the performance depends on the number of substructure attributes: it is best for 500 substructures (error: 0.5701), slightly worse for 100 substructures (0.5744) and clearly worse for 1000 substructures (0.5897). This indicates that overfitting is an important issue for the set-up with 1000 substructures. Consequently, adding biological information in this case only increases the prediction error further to 0.5969 and 0.6054.

With 100 and 500 substructure features, adding biological information as correlation coefficients generally reduces the error. However, overfitting is an issue as well: using 7 representative cell lines performs better than using all 21 training cell lines in both cases, in the latter case even by a large margin (0.5694 vs. 0.5596). If one uses single genes as features, the performance depends very strongly on the number of features. There is an improvement for 500 genes (0.5637), but significant deterioration for 100 or 1000 gene attributes. Apparently, overfitting is again an issue. Overall, using selected reference cell lines to represent biological information seems to yield the best results.

A few of the model's prediction rules produced by the learner are given in table 2. Rule 2 covers 2322 examples and states that for a substance where the substructures `c(:c(:c(:c(:c))))(:c(:c(:c(:n(:c)))))` and `c(:c(:c(:c(-O))))(:c(-C(-C(-C(-C))))(:c))` (in SMARTS format, see

Table 1. Quantitative results from applying Cubist to the NCI data. We present results for varying numbers of compound features (100, 500, 1000): In each case the first row gives the reference result without the use of biological information. The remaining row show results for varying representations of the biological information. The third column states the mean absolute error. The fourth column gives the value of the t statistic from a paired t-test on the test instances, followed by the associated p-value. The sixth column states whether the outcome is a significant win or loss when compared to the reference result without biological information. The final column shows the runtime of Cubist on a Pentium IV CPU with 2.8 GHz.

# Sub-structures	Biological Information	Mean Abs. Error	t-Stat	p-value	sign. Win/ Loss	CPU Time
100	none	0.5744	-	-		79.1 s
100	cell lines	0.5741	-1.83	$p \approx 0.034$	+	162.1 s
100	tissue	0.5735	-7.32	$p < 0.001$	+	116.5 s
500	none	0.5701	-	-		597.5 s
500	single genes (100)	0.6319	78.49	$p < 0.001$	-	1048.2 s
500	single genes (500)	0.5637	-34.41	$p < 0.001$	+	4059.9 s
500	single genes (1000)	0.5826	33.83	$p < 0.001$	-	10040.0 s
500	cell lines	0.5694	-5.91	$p < 0.001$	+	760.4 s
500	tissue	0.5596	-51.13	$p < 0.001$	+	674.2 s
1000	none	0.5897	-	-		1186.2 s
1000	cell lines	0.5969	25.45	$p < 0.001$	-	1334.7 s
1000	tissue	0.6054	40.03	$p < 0.001$	-	1157.6 s

`http://www.daylight.com/dayhtml/doc/theory/theory.smarts.html` for a description) are present the GI_{50} value for a cell line is 4.6566 plus 0.29 times the correlation of this cell line with cell line `SK-MEL-28` plus 0.17 times the correlation with cell line `UO-31`. Rule 9 works similarly for 7325 examples. The GI_{50} value for a substance/cell line combination can be predicted as 4.8483 plus 0.19 times the correlation with cell line `SK-MEL-28` if the substance contains the substructures `c(:c(:c(:c(:c(:c(:c))))))-(:c(:c(:n(:c(:c(:c))))))` while the substructures `c(-C(-C))-(:c(-C(-C)(=O))(:c))` and `c(:c(:c(-O)))(:c(:c(-C(-C))(:c)))` are missing. Rules can also uncover more complex contributions from the biological features. For instance, in rule 23 (covering 4911 examples) in the presence of the two substructures, `C(-C(-C)(-C))(-C(-C)(-O(-C)))` and `C(-C(-C(-C)))(-O(-C(-C(-C))))`, the GI_{50} prediction depends on the correlations with cell lines `UO-31` (-0.43 times), `SK-MEL-28` (0.38 times), `OVCAR-8` (0.35 times), `SW-620` (0.4 times) and `MOLT4` (0.23 times), which makes use of five of the seven biological features present in the data.

The rules presented here clearly show that the use of appropriately represented biological information can enhance prediction power and complete structure-based information.

Table 2. Sample rules from Cubist. The rules are part of the model for 500 substructures and 7 selected tissues. On the left-hand side of the rules (the antecedent), we have tests for the occurrence or non-occurrence of substructures (in SMARTS format). On the right-hand side, we have linear models on variables representing the correlation with the reference cell lines.

```
Rule 2: [2322 cases, mean 4.8189, range 3.456 to 10, est err 0.4961]
    if
        c(:c(:c(:c(:c)))(:c(:c(:c(:n(:c))))) = 1
        c(:c(:c(:c(-O))))(:c(-C(-C(-C(-C))))(:c)) = 1
    then
        GI50 = 4.6566 + 0.29 SK-MEL-28 + 0.17 UO-31

Rule 9: [7325 cases, mean 4.9445, range 2.903 to 10, est err 0.5976]
    if
        c(-C(-C))(:c(-C(-C)(=O))(:c)) = 0
        c(:c(:c(:c(:c(:c(:c))))))(:c(:c(:n(:c(:c(:c)))))) = 1
        c(:c(:c(-O)))(:c(:c(-C(-C))(:c))) = 0
    then
        GI50 = 4.8483 + 0.19 SK-MEL-28

Rule 23: [4911 cases, mean 6.2986, range -1.02 to 13, est err 1.3271]
    if
        C(-C(-C)(-C))(-C(-C)(-O(-C))) = 1
        C(-C(-C(-C)))(-O(-C(-C(-C)))) = 1
    then
        GI50 = 5.9707 - 0.43 UO-31 + 0.38 SK-MEL-28 + 0.35 OVCAR-8 + 0.4 SW-620
        + 0.23 MOLT-4
```

5. Conclusion and Outlook

In this paper, we investigated whether the inclusion of gene expression data can improve a predictive model for growth inhibition learned from the NCI DTP human tumor cell line screening data. Experiments showed that simply relying on individual gene expressions does not necessarily improve, but might degrade performance in predictive modeling of growth inhibition. To show statistically significant improvements over using chemical information alone, a suitable representation of the biological information has to be found.

In future work, we plan to take advantage of the rich structure in the various parts of the input data. For instance, there are classes of substances sharing the same mechanism of action or physico-chemical properties or groups of genes (functional modules) that belong together. Our approach did not yet address these interrelationships. The goal should be to find these groups automatically in the data: Learning mechanisms should be able to recognize and handle subgroups in sets of attributes as well as examples and ultimately use them for prediction.

Appendix A: Data Flow in the Study

For the first batch of experiments (individual genes as features), the parameters to obtain varying numbers of features are *NrFreeTrees* and *NrGenes*:

(1) Sampling

  (a) $(\sqrt{2}-1)\cdot 100\%$ of the cell lines are randomly sampled (without replacement) for testing, the rest is used for training.
  (b) $(\sqrt{2}-1)\cdot 100\%$ of the compounds are randomly sampled (without replacement) for testing, the rest is for training.

(2) Find frequent substructures in training compounds: apply FreeTreeMiner to find frequently occurring substructures in the molecular graphs of the training compounds.

(3) Rank and select significant frequent substructures

  (a) Rank the frequent substructures according to the minimum p-value over all the training cell lines they were tested on. Note that not all compounds were tested on all cell lines.
  (b) Select the first *NrFreeTrees* from the sorted list of substructures.

(4) Reformulate compounds in terms of substructures

  (a) Reformulate the training compounds in terms of the selected substructures. A substructural feature is set to one if the substructure is contained in the molecular graph, and set to zero otherwise.
  (b) Analogously, reformulate the test compounds in terms of the selected substructures.

(5) Rank and select genes

  (a) Rank all genes according to their highest variance on the training cell lines.
  (b) Select the first *NrGenes* from the list of genes.

(6) Project on selected genes

  (a) Project the training cell lines onto the selected genes. In other words, we keep only the *NrGenes* genes with the highest variance to describe the cell lines.
  (b) Project the test cell lines onto the selected genes.

(7) Join the training compounds, the training cell lines, and the GI_{50} values to obtain the training set. Note that not all compounds were tested on all cell lines. Therefore, the join involves a look-up in the GI_{50}s table for each pair of a compound and a cell line.

(8) Join the test compounds, the test cell lines, and the GI_{50} values to obtain the test set. This is the same procedure as for the training set. In this way, the training set contains approximately two thirds of the instances, and the test set approximately one third. Also notice that in this way no compound or cell line in the test set is used for training.

(9) Train Cubist on the training set and test it on the test set.

The data flow in the second batch of experiments (correlations with reference cell lines as features) differs slightly from the one before (see also Fig. 3), in steps 5 to 9. The only parameter is the number of substructures (*NrFreeTrees*):

(5) Correlate training cell lines: compute a matrix with the Pearson correlation coefficient between all pairs of cell lines in the training cell lines. Thus, the correlation between a cell line and another is used as a feature to describe the former.

(6) Correlate test cell lines with training cell lines: a matrix containing the Pearson correlation coefficient between each test cell line and each training cell line is computed. The correlation of a test cell line with a training cell line is used to describe the former.

(7) Join the training compounds, the training cell line correlations, and the GI_{50} values to obtain the training set.

(8) Join the test compounds, the test cell line correlations, and the GI_{50} values to obtain the test set.

(9) Train Cubist on the training set and test it on the test set.

References

1. M.R. Boyd, in : A.B. Teicher (ed.), *Cancer Drug Discovery and Development, Vol.2; Drug Development; Preclinical Screening, Clinical Trial and Approval*, Humana Press, 23-43, (1997).
2. P.E. Blower, C. Yang, M.A. Fligner, J.S. Verducci, L. Yu, S. Richman, J.N. Weinstein, *Pharmacogenomics J.* **2(4)**, 259-271 (2002).
3. X. Fang, L. Shao, H. Zhang and S. Wang, *J. Chem. Inf. Comput. Sci.* **44**, 249-257 (2004).
4. A.A. Rabow, R.H. Shoemaker, E.A. Sausville and D.G. Covell, *J. Med. Chem.* **45**, 818-840 (2002).
5. L.M. Shi, Y. Fang, J.K. Lee, M. Waltham, D.T. Andrews, U. Scherf, K.D. Paul and J.N. Weinstein, *J. Chem. Inf. Comput. Sci.* **40**, 367-379 (2000).
6. A. Wallqvist, A.A. Rabow, R.S. Shoemaker, E.S. Sausville and D.G. Covell, *Bioinf.* **19(17)**, 2212-2224 (2003).
7. J.N. Weinstein, Y. Pommier, *C.R.Biologies* **326**, 909-920 (2003).
8. D.T. Ross, U. Scherf, M.B. Eisen, C.M. Perou, C. Rees, P. Spellman, V. Iyer, S.S. Jeffrey, M. Van de Rijn, M. Waltham, A. Pergamenschikov, J.C.F. Lee, D. Lashkari, D. Schalon, T.G. Myers, J.N. Weinstein, D. Botstein, P.O. Brown, *Nature Genetics* **24**, 227-235 (2000).
9. J.E. Staunton, D.K. Slonim, H.A. Coller, P. Tamayo, M.J. Angelo, J. Park, U. Scherf, J.K. Lee, W.O. Reinhold, J.N. Weinstein, J.P. Mesirov, E.S. Lander, T.R. Golub, *PNAS* **98(19)**, 10787-10792 (2001).
10. U. Scherf, D.T. Ross, M. Waltham, L.H. Smith, J.K. Lee, L. Tanabe, K.W. Kohn, W.C. Reinhold, T.G. Myers, D.T. Andrews, D.A. Scudiero, M.B. Eisen, E.A. Sausville, Y. Pommier, D. Botstein, P.O. Brown, J.N. Weinstein, *Nature Genetics* **24**,236-244 (2000).
11. RuleQuest Research (2005). `http://www.rulequest.com/cubist-info.html`
12. U. Rückert, S. Kramer, *Proc. ACM Symp. on Appl. Comp. (SAC 2004)*, 564-570 (2004).
13. S. Sommer, *Lazy Structure-Activity Relationships*, Diploma Thesis, TU München (2005).